T0234676

# Lecture Notes in Computer Science     14299

## Founding Editors

Gerhard Goos
Juris Hartmanis

## Editorial Board Members

Elisa Bertino, *Purdue University, West Lafayette, IN, USA*
Wen Gao, *Peking University, Beijing, China*
Bernhard Steffen®, *TU Dortmund University, Dortmund, Germany*
Moti Yung®, *Columbia University, New York, NY, USA*

The series Lecture Notes in Computer Science (LNCS), including its subseries Lecture Notes in Artificial Intelligence (LNAI) and Lecture Notes in Bioinformatics (LNBI), has established itself as a medium for the publication of new developments in computer science and information technology research, teaching, and education.

LNCS enjoys close cooperation with the computer science R & D community, the series counts many renowned academics among its volume editors and paper authors, and collaborates with prestigious societies. Its mission is to serve this international community by providing an invaluable service, mainly focused on the publication of conference and workshop proceedings and postproceedings. LNCS commenced publication in 1973.

Moti Yung · Chao Chen · Weizhi Meng
Editors

# Science of Cyber Security

5th International Conference, SciSec 2023
Melbourne, VIC, Australia, July 11–14, 2023
Proceedings

Springer

*Editors*
Moti Yung 🆔
Columbia University
New York, NY, USA

Chao Chen 🆔
RMIT University
Melbourne, VIC, Australia

Weizhi Meng 🆔
Technical University of Denmark
Kongens Lyngby, Denmark

ISSN 0302-9743        ISSN 1611-3349 (electronic)
Lecture Notes in Computer Science
ISBN 978-3-031-45932-0        ISBN 978-3-031-45933-7 (eBook)
https://doi.org/10.1007/978-3-031-45933-7

© The Editor(s) (if applicable) and The Author(s), under exclusive license
to Springer Nature Switzerland AG 2023

This work is subject to copyright. All rights are reserved by the Publisher, whether the whole or part of
the material is concerned, specifically the rights of translation, reprinting, reuse of illustrations, recitation,
broadcasting, reproduction on microfilms or in any other physical way, and transmission or information
storage and retrieval, electronic adaptation, computer software, or by similar or dissimilar methodology now
known or hereafter developed.
The use of general descriptive names, registered names, trademarks, service marks, etc. in this publication
does not imply, even in the absence of a specific statement, that such names are exempt from the relevant
protective laws and regulations and therefore free for general use.
The publisher, the authors, and the editors are safe to assume that the advice and information in this book
are believed to be true and accurate at the date of publication. Neither the publisher nor the authors or the
editors give a warranty, expressed or implied, with respect to the material contained herein or for any errors
or omissions that may have been made. The publisher remains neutral with regard to jurisdictional claims in
published maps and institutional affiliations.

This Springer imprint is published by the registered company Springer Nature Switzerland AG
The registered company address is: Gewerbestrasse 11, 6330 Cham, Switzerland

Paper in this product is recyclable.

# Preface

This volume contains the papers presented at the 5th International Conference on Science of Cyber Security (SciSec 2023), which was held as a hybrid conference during 11–14 July, 2023, in Melbourne, Australia.

SciSec is an annual international forum for researchers and industry experts to present and discuss the latest research, trends, breakthroughs, and challenges in the domain of cybersecurity. This new forum was initiated in 2018 and aims to catalyze the research collaborations between the relevant communities and disciplines that can work together to deepen our understanding of, and build a firm foundation for, the emerging Science of Cyber Security.

This year we received 60 submissions from around the world. Each paper was reviewed by at least three Program Committee members in a single-blind process. At last, we accepted 21 full papers with an acceptance rate of 35%. Also, we accepted 6 short papers based on their quality. The selected Best Paper Award went to "ACDroid: Detecting Collusion Applications on Smart Devices" by Ning Xi, Yihuan He, Yuchen Zhang, Zhi Wang, and Pengbin Feng, and the selected Best Student Paper Award went to "A Graphical Password Scheme based on Rounded Image Selection" by Xinyuan Qin and Wenjuan Li. The conference program also included two invited keynote talks: the first keynote from Moti Yung, titled "Methodology for Secure Systems Deployment: Science of Security for Evolving Ecosystems", and the second keynote from Allen Au, titled "The Role of Cryptography in Blockchains". Also, we had one invited talk from Kaitai Liang, titled "Searchable Symmetric Encryption and its attacks".

We would like to thank all of the authors of the submitted papers for their interest in SciSec 2023. We also would like to thank the reviewers, keynote speakers, and participants for their contributions to the success of SciSec 2023. Our sincere gratitude further goes to the Program Committee, the Publicity Committee, and the Organizing Committee, for their hard work and great efforts throughout the entire process of preparing and managing the event.

We hope that you will find the conference proceedings inspiring and that they will further help you in finding opportunities for your future research.

July 2023

Moti Yung
Chao Chen
Weizhi Meng

# Organization

## Steering Committee

Guoping Jiang      Nanjing University of Posts and
Telecommunications, China
Feng Liu      Institute of Information Engineering, Chinese
Academy of Sciences, China
Shouhuai Xu      University of Colorado Colorado Springs, USA
Moti Yung      Google and Columbia University, USA

## General Chairs

Yang Xiang      Swinburne University of Technology, Australia
Qiang Tang      University of Sydney, Australia
Feng Liu      Institute of Information Engineering, Chinese
Academy of Sciences, China

## PC Chairs

Moti Yung      Google Inc, USA
Chao Chen      RMIT, Australia
Weizhi Meng      Technical University of Denmark, Denmark

## Publicity Chairs

Xingliang Yuan      Monash University, Australia
Jun Shao      Zhejiang Gongshang University, China
Bo Chen      Michigan Technological University, USA

## Publication Chairs

Maggie Liu      RMIT University, Australia
Wenjuan Li      Hong Kong Polytechnic University, China

# Program Committee Members

| | |
|---|---|
| Richard Brooks | Clemson University, USA |
| Bo Chen | Michigan Technological University, USA |
| Hung-Yu Chien | National Chi Nan University, Taiwan |
| Chun-I Fan | National Sun Yat-sen University, Taiwan |
| Dieter Gollmann | Hamburg University of Technology, Germany |
| Thippa Reddy Gadekallu | VIT Vellore, India |
| Yingjiu Li | University of Oregon, USA |
| Wenjuan Li | Hong Kong Polytechnic University, China |
| Weizhi Meng | Technical University of Denmark, Denmark |
| Lingguang Lei | IIE, Chinese Academy of Science, China |
| Hiroaki Kikuchi | Meiji University, Japan |
| Weizheng Wang | City University of Hong Kong, China |
| Lingyu Wang | Concordia University, Canada |
| Toshihiro Yamauchi | Okayama University, Japan |
| Chuan Yue | Colorado School of Mines, USA |
| Leo Yu Zhang | Deakin University, Australia |
| Zhenpeng Lin | Northwestern University, USA |
| Christos Xenakis | University of Piraeus, Greece |
| Leandros Maglaras | De Montfort University, UK |
| Fuchun Guo | Wollongong University, Australia |
| Taeho Jung | University of Notre Dame, USA |
| Nora Cuppens-Boulahia | École Polytechnique de Montréal, Canada |
| Steven Furnell | University of Nottingham, UK |
| Ding Wang | Nankai University, China |
| Edgar Weippl | University of Vienna & SBA Research, Austria |
| Brajendra Panda | University of Arkansas, USA |
| Hélder Gomes | Universidade de Aveiro, Porto |
| Yvo Desmedt | University of Texas at Dallas, USA |
| Jun Shao | Zhejiang Gongshang University, China |
| Willy Susilo | University of Wollongong, Australia |
| Lei Xue | Sun Yat-sen University, China |
| Qingni Shen | Peking University, China |
| Joaquin Garcia-Alfaro | Télécom SudParis, France |
| Jianting Nin | Singapore Management University, Singapore |
| Xiaofeng Chen | Xidian University, China |
| Rongmao Chen | National University of Defense Technology, China |
| Songqing Chen | George Mason University, USA |
| Fan Zhang | Zhejiang University, China |
| Kun Sun | George Mason University, USA |

Kwok Yan Lam          Nanyang Technological University, Singapore
Carlo Blundo          Università degli Studi di Salerno, Italy
Joachim Posegga        University of Passau, Germany

# Contents

## AI for Security

## Threat Detection and Analysis

## Web and Privacy Security

**Cryptography and Authentication II**

**Advanced Threat Detection Techniques and Blockchain**

**Workshop Session**

# Network and System Security

# ACDroid: Detecting Collusion Applications on Smart Devices

Ning Xi, Yihuan He, Yuchen Zhang(✉), Zhi Wang, and Pengbin Feng

School of Cyber Engineering , XiDian University, Xi'an, Shaanxi, China
nxi@xidian.edu.cn, zhangyuchen@stu.xidian.edu.cn

**Abstract.** With the continuous innovation of artificial intelligence technology, more and more smart devices are being applied to emerging industrial Internet and IOT (Internet of Things) platforms. Most of these smart devices are designed based on the Android framework, and the security of Android applications is particularly important for these smart devices. To facilitate communication between applications, ICC (Inter-Component Communication) is widely used in Android. While bringing convenience to users, it also brings the risk of privacy leakage and privilege escalation. In this way, two or more applications can collude and thereby evade detection by tools that analyze the security of a single application. To defend against this attack, we propose a machine learning-based static analysis method and design and implement ACDroid. ACDroid uses static analysis to obtain inter-application collaboration characteristics, including inter-application ICCs and dangerous permission group combinations. The deep learning algorithm is then used for efficient classification to detect collusion attacks. ACDroid improves the detection performance of existing research focusing on single permission features via constructing synergistic components. We validate our tool by conducting experiments on over 10,000 real-world applications. Compared with state-of-the-art approaches, our method expresses superior performance in collusion attack detection.

**Keywords:** Smart Devices · Inter-app communication · Collusion · Android Security · Machine Learning

## 1 Introduction

Application collusion is an emerging threat to Android based smart devices. In app collusion, two or more apps collude in some manner to perform a malicious action that they are unable to do independently. Detection of colluding apps is a challenging task. With the continuous development of the information technology industry, the use of smart devices is exploding. As the widely operating system on these smart devices, Android can handle various user information such as photos, finances, and location. To fulfill the goal of collaborative work, Android applications can interact between components and inter-applications through ICC. While this cooperative working mode is advantageous, it also introduces a

© The Author(s), under exclusive license to Springer Nature Switzerland AG 2023
M. Yung et al. (Eds.): SciSec 2023, LNCS 14299, pp. 3–22, 2023.
https://doi.org/10.1007/978-3-031-45933-7_1

new attack approach called the collusion attack. The attack enables numerous programs to generate longer call pathways to accomplish the malicious operation, with each participating portion potentially eluding individual application security detection.

The emergence of collusion attacks has brought new challenges to Android security. Because most current Android malware detection techniques consider malware applications to conduct attacks independently, only individual application security detection is performed. For example, XmanDroid [5] monitors communication links between applications on Android at runtime. It then validates them against security rules defined in the system policy to detect and block confused deputy attacks and collusion attacks. Bhandari et al. [2] proposed a framework to check the existence of connectivity between two applications through static analysis and add edges accordingly. Then check whether these edges violate the defined collusion policy, so as to cross-determine whether there is a collusion attack. DIALDroid [4] uses static analysis to obtain inter-application ICCs and uses fine-grained security policies to detect sensitive inter-application ICCs to calculate whether there are collusion data leaks or privilege escalations. Elish et al. [9] propose a scalable and efficient method to screen applications for possible collusion. The risk level of conspiracy attacks between applications is determined by extracting the flow feature of inter-application communication through static analysis and then classifying it according to a predefined security policy.

The biggest challenge of these policy analysis-based methods is to identify policy sets, which require extensive relevant domain knowledge and be as comprehensive as possible. The policy set should not be too strict, which may produce false positives, but it should not be too flexible, leading to more false negatives. This tough and loose metric is not straightforward, and many uncertainties exist. Currently, these policies are specific or manually defined. Therefore, it is likely that certain application information, which may be potentially important for improving the determination accuracy, is ignored due to its lack of comprehensiveness. To overcome these limitations, we propose a machine learning-based method for collusion detection.

We describe ACDroid as a tool for efficiently analyzing inter-application security. ACDroid is an inter-application collusion detection tool that statically analyzes the ICCs between applications and obtains collaborative permission features through algorithms, which are then efficiently classified and detected by machine learning. These applications may collude, resulting in data leakage or abuse of permissions. Our contributions are summarized as:

- For the first time, we built the Android collusion assault dataset. By collecting applications from the same manufacturer or developer, we obtained more inter-application communication data and constructed a dataset suitable for collusion research.
- We propose and implement an algorithm for creating collusion research characteristics based on the combination of permission groups. Incorporating the synergistic features is promising in collusion detection.

- We compared with various traditional machine learning algorithms and deep learning algorithms and obtained the algorithm with the best performance form for collusion analysis.
- We implemented ACDroid to extract collaborative features from applications through static analysis and finally use machine learning to classify. It could achieve the automatic and efficient detection of collusion attacks.

The remainder of the paper is organized as follows: Sect. 2 describes the collusion attack model, Sect. 3 details our approach and tools, Sect. 4 evaluates ACDroid from multiple aspects, and Sect. 5 discusses our work and proposes limitations. Finally, the paper concludes with an introduction to related research and summarizes the outcome of this paper.

## 2    Security Model

Through application collusion, the malicious purpose of one malware can be broken down into multiple applications to complete. This way, a more stealthy attack can achieve, avoiding security detection. As shown in Fig. 1, multiple seemingly benign applications conspire to accomplish the malware's attack mission. Collusion applications can communicate with each other. During the communication process, each application can apply permissions to access resources or implement system features so that multiple collusion applications can jointly perform malicious tasks.

The malware in Fig. 1(a) reads the address book through permission READ_CONTACTS and sends the read private information to the attacker through permission SEND_SMS. In Android security checks, the permissions and specific actions of the malware in 1(a) can be easily detected and therefore judged as malicious. APP A and APP B, as depicted in 1(b), cooperate to complete the attack task of the first malicious application. App A needs to read the contact with permission READ_CONTACTS and pass the read information to App B via inter-application communication. After that, application B sends the attacker a message containing the data. App A and B split the malware's behavior into multiple parts. Then they completed some of them respectively, which achieved the purpose of evading the detection of Android security analysis tools.

Collusion in multiple applications to achieve malicious attacks reduces the risk of individual applications being identified, thereby weakening the capabilities of various Android security analysis tools. In our research, the malicious purposes that Android applications collude to achieve can be mainly classified into privilege escalation or privacy leakage.

**Privilege Escalation in Collusion Scenarios:** Android devices enforce permission protection mechanisms to regulate user applications' data access. Privilege escalation enables applications to bypass privilege protection mechanisms by delivering permission between applications. Specifically, in the collusion scenario, two or more applications with limited permissions communicate with each other to obtain unauthorized resources.

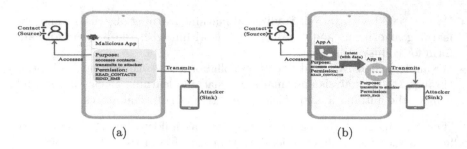

**Fig. 1.** Multiple applications conspiracy to replace a single malware implementation attack model

As shown in Fig. 1(b), App B does not have permission to access the contact but obtains the address book information because it has achieved privilege escalation through collusion.

**Privacy Leakage in Collusion Scenarios:** In the collusion scenario, sensitive data is sent as a source to other applications or even outside the mobile device through a secret path (without the user's consent), resulting in a privacy leak.

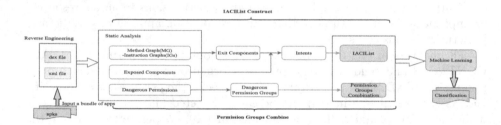

**Fig. 2.** The Overview of ACDroid

## 3   Framework Design

This section mainly introduces our approach and the Android collusion detection tool ACDroid, which statically analyzes a set of applications and then determines whether there is a collusion risk through machine learning. As shown in Fig. 2, our approach is mainly composed of three parts: (1) apply static analysis to extract an Inter-Application Communication Intent List (IACIList) for each bundle of applications. (2) permission groups combine to analyze dangerous permission groups and generate collaborative features by the algorithm for sample groups. (3) The machine learning part uses machine learning algorithms to judge the collusion risk of a sample according to its permissions and inter-application communication.

In the first part, static analysis takes a bundle of android applications (APK files) as input. First, we can obtain the dex files and manifest files through reverse engineering. Analyzing the manifest file, we can get the basic structure information of the single application (including the permissions and exposed components). Then we further explore the dex file to obtain the inter-application communication intents and the Inter-Application Communication Intent List (IACIList) from the bundle of applications.

The second part aims to get the result of the combination of dangerous permissions for each application. Instead of examining permissions to analyze the security of a single application, we mainly focus on collusion attacks involving multiple applications. In this attack method, the application colludes to leverage their authorized permissions to interact with each other to achieve privilege escalation. Therefore, this part introduces the principle of permission group combination and designs an algorithm for obtaining the result of permission group combination between applications.

The machine learning part is to help us better discover the deep connection between permission combinations and intents in the collusion scenario. Through experiments, we found that there are many combinations of permission groups, and each intent involves a lot of permissions. The final experimental results also verify that intent and permission can predict the risk of collusion.

In the following, we will describe the details of our approach.

## 3.1   IACIList Construct

Our static analysis consists of two phases: 1) single application information extraction and 2) ICCs extraction. Finally, our static analysis will get the Inter-Application Communication Intent List (IACIList).

**Single Application Information Extraction.** The input of this part is the APK files of a bundle of applications. We first extract dex and manifest files from these APK files via reverse engineering.

Firstly, analyze the manifest file to generate, for each application, a table of exposed components. These components serve as an entry point for inter-application communication, and the exposed component could directly communicate with other external components. When a component has an inter-filter or its exported status is true, it is considered an exposed component. We generate an entry table for each application, which records the exposed component information. In this phase, we also extract the permissions of each application for use in the following combination of permissions.

The dex file contains all the code of one application, so it is the critical file for analysis. In practice, the application's source code need not be obtained; we can perform security analysis on the compiled dex file of the applications. So we perform static analysis based on Dalvik bytecode, generate Method Graph for each application through Class-Loader Virtual Machine (CLVM), and then create Instruction graphs for each method.

To determine the exit points of inter-application communication, we define some APIs as Exit APIs. Explicit intent uses Exit APIs such as setClassName, setComponent, and setClass. Because this type of intent requires specifying the name of the connected component. Implicit intents must set the action to match the intent-filter and use setAction as the Exit API. We traverse all Instruction graphs nodes to find Exit APIs and record the received component information by performing data flow analysis on the Instruction graph. If it is an Implicit intent, additional information needs to be recorded.

Through this phase of the processing, we get each application's exposed component list, exit component list, and dangerous permissions.

**ICCs Extraction.** For each application's exit component list, we iterate over their target components to match other applications' exposed components. The intent connection obtained according to the component name is marked as explicit intent, and the communication determined by the intent-filter is marked as implicit intent. When matching implicit intents, other information in the intent-filter must also be consistent with what the ICC holds. Through this phase, ACDroid creates an inter-application communication intent list (IACIList) for the input applications. The information structure of each intent in IACIList is presented in Table 1.

**Table 1.** Describe the attributes of intent in IACIList

|        | Attribute | Description |
|--------|-----------|-------------|
| Intent | Sender | The name of the application which sends the intent |
|        | Target | The name of the application which the intent is directed to |
|        | Category | The intent is explicit or implicit |
|        | Action | The action of implicit intent |
|        | Data | The data of the intent |

Some studies have shown that directly specifying a component's explicit intent is more dangerous because its objectives are clear and targeted; thus, some previous studies only analyzed the explicit intent [1,9]. But in our study, both intentions have significant research value and lead to successful inter-app communication. Ignoring any of them is incomplete. In addition, implicit intent is more commonly used in the IAC than explicit intent. Studying both explicit and implicit intent is necessary, and that is precisely what we are doing.

## 3.2   Permission Groups Combine

Android permission mechanism is continuously upgraded to improve security protection [10]. Previous studies mainly focus on the lack of authorization verification during inter-app communication. With the continuous upgrade of the authorization mechanism, these problems have basically been solved. However, in the current authorization mechanism, there still exists collusion risk among applications. Therefore, we propose new ideas based on the combination of dangerous permission groups. According to the collusion scenario mentioned above, each application only needs to provide partial permission assistance to achieve malicious purposes. It is helpful to extract the combination of dangerous permission groups as behavior features.

Collusion attacks often require permission to collaborate, so we propose permission combination as an analysis feature. Below we describe the situation of permission in collusion:

**Definition 1.** $Z$ represents the set of Android applications, P represents the set of all dangerous permissions in Android, $a_i$ represents an action of the Android application, and $p_i$ represents the permission corresponding to $a_i$. All actions of application $z \in Z$ are denoted by $Action(z)$, All permissions of application are indicated by $P(z)$, the cooperative work between two applications is denoted by $U(z_i, z_j)$, and $L$ represents the existing communication.

$$U(z_i, z_j) = Action(z_i)LAction(z_j) \tag{1}$$

**Definition 2.** The minimum sequence of actions required to implement a malicious attack $F(X)$ is expressed as $m$. At the same time, the minimum permissions must be obtained $P(m)$. If some additional sequence of actions can also realize the function $F(X)$, these actions are considered the extended sequence of $m$, counted as $M*$.

$$P(m) \to m \tag{2a}$$
$$m \to F(X) \tag{2b}$$
$$M* \to F(X) \tag{2c}$$
$$m \in M* \tag{2d}$$
$$P(m) \in P(M*) \tag{2e}$$

$\exists z_1, z_2, z_m \in Z$, when

$$Action(z_1) \notin M* \tag{3a}$$
$$Action(z_2) \notin M* \tag{3b}$$
$$Action(z_m) \in M* \tag{3c}$$

At this point, consider $z_m$ malware that can cause F(X). $z_1$ and $z_2$ as benign applications. However, $z_1$ and $z_2$ are benign in a single detection but can achieve F(X) in a collusion scene. At this time, the collusion $z_1$ and $z_2$ can be expressed as:

$$U(z_1, z_2) = Action(z_1)LAction(z_2) \tag{4a}$$
$$Action(z_1)LAction(z_2) \in M* \tag{4b}$$
$$(P(z_1) \cup P(z_2)) \in P(M*) \tag{4c}$$

In the above description of collusion scenarios, every single application can evade the security checks of a single application. Simultaneously, we also find that the union of application permissions must contain the permission set $P*$ that realizes the malicious attack $F(X)$. As a result, the permission groups combine algorithm is described as follows:

**Definition 1.** Each device is a tuple $X< x_i, ..., x_j >$, where $x_i \neq x_j$ indicates different applications installed on the device $X$. Meanwhile, each application $x_i$ has its own dangerous permissions $D_i$.

In the collusion scenario, the application can achieve the purpose of the attack by privilege escalation. We group each app's permissions to derive dangerous permissions groups involving privacy or vital functions. To implement the algorithm, the dangerous permission groups among the applications are compared and divided into the same and different dangerous permission groups.

**Definition 2.** In device $X$, the application $x_i$ has a set of dangerous permissions $D_i$, and another application $x_j$ has a set of dangerous permissions $D_j$. The same dangerous permissions between $x_i$ and $x_j$ are recorded as set $SP_{<i,j>}$, $x_i$ different permissions from $x_j$ are recorded as set $DP_1< i, j >$, $x_j$ different permissions from xi are recorded as set $DP_2< j, i >$.

According to definition 1 and definition 2, we can get the following mathematical expressions.

$$SP_{<i,j>} = D_i \cap D_j \tag{5a}$$
$$DP_1< i, j > = D_i \setminus D_j \tag{5b}$$
$$DP_2< j, i > = D_j \setminus D_i \tag{5c}$$

**Definition 3.** To find the complete combination structure of $D_i$ and $D_j$, we record the set $Sub_{<i,j>,n}$, n represents the combined result of n elements. $Sub_{<i,j>}$ is the final output result, which contains the set of all combinations. To achieve the purpose of a collusion attack, a set of applications needs at least one combination of dangerous permissions that cannot be formed by a single application, as shown in Fig. 2. This can be interpreted as a group of applications that can be directly filtered out with no combined results when one of the applications does not contain dangerous permissions. Through our Algorithm 1, get the output result $Sub< i, j >$

### 3.3 Machine Learning

With the rapid development of emerging technologies, the volume of data is also increasing. It makes the application of machine learning more widespread and

---

**Algorithm 1.** Permissions Combination

---

**Input:** $D_i$:a set of dangerous permissions for the app $x_i$;

$D_j$:a set of dangerous permissions for the app $x_j$;

$|D_i|$: the number of permissions in the set $D_i$;

**Output:** combinations $Sub_{<i,j>}$

1: **if** $|D_i|==0$ or $|D_j|==0$ **then**
2:     $Sub_{<i,j>} \leftarrow 0$
3: **else**
4:     $SP_{<i,j>} = D_i \cap D_j$
5:     $DP_1<i,j> = D_i \setminus D_j$
6:     $DP_2<j,i> = D_j \setminus D_i$
7:     **if** $DP_1<i,j> == null$ or $DP_2<j,i> = D_j == null$ **then**
8:         $Sub_{<i,j>} \leftarrow 0$
9:     **else**
10:         $Sub_{<i,j>,2} = \{(a_1,a_2)|a_1 \in DP_1<i,j> \wedge a_2 \in DP_2<j,i>\}$
11:         $m = (|SP_{<i,j>}| + |DP_1<i,j>| + |DP_2<j,i>|)$
12:         **for** $n = 3, n < m+1, n++$ **do**
13:             $(a_n \in SP_{<i,j>}) \vee (a_n \in DP_1<i,j>) \vee (a_n \in DP_2<j,i>)$
14:             $a_1 \neq a_2 \neq ... \neq a_i$
15:             $Sub_{<i,j>,n} = \{(Sub_{<i,j>,n-1}, a_n)\}$
16:         **end for**
17:         $Sub_{<i,j>} \leftarrow \bigcup_{n=2}^{m} Sub_{<i,j>,n}$
18:     **end if**
19: **end if**
20: **return** $Sub_{<i,j>}$

---

also brings enormous challenges to data processing. In recent years, there have been endless machine learning applications on Android, but most of these studies are focused on single applications [7,14,18]. Therefore, there is much room for improvement in the application of machine learning in solving the IAC problem. The workflow of the machine learning part within ACDroid is shown in Fig. 3.

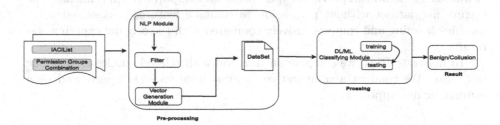

**Fig. 3.** The overview of machine learning part some words wrong in the picture

**Processing.** To expand the applicability of our approach, we have included one-class classification algorithms such as SVDD, DeepSVDD [15], binary classification algorithm SVM, LSTM, TextCNN [11], TextRNN [13] and so on. These algorithms could predict a correlation score based on the numeric feature vector outputted from the previous step.

Our evaluation criteria include True Negative Rate (TNR), Precision, Accuracy Rate, F-score and Recall Rate, which measure the performance of the classification algorithm in our data set. The one-class classification algorithm is thought to be more suitable for real-world anomaly detection scenarios, but the binary classification algorithm performs better via our experiment.

# 4     Empirical Evaluation

This section assesses our approach's effectiveness in classifying collusion and benign applications. We evaluate our system by answering the following research questions:

RQ1: What are the statistical results of Inter-App Communication analysis in real-world Android applications? (Section IV-A)

RQ2: What is the performance of our approach? (Section IV-B)

RQ3: What is the effect of using permission group combinations as machine learning features? (Section IV-C)

RQ4: What are the merits of our approach compared with other research? (Section IV-D)

To confirm RQ1, we constructed an application dataset suitable for collusion research from public platforms. We conduct our experiments using real-world Android apps, presenting the results of our inter-app communication analysis with data.

To answer RQ2, we apply our data to different machine learning algorithms for multiple independent experiments and then compare the performance of these algorithms. Finally, the results of machine learning experiments are summarized and analyzed.

To prove the importance of permission group combination as our machine learning feature (RQ3), we selected the machine learning algorithm that was evaluated as the best in previous experiments for comparative experiments. The feature file dataset without permission to combine features is classified using machine learning and comprehensively compared with the original experimental results.

To illustrate RQ4, we compare our method with related studies on a public test suite. We conduct a comparative analysis from various aspects and then summarize our approach.

## 4.1     RQ1: Our Data Set and Effectiveness of Static Analysis

Unlike the researchers in [9], who wrote the collusive applications themselves, all the applications in our dataset are from the application platform to be more convincing. We built our data set by exploiting several well-known Android application repositories (like Google Play and Baidu App Store) composed of legitimate Android applications. There are about 5000 of these applications, including about 2000 capable of ICC communication, Table 2 illustrates the distribution of applications from the data set used in our experiments, showing that they are sufficiently diverse from different categories. Data is divided into training and test sets, and their distributions are basically the same.

**Table 2.** Distribution the applications from our data-set

| App type | Date Sets | | |
|---|---|---|---|
| | Training set | Test set | Tall |
| Weather | 0.7% | 1.0% | 1.7% |
| Music& audio | 1.3% | 1.4% | 2.7% |
| Business tools | 4.2% | 3.6% | 7.8% |
| Entertainment | 6.9% | 6.2% | 13.1% |
| Efficient Office | 7.5% | 7.8% | 15.3% |
| Photography | 5.3% | 4.6% | 9.9% |
| Shopping | 8.6% | 9.5% | 18.1% |
| System Tool | 16.8% | 13.9% | 30.7% |
| Game& Auxiliary | 11.5% | 17.4% | 28.9% |
| Health& Medical | 1.9% | 3.3% | 5.2% |
| Learning& Education | 1.9% | 4.2% | 6.1% |
| Books& News | 7.3% | 6.5% | 13.8% |
| Social Networking | 10.8% | 9.6% | 20.4% |
| Finance | 3.5% | 3.0% | 6.5% |
| Domestic Services | 7.9% | 8.8% | 16.7 % |

We download applications from the same manufacturer or developer from the application market, and the combination of the same manufacturer is used as a training set. To account for the more frequent use of ICC links between applications of the same manufacturer for cross-application communication, we compared applications in the training set by shuffling and randomly combining them into pairs of applications. The red dots in Fig. 4 represent application pairs connected according to the same manufacturer, and the green dots represent randomly combined application pairs. We set a gray horizontal slice when the number of ICC connections is 20. It is not difficult to find that the red points are much more than the green points, and the ICC links are also more numerous.

To be more intuitive, we further use CDF (Cumulative Distribution Function) maps to illustrate. Figure 5 is the CDF diagram of the data in Fig. 4, which is the integral of the probability density function. It can completely describe the probability distribution of the number of ICC links. It can be seen from the CDF diagram that the ICC links in the application pairs matched by the same manufacturer are more evenly distributed and more numerous than random matches.

**Fig. 4.** Describe the comparison of ICC links obtained by different matching methods

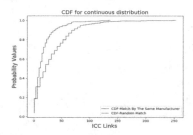

**Fig. 5.** The CDF of ICC links obtained by different matching methods

Experimental data show that our constructed dataset is suitable for collusion research. ACDroid performs well in extracting ICC links for inter-application communication, and our dataset feature extraction can meet the requirements for further machine-learning classification.

### 4.2 RQ2: Performance

In this research question, we will introduce the performance of our approach at runtime. Measure ACDroid true positives (TP), true negatives (TN), false positives (FP), and false negatives (FN) based on what we manually label. Using the TP, TN, FP, and FN, further calculate the TNR, Recall, Precision, F-score, and Accuracy. We will conduct experiments and evaluate various classification algorithms on the data set to choose the most suitable machine-learning method for our approach. The experimental process and evaluation results are introduced in the following two sections.

**One-Class Classification.** In the course of the experiment, combined with relevant knowledge and our data set, it was found that communication between apps is widespread. Although there are some insecure situations, safe communication still accounts for the majority. In other words, the ratio of safe examples to unsafe examples during applications of IAC is seriously out of balance, regardless of the data set or real-world situation. As a result of imbalanced positive and negative samples, our first thought is to focus on the security category with a large percentage. We believe that one-class classification algorithms may perform well in collusion scenarios. We evaluated a range of one-class classifiers, and we finally focused on the support vector data description (SVDD). SVDD is a one-class classification approach expanded based on the principles of the SVM, which also has the characteristics of the boundary method

**Table 3.** The Performance of One-Class Classification Algorithms

| Algorithms | TNR | Recall | Precision | F-score | Accuracy |
|---|---|---|---|---|---|
| SVDD | 99.37% | 10.00% | 92.86 | 18.06% | 59.17% |
| Deep-SVDD | 100.00% | 10.87% | 100.00% | 31.51% | 65.25% |

In the case of SVDD, our experiment did not perform satisfactorily. Therefore, a new one-class classification algorithm based on SVDD, Deep SVDD [15], by deploying a deep neural network, has been further carried out in the place of the kernel function. The neural network was trained to extract a data-dependent representation, removing the need to choose an appropriate kernel function by hand. We list the experimental results of our data set on the one-class classification algorithms in Table 3.

Although we are full of expectations for the one-class classification algorithm, their performance on our data set is unsatisfactory. So we further conducted experiments on the Binary classification algorithms.

**Binary Classification.** For the experiments in this part, we will use a classic binary classification algorithm. Since we are processing our features based on natural language processing, the experiments are more adapted to algorithms that perform well in text classification. Among many binary classification algorithms, we chose SVM, textCNN, textRNN, and LSTM to conduct our experiments.

**Table 4.** The Performance of Different Binary Classification

| Algorithms | TNR | Recall | Precision | F-score | Accuracy |
|---|---|---|---|---|---|
| SVM | 35.59% | 56.90% | 8.8% | 15.24% | 37.69% |
| LSTM | 100.00% | 96.55% | 100.00% | 98.25% | 99.66% |
| TextCNN | 99.62% | 74.14% | 95.56% | 83.49% | 97.11% |
| TextRNN | 99.62% | 31.03% | 90.00% | 46.15% | 92.87% |

Table 4 shows the evaluation results of different binary classification algorithms when they are tuned to the optimal state. It is not difficult to find that the performance of LSTM is better than other experimental algorithms on each evaluation data. Compared with LSTM, the two one-class classification algorithms in Table 3 have good judgment accuracy for positive examples but are very insensitive to negative classes. Some traditional machine learning effects in the experiment are even more unsatisfactory, so we haven't shown them. Although TextCNN performs well in Table 4, it is still slightly inferior to LSTM. From multiple aspects of evaluation, our method finally chooses the deep learning algorithm LSTM as the classification algorithm of the machine learning part.

By tweaking the hyperparameters during our machine learning training. Different hyperparameters have different effects on the model. We mainly focus on hidden size, batch size, and layer. The variations in the evaluation results of three distinct deep learning algorithms during the tuning procedure are depicted in Fig. 6, Appendix 8. Search for optimal hyperparameters for each model algorithm separately so that all results in Fig. 6 are based on optimal hyperparameters. Other hyperparameters may have been adjusted in the experiment but remained unchanged in the end, not all of which are listed here. After tuning the epoch, we find that each algorithm achieves slightly higher test accuracy. Based on the results of the experiments, Fig. 2 depicts the convergence of each algorithm that achieved reliable performance. When the epoch was adjusted to about 30, we found that all three algorithms converged and reached stability. Through the performance comparison of the three algorithms, it is not difficult to see that the F-score and Recall showed in Fig. 6(b) and Fig. 6(d), LSTM is much better than the other two algorithms. This demonstrates that LSTM performs well for both positive and negative examples and is appropriate for the dataset.

We next describe how stable the performance of LSTM is In our classification scene when hyperparameter values are varied. Figure 7, Appendix 8 shows the performance of LSTM for different epochs, hidden layers, hidden sizes, and batch sizes. Accuracy and TNR are relatively smooth for all hyperparameters, which has a certain relationship with the proportion of negative samples in our application scenario. It shows that LSTM's judgment accuracy rate for negative samples is stable and high. In contrast, Recall and F-score have large oscillations in the tuning process of all hyperparameters. Significantly when the hidden layers exceed 5, the model is obviously overfitting. However, after we get the optimal LSTM model, we compare it with other algorithms. As shown in Fig. 6, we still can observe that the LSTM curve is higher than the curves of TextCNN and TextRNN.

### 4.3    RQ3: Effectiveness of Permission Groups Combine

In this experiment, our purpose is to prove the validity of the permission group combination. In the comparison experiment, to form a more intuitive comparison, we removed the permission group combination feature in the feature extraction part.

In the previous chapters, we have introduced the necessity of permission group combinations. During the experiment, we used the best-performing LSTM model. Without changing other hyperparameters, we continuously adjust the epoch to conduct comparative experiments, and the results are shown in Fig. 8, Appendix 8. The epochs at which the two datasets reach convergence are basically the same, meaning the training speed is not much different.

Throughout the experimental performance curves of Fig. 8, the solid red line is consistently higher than the blue dotted line. All performance functions are better for datasets with permission group combinations. We can find that the Accuracy and F-Score of our original feature datasets, which has the feature permission groups combination are better for classification under the same conditions. In other words, the combination of permission group features can effectively improve our classification effect. Adding permission group combinations can help distinguish collusive applications from benign applications.

## 4.4  RQ4: Comparison with Alternative Approaches

To evaluate the effectiveness of our approach, we conducted experiments using a benchmark test suite related to inter-application communication. The main reason is that the analysis results of these benchmark suites can be publicly queried and confirmed. We run different analytical tools on these public test sets to identify collusion and compare their verdicts with our method.

The availability of Android benchmark suites has increased with the development of Android security. However, our research options are still limited since very few of them can be used for collusion research [3]. For the above reasons, we used the benchmark suites, including DroidBench 3.0 and DroidBench (IccTA), which contain test cases for inter-app communication experiments. We have chosen 17 benchmark apps from them that have inter-application communication as indicated by the benchmark providers.

**Table 5.** ACDroid Experimental Results on DroidBench 3.0

| Total Apps | # Total App Pairs | # Collusion App Pairs | # Benign App Pairs |
|---|---|---|---|
| 11 | 55 | 9 | 46 |

DroidBench 3.0 is an open test suite for evaluating the effectiveness of tools for Android apps, which comprises 190 test cases as of December 2021. We conduct experiments using 11 inter-application communication benchmark applications provided by DroidBench 3.0. We randomly paired these applications into 55 application pairs and displayed the classification results of the above test application pairs in Table 5 according to the developer's instructions. Of the 55 application pairs, only nine were collusive. This result will be compared with the experimental results of each tool. DroidBench (IccTA) is the IccTA branch of the DroidBench, which contains three pairs of test cases for inter-application communication analysis. Due to the small number of test cases available, we did not perform further matching but directly used these test cases.

The experimental results are described in detail in Table 7, Appendix 8. Covert performed poorly on DroidBench 3.0, missing all collusion detections and providing the highest false positives. This happens mainly because Covert focuses on explicit Intent, therefor the code covered is not comprehensive. Because IccTA+APK Combiner must combine applications before testing, the combination of numerous applications fails, causing it to crash, and no subsequent experiments could be performed. As a result, IccTA+Apk Combiner cannot be employed in most circumstances, and its experimental results are skewed. Experimental results show that its good performance seems limited to its dataset. DIALDroid filters out ICC communications that are considered insensitive to reduce computational complexity. At the same time, to avoid profiling into deadlocks, the analyzer profiles the application for a maximum of 20 min. DAILDroid's storage filtering and analysis time limit improve computational efficiency and bring accuracy problems. In addition, DAILDroid does not accurately match whether the data in the Intent meets the requirements of the receiving component, which also brings false positives. Table 6 shows the comparison results of our approach with other methods on two benchmarks. Compared with the other three methods, ACDroid performs better.

**Table 6.** Compare ACDroid with Other Approach

|           | COVERT | # IccTA +APK Combiner | DIALDroid | **ACDroid** |
|-----------|--------|------------------------|-----------|-------------|
| **Precision** | 17.65% | 13.04% | 71.43% | 92.31% |
| **Recall** | 25.00% | 25.00% | 83.33% | 100% |
| **F-score** | 20.69% | 17.14% | 76.92% | 96% |

## 5  Discussion and Limitations

### A. Inter-App Communication Analysis

In our study, only the colluding applications that communicate with each other through the intents are considered. However, in practical applications, inter-application communication can also be performed by content providers, shared preferences, external storage, etc.

Although we have tried our best to obtain a comprehensive analysis of explicit and implicit intents to avoid ignoring implicit intentions like [9]. There are still many IAC methods besides intents. In the forward, our research will face solving the problem of a broader range of IAC communication methods. Meanwhile, our IAC analysis employs a static approach. So like other static analysis-based methods, our method also has limitations when analyzing programs that use dynamic code loading or obfuscation techniques. To overcome these problems, dynamic analysis techniques will be introduced in future research to improve the whole project.

## B. ICC Data Flow Analysis

Our approach is based on analyzing DEX files and does not work at the java level like soot and the soot-based tool [12], which makes it difficult for us to conduct a comprehensive data flow analysis. As a result, we sometimes cannot fully retrieve the content provided by ICCs. We believe that the transmission content is arbitrary information, resulting in some false positive samples. For follow-up research, we plan to use Flowdroid for IDE analysis and combine cross-application communication to improve data flow.

## 6   Related Work

Malware collusion is a more emerging threat to Android security than other attack methods. Related technical research is growing and improving. We will analyze the existing advanced search from three main aspects: inter-app analysis, application collusion analysis, and machine learning search on Android.

**Inter-App Analysis.** ComDroid [8] is a tool that evaluates Android applications for potential inter-application communication vulnerabilities. Despite the fact that its study focuses on Inter-Application Communication, it merely detects specific application components and tracks potential component and Intent vulnerabilities, but it does not validate the existence of attacks. The developer is responsible for reviewing the applications and ensuring the veracity of the warnings. PermissionFlow [16] is an automated rule-based static pollution analysis tool that automatically generates rules specifying how permissions leak to unauthorized applications. Its research focuses on permission collusion, confused deputy, and intent spoofing attacks.

**Collusion Analysis.** Xmandroid [5] is a dynamic framework proposed by Bugiel et al. that allows the monitoring of communication links between applications on Android at runtime, extending Android's reference monitor. It mainly targets two types of privilege escalation attacks (confused deputy attacks and collusion attacks) and is validated against the security rules defined in the system policy. Since the defined system policies are not well adapted to the application scenarios, XManDroid may have a large number of false positives when the application has too many privileges. Elish et al. presented an end-to-end and static analysis algorithm to screen apps for possible collusion [9]. Based on features extracted from ICC flows between applications, a risk classification method is designed to calculate the risk of ICC between applications. Thereby, it is determined whether there is a risk of collusion between end-to-end application pairs. Unfortunately, the research of Elish et al. only focuses on explicit intent, so it cannot fully implement collusion detection. 2020malicious [6]

**Machine Learning Research on IAC.** Xu et al. proposed ICCDetector [17] for malware detection. The ICC features are extracted using static analysis, and the SVM algorithm is used to train the detection model for machine learning classification. Based on the ICC model, five additional malware categories are presented, and finally, identified malware will be sorted into the relevant categories. However, ICCDetector can only identify a single application for classification and cannot detect collusion.

# 7   Conclusions and Future Work

We propose a machine learning-based collusion detection system that combines static analysis and machine learning techniques to detect collusion attacks. Static analysis is used to acquire inter-application ICCs and individual application permissions, and algorithms are designed to construct risky permission group combinations. To gain cooperative features, increase the detection accuracy of collusion attacks, and propose a novel idea for collusion attack detection. We experimented with various machine learning algorithms, confirming which worked best for collusion attack categorization scenarios. Experiments using publicly available test suites illustrate the efficacy of our strategy. In future work, we explore including the dynamic analysis approach to address the inherent flaws of static analysis and improve feature extraction.

**Acknowledgements.** This work was supported by the Major Research plan of the National Natural Science Foundation of China (Grant No. 92267204, 92167203), Natural Science Basis Research Plan in Shaanxi Province of China (Grant No. 2022JM-338).

# 8   Appendix

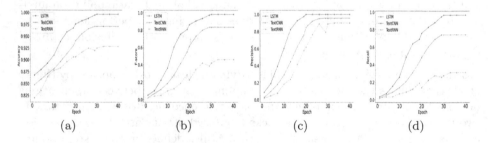

**Fig. 6.** Comparison of different algorithms

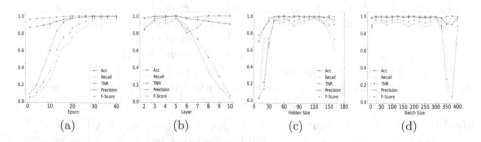

**Fig. 7.** Performance for LSTM as a function of four hyperparameters: epoch(a), hidden layers (b), hidden size (c) and batch size (d).

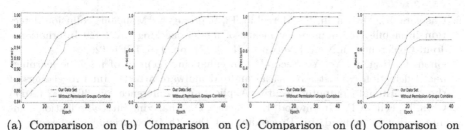

(a) Comparison on Accuracy    (b) Comparison on F-Score    (c) Comparison on Precision    (d) Comparison on Recall

**Fig. 8.** Comparison of the classification effects of using feature permission groups combination (Color figure online)

**Table 7.** Compare ACDroid with Other Approach on DroidBench

| Source Apps | Destination Apps | # Inter-App ICC Links | Collusion | COVERT | # IccTA + APK Combiner | DIALDroid | ACDroid |
|---|---|---|---|---|---|---|---|
| DroidBench 3.0 | | | | | | | |
| DeviceId_Broadcast1 | Collector | 1 | √ | ○ | ○ | ☑ | ☑ |
| DeviceId_ContentProvider1 | Collector | 1 | √ | ○ | ○ | ☑ | ☑ |
| DeviceId_OrderedIntent1 | Collector | 1 | √ | ○ | ○* | ☑ | ☑ |
| DeviceId_Service1 | Collector | 1 | √ | ○ | ○* | ○ | ☑ |
| Location1 | Collector | 1 | √ | ○ | ○* | ☑ | ☑ |
| Location_Broadcast1 | Collector | 1 | √ | ○ | ○* | ☑ | ☑ |
| Location_Service1 | Collector | 1 | √ | ○ | ○* | ○ | ☑ |
| SendSMS | Echoer | 1 | √ | ○ | ○* | ☑ | ☑ |
| StartActivityForResult1 | Echoer | 1 | √ | ○ | ○* | ☑ | ☑ |
| Benign app pairings(46) | | | × | ☒14 | ☒20* | ☒4 | ☒1 |
| DroidBench(IccTA) | | | | | | | |
| startActivity1_source | startActivity1_sink | 1 | √ | ☑ | ☑ | ☑ | ☑ |
| startSevice1_source | startService1_sink | 1 | √ | ☑ | ☑ | ☑ | ☑ |
| sendbroadcast1_source | sendbroadcast1_sink | 1 | √ | ☑ | ☑ | ☑ | ☑ |

* √ = collusion occurs, ×= Collusion did not occur, ○ = a missing report, * = can't tested due to crash, ☑ = a correct test result, ☒ = a false result

# References

1. Bagheri, H., Sadeghi, A., Garcia, J., Malek, S.: COVERT: compositional analysis of android inter-app permission leakage. IEEE Trans. Software Eng. **41**(9), 866–886 (2015)
2. Bhandari, S., Laxmi, V., Zemmari, A., Gaur, M.S.: Intersection automata based model for android application collusion. In: 2016 IEEE 30th International Conference on Advanced Information Networking and Applications (AINA), pp. 901–908. IEEE (2016)
3. Blasco, J., Chen, T.M.: Automated generation of colluding apps for experimental research. J. Comput. Virol. Hacking Tech. **14**(2), 127–138 (2018)
4. Bosu, A., Liu, F., Yao, D., Wang, G.: Collusive data leak and more: large-scale threat analysis of inter-app communications. In: Proceedings of the 2017 ACM on Asia Conference on Computer and Communications Security, pp. 71–85 (2017)
5. Bugiel, S., Davi, L., Dmitrienko, A., Fischer, T., Sadeghi, A.R.: XmanDroid: a new android evolution to mitigate privilege escalation attacks. Technische Universität Darmstadt, Technical Report TR-2011-04 (2011)

6. Casolare, R., Martinelli, F., Mercaldo, F., Santone, A.: Malicious collusion detection in mobile environment by means of model checking. In: 2020 International Joint Conference on Neural Networks (IJCNN), pp. 1–6. IEEE (2020)

7. Chen, L., Hou, S., Ye, Y.: SecureDroid: enhancing security of machine learning-based detection against adversarial android malware attacks. In: Proceedings of the 33rd Annual Computer Security Applications Conference, pp. 362–372 (2017)

8. Chin, E., Felt, A.P., Greenwood, K., Wagner, D.: Analyzing inter-application communication in android. In: Proceedings of the 9th International Conference on Mobile Systems, Applications, and Services, pp. 239–252 (2011)

9. Elish, K.O., Cai, H., Barton, D., Yao, D., Ryder, B.G.: Identifying mobile inter-app communication risks. IEEE Trans. Mob. Comput. **19**(1), 90–102 (2018)

10. He, Y., Li, Q.: Detecting and defending against inter-app permission leaks in android apps. In: 2016 IEEE 35th International Performance Computing and Communications Conference (IPCCC), pp. 1–7. IEEE (2016)

11. Kim, Y.: Convolutional neural networks for sentence classification (2014). https://doi.org/10.48550/ARXIV.1408.5882. https://arxiv.org/abs/1408.5882

12. Li, L., et al.: IccTA: detecting inter-component privacy leaks in android apps. In: 2015 IEEE/ACM 37th IEEE International Conference on Software Engineering, vol. 1, pp. 280–291. IEEE (2015)

13. Liu, P., Qiu, X., Huang, X.: Recurrent neural network for text classification with multi-task learning. arXiv preprint arXiv:1605.05101 (2016)

14. Mahindru, A., Sangal, A.: MLDroid-framework for android malware detection using machine learning techniques. Neural Comput. Appl. **33**(10), 5183–5240 (2021)

15. Ruff, L., et al.: Deep one-class classification. In: International Conference on Machine Learning, pp. 4393–4402. PMLR (2018)

16. Sbîrlea, D., Burke, M.G., Guarnieri, S., Pistoia, M., Sarkar, V.: Automatic detection of inter-application permission leaks in android applications. IBM J. Res. Dev. **57**(6), 10–1 (2013)

17. Xu, K., Li, Y., Deng, R.H.: ICCDetector: ICC-based malware detection on android. IEEE Trans. Inf. Forensics Secur. **11**(6), 1252–1264 (2016)

18. Zhu, D., Jin, H., Yang, Y., Wu, D., Chen, W.: DeepFlow: deep learning-based malware detection by mining android application for abnormal usage of sensitive data. In: 2017 IEEE Symposium on Computers and Communications (ISCC), pp. 438–443. IEEE (2017)

# DomainIsolation: Lightweight Intra-enclave Isolation for Confidential Virtual Machines

Wenwen Ruan[1,2], Wenhao Wang[1,2(✉)], Shuang Liu[3], Ran Duan[3], and Shoumeng Yan[3]

[1] State Key Laboratory of Information Security, Institute of Information Engineering, CAS, Beijing, China
{ruanwenwen,wangwenhao}@iie.ac.cn
[2] University of Chinese Academy of Sciences, Beijing, China
[3] Ant Group, Hangzhou, China
{ls123674,dr264275,shoumeng.ysm}@antgroup.com

**Abstract.** In recent years, there has been a rise in the use of confidential computing as a new computing paradigm that enables privacy-preserving computation on sensitive and regulated data. This approach relies heavily on hardware-based Trusted Execution Environments (TEE), which establish isolated regions for data processing within a protected CPU region. Currently, a variety of TEEs (such as p-enclave in HyperEnclave and AMD SEV) support privilege separation and running a fully-fledged operating system within the confidential Virtual Machines (VMs). However, running a fully-fledged operating system inevitably increases the trusted computing base (TCB), making it challenging to conduct security verification.

To address the problem, this paper studies the cases when complex OS services (such as device drivers and networking etc.) are removed from the confidential VM, and presents DomainIsolation, a page table based lightweight and efficient isolation scheme. We show that DomainIsolation enhances both the security and performance of enclave applications through several case studies, including confinement for untrusted libraries, fine-grained data protection, and fast communication. We have integrated DomainIsolation with the Occlum library OS, Enarx and ported several real-world applications. The evaluations on common benchmarks and applications (such as NBench, Lighttpd, Redis and OpenSSL) show that DomainIsolation only introduces a low overhead (<2% in most cases).

## 1 Introduction

Data protection typically involves three stages: storage, in transmission, and in use. Currently, encryption is commonly used to protect data during storage

This work is supported by the National Natural Science Foundation of China (Grant No. 62272452), Ant Group and Henan Key Laboratory of Network Cryptography Technology (No. LNCT2020-A03).

© The Author(s), under exclusive license to Springer Nature Switzerland AG 2023
M. Yung et al. (Eds.): SciSec 2023, LNCS 14299, pp. 23–41, 2023.
https://doi.org/10.1007/978-3-031-45933-7_2

and in transmission, while protecting data in use is becoming an increasingly important issue, which is generally referred to as confidential computing. In this paper, we concentrate on privacy-preserving computations with Trusted Execution Environment (TEE). TEEs provide hardware-enforced memory partitions where sensitive data can be securely processed. Existing TEE designs support different levels of TEE abstractions, such as process-based (Intel's Software Guard eXtensions (SGX [24]), virtualization-based (AMD SEV [19]), separate worlds (ARM TrustZone [5]), and hybrid (Keystone [21]).

Intel's SGX is the most representative process-based TEE, which provides isolated regions (called enclaves) within the process address space. As such, a normal application needs to be partitioned into trusted and untrusted parts before SGX can protect it. The SGX programming model minimizes the trusted computing base (TCB), which only includes the CPU and the trusted code itself. However, it is difficult to protect existing applications without extensive code refactoring. For this purpose, Library Operating System (LibOS), such as Graphene-SGX [30] and Occlum [29], is introduced to run existing applications with minimum modifications in the enclaves.

Unfortunately, Intel SGX only provides a *monolithic protection model* within an enclave, such that all the enclave code runs in a single address space and has the same protection domain. This creates a challenge for developers who require isolation between their own code and third-party libraries. Within Intel SGX, third-party libraries linked with user enclave code run in the same enclave, which may make them susceptible to data leaks or remote code execution attacks. One example is the HeartBleed attack in the OpenSSL library [33]. A small bug in processing heartbeat messages may allow the attackers to extract information from arbitrary freed buffers within the enclave that were linked with the OpenSSL library. As a result, both hardware and software changes have been proposed to provide efficient intra-enclave isolation. However, hardware-based methods (such as Nested Enclave [28] and LIGHTENCLAVE [13]) require hardware modifications and cannot be applied to existing commercial CPUs, while software-based solutions (such as WARM [2] and Occlum [29]) are evidently slower, e.g., as shown in [32], Occlum and WAMR are about 5.21x and 1.51x slower on average than the baseline on the same machine learning tasks.

In contrast to process-based TEE, VM-based TEEs, such as AMD SEV [19] and privileged enclave of HyperEnclave (i.e., p-enclave) [18], support privilege separation within TEEs. They are capable of supporting in-enclave page table management and exception handling, and can run fully-fledged operating systems inside the TEEs. As a result, existing applications can be protected by VM-based TEEs without huge modifications. However, introducing a complete OS inside the TEE will cause an expanding TCB and bring much bigger attack surfaces. In this paper, we study the following research question:

*Can we support efficient and fine-grained intra-TEE isolation for virtualization based TEEs without significantly expanding the TCB?*

To answer the question, this paper focuses on existing designs where complex OS services (such as device drivers and networking etc.) are removed from the confidential VM. The most representative designs include the Occlum LibOS with HyperEnclave [18] and Enarx [1] with AMD SEV. However, both of them only adopt the monolithic model which does not support fine-grained intra-enclave isolation[1], and Occlum even does not take advantage of the privilege separation feature at all. For this purpose, we propose a lightweight internal isolation scheme (called DomainIsolation) to support *fine-grained intra-enclave isolation among different security domains*. Specifically, a trusted software module (called DomMng) runs at the higher privilege level (i.e., guest ring 0), which maintains a shadow page table for every security domain running at the lower privilege level (i.e., guest ring 3), and supports the management of security domains with page table based memory isolation. Furthermore, we demonstrate that DomainIsolation enhances both the security and performance of enclave applications through several case studies, including confinement for untrusted libraries, fine-grained data protection, and fast communication.

We have implemented the prototype of DomainIsolation on Occlum with HyperEnclave and Enarx with AMD SEV respectively. The evaluations show that domain switches take 1330 and 1400 CPU cycles respectively, which is significantly lower then switches between different enclaves (9700 and 5360 CPU cycles). The evaluations of real-world applications shows that the performance overhead in most cases is less than 2%. As a result, for the first time we show that the proposed design achieves fined-grained intra-enclave isolation, great compatibility with existing applications without significantly increasing the TCB, and low performance overhead.

In summary, the paper makes the following contributions.

- *Lightweight and efficient intra-enclave isolation.* We present DomainIsolation for the management of security domains to support fine-grained intra-enclave isolation for confidential VMs.
- *Case studies.* We show that DomainIsolation enhances both the security and performance of enclave applications through several case studies, including confinement for untrusted libraries, fine-grained data protection, and fast communication.
- *Implementations and evaluations.* We implement a prototype of DomainIsolation which provides full compatibility with existing applications through customized modifications to Occlum and Enarx. The evaluations show that the design only incurs a minimal overhead (about 2%).

## 2    Background

### 2.1    Trusted Execution Environment (TEE)

Trusted Execution Environment (TEE) technology provides hardware-based secure partitions that ensure the integrity and confidentiality of sensitive code and data. These secure partitions can only be accessed by authorized entities,

---

[1] Since they support the same process-based TEE model as Intel SGX, we still refer to the TEE's isolated regions as enclaves for simplicity.

even though untrusted privileged software is still responsible for managing available resources such as CPU and memory. Remote attestation is a critical feature of TEEs that allows remote users to verify the genuineness of the TEE and ensure that it has not been compromised by malicious operators. We introduce existing TEEs in terms of their provided TEE abstractions as follows.

**Process-Based TEEs.** Currently, the TEE that gets the most attention from both the industry and academia is Intel's Software Guard eXtensions (SGX) [24]. SGX is an x86 extended instruction set that supports the management (such as creation, enter, exit, and destruction) of *enclaves*, where the enclave is a region within the *process*'s address space for the safe processing of sensitive code and data. Applications are divided into the trusted and untrusted parts, and only the trusted part needs to be protected by SGX. Although the process-based TEE represented by SGX has significantly reduced the Trusted Computing Base (TCB), determining the partition area and the workload required for the partition can be time-consuming for complex applications. Moreover, SGX cannot handle system calls (syscalls) inside the enclave, so it is needed to switch to untrusted operating systems (OSes) when processing syscalls. This can result in significant performance losses when the OS completes the syscall and then returns to enclaves. To address these challenges, researchers have proposed a solution that places LibOS in SGX. In addition to essential system services such as memory management and file systems, LibOS provides POSIX abstraction for applications, reducing their dependence on the untrusted world and facilitating the transplantation of existing applications. Gramine [30] and Occlum [29] are currently the most widely used SGX LibOSes. Compared with Gramine which is mostly written in C, Occlum is written using the memory-safe language Rust.

**VM-Based TEEs.** Process-based TEEs do not provide privilege separation inside enclaves. In recent years, VM-based TEEs, such as those offered by AMD SEV [19], Intel TDX [17], IBM PEF [15], ARM CCA [6], and p-enclave in Hyper-Enclave [18], are paving the way for the future of confidential cloud computing. One key advantage of these TEEs is their ability to securely run fully-fledged guest operating systems, such as common Linux distributions like Ubuntu, without requiring modifications to existing toolchains or applications. This makes it possible to run complex software like language runtimes, deep learning frameworks, and device drivers within the TEE. To support the functionality of the TEE, the host software (QEMU and KVM) and guest operating system cooperate in a para-virtualization manner to securely manage VM resources.

### 2.2 Enclave Abstraction Model

The intricate architecture of OSes (such as Linux) poses a significant challenge to running fully-fledged guest OSes within TEEs, which can potentially expose numerous attack vectors. The attackers can leverage these vulnerabilities to circumvent TEE protections, therefore enabling the theft of sensitive information or compromise of TEE's integrity. To reduce the attack surface, this paper studies the support of process-based *enclave abstraction model* within VM-based TEEs, which is explained below (Fig. 1).

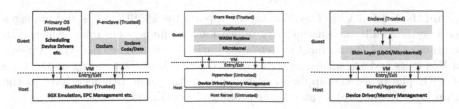

(a) Occlum with HyperEn-  (b) Enarx with AMD  (c) Unified enclave abstrac-
clave                      SEV                  tion model

**Fig. 1.** Enclave abstraction model.

**Occlum with HyperEnclave.** HyperEnclave is constructed by introducing a tiny *trusted* hypervisor – RustMonitor, which operates in the host mode [18]. RustMonitor creates a normal VM to host the untrusted primary OS, and a separate VM for each enclave. HyperEnclave supports SGX's enclave abstraction model by emulating SGX instructions during the transitions between the enclave and primary OS. With the support of Rust-SGX toolchains [31] and Occlum [29] on top of HyperEnclave, existing SGX applications can run on HyperEnclave with only minimal or no code modifications (Fig. 1a).

Despite running in a VM environment, the enclave does not run a complicated OS and does not utilize any OS services. Specifically, the enclave's page tables are managed by RustMonitor, and the enclave follows a one-to-one mapping between the guest virtual address and guest physical address. As a result, the entire enclave runs in a monolithic model.

**Enarx with AMD SEV.** Enarx is an open source project that enables applications to run within TEEs without rewriting for particular platforms (such as Intel SGX and AMD SEV) [1]. On AMD SEV, Enarx provides a four-layer runtime stack (Fig. 1b). The untrusted kernel and hypervisor provide memory management for processes within the Enarx Keep[2]. Inside the trusted VM, Enarx provides a microkernel running at guest ring-0 and a WebAssembly (WASM) runtime running at guest ring-3 that provides the runtime environment for applications within the Enarx Keep. The application layer is the workload provided by the users to run within the Enarx Keep, which could be written in high-level languages like C, C++ and Rust and compiled to WASM bytecode.

**Unified Enclave Abstraction Model.** In this paper, we propose a unified enclave abstraction model for VM-based TEEs as shown in Fig. 1c. Our design of the lightweight in-enclave isolation scheme on top of the unified enclave abstraction model will be introduced in Sect. 3.

Specifically, the model involves the following components. The first two components are the host kernel and hypervisor, both of which may be either trusted or untrusted. To enable the running of existing applications without any required modifications, a shim layer (e.g., Occlum LibOS and micro-kernel) is included

---

[2] Keep is Enarx's term for TEE instances, such as enclaves.

within the trusted VM. The shim layer supports the POSIX APIs and manages basic system calls such as memory allocation and file system management. To minimize the TCB size, most system calls initiated within the trusted VM result in VM exits and are subsequently forwarded to the host. Depending on the type of TEE platforms, the application may run at the same privilege level as the shim layer (p-enclave) or at a lower privilege level (Enarx).

### 2.3   Assumptions and Threat Model

This paper proposes lightweight intra-enclave isolation scheme for the above-mentioned enclave abstraction model on top of existing TEEs. Therefore, we assume that the underlying TEE platforms (i.e., HyperEnclave and AMD SEV) as well as the shim layer (i.e., Occlum LibOS and Enarx) are trustworthy. Our design relies on the TEE designs to prevent unauthorized access to the enclave from external parties, including privileged software attacks [35], cold boot attacks [14], and DMA attacks [22]. Similarly, side channel attacks (such as Spectre [20], micro-architecture attacks [11]) and denial-of-service attacks are considered out of scope for this paper.

We assume that multiple *security domains* may exist in the enclave, which is common for complex applications protected by the enclave that may involve untrusted third-party libraries. Our design introduces a lightweight domain manager (called DomMng) running at the guest VM's privileged level (called kernel mode), which is trusted and provides intra-enclave isolation among these security domains. The security domains run at the guest VM's user level (called user mode). In this case, vulnerabilities may exists in certain security domain (e.g., the one that hosts third-party libraries) and could be exploited by the attacker to take full control of the security domain. In this case, the attacker intends to break the domain isolation, and tries to tamper with the execution or to access the secret of other security domains.

## 3   Design

### 3.1   Overview

Modern applications are too complex to make them secure, especially when integrating existing untrusted third-party libraries to reduce development complexity. Such libraries often contain security vulnerabilities that can compromise the security of the entire application. One notable example is the HeartBleed vulnerability in OpenSSL [33], which allowed attackers to exploit a buffer overflow in processing heartbeat messages to access the server's in-memory secrets, such as private keys, user sessions, and passwords.

In this paper, we focus on providing fine-grained protection for applications under the enclave abstraction model outlined in Sect. 2.2. Existing designs only support the monolithic protection model, meaning that all enclave code and data run in the same address space. This leaves applications vulnerable to attacks that exploit vulnerabilities in third-party libraries, potentially leading to secret breaches or control flow manipulation. To address these issues, a strong isolation mechanism is essential. By compartmentalizing a large application into smaller pieces and mapping them to different protection domains, developers can confine any unintended damages caused by malicious or misbehaving components.

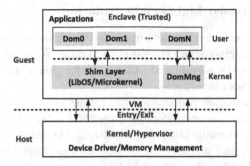

**Fig. 2.** `DomainIsolation` architecture under the enclave abstraction model.

In particular, we introduce the concept of a *security domain* as the smallest isolation unit within the enclave. Different security domains within the enclave are fully isolated from each other, providing an additional layer of protection against potential attacks. The architecture design of the domain isolation scheme (called `DomainIsolation`) is shown in Fig. 2. The enclave application runs in user mode and is divided into multiple security domains. The shim layer is trusted and runs in the kernel mode. We have introduced a small trusted domain manager (DomMng) that operates in kernel mode. DomMng is designed to enforce memory isolation among security domains (see Sect. 3.2). It is also responsible for managing security domains, including domain creation and switching (see Sect. 3.3). In this paper, we enforce the following two kinds of intra-enclave domain isolation:

- *Security domain-DomMng isolation*, where the DomMng runs in the kernel mode and security domains run in the user mode. The application in the security domains cannot tamper with the functionalities of the DomMng and the security domains are not trusted by DomMng.
- *Cross-security-domain isolation*, where the attacker who has compromised certain security domains does not have access to other security domains or the entire address space of the enclave.

Additionally, the design supports memory sharing across security domains. For instance, in an application serving multiple users, each user's security-sensitive data and code execute independently in their own security domain, while the desensitized data and shared libraries run in a shared security domain. This memory sharing also facilitates secure and efficient communication across security domains, which is much more efficient than cross-enclave communication when the security domains are isolated using multiple enclaves.

Since both of the existing designs are built upon AMD platforms (Hyper-Enclave and SEV), in the rest of paper we introduce our detailed design and implementation with x86/64 notations.

## 3.2 Page-Table-Based Lightweight Isolation

In the monolithic protection model supported by existing designs, all security domains within the enclave share the same page table. As a consequence, security domain $a$ can access the memory of other security domains, which can lead to security vulnerabilities and attacks.

To address this issue, we propose a more robust isolation mechanism – establishing an *independent security domain page tables* for each security domain. To support security domain-DomMng isolation, DomMng runs at the kernel mode of the enclave, and memory areas used by DomMng are marked as privileged, preventing illegal access attempts from user security domains. Additionally, DomMng marks the memory areas used by page tables as privileged, ensuring that user security domains cannot directly modify the page table. To enforce cross-security-domain isolation, we create different page tables for different security domains in the enclave. As a result, the security domain in user mode can only access the allocated memory area, whereas memory areas belonging to other security domains are marked as unmapped and inaccessible.

**Security Domain-DomMng Isolation.** To support security domain-DomMng isolation, we place DomMng in kernel mode, while security domains run in user mode. Switches between user mode and kernel mode can be triggered through system calls, interrupts, and exceptions. In this paper, we use system calls to switch the processor between kernel mode and user mode (see Sect. 3.3).

Specifically, to support memory isolation between the security domain and DomMng, we utilize the User/Supervisor (U/S) bit in the page table entries (PTEs). Specifically, the third bit in the PTE is the U/S bit, which indicates the access permission of the memory page regarding the privilege level. If the bit is set, code running in both user and kernel modes can access the page; otherwise, only the code running in kernel mode can access the page. Therefore, DomMng sets all the physical pages allocated as Supervisor during the initialization of the enclave. During the construction of a security domain, DomMng allocates memory for the security domain and creates a security domain page table. The allocated memory area is set to User in the security domain page table, ensuring that the security domain can only access the allocated memory region.

**Cross-Security-Domain Isolation.** In the enclave abstraction model, the host kernel and hypervisor allocate physical pages for the enclave. DomMng is responsible for memory management within the enclave. After the enclave VM is launched and the system enters the kernel mode, DomMng constructs a VM page table, which manages the entire internal memory in the enclave.

The security domain's page table is created during the process of establishing the security domain. To handle the synchronization problem between the VM page table and the security domain page table, DomMng establishes the same mapping in both tables when it allocates memory for the security domain. However, all physical pages are marked as Supervisor in the VM page table, while the mapping in the security domain is set to User. With this configuration, DomMng ensures that user-mode programs can only access the allocated memory regions through the security domain page table. When the security domain attempts to access an unallocated memory region, a page fault is triggered immediately, and control is transferred back to DomMng.

**Userspace Memory Management.** The userspace memory area within the enclave's address space is managed as a memory pool, which is divided into multiple virtual memory area (VMA) chunks. Each VMA chunk contains a virtual memory area and is assigned to a specific security domain. DomMng ensures that memory areas allocated to security domains do not overlap by managing these memory chunks.

**Shared Memory Among Security Domains.** Data transmission among different security domains is achieved through shared memory. Specifically, the shared memory area of security domains $a$ and $b$ is mapped to their respective security domain page tables and marked as User. However, the memory area is not mapped to the page table of other security domains. For public services, we provide a shared security domain, whose memory is mapped to all security domains' page tables and marked as User. This enables efficient and secure communication across different security domains.

### 3.3 Security Domain Life Cycle

**Bootstrapping the Enclave Execution.** After allocating the necessary resources for the enclave VM, the host triggers a VM entry (using the VMRUN instruction) to enter the shim layer within the enclave. The shim layer then initiates the bootstrap process, which includes setting up the enclave's page table (VM page table), as well as the stack and heap regions. In existing designs that use the monolithic protection model, the shim layer then hands control over to the user application's entry point, which starts the execution of user applications.

To support memory isolation among security domains, the shim layer hands control over to the DomMng before passing control to user domains. DomMng then creates separate page tables and prepares the heap and stack regions for each secure domain. After setting up the page tables, DomMng directs the execution to the user-mode entry point of the secure domain.

**Switches Between Security Domain and the Host.** When the execution needs to switch from the security domain to the host, such as when handling interrupts or system calls, the current execution context is automatically saved in the Virtual Machine Control Block (VMCB) by hardware. When re-entering the enclave, the CPU resumes the computation that the security domain was performing. These switches do not require the involvement of DomMng since the Current Privilege Level (CPL) in the VMCB records the privilege level at which the security domain executed. The security domain's state is then restored according to its register context.

**Switches Between Security Domains.** We provide a system call interface inside DomMng that enables a security domain to switch to another specified security domain. For example, when security domain $a$ intends to call functions in security domain $b$, it first needs to switch to security domain $b$. For this purpose, security domain $a$ can leverage the security domain switch function provided by DomMng, giving the unique identification of security domain $b$ as the input parameter. Upon entering kernel mode, DomMng saves the context of security domain $a$, loads the security domain $b$'s page table, flushes the TLB entries corresponding to domain $a$, and loads the context of security domain $b$ into the registers. Currently, we do not impose any security checks or restrictions during domain switches. It is the programmer's responsibility to ensure that domain switches are legitimate.

## 4   Implementations

We have implemented the prototype of DomainIsolation on AMD platforms that support Occlum with HyperEnclave and Enarx with AMD SEV respectively. The details are presented as follows.

### 4.1   Supporting Occlum on HyperEnclave

In the original design, Occlum leverages Memory Protection Extensitons (MPX) [34] for domain isolation in the enclave's address space. However, MPX has been deprecated by Intel due to possible defects [25]. DomainIsolation revives the support of the internal isolation in the enclave. The prototype has three major components: (1) Integrating DomainIsolation with Occlum; (2) system call handler; and (3) page table management for security domains.

**Integrating DomainIsolation with Occlum.** To confine untrusted security domains, the trust RustMonitor initializes the VMCB of the enclave VM to ensure the enclave's entry point (called OENTRY) points to the starting address of the shim layer (i.e., Occlum). RustMonitor also marks the CPL as 0 so that Occlum runs in the privilege mode. Before switching to the userspace security domain, Occlum invokes DomMng, who is responsible for the management of security domains.

DomMng requests memory from the enclave's memory pool, and manages the allocated memory to every security domain. It uses a red-black tree data structure to ensure efficient memory management and record memory in use. After the security domains' page tables are set up, DomMng invokes the task-switching code (see Fig. 3a). Within this function, DomMng saves the thread local storage of the kernel and switches to the userspace fsbase and stack.

Afterwards, DomMng establishes the security domain's page table (see "page table management" for details) and finally executes the sysretq instruction, which switches the CPU to the userspace security domain.

```
                                         1  ;entry point of syscall from
                                                standard library
                                         2  __dommng_syscall_linux_abi:
                                         3      push %rcx
                                         4      syscall
                                         5      ret
 1 ; save kernel fsbase                  6
 2 rdfsbase %r10                         7  __dommng_syscall_entry:
 3 movq %r10, TASK_KERNEL_FS             8      ; save rsp in r11
 4 ; use user fsbase                     9      movq %rsp, %r11
 5 movq %TASK_USER_FS, %r10             10      ; switch to the kernel stack
 6 wrfsbase %r10                        11      movq TASK_KERNEL_RSP, %rsp
 7 ; use user stack                     12      ;save the CPU state
 8 movq TASK_USER_RSP, %rsp             13      push %rcx ;save %rip
 9 ; run user code                      14      push %r11 ;save %rsp
10 movq TASK_USER_ENTRY_ADDR, %rcx      15      ...
11 ; change into user security mode     16      ; use kernel fsbase
      page table                        17      movq TASK_KERNEL_FS, %r11
12 movq TASK_USER_CR3, %cr3             18      wrfsbase %r11
13 ; use sysretq to jump into the user  19      ;change to the VM page table
      code and ring3                    20      movq TASK_KERNEL_CR3, %cr3
14 sysretqs                             21      call syscall_handler
```

|                (a) Task switching                |            (b) System call handler             |

**Fig. 3.** Code snippets of integrating DomMng with Occlum.

Security domains are specified by programmers, and each should be separate process. Third-party libraries and components that contain sensitive information should also be isolated. The number of security domains is not limited, but it is important to note that when a large number of security domains are allocated within an enclave, the protection granularity will become very small. At the same time, the switch of security domains will become very frequent, which leads to TLB refreshes and memory access delay, resulting in performance degradation.

**System Call Handler.** In the monotonic protection model supported by the original Occlum, all system calls have been removed with the modified Occlum-specific standard libraries (i.e., musl-libc). To support security domain switches by providing a system call based interface, we consider two common ways that an enclave application may use to invoke system calls. The first way is to use assembly code directly (i.e., asm ("syscall")). The second way is to invoke APIs provided by standard libraries such as glibc.

To handle the two aforementioned situations, we have implemented two entry points for entering kernel mode. The first entry point is __dommng_syscall_linux_abi, which is invoked at the system call entry point of standard libraries. The second entry point is __dommng_syscall_entry, and its address is stored in the model-specific register LSTAR. When the syscall instruction is triggered, the CPU jumps to the __dommng_syscall_entry entry point. This entry point saves the context, switches to the kernel stack, and processes the system call (see Fig. 3b).

**Page Table Management for Security Domains.** In HyperEnclave, the VM page table is managed by RustMonitor. To enable DomMng to manage the page table, we added a layer of linear mapping in RustMonitor for the physical pages where the page table resides. With this layer of linear mapping, DomMng can directly add a fixed offset value to the physical address pointed to by the PTEs to obtain the corresponding virtual address and access the corresponding memory area through this virtual address. In the current design of HyperEnclave, specifically p-enclave, all memory mapped by the PTEs in the enclave's page table is marked as privileged.

During the initialization of a security domain, DomMng performs a deep copy of the enclave's page table to establish a new page table for the security domain. However, since the memory is marked as privileged (as mentioned above), it is not currently accessible by the userspace security domains. When allocating memory (such as code, heap, and stack regions) from the memory pool to a specific security domain, DomMng updates the corresponding PTEs to make it accessible to the security domain. To achieve this, we maintain a hash table that records each virtual page in the enclave and the position where the corresponding PTE is located. DomMng looks up the hash table for the allocated memory to locate the PTEs and sets the U/S bit of the PTEs accordingly.

### 4.2   Supporting Enarx on AMD SEV

Most of the implementations used for Occlum can also be applied to Enarx. In this section, we provide a brief discussion on some key differences.

In terms of system call handling, Enarx offers the WASM runtime (wasmtime [8]), which supports the standard WebAssembly System Interface. This means that Enarx can run user applications written in RUST, C, and C++, compiled to WASM bytecode. As a result, there is no need to provide additional entry points for system calls besides __dommng_syscall_entry.

Additionally, Enarx manages the application's own page table. When a security domain is initialized, a new blank page table is created. Then, at the time of memory allocation, the mapping for the allocated memory is added to the security domain's page table.

### 4.3   Case Studies

In this section, we provide three case studies as presented in Fig. 4, to explain how DomainIsolation improves upon the monolithic protection model.

**Confinement.** We demonstrate the effectiveness of DomainIsolation in confining untrusted third-party libraries using OpenSSL [26]. We tested OpenSSL-1.0.1e [27] which is vulnerable to HeartBleed in Occlum. In the monolithic protection model of the original Occlum, the server code and OpenSSL share the entire address space. However, with DomainIsolation, we placed OpenSSL and

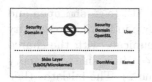
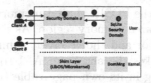
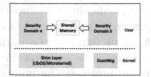

(a) Confinement             (b) Data protection       (c) Fast communication

**Fig. 4.** Case studies.

the server code in different security domains, effectively isolating them from each other. Our tests confirmed that the vulnerable OpenSSL code could not access sensitive data in the server's security domain, thus demonstrating the effectiveness of `DomainIsolation` in confining untrusted libraries.

**Fine-Grained Data Protection.** In this case study, we consider the scenario that a specific security domain provides shared SQLite services to other security domains, and *only the SQLite security domain has accesses to the clients' request and data.*

Besides the SQLite security domain, we implemented two security domains $a$ and $b$ in the enclave, providing services to two remote client $A$ and $B$. To handle the request, ❶ firstly, client $A$ sends an encrypted request to security domain $a$ with a pre-shared secret key. ❷ Security domain $a$ decrypts the request and filters out the client's private data. ❸ Security domain $a$ requests service from the SQLite security domain and sends relevant query requests. ❹ The SQLite security domain queries the database and ❺ returns the results to security domain $a$, ❻ which then encrypts the results and sends them to the client $A$.

**Fast Communication.** In this case study, we introduce the concept of a shared security domain, which is accessible by other security domains by setting the U/S bit in the corresponding PTEs. As the enclave is protected from the untrusted world, security domains can establish a fast communication channel through the shared security domain without the need of exiting the enclave.

```
1 // Initiator initiazes a shared
     memory
2 fd = shm_open("shm_file", O_CREATE|
     O_RDWR, 0777);
3 // Set the shared memory size
4 ftruncate(fd, size);
5 // Mapping the memory in the
     security domain
6 data = mmap(NULL, size, PROT_READ|
     PROT_WRITE, MAP_SHARED, fd, 0);
7 sprintf(data, "share memory%d\n",
     fd);
```

```
1 // security domain receiver
2 fd = shm_open("shim_file", O_RDWR,
     0777);
3 data = mmap(NULL, size, PROT_READ,
     MAP_SHARED, fd, 0);
4 // read data
5 printf(data);
```

(a) Initiator initializes shared memory.    (b) Receiver reads the shared memory.

**Fig. 5.** Example code for fast communication.

Specifically, we support the POSIX shared-memory API (e.g., `shm_open` and `shm_unlink`) to provide a shared memory region. As shown in Fig. 5, two security domains, namely the initiator and the receiver, communicate with the shared memory. The initiator invokes the `shm_open` API to create a shared memory region and sets the mapping in its own page table (line 2). The initiator then writes the data to the shared memory (line 7). On the other hand, the receiver opens the shared memory and maps it in its own page table using the `shm_open` API (line 2). Finally, the receiver can read the data from the shared memory (line 5).

## 5   Performance Evaluation

In this section, we present the performance evaluation of `DomainIsolation`. We conducted both micro-benchmarks and macro-benchmarks of `DomainIsolation` on the HyperEnclave platform equipped with AMD R3-5300G CPU and 16 GB RAM. The primary OS is Ubuntu 18.04 LTS with Linux kernel 5.3.0-28-generic. We used the HyperEnclave artifact evaluation version [16] and Occlum version 0.27.2. Since there is no public available benchmarks for Enarx, we only conducted micro-benchmarks on the Enarx platform equipped with a dual-socket AMD EPYC 7543 CPU and 64 GB RAM.

### 5.1   Micro-Benchmarks

**Table 1.** Micro-benchmarks (latencies are measured in CPU cycles).

| Platform | Security domain→DomMng | Security domain→security domain |
|---|---|---|
| Occlum/HyperEnclave | 166 | 1330 |
| Enarx/SEV | 140 | 1400 |

**Switches Between Security Domain and DomMng.** We conducted a latency measurement of switches between security domains and DomMng on both HyperEnclave and Enarx platforms. To obtain the latency values, we executed an empty system call with no explicit parameters 100,000 times and calculated the median value. The results, presented in Table 1, indicate that the switches took 166 and 140 CPU cycles for HyperEnclave and Enarx, respectively. It is worth noting that these latency values are significantly lower than those required for a world switch when domain isolation is enforced using separate enclaves. For instance, switching between the enclave VM and the host in HyperEnclave requires 9700 and 5360 CPU cycles, respectively.

**Switches Between Security Domains.** DomMng offers a new system call named `sd_switch` that enables switching between security domains by adding it to the system call table. In this evaluation, we created two security domains and invoked the sd_switch system call 100,000 times. We then calculated the median value, which is presented in Table 1. The results indicate that the switch took 1330 clock cycles on HyperEnclave and 1400 clock cycles on Enarx. During security domain switches, the time spent mainly involves context switches and page tables switches. In real-world applications, the overhead may also include additional TLB misses caused by context switches.

## 5.2 Real-World Workloads

In this section, we select four real-world applications for evaluation: an algorithm benchmark package NBench [23], a lightweight web server Lighttpd [3], a popular database Redis [4] and a simple echo server based on OpenSSL [26] as representations for the CPU-intensive, I/O-intensive, and memory-intensive task for DomainIsolation. We measured the performance on both DomMng and vanilla Occlum, and the baseline we refer to is vanilla Occlum running on the unmodified HyperEnclave, while the DomMng runs in the modified HyperEnclave for comparison.

**NBench.** NBench measures the performance of the CPU, FPU, and memory system of the system. As presented in Fig. 6a of Appendix A, the performance of DomMng is consistent with vanilla Occlum without obvious performance loss. In the test of the numeric sort function, DomMng is slightly higher than the baseline.

**Lighttpd.** We ran a Lighttpd (v1.4.40) server on HyperEnclave and vanilla Occlum. We used the Apache HTTP benchmark tool and generated 100 concurrent clients over the local loopback to obtain pages of various sizes to evaluate throughput. As presented in Fig. 6b of Appendix A, in terms of throughput, DomMng has a slightly 2% loss compared with vanilla Occlum.

**Redis.** We used Redis to evaluate the performance of memory and I/O-intensive scenarios. We ran a Redis server on DomMng and vanilla Occlum, and used YCSB [10] workload A. In the evaluation, 50000 records were loaded into the server, which is in total 50MB of data, and then performed 100,000 operations from 20 clients over the local loopback, while the request frequency was gradually increased and the latency under different throughput rates was calculated. As shown in Fig. 6c of Appendix A, DomainIsolation has a slight delay increase when reaching the same throughput rate, and can reach 97.5% of the baseline. It is worth noting that when the throughput reaches 15 kop/s, the delay caused by DomMng is doubled. The reason for the high delay is that DomMng also triggers the switch of VM page table and security domain page table during the system call, which leads to TLB refresh and causes memory access delay.

**OpenSSL.** We have built a simple Echo server and Echo client service. The server and client code is modified base on the mongoose demo [9], and the OpenSSL code is the vulnerable version [27]. We set the OpenSSL library and the security-sensitive server code to run in two separated security domains. The server opens port 8443, and the client sends data to the server every 5 s. The data increases from 256 Bytes to 256KB. We record the time server takes to receive the data from client and resend the data to client. We also conduct the same experiment on vailla Occlum. As shown in the Fig. 6d of Appendix A, when transmitted data is less than 2KB, there is almost no loss. While the data is greater than 2KB, the evaluation shows that the overhead is about 9%, which is caused by the function calls between the service code and OpenSSL, incurring extra security domain context switches and TLB flushes.

## 6   Related Works

**Hardware-Based Intra-enclave Isolation on Intel SGX.** Intel Memory Protection Keys (MPK) is a new hardware primitive that enables thread-local permission control on groups of pages without modifying page tables. With MPK, an application can partition its virtual address space into 16 memory domains. LIGHT-ENCLAVE [13] introduces a non-intrusive expansion scheme for SGX that can safely utilize the MPK features in the untrusted OS, with an overhead of up to 4%. Nested Enclave [28] introduces Multi-Level Security (MLS) [7] to the current TEE model, providing hierarchical security domains. However, both LIGHT-ENCLAVE and Nested Enclave require hardware modifications, making it challenging to apply them to commercial CPUs. In contrast, DomainIsolation provides intra-enclave isolation using existing hardware.

**Supporting SGX Model on AMD SEV.** vSGX [36] provides a virtualized execution environment for SGX enclaves on the AMD SEV platform using software, while maintaining the same security guarantees. In vSGX, the SGX enclave is placed within a SEV-protected enclave VM (EVM), and a microkernel is introduced within the EVM for memory and thread management. The untrusted portion of the application runs in the App VM (AVM), where users can install an OS and invoke SGX commands to build the EVM. Enclavisor [12] proposes a micro-kernel that runs in guest supervisor mode to balance security, performance, and flexibility in cloud computing. Enclavisor is designed to manage enclave instances running in guest user mode. Enclavisor's isolation is between multiple enclaves, while DomainIsolation implements intra-enclave isolation. Although both use page table isolation, Enclavisor does not involve security domain switching and other such features in DomainIsolation. These designs also follow the monolithic protection model in the enclave, and we anticipate that DomainIsolation can be similarly applied, although this is left as future work.

## 7   Conclusion

In this paper, we have presented DomainIsolation, a lightweight and efficient intra-enclave isolation scheme. We have integrated DomainIsolation with the Occlum library OS on the HyperEnclave platform and Enarx on the AMD SEV platform, demonstrating its effectiveness in preventing potential information leaks. Our evaluation of real-world applications shows that `DomainIsolation` introduces only a slight performance overhead, making it a practical solution for protecting sensitive data within enclaves. Furthermore, our approach is flexible and can be adapted to different hardware environments, providing a scalable solution for secure computing.

## A  Appendix

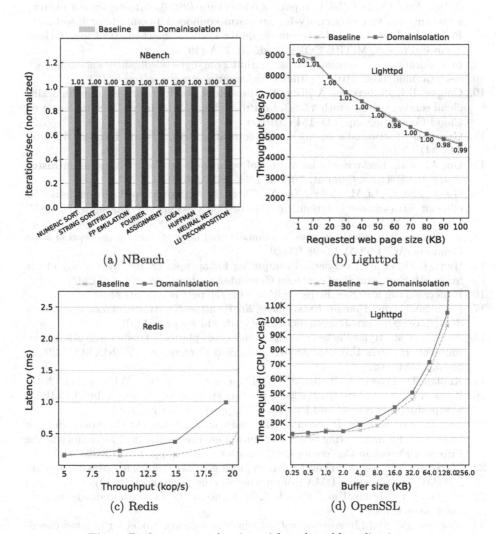

(a) NBench                          (b) Lighttpd

(c) Redis                           (d) OpenSSL

**Fig. 6.** Performance evaluation with real-world applications.

# References

1. Enarx: Confidential computing with webassembly. https://github.com/enarx/enarx/. Accessed 28 Mar 2023
2. Webassembly micro runtime. https://github.com/bytecodealliance/wasm-micro-runtime/. Accessed 13 Oct 2022
3. Lighttpd (2021). https://www.lighttpd.net
4. Redis (2022). https://redis.io
5. Alves, T., Felton, D.: Trustzone: integrated hardware and software security. White paper (2004)
6. ARM: Arm Confidential Compute Architecture (2020). https://www.arm.com/why-arm/architecture/security-features/arm-confidential-compute-architecture
7. Bell, D.E., LaPadula, L.J.: Secure computer systems: mathematical foundations. Technical report, MITRE CORP, Bedford, MA (1973)
8. bytecodealliance: wasmtime. https://github.com/bytecodealliance/wasmtime
9. cesanta: mongoose. https://github.com/cesanta/mongoose/tree/6.15/
10. Cooper, B.F., Silberstein, A., Tam, E., Ramakrishnan, R., Sears, R.: Benchmarking cloud serving systems with YCSB. In: Proceedings of the 1st ACM Symposium on Cloud Computing, pp. 143–154 (2010)
11. Gruss, D.: Software-based microarchitectural attacks. IT-Inf. Technol. **60**(5–6), 335–341 (2018)
12. Gu, J., et al.: Enclavisor: a hardware-software co-design for enclaves on untrusted cloud. IEEE Trans. Comput. **70**(10), 1598–1611 (2020)
13. Gu, J., Zhu, B., Li, M., Li, W., Xia, Y., Chen, H.: A hardware-software co-design for efficient intra-enclave isolation. In: 31st USENIX Security Symposium (USENIX Security 2022), pp. 3129–3145 (2022)
14. Halderman, J.A., et al.: Lest we remember: cold-boot attacks on encryption keys. Commun. ACM **52**(5), 91–98 (2009)
15. Hunt, G.D., et al.: Confidential computing for openpower. In: Proceedings of the Sixteenth European Conference on Computer Systems, pp. 294–310 (2021)
16. HyperEnclave: atc22-ae. https://github.com/HyperEnclave/atc22-ae
17. Intel: Intel Trust Domain Extensions (2020). https://software.intel.com/content/dam/develop/external/us/en/documents/tdxwhitepaper-v4.pdf
18. Jia, Y., et al.: HyperEnclave: an open and cross-platform trusted execution environment. In: 2022 USENIX Annual Technical Conference (USENIX ATC 2022), pp. 437–454 (2022)
19. Kaplan, D., Powell, J., Woller, T.: AMD memory encryption. White paper (2016)
20. Kocher, P., et al.: Spectre attacks: exploiting speculative execution. In: 2019 IEEE Symposium on Security and Privacy (SP), pp. 1–19. IEEE (2019)
21. Lee, D., Kohlbrenner, D., Shinde, S., Asanović, K., Song, D.: Keystone: an open framework for architecting trusted execution environments. In: Proceedings of the Fifteenth European Conference on Computer Systems, pp. 1–16 (2020)
22. Markettos, A.T., et al.: Thunderclap: exploring vulnerabilities in operating system IOMMU protection via DMA from untrustworthy peripherals (2019)
23. Mayer, U.F.: Linux/Unix nbench (2017). https://www.math.utah.edu/mayer/linux/bmark.html
24. McKeen, F., et al.: Innovative instructions and software model for isolated execution. Hasp Isca **10**(1) (2013)
25. Oleksenko, O., Kuvaiskii, D., Bhatotia, P., Felber, P., Fetzer, C.: Intel MPX explained: a cross-layer analysis of the intel MPX system stack. Proc. ACM Measur. Anal. Comput. Syst. **2**(2), 1–30 (2018)

26. OpenSSL: OpenSSL. https://www.openssl.org/
27. OpenSSL: OpenSSL-1.0.1e. https://ftp.openssl.org/source/old/1.0.1/openssl-1.0.1e.tar.gz
28. Park, J., Kang, N., Kim, T., Kwon, Y., Huh, J.: Nested enclave: supporting fine-grained hierarchical isolation with SGX. In: 2020 ACM/IEEE 47th Annual International Symposium on Computer Architecture (ISCA), pp. 776–789. IEEE (2020)
29. Shen, Y., et al.: Occlum: secure and efficient multitasking inside a single enclave of Intel SGX. In: Proceedings of the Twenty-Fifth International Conference on Architectural Support for Programming Languages and Operating Systems, pp. 955–970 (2020)
30. Tsai, C.C., Porter, D.E., Vij, M.: Graphene-SGX: a practical library OS for unmodified applications on SGX. In: 2017 USENIX Annual Technical Conference (USENIXATC 2017), pp. 645–658 (2017)
31. Wang, H., et al.: Towards memory safe enclave programming with rust-SGX. Proceedings of the 2019 ACM SIGSAC Conference on Computer and Communications Security (2019)
32. Wang, W., Liu, W., Chen, H., Wang, X., Tian, H., Lin, D.: Trust beyond border: lightweight, verifiable user isolation for protecting in-enclave services. IEEE Trans. Dependable Secure Comput. (2021)
33. Wikipedia: Heartbleed. https://en.wikipedia.org/wiki/Heartbleed
34. Wikipedia: Intel MPX. https://en.wikipedia.org/wiki/Intel_MPX
35. Xu, Y., Cui, W., Peinado, M.: Controlled-channel attacks: deterministic side channels for untrusted operating systems. In: 2015 IEEE Symposium on Security and Privacy, pp. 640–656. IEEE (2015)
36. Zhao, S., Li, M., Zhangyz, Y., Lin, Z.: vSGX: virtualizing SGX enclaves on AMD SEV. In: 2022 IEEE Symposium on Security and Privacy (SP), pp. 321–336. IEEE (2022)

# Keeping Your Enemies Closer: Shedding Light on the Attacker's Optimal Strategy

Weixia Cai[1,2], Huashan Chen[1](✉), and Feng Liu[1,2]

[1] Institute of Information Engineering, Chinese Academy of Sciences, Beijing, China
{caiweixia,chenhuashan,liufeng}@iie.ac.cn
[2] School of Cyber Security, University of Chinese Academy of Sciences, Beijing, China

**Abstract.** Realistically simulating a human attacker can effectively help the defender identify security weaknesses in the network. One important factor that affects the attacker's strategy is the various human characteristics. In this paper, we develop an attack engine, dubbed Attacker-Patience-Experience-Curiosity or APEC for short, to model the attacker's strategy under uncertainty. The proposed model is based on the Partially Observable Markov Decision Process (POMDP) model, taking three familiar characteristics of the attacker into consideration, including: (i) patience towards the target network; (ii) experience with attack tools; (iii) curiosity to develop new attack tools. These characteristics are modeled into the state space, action space, transition function, and reward function in the POMDP model. We further propose the betrayal principle, sunk cost, and "silence speaks volumes" to demonstrate how the attacker's characteristics affect its strategy, and why the attacker's strategy is changed at some specific points. We evaluate the effectiveness of the proposed model over two realistic network scenarios and draw several useful insights.

**Keywords:** Optimal strategy · Human characteristics · Non-deterministic planning · POMDP

## 1 Introduction

With the increasingly widespread use of the internet, both individuals and organizations become more vulnerable to cyber attacks.

A promising approach to effectively identify security vulnerabilities in advance is conducting attacks in the same way that a real attacker would do. As the saying goes, "Keep your friends close, but your enemies closer". This has led to the emergence of various technologies to model cyber attacks, such as attack graphs [1,2], cyber kill chains [3], and automated penetration testing [4].

However, most previous work [5] has concentrated on representing and analyzing the complex attack process, with little attention paid to the investigation of the attacker's human characteristics that have a significant impact on cyber

© The Author(s), under exclusive license to Springer Nature Switzerland AG 2023
M. Yung et al. (Eds.): SciSec 2023, LNCS 14299, pp. 42–59, 2023.
https://doi.org/10.1007/978-3-031-45933-7_3

attacks. In reality, attackers with different intentions, skill levels, and preferences would adopt distinct attack strategies [6]. These distinct attack strategies may lead to entirely different attack results. Furthermore, from the defender's perspective, an important pre-requisite for employing a successful defense strategy is the capability to accurately analyze and forecast the attacker's strategy, which can only be achieved when we absolutely throw light on the influence of human characteristics on the attacker's strategy. To the best of our knowledge, by now, there is no relevant studies focusing on the modeling and analysis of the attacker's human characteristics.

To address the above-mentioned issue, this work takes the first step to mine the relations between the attackers' human characteristics and their attack strategies. We abstract the attacker's characteristics into three features: (i) Patience towards target networks, which refers to the amount of time an attacker is willing to spend on network penetration. For example, advanced persistent threat (APT) actors often patiently lurk for years before reaching their ultimate goal, whereas script kiddies usually undertake one-step attacks and give up easily [7]. (ii) Experience with attack tools. A professional attacker usually creatively uses a variety of sophisticated attack tools to reach the goal, while a novice attacker may only master some user-friendly attack tools. (iii) Curiosity to develop new attack tools. Developing a zero-day exploit is extremely tempting for most attackers, but the success rate is often exceedingly low. Attackers with more curiosity are more likely to explore new techniques, while others may prefer ready-made ones.

When looking for an optimal attack strategy, an attacker considers two factors: the configuration of the target network and the accessible attack tools. Depending on whether the target network states are fully observable or the actions of attack tools are deterministic, studies can be divided into deterministic planning [1, 2, 8] and non-deterministic planning [9–12]. The current state-of-the-art methods are based on POMDP in academic research [13, 14], which fits well for modeling the target network scenarios under uncertainty. Inspired by the work [13], we propose the Attacker-Patience-Experience-Curiosity (APEC) model, which represents the attacker characteristics based on POMDP, allowing us to analyze the optimal attack strategy under uncertainty.

In order to reveal the underlying reasons for the attacker's strategy, we further propose the concepts of betrayal principle, sunk cost, and "silence speaks volumes" to demonstrate how the attacker's characteristics affect their behavior, and why the attacker's strategy is changed at some point. To the best of our knowledge, this is the first attempt towards explaining the attacker's strategy from a theoretical point of view. We evaluate the effectiveness of our approach through two experimental scenarios.

To summarize, we make the following contributions in this work:

– We identify three inherent human characteristics of cyber attackers that have a significant impact on attack strategies, which unfortunately have not been explored in state-of-the-art works.

- We propose the APEC model that incorporates the attacker's characteristics into the state space, action space, transition function, and reward function in POMDP for analyzing the attacker's strategy.
- We originally propose three novel concepts, which would greatly help to find out the reasons behind the attacker's strategy in real security incidents.
- We conduct experiments over realistic network scenarios to evaluate the effectiveness of the proposed APEC model and draw several useful insights.

The rest of the paper is organized as follows. Section 2 reviews related prior work. Section 3 presents our approach. Section 4 describes the experiments. Section 5 presents and discusses the results. Section 6 concludes the paper.

## 2  Related Work

### 2.1  Optimal Attack Strategy

Attack trees/graphs were previously a prominent method to represent a space of sequential attacks for planning an attack path against a specific network, by modeling network attacks in terms of pre-conditions and post-conditions. However, as the fact that cyber attacks are more sophisticated and targeted, these attacks frequently involve a long series of exploits, and it is difficult for humans to create attack trees or graphs [13].

In order to reduce reliance on highly knowledgeable and experienced experts, paper [8] was the first to integrate attack planning into the field of AI (artificial intelligence) planning and present a solution based on the Plan Domain Definition Language (PDDL). It then has achieved great success in the commercial product "Core Impact" [11]. More autonomous attack tools have been introduced to find the optimal attack strategy, and identify critical nodes [15–18].

However, those planning methods require static network settings and full observation of system states. In actual attack planning, attackers unavoidably need to deal with uncertain information [19], where the uncertainty is caused by partial observability of the network and lack of reliability of attack tools. For the former, given the fact that a large number of personal computers are linked from home, and various defense tools becoming part of an enterprise network, making it difficult to have a static picture of the network topology or the precise configuration of each device in the target network. For the latter, limited by the attacker's characteristics, whether the vulnerability can be successfully exploited is probabilistic.

To solve the above problems, in paper [10], the authors adopted POMDP to formalize the attack process under uncertainty. Inspired by this work, studies [11–13] extended sources of uncertainty from an attacker perspective, but none of them are related to the attacker's characteristics. Although these approaches can simulate attackers more realistically than deterministic planning, they still presume that all attackers have the same intention, skill level, and preferences. And our work presents an approach to analyze the optimal attack strategies that take attackers' characteristics into consideration.

## 2.2   Attackers' Characteristics Affect Task Behaviors

Some interesting attempts have been made that take human factors into account when given attackers' optimal attack strategy. In [20], the authors evaluated the entire state of a network by emulating real adversaries, and tested four different types of decision engines - one that chooses randomly, one built on a finite-state machine, one based on greedy action selection, and one that uses a planning method. In [21], the authors considered a finite collection of attacker types, and each type of attacker has different coefficients of influence on their true strategy. In [22], the authors proposed classifying attackers into three categories - individuals, organized groups, and intelligence agencies - based on various attributes such as the quantity of the malefactors, their motives, and their goals. However, these works can not explain why and how attackers' characteristics affect their strategy.

## 2.3   Feasibility

There are various approaches used to further clarify the type of attackers, such as using the attack samples and the attacker's packet capture (PCAP) data to capture attacker behavioral patterns. In [23], the authors analyzed a rich competition data set and gained a detailed timeline of events to characterize and describe attacker behaviors. In [24], the authors proposed a model that is capable of capturing attacker behavioral patterns. The input to their classifiers is a derived set of features that can be obtained from network traffic. On the basis of these studies, our attack engine will work in practice.

# 3   Model Description

In this paper, we propose a POMDP-based attack engine called APEC, which takes attackers' characteristics into consideration. The attackers' characteristics towards the target system can be summarized into three characteristics: 1) patience towards the target network, 2) experience with attack tools, and 3) curiosity to develop new attack tools. These characteristics will directly affect the optimal action strategy. The proposed APEC functioning diagram is detailed in Fig. 1.

## 3.1   POMDP

We use the POMDP framework to model the attack sequential decision process and obtain an optimal strategy that maximizes its expected total discounted reward.

Formally POMDP is specified as a tuple $< S, A, T, Z, O, R, b_0, \gamma >$. At each time period, an agent is in a state $s \in S$, takes an action $a \in A$, converts to the next state $s' \in S$ with probability $T(s'|s, a)$, perceives an observation $z \in Z$ with probability $O(z|s', a)$, and receives a reward $R(s, a)$. The probabilities $T(s'|s, a)$

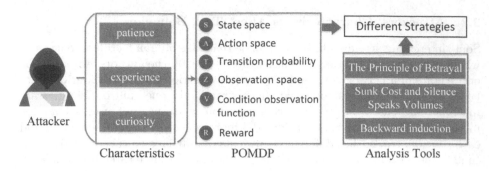

**Fig. 1.** APEC Functional Diagram

and $O(z|s', a)$ are all conditional probability functions that indicate uncertainty in the effect of action and sense. The discount factor $\gamma \in [0, 1]$ determines how much immediate rewards are preferred over longer-term rewards. If an agent only concerns with the immediate rewards, then $\gamma$ can be set to 0, implying that future rewards do not have to be taken into account in the action decision; on the other hand, if the agent regards future rewards as equally essential as current benefits, set $\gamma$ equals to 1. Finally, $b_0$ denotes the initial belief, which is a probability distribution over states at the beginning.

In a POMDP issue, the goal is to find an optimal strategy $\pi^*$ that maximizes the expected total discounted reward over an infinite horizon, i.e. $\pi^* = \arg\max_\pi V^\pi(b_0)$, where $V^\pi(b_0) = \mathbb{E}\left[\sum_{t=0}^\infty \gamma^t R(s_t, a_t)\right] = \sum_{t=0}^\infty \gamma^t \mathbb{E}\left[R(s_t, a_t)|b_0, \pi\right]$. The actions $a_0, a_1, a_2...$ are guided by strategy $\pi$.

A strategy $\pi$ is a decision tree, which is a tree-like description of decisions and their potential consequences. Uncertainty about states and actions leads to solving the POMDP as finding an "optimal tree", rather than finding the "shortest path" from the start node to the goal node when the target network is static and fully observable. In a POMDP issue, for each state, the agent has multiple optional actions, each of which leads to different states with different probabilities. Therefore, the recursively calculate expected total discounted reward for each state can be expressed as $V_t(s, b_t) = R(s, a) + \gamma \sum_{s' \in S} T(s'|s, a) V_{t+1}(s', b_{t+1})$ [25].

## 3.2   POMDP-Based APEC

Now we can extend the POMDP framework to accommodate attackers' characteristics, such as patience, experience, and curiosity.

The attacker's patience with the target network is directly related to the expected reward from an attack. Two considerations are taken into account from this point of view. The first is *patience discount factor* $\gamma_p$. It replaces the position of the original discount factor as introduced in Sect. 3.1, which specifies the agent's preference among short-term or long-term rewards and also contributes to the convergence of the POMDP calculation. However, its practical significance in influencing decision strategies is neglected, typically employed just for convergence (by setting it equal to 0.95). In [27], the authors took the discount rate into account, but its precise meaning was not specified, and the impact of its change was not examined. Our work strives to fill this gap by trying to figure out when and why the attack strategy should be changed under different patience. The second is *patience time horizon* $T_p$, which represents the maximum steps of attacks that the attacker is willing to spend. This factor will affect the horizon of the expected total discounted reward and the final decision at time $T_p$.

The attacker's experience with tools determines the tool utilization success rate. Even a single symbolic error in the attack payload could result in the failure of the attack. This observation enables us to represent attackers' behavior as a variable noise. An attack with a high noise level indicates that there is a lot of interference, so the attack success rate is low. We assume that an attacker would not intentionally insert noise in this paper. Depending on the type of action, the noise of the action is also different. This study investigates noise generated by two types of activities ($E : \{noise_1, noise_2\}$). For exploits, the $noise_1$ represents the probability that the value of a state variable fails to change after the action. As for detection, such as port scanning, the $noise_2$ of this kind of action represents that an attacker will update the belief function of the system state with probability $(1 - noise_2)$.

The curiosity about the target system and the temptation to mine zero-day vulnerabilities will affect the information-gathering process for different attackers. Attackers with more curiosity would try more actions and get more information about the target system. With more information at hand, more actions are possible. Here we set two additional actions $C : \{ac_1, ac_2\}$ for attackers with more curiosity. Action $ac_1$ is a general vulnerability that can be exploited on specific systems and ports. Action $ac_2$ is a zero-day vulnerability that can bypass some restrictions and directly attack the target host. These two kinds of actions have different probabilities being available in practice, and the discovery of zero-day vulnerabilities will be more difficult. Since the success rate of the curiosity action is difficult to fix in realistic scenarios, the decision on when to perform outreach to new actions will be discussed in this framework.

To sum up, we define the POMDP-Based attack engine APEC as

$$P_p = < S_C, A_C, T_{E,C}, Z_C, O_{E,C}, R_C, b_0, \gamma_p, T_p > \tag{1}$$

where:

- $S_C = G \times T \times K_1 \times ... \times K_{n_s} \times (C_1 \times ... \times C_{m_s})$ is a set of states, where $G$ represents if the goal of the attack is reached, $T$ represents if the attacker gave up the attack, $K_i$ for $i \in [1, n_s]$ represents the original component of a target system, and $C_j$ for $j \in [1, m_s]$ represents the new component required by curiosity attacks from additional actions set $C$. Both $G$ and $T$ are fully observable, while $K_i$ and $C_j$ are partially observable.
- $A_C = \{Exploit_1, ..., Exploit_{n_a}, Scan_1, ..., Scan_{m_a}, Terminate, C\}$ is a set of actions, where $Exploit_i$ represents the vulnerability exploitation, $Scan_j$ rep-

resents the action to detect the configuration of the target system, *Terminate* represents the attacker gave up the at $T$ state, and $C$ are additional actions out of curiosity.

– $T_{E,C}$ is a set of condition transition probability between states, where $T_{ij}$ represents the transition probability of state $i$ to state $j$ under a certain action. The experience of different attackers will have different $noise_1 \in E$, and the transition probability will change accordingly. An action corresponds to a state condition transition matrix, then each added action in $C$ will also have a corresponding transition matrix.

– $Z_C$ and $O_{E,C}$ are the observations and conditional observation functions, where experience with $noise_2 \in E$ will affect the conditional observation functions, and curiosity may have new observation functions.

– $R_C = S \times A_C$ is the reward function, getting a reward $R_{ij}$ after taking $action_j \in A_C$ at $state_i$.

– $b_0$ denotes the initial belief and refers to the initial belief about the target network states. In our experiments, the attackers start with the attack tools they currently have, and the initial belief about the target network is uniformly distributed.

### 3.3   The Principle of Betrayal

This subsection will introduce the principle of betrayal to analyze why changes in the *patience discount factor* ($\gamma_p$) lead to different strategies, how the $\gamma_p$ affects the strategy selection and what the critical point for strategy change is.

Suppose $\pi_1 = \{a_{11}, a_{12}, ...a_{1T}, ...\}$ and $\pi_2 = \{a_{21}, a_{22}, ..., a_{2T}, ...\}$ are the optimal strategies for *patience discount factor* equal to $\gamma_1$ and $\gamma_2$ respectively. Time $T$ is the first time that strategy $\pi_1$ is unequal to $\pi_2$, i.e. $a_{1i} = a_{2i}$ for $i < T$, and $a_{1i} \neq a_{2i}$ at time $i = T$. Assuming other conditions remain the same, it can be concluded that the *patience discount factor* is the reason why strategy $a_{1T}$ switches to $a_{2T}$ at moment $T$. And the principle of betrayal is motivated by the fact that the immediate payment from action betrayal at time $T$ is greater than the expected future rewards.

$$r(a_{2T}) - r(a_{1T}) \geq \gamma_2(V_{\pi_1-future} - V_{\pi_2-future}) \tag{2}$$

where $r(a_{iT})$ represents the immediate reward with strategy $a_{iT}$, $V_{\pi_i-future}$ represents the total future discounted rewards beginning at time $T + 1$. It's worth noting that the above formula is guaranteed to be correct, as strategy $\pi_1$ is the optimal strategy in the $\gamma_2$ situation.

Correspondingly, a similar inequality can be drawn from the fact that in the $\gamma_1$ situation, at time $T$, action $a_{1T}$ was preferred over action $a_{2T}$, as shown in formula 3.

$$r(a_{1T}) - r(a_{2T}) \geq \gamma_1(V_{\pi_2-future} - V_{\pi_1-future}) \tag{3}$$

The formula 2 and 3 tell the driving force and critical point for strategy changes, but how to calculate the total future discounted rewards of an attack strategy is a big problem. Recall the introduction to POMDP in Sect. 3.1, the immediate reward at time $T$ is $R(s, a)$, the total future discounted reward is $\gamma \sum_{s' \in S} T(s'|s, a_t)V_{t+1}(s', b_{t+1})$. And according to the principle of betrayal, the following formula 4 can be obtained based on the formula 2 (only discuss on $\gamma_2$ for convenience):

$$R(a_{2T}) - R(a_{1T}) \geq \gamma_2 \left[ \sum_{s' \in S} T(s'|s, a_{1T})V_{T+1}(s', b'_{t+1}) - \sum_{s'' \in S} T(s''|s, a_{2T})V_{T+1}(s'', b''_{t+1}) \right] \tag{4}$$

So, by solving the POMDP problem, we could approximate the total future discounted rewards of an attack strategy.

### 3.4   Sunk Cost and Silence Speaks Volumes

A sunk cost is a cost that has already been incurred and cannot be recovered. The problem is how to avoid the sunk costs caused by mistakes in decision-making.

Assume that the cost of scanning is $Cost_{scan}$ and the expected reward of an exploit is $R(s, a_i)$. If $R(s, a_i) \geq Cost_{scan}$ holds, the attacker would prefer to choose exploit $a_i$ rather than scan. If the action fails, the attacker will be unable to obtain any information about the target system, and the cost of the exploit will become a sunk cost. In this case, the $R(s, a_i) \geq Cost_{scan}$ holds true for the following time, and the attacker will choose to exploit again, which may also fail.

However, countless failed attacks can reveal something, where failure is caused by (i) The target system configuration failing to match the preconditions of the attack, (ii) the attacker's lack of expertise with the tool. For the former, no matter how many attacks are launched, they will always result in failure. For the latter, the probability of failing after repeated attempts is quite small. Even if the tool used by the attacker only has a $success_{rate}$ of 0.1, then the probability of failing after fifty attempts is $(1 - 0.1)^{50}$ less than 0.01. This probability is small enough to persuade the attacker that the former is the cause of the failure. This is called "silence speaks volumes", silence expresses more than words do.

Therefore, repeated failures can be exploited by the attacker to adapt its strategy. It should be clear that "silence speaks volumes" would already exist in the APEC framework due to the state transition probabilities, but it does not provide an intuitive indication of its result. Rather than just presenting the optimal strategy without explanation, the "silence speaks volumes" provides the attacker with practical advice on the action's empirical validity, which is another perspective to assess the effect of attackers' experience.

### 3.5   Backward Induction

Backward induction is the process of reasoning backward in time, from the end to determine a sequence of optimal actions. This method is often adopted in a limited and repetitive game process, and actual cyber attacks are exactly such a process. In cyber attacks, the attack's time and resources are limited, and the tightly constrained constant $T_p$ can be used to anticipate the maximum expected discount reward in a finite time.

Suppose $T_p$ is the number of actions of an attacker. At the time $T_p$, the attacker will choose an action with the biggest reward, i.e. $a_{T_p}^* = \arg\max\limits_a R(s,a)$. The reward of each action $a$ is shown in the Table 1. Where $r_{ij}$ is the expected reward of action $a_i$ at state $s_j$ with success rate $success_{ij}$.

**Table 1.** Expected Reward of Attacks

| | $s_1$ | $s_2$ | $s_3$ | | $s_M$ | $R(s,a)$ |
|---|---|---|---|---|---|---|
| $a_1$ | $r_{11}$ | $r_{12}$ | $r_{13}$ | | $r_{1M}$ | $\sum\limits_{j=1}^{M} b_j r_{1j}$ |
| $a_2$ | $r_{21}$ | $r_{22}$ | $r_{23}$ | | $r_{2M}$ | $\sum\limits_{j=1}^{M} b_j r_{2j}$ |
| $a_3$ | $r_{31}$ | $r_{32}$ | $r_{33}$ | | $r_{3M}$ | $\sum\limits_{j=1}^{M} b_j r_{3j}$ |
| ... | ... | ... | ... | ... | ... | ... |
| $a_N$ | $r_{N1}$ | $r_{N2}$ | $r_{N3}$ | | $r_{NM}$ | $\sum\limits_{j=1}^{M} b_j r_{Nj}$ |

From this point of view, action $a_i$ would be taken by attackers when the $\sum\limits_{j=1}^{M} b_j r_{ij}$ is the maximum expected reward. A more special case is when $r_{ij}$ is the largest reward among others, and the belief in state $s_j$ is $b_j = 1$ at time $T_p$, then the optimal action is $a_i$. Therefore, by the backward induction, all actions before time $T_p$ should make belief $b_j$ more clear. If the belief $b_j$ in state $s_j$ is still small, then turn to $b_{j'}$ with the second largest gain, and so on.

## 4    Experiment Scenarios

We have evaluated our approach in two different scenarios. The difference lies in the attacker's actions and the minimum number of actions to reach the goal. The first is based on studies [10,13], where there are two types of actions: scans and exploits, and the minimum of action to reach the goal (*MinStep*) is one step. The second is based on studies [13,28], where all actions are exploits and *MinStep* is five steps. Table 2 shows the POMDP size for each scenario. The rest of this section presents a brief introduction to these two scenarios.

### 4.1    Experiment Scenario 1

Scenario 1 involves a single host, with the goal of controlling the host. There are 12 exploit actions, two of which are additional actions out of curiosity. Their

**Table 2.** POMDP problem size and rewards for each scenario and agent.

| Scenario | $S$ | $A$ | $Z$ | $S_C$ | $A_C$ | $Z_C$ | MinSteps |
|----------|-----|-----|-----|-------|-------|-------|----------|
| 1 | 3072 | 20 | 6 | 3072 | 21 | 6 | 1 |
| 2 | 256 | 7 | 2 | 256 | 8 | 2 | 5 |

preconditions are shown in Table 3. Each of these actions is independent, i.e. the order in which actions are carried out has no bearing on the outcome. In Scenario 1 (excluding $C$), there are 10 exploits, 8 ports, and 3 operation systems, so the state space $|S| = (G = 2) \times (T = 2) \times (K_1 = 2) \times (K_2 = 2) \times (K_3 = 2) \times (K_4 = 2) \times (K_5 = 2) \times (K_6 = 2) \times (K_7 = 2) \times (K_8 = 2) \times (K_9 = 3) = 3072$, the action space $|A| = (exploits = 10) + (portscan = 8) + (OSscan = 1) + Terminate = 20$, and the observation space is $|O| = 2 \times 3 = 6$ for each observation consists of port = [open, closed], operation systems (OS) = [linux, windows, openBSD]. The same method can be used for scenarios with additional actions. The additional actions in this article are added one at a time for easy comparison. In line with the original study [27], the cost for both port scans and exploit actions is 10, and 50 for an OS scan. If the target host was successfully controlled, the reward is 9000. The problem of Scenario 1 is how to plan the sequence of actions of attack, port scan, and OS scan to maximize the expected reward when the initial belief of the target system state is $b_0$.

**Table 3.** Exploits for Scenario 1 [13]

| Attack Action | Target OS | Target Port |
|---------------|-----------|-------------|
| vsftpd-234 | linux | 21 |
| IIS-ftp | windows | 21 |
| openBDSD-ssh | openBSD | 22 |
| IIS-http | windows | 80 |
| apache-openBSD | openBSD | 80 |
| rsh | linux | 512 |
| unreal-ircd-3281 | linux | 3632 |
| manageEngine9 | windows | 8020 |
| tomcat | windows | 8282 |
| elastic-search | windows | 9200 |
| C: $action_1$ | windows | 512 |
| C: $action_2$ | openBSD | 80 |

## 4.2  Experiment Scenario 2

The second scenario is starting from the Normal operation, with the goal of stealing confidential data, as shown in Fig. 2. Action $a_7, a_8$ in the red line are additional actions out of curiosity. In Scenario 2, there are 8 states, each of which has two conditions: successful or not. So the state space $|S| = 2^8 = 256$, the action space $|A| = 7$, and the observation space is $|O| = 2$ for each observation consists of [success, failure]. And the cost for both port scans and exploit actions is 10. If the confidential data was successfully stolen, the reward is 999. Different from the independence assumption of actions in scenario 1, the actions in scenario 2 have a sequential relationship. It means that if the previous action fails, the next action will fail as well. For example, if action $a_5$ fails, action $a_6$ would not succeed as well.

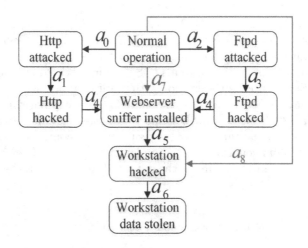

**Fig. 2.** Graph representation of scenario 2

## 4.3   Experimental Setup

The effectiveness of our work was tested on a machine running an Inter Core CPU i5 at 1.60 Ghz with 8GB RAM, by the Approximate POMDP Planning Toolkit (APPL) [29]. APPL is an efficient C++ toolbox for solving approximate POMDP. Each experiment was simulated 1000 times until the planning time of 210 s or the target precision of 0.01 was met. The initial belief of Scenario 1 was uniformly distributed over the possible state while starting from the state of "Normal operation" for Scenario 2.

An attacker can repeat the attack after successfully reaching the goal, but the reward is only paid on the first successful attempt. It is thus meaningless to repeat the operation, for which the reward is zero and the operation's cost must be paid. However, the agent can only learn this apparent knowledge after several failed attempts. To make this more efficient, we set up the first successful attacks to trigger the termination state and receive a quick learning reward. This setting is natural for a successful attack and is equivalent to the termination of the attack.

# 5    Results and Discussion

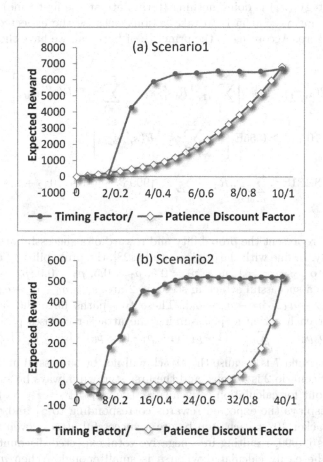

**Fig. 3.** APEC performance with *patience discount factor* at an interval of 0.05. (a) and (b) show results for Scenario 1 and Scenario 2, respectively.

## 5.1    The Effect of Attacker's Patience

The performance of the APEC with respect to the *patience discount factor*($\gamma_p$) and *patience time horizon*($T_p$) is shown in Fig. 3. The expected reward is a concave function with regard to the $\gamma_p$, while an approximated convex function with respect to the $T_p$. The performance in terms of the expected discount reward increases rapidly after the $\gamma_p$ exceeds 0.6. When comparing steps required for the effective attack in (a) and (b), the number of steps required for the successful attack in (b) is larger, and correspondingly the process is terminated sooner with a large $\gamma_p$ due to the difference in the *MinStep* required for the two scenarios.

Here we apply the betrayal principle to explain why strategy changes at $\gamma = 0.55$ in Scenario 2. The expected reward is zero when $\gamma_1 = 0.5$, indicating that the strategy $(\pi_1)$ is doing nothing (terminate at the first time). While the strategy $(\pi_2)$ for $\gamma_2 = 0.55$ is to take actions because the expected reward is greater than zero. According to the principle of betrayal, we have the formula 7.

$$R(s, \pi_2) - R(s, \pi_1) \geq \gamma_2 \mathbb{E}\left[\sum_{t=1}^{\infty} \gamma_1^{t-1} R(s_t', \pi_1) - \sum_{t=1}^{\infty} \gamma_2^{t-1} R(s_t'', a_{2t})\right] \tag{5}$$

$$-10 - 0 \geq 0.55\mathbb{E}\left[0 - \sum_{t=1}^{\infty} \gamma_2^{t-1} R(s_t'', a_{2t})\right] \tag{6}$$

$$18.1818 \leq \sum_{n=4}^{\infty} \mathbb{P}_n R_n = \sum_{n=4}^{\infty} \mathbb{P}_n \left[999\gamma_2^n - (10 + 10\gamma_2 + \ldots + 10\gamma_2^{n-1})\right] \tag{7}$$

where $p_n$, $R_n$ represent the probability and reward of a successful attack at step $n$, respectively. In line with the original study [28], the probability of a successful attack is set to $p_0 = p_2 = 1$, $p_1 = p_3 = 0.5$, $p_4 = 0.5$, $p_5 = 0.9$, $p_6 = 0.999$. The approaches to a successful attack in scenario 2 are $[s_1 = a_0 - a_1 - a_4 - a_5 - a_6]$ and $[s_2 = a_2 - a_3 - a_4 - a_5 - a_6]$. These two paths have the same chance of success for each action step, assuming the attacker chooses path $s_1$, hence $\mathbb{P}_n = p_0 p_1 p_4 p_5 p_6 \sum_{i+j+k+l=n-4} (1-p_1)^i (1-p_4)^j (1-p_5)^k (1-p_6)^l$. The sum starts at n = 4 in formula 7 is because the attack will not be successful until $n = 4$ for *MinStep* in Scenario 2 is 5 steps, and the first step will always be successful, so it was taken out for calculation convenience. (i.e., $\mathbb{P}_1 = \mathbb{P}_2 = \mathbb{P}_3 = 0$).

Table 4 displays the expected rewards corresponding to $\gamma_1$ and $\gamma_2$. Since it costs the attacker 10 at each step before a successful attack, even though the success bonus is 999, resulting in a negative total expected discounted reward. The $\mathbb{P}_n$ can be easily calculated when $n$ is small enough. When $n \geq 10$, the probability $\mathbb{P}_n$ decreases faster than the rate at which $R_n$ decreases, therefore $\mathbb{P}_n R_n$ is monotonically increasing and eventually converges to zero. Based on this assessment, we set the probability $\mathbb{P}_9$ to the sum of the remaining probabilities, i.e. $\mathbb{P}_9 = \sum_{n \geq 9} \mathbb{P}_n$.

The expected reward of the strategy of doing nothing $(\gamma_1)$ and the strategy of always taking action $(\gamma_2)$ are 6.8967 and 20.2972, respectively, satisfying $6.6967 < 18.1818 < 20.2972$. This allows us to observe how $\gamma$ changes attack strategy. That is, the immediate payoff of the strategy of always taking action is greater than the expected reward of doing nothing at $\gamma = 0.55$, resulting in strategy betrayal.

Here we present an example to show how the principle of betray guides strategy change, which is consistent with the strategy obtained by APEC. But it is not sufficient to illustrate such a generic principle with just one single, arbitrarily chosen numerical example. The challenge lies in the complexity of

**Table 4.** expected rewards for $\gamma_1$ and $\gamma_2$ in Scenario 2

| n | $\gamma_2 = 0.5$ | | $\mathbb{P}_n$ | $\gamma_2 = 0.55$ | |
|---|---|---|---|---|---|
| | $R_n$ | $\mathbb{P}_n R_n$ | | $\mathbb{P}_n R_n$ | $R_n$ |
| 4 | 43.6875 | 9.8199 | 0.2248 | 16.0098 | 71.2260 |
| 5 | 11.8438 | 2.9311 | 0.2475 | 7.2200 | 29.1743 |
| 6 | −4.0782 | −0.7893 | 0.1936 | 1.1702 | 6.0459 |
| 7 | −12.0391 | −1.5881 | 0.1319 | −0.8805 | −6.6748 |
| 8 | −16.0195 | −1.3384 | 0.0835 | −1.1422 | −13.6711 |
| 9 | −18.0098 | −2.1384 | 0.1187 | −2.0802 | −17.5191 |
| | | **6.8967** | | **20.2972** | |

computing future rewards, which is outside the scope of this paper and can be left for future research.

**Insight 1.** *The principle of betray can explain reasonably why the attacker gives up at the beginning of an attack and provide a specific threshold of patience when the attacker would change the strategies.*

### 5.2 The Effect of Attacker's Experience

As can be seen from Table 5, as $noise_1$ becomes more intrusive to exploit vulnerabilities, the expected reward decreases gradually at first, only to plummet dramatically later. Furthermore, the algorithm gets more difficult to converge and becomes bogged down in repeating actions when there is a lot of noise. Scenario 1 shows a considerable drop in performance when $noise_1 \geq 0.75$, the results show a significant decline in performance, with agents performing the same subset of actions hundreds of times. Finally give up the attack when $noise_1 = 1$, and the expected reward turns to zero.

On the other hand, changing the variable of observation $noise_2$ has little impact on the expected reward, which is symmetric around $\gamma_2 = 0.5$. When the noise of observation is $noise_2 = 0.5$, the action of observation no longer has any useful feedback, and the cost becomes a sunk cost. In such a situation, the attacker will choose to take exploitation and apply "silence speaks volumes" to see whether the configuration matches. In our experiment, when the preconditions of the vulnerability are met, the success rate of the vulnerability exploitation is 0.95, regardless of the observed noise variation. The probability $(1 - 0.95)^n$ that all $n$ attempts of the same exploit fail will persuade the attacker that the failure is due to the target system failing to meet the preconditions of the vulnerability. After then, the attacker will choose another weakness to exploit until all vulnerabilities have been exploited. Scenario 1 and Scenario 2 has similar result, we only show the result of Scenario 1 here.

**Table 5.** APEC performance with $noise_1$, $noise_2$ at an interval of 0.05, $\in [0.05, 1]$.

| interval | $noise_1$ | | $noise_2$ | |
|---|---|---|---|---|
| | Expected reward | Time | Expected reward | Time |
| 0.05 | **6805.74** | 108.11 | **6805.74** | 108.11 |
| 0.10 | **6769.33** | 212.85 | **6799.51** | 68.00 |
| 0.15 | **6729.42** | 202.84 | **6793.28** | 57.31 |
| 0.20 | **6684.4** | 210.10 | **6787.06** | 67.38 |
| 0.25 | **6635.03** | 213.56 | **6780.83** | 69.61 |
| 0.30 | **6579.86** | 212.55 | **6774.61** | 76.22 |
| 0.35 | **6517.16** | 211.16 | **6768.39** | 104.00 |
| 0.40 | **6445.77** | 210.16 | **6762.17** | 110.45 |
| 0.45 | **6364.20** | 212.63 | **6755.95** | 140.28 |
| 0.50 | **6267.69** | 212.34 | **6749.74** | 68.50 |
| 0.55 | **6156.17** | 211.14 | **6755.95** | 125.59 |
| 0.60 | **6019.80** | 213.53 | **6762.17** | 112.81 |
| 0.65 | **5859.04** | 211.81 | **6768.39** | 99.19 |
| 0.70 | **5659.39** | 211.70 | **6774.61** | 71.03 |
| 0.75 | **5402.87** | 212.14 | **6780.83** | 63.86 |
| 0.80 | **5062.10** | 212.14 | **6787.06** | 63.13 |
| 0.85 | **4568.50** | 212.88 | **6793.28** | 63.58 |
| 0.90 | **3852.07** | 210.74 | **6799.51** | 63.30 |
| 0.95 | **2615.64** | 211.05 | **6805.74** | 89.41 |
| 1.00 | **0** | 0.25 | **6815.89** | 6.61 |

**Insight 2.** *Exploit actions make more sense than scans. This is because exploits contribute directly to the final rewards.*

**Insight 3.** *The observation noise has little impact on the expected reward. Instead of checking for invalid observations, the attacker can utilize the "Silence Speaks Volumes" to check whether the preconditions for the exploit match.*

## 5.3   The Effect of Attacker's Curiosity

Figure 4 illustrates the relationship between the probability of success and the expected reward of adding additional action to Scenario 2. As the probability of the additional action succeeding rises, so does the expected reward. When the success rate of the additional action is low, the expected reward is extremely similar to the expected reward without the additional action, implying that the additional action is discarded when developing an action strategy. The expected reward increases when the success rate exceeds a certain threshold. And increased expected reward far outweighs the expected reward of no additional action, indicating that the action has been used as a priority planning strategy.

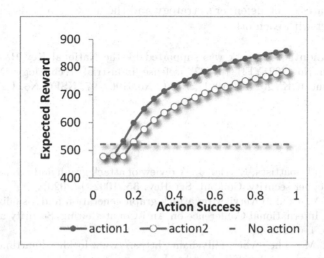

**Fig. 4.** The expected reward of adding additional actions in Scenario 2

**Insight 4.** *Even though the success rate of a curiosity action is difficult to fix in realistic scenarios, our work provides the attacker with a threshold on whether to explore new attacks.*

**Insight 5.** *Exploring new attack tools is highly encouraged. In our case, the threshold of success rate is lower than 0.2.*

## 6   Summary and Future Work

How to simulate an attack with the attacker's characteristics under uncertainty is an important problem. This paper presents a POMDP-based attack engine called APEC to study the attacker's characteristics when generating an attack strategy. We abstract the attacker's characteristics into three features: (i) Patience towards the target network; (ii) Experience in the use of attack tools; and (iii) Curiosity to develop new attack tools. Based on POMDP, these three characteristics of the attacker are modeled into the state space, action space, transition function, and reward function. The effectiveness of the method is evaluated by two experimental scenarios, in which the attackers with different intentions, skill levels, and preferences adopt distinct attack strategies. The proposed concepts of betrayal principle, sunk cost, and "silence speaks volumes" thoroughly demonstrate how the attacker's characteristics influence their performance, as well as why and when the attack strategy should be changed. The results are instructive for the defenders to protect their network in advance.

In this paper, the target network is considered to be static, and the defensive activities are ignored. Our future work will concentrate on the intelligent decision-making for both the attacker and the defender, meaning that the attacker's strategy, the defender's strategy and the network are all dynamic and would interact with each other.

**Acknowledgment.** This work was supported by the National Key R&D Program of China with No. 2021YFB3101402, Defense Industrial Technology Development Program (Grant JCKY2021906A001), NSFC No.61902397, NSFC No. U2003111 and 61871378.

# References

1. Lallie, H.S., Debattista, K., Bal, J.: A review of attack graph and attack tree visual syntax in cyber security. Comput. Sci. Rev. **35**, 100219 (2020)
2. Shengwei, Y., et al.: Overview on attack graph generation and visualization technology In: International Conference on Anti-Counterfeiting, Security and Identification, pp. 1–6 (2013)
3. Lockheed, M.: The cyber kill chain. https://www.lockheedmartin.com/en-us/capabilities/cyber/cyber-kill-chain.html
4. Al Shebli, H.M.Z., Beheshti, B.D.: A study on penetration testing process and tools. In: 2018 IEEE Long Island Systems, Applications and Technology Conference, pp. 1–7 (2018)
5. Vyas, S., Hannay, J., Bolton, A., Burnap, P.P.: Automated cyber defence: a review. arXiv (2023)
6. Kadivar, M.: Cyber-attack attributes. Technol. Innov. Manage. Rev. 22–27 (2014)
7. Chapple, M., Seidl, D.: CompTIA Security+ Study Guide: Exam SY0-601. 8th edn. ISBN-13 is 978–1119736257, Sybex (2021)
8. Boddy, M.S., Gohde, J., Haigh, T., Harp, S.A.: Course of action generation for cyber security using classical planning. In: The Fifteenth International Conference on Automated Planning and Scheduling, pp. 12–21 (2005)
9. Hoffmann, J.: Simulated penetration testing: from "Dijkstra" to "Turing Test++". In: The Twenty-Fifth International Conference on Automated Planning and Scheduling, pp. 364–372 (2015)
10. Sarraute, C., Buffet, O., Hoffmann, J.: POMDPs make better hackers: accounting for uncertainty in penetration testing. In: The Twenty-Sixth AAAI Conference on Artificial Intelligence, pp. 1816–1824 (2012)
11. Sarraute, C., Richarte, G., Lucángeli Obes, J.: an algorithm to find optimal attack paths in nondeterministic scenarios. In: The 4th ACM Workshop on Security and Artificial Intelligence, pp. 71–80 (2011)
12. Shmaryahu, D., Shani, G., Hoffmann, J., Steinmetz, M.: Simulated penetration testing as contingent planning. In: The Twenty-Eighth International Conference on Automated Planning and Scheduling, pp. 241–249 (2018)
13. Schwartz, J., Kurniawati, H., El-Mahassni, E.: "POMDP + information-decay: incorporating defender's behaviour in autonomous penetration testing. In: The Thirtieth International Conference on Automated Planning and Scheduling, pp. 235–243 (2020)
14. Holm, H.: Lore a red team emulation tool. IEEE Trans. Dependable Secure Comput. **20**(2), 1596–1608 (2023)

15. Randhawa, S., Turnbull, B., Yuen, J., Dean, J.: Mission-centric automated cyber red teaming. In: The 13th International Conference on Availability, Reliability and Security, pp. 1–11 (2018)
16. Vats, P., Mandot, M., Gosain, A.: A comprehensive literature review of penetration testing & its applications. In: 8th International Conference on Reliability, Infocom Technologies and Optimization, pp. 674–680 (2020)
17. Maeda, R., Mimura, M.: Automating post-exploitation with deep reinforcement learning. Comput. Secur. **100**, 102–108 (2020)
18. Alsaheel, A., et al.: ATLAS: a sequence-based learning approach for attack investigation. In: The 30th USENIX Security Symposium, pp. 3005–3022 (2021)
19. Li, J., Ou, X., Rajagopalan, R.: Uncertainty and risk management in cyber situational awareness. In: Uncertainty and Risk Management in Cyber Situational Awareness (2010)
20. Applebaum, A., Miller, D., Strom, B., Korban, C., Wolf, R.: Intelligent, automated red team emulation. In: The 32nd Annual Conference on Computer Security Applications, pp. 363–373 (2020)
21. Erik, M., Mohammad, R., Demosthenis, T.: A POMDP approach to the dynamic defense of large-scale cyber networks. IEEE Trans. Inf. Forensics Secur. **13**, 2490–2505 (2018)
22. Abomhara, M., Køien, G.M.: Cyber security and the internet of things: vulnerabilities, threats, intruders and attacks. Cyber Secur. 65–88 (2015)
23. Munaiah, N., Rahman, A., Pelletier, J., Williams, L., Meneely, A.: Characterizing attacker behavior in a cybersecurity penetration testing competition. In: ACM/IEEE International Symposium on Empirical Software Engineering and Measurement, pp. 1–6 (2019)
24. Gabrys, R., et al.: Emotional state classification and related behaviors among cyber attackers. In: Proceedings of the 56th Hawaii International Conference on System Sciences, pp. 846–855 (2023)
25. Åström, K.J.: Optimal control of Markov processes with incomplete state information. J. Math. Anal. Appl. **10**, 174–205 (1965)
26. Perry, I., et al.: Differentiating and predicting cyberattack behaviors using LSTM. In: IEEE Conference on Dependable and Secure Computing, pp. 1–8 (2018)
27. Ghanem, M.C., Chen, T.M.: Reinforcement Learning for efficient network penetration testing. Article 6 (2020)
28. Lye, K.-W., Wing J.M.: Game Strategies in Network Security. vol. 4, pp. 1615–5262. Springer, Cham (2005)
29. Shridhar, M, Panpan, C.: Efficient point-based POMDP planning by approximating. Accessed 10 Nov 2021

# Cyber Attacks Against Enterprise Networks: Characterization, Modeling and Forecasting

Zheyuan Sun[1], Maochao Xu[2], Kristin M. Schweitzer[3], Raymond M. Bateman[3],
Alexander Kott[3], and Shouhuai Xu[4(✉)]

[1] Department of Computer Science, University of Texas at San Antonio, San Antonio, USA
[2] Department of Mathematics, Illinois State University, Normal, USA
[3] DEVCOM Army Research Laboratory, Adelphi, USA
[4] Department of Computer Science, University of Colorado Colorado Springs,
Colorado Springs, USA
sxu@uccs.edu

**Abstract.** Cyber attacks are a major and routine threat to the modern society. This highlights the importance of forecasting (i.e., predicting) cyber attacks, just like weather forecasting in the real world. In this paper, we present a study on characterizing, modeling and forecasting the number of cyber attacks at an aggregate level by leveraging a high-quality, publicly-available dataset of cyber attacks against enterprise networks; the dataset is of high quality because more than 99% of the attacks were examined and confirmed by human analysts. We find that the attacks exhibit high volatilities and burstiness. These properties guide us to design statistical models to accurately forecast cyber attacksand draw useful insights.

**Keywords:** Cybersecurity data analytics · attack forecasting · attack prediction · burstiness · cyber threats · cybersecurity dynamics · statistical models

## 1 Introduction

The importance of forecasting cyber attacks (or attack events) is well appreciated because it can enable proactive cyber defense, similar to how weather forecasting helps us in planning our daily activities. For example, being ale to forecast cyber attacks against a network will give cyber defenders useful information in planning defenses [32]. Moreover, the forecasting capability allows the defender to dynamically adjust the allocation of defense resources [3,66,67], including both human analysts who need to examine the alerts triggered by defense tools [22] and sensor deployments when the prediction is geared toward specific type of attacks. Moreover, when the predicted number of attacks is high but the detected number of attacks is low, it hints that the defense may not be effective and/or the attacks may be new. Although there are studies on forecasting cyber attacks (e.g., [12,18,43,44,55,66,67]), these studies are limited primarily by the quality of the datasets they leverage because they are often collected from *non-enterprise networks* (e.g., cyber honeypots). This is not surprising because high-quality cyber attack data against production/enterprise networks is sensitive.

In this paper, we leverage a dataset which contains some aggregated information about cyber attacks against enterprise networks, rather than honeypots. The dataset

© The Author(s), under exclusive license to Springer Nature Switzerland AG 2023
M. Yung et al. (Eds.): SciSec 2023, LNCS 14299, pp. 60–81, 2023.
https://doi.org/10.1007/978-3-031-45933-7_4

describes weekly-binned cyber attacks, but not the individual attacks. This dataset is of high quality in the sense that more than 99% of the attacks are confirmed by analysts.

**Our Contributions**. We make three contributions. First, we analyze three time series derived from the dataset: (i) the weekly-binned *number* of attacks, referred to as $X_t$; (ii) the weekly *average* attack report length, referred to as $Y_t$; and (iii) the weekly *total* attack report length, referred to as $Z_t$ and derived from $X_t$ and $Y_t$. Note that $Y_t$ and $Z_t$ are analyzed here for the first time. Second, we show that these time series exhibit high *volatilities* and *burstiness*, meaning that they cannot be modeled by simple stochastic process models (e.g., Poisson). Third, we show that these time series can be accurately modeled by an ARIMA+GARCH model, where ARIMA stands for "AutoRegressive Integrated Moving Average" and can model the dynamic mean, and GARCH stands for "Generalized AutoRegressive Conditional Heteroskedasticity" and can model the burstiness. Moreover, we show that the ARIMA+GARCH model can forecast the number of attacks one week ahead of time. These allow us to draw a number of useful insights, such as: (i) the average attack report length reflects attack sophistication but not attack newness; (ii) new attacks are not necessarily sophisticated; and (iii) burstiness in the average attack report length suggests that sophisticated attacks are seen often.

**Related Work**. The first study on forecasting cyber attacks based on real-world datasets may be [12], which leverages a dataset collected at a campus network from 6/14/2001 to 3/14/2007. The study *heuristically* uses the ARIMA model without giving statistical justification. Another study [18] forecasts distributed denial-of-service attack rate based on data collected by a network blackhole (not enterprise networks). These datasets were not publicly available. The dataset we analyze has been studied in [3], which was influenced by [66] but only considered the aforementioned time series $X_t$. Going beyond [3], we consider time series $X_t, Y_t, Z_t$, which allows us to draw more insights.

To our knowledge, the problem of forecasting cyber attacks was not systematically investigated until [66], which proposed a systematic *gray-box* framework for data-driven modeling and forecasting cyber attacks. The term "gray-box" means: first characterizing the statistical properties exhibited by the data, and then building models that can accommodate these properties to forecast cyber attacks. The gray-box nature explains why the resulting statistical models can predict, contrasting the *black-box* nature of machine learning, especially deep learning, models. The framework [66] has been extended to forecast cyber attacks with extreme values [43,67], investigate the effectiveness of cyber defense early-warning [55], forecast the distribution of multivariate cyber attack rates while accommodating the dependence between them [44], characterize the cyber threat posture [65], investigate the prediction upper bound of cyber attack rates [9], and forecast cyber data breach incidents [20,56]. The present study can be seen as an extension of the gray-box framework [66] to accommodate newly identified statistical properties exhibited by cyber attacks against enterprise networks (i.e., high volatilities and burstiness). Related to this statistical approach, deep learning has been applied to forecast this type of data [19] as well as causality-based forecasting [53].

At a higher level of abstraction, cyber threats forecasting, including the "grey-box" framework [66], belongs to cybersecurity data analytics, which is integral to Science of

Cyber Security [64] and one pillar of the broader Cybersecurity Dynamics framework [58,59,61,62]. This broader framework is driven by cybersecurity metrics and quantification [6,10,15,38,42,63], including the quantification of attack and defense capabilities (e.g., cyber social engineering attacks [39–41,49]). Another pillar of the framework aims to model the evolution of the global cybersecurity state incurred by cyber attack-defense-use interactions, where "global" highlights the perspective of looking at a network (e.g., enterprise wide, nation wide, or even the entire cyberspace) as a whole. This pillar has led to many significant results (e.g., [7,8,23,25,26,34,35,57,60,68]).

**Paper Outline.** Section 2 describes the dataset and its cybersecurity implications. Section 3 characterizes the time series. Section 4 presents our model fitting and forecasting results. Section 5 concludes the paper. To improve the readability of the paper, Appendix A reviews the statistical knowledge that is used in this study.

# 2   Dataset and Its Cybersecurity Implications

## 2.1   Data Description

The dataset [2] is collected by a Network Defense Service Provider. It contains attack events against multiple enterprise networks managed by the Provider. These attack events are first detected by cyber defense tools and then more than 99% of them are manually examined and confirmed by human analysts (i.e., false positives are eliminated, while noting that the false-positive rate is not available). This means that at least 99% of the cyber attacks in the dataset are real attacks (i.e., true positives), while noting that the dataset may not contain all attacks (i.e., missing the false negatives). This does not invalidate the value of the present study on forecasting attacks based on true positives because the predictions can help defenders allocate resources to deal with detectable attacks. To our knowledge, this is the only publicly available dataset on cyber attacks against enterprise networks (rather than honeypots and network blackholes).

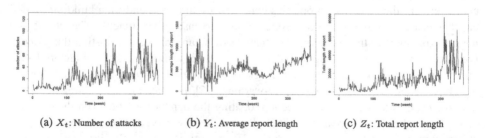

(a) $X_t$: Number of attacks         (b) $Y_t$: Average report length         (c) $Z_t$: Total report length

**Fig. 1.** Plots of time series $X_t, Y_t, Z_t$, where the $x$-axis represents time (unit: week).

The dataset contains 9,302 cyber attacks over a period of 366 weeks (or 7 years), but the precise data collection time is not given (except that it was after year 2000). For each attack, an *attack report* was written by a human analyst; unfortunately, no detailed information on attack reports is given. The dataset consists of two time series:

(i) the weekly-binned number of attacks, denoted by $X_t$ for $t = 1, \ldots, 366$; (ii) the weekly-binned average length of attack reports, denoted by $Y_t$ for $t = 1, \ldots, 366$. We propose deriving the weekly-binned *total* length of attack reports, denoted by $Z_t$, as follows: the weekly-binned total length of attack reports, denoted by $Z_t = X_t \cdot Y_t$ for $t = 1, \ldots, 366$. Figure 1a-1c respectively plot $X_t$, $Y_t$, and $Z_t$.

## 2.2 Attack Report Length, Attack Newness and Attack Sophistication

**The Notions of Attack Newness and Attack Sophistication.** We hypothesize that the attack report length may reflect (i) *attack newness*, meaning that a new attack that was not seen before would need to be documented in details, leading to a long report, and/or (ii) *attack sophistication*, meaning that a sophisticated attack would need to be documented in details, also leading to a long report. To (in)validate this, we are allowed to have access to a random sample of 100 attacks (rather than all the 9,302 attacks) in terms of the following three attributes: (i) attack report length, which is in the number of bytes but not the report itself; (ii) attack newness, which is subjectively labeled by a human analyst as "low", "medium" or "high"; (iii) attack sophistication, which is also subjectively labeled by a human analyst as "low", "medium" or "high". Among the 100 attacks, 14, 19, and 67 are respectively labeled as high-, medium-, and low-newness; whereas, 82, 15, and 3 are respectively labeled as high-, medium-, and low-sophistication. Since there are only 3 low-sophistication attacks (which are too few to make statistical sense), we combine medium- and low-sophistication attacks into one category, also called low-sophistication attacks in contrast to high-sophistication ones. This leads to 82 high-sophistication attacks and 18 low-sophistication attacks. Unfortunately, we are not authorized to share the data on these 100 attacks.

Given that newness and sophistication are subjectively labeled, we examine whether or not there is a dependence between the labels corresponding to them. For this purpose, we first use the Chi-square test, which is based on the contingency table [14] of the newness and sophistication labels. The Chi-square test result is 41.432 with a $p$-value $1.007e-09$, suggesting that there is a degree of dependence between these two notions. In order to characterize the dependence, we code the labels "low" as '0', "medium" as '1', and "high" as '2'; we then use the Kendall's rank correlation test [29] on these coded values. The estimated Kendall's tau is $-0.5089$ with a $p$-value $1.474e-07$, suggesting that there is a negative dependence between attack newness and attack sophistication.

**Insight 1.** *A high-newness attack can have a low sophistication, namely that a new attack is not necessarily sophisticated.*

**Relationship between Attack Report Length, Attack Newness, and Attack Sophistication.** Figure 2a is the boxplot of the attack report length vs. attack newness based on the 100 samples mentioned above. We observe: (i) the boxplots of medium- and high-newness attacks are similar; (ii) the medians of attack report lengths of low-, medium-, and high-newness attacks are similar; (iii) there are very long attack reports (i.e., outliers) for some low-newness attacks; and (iv) the variability among attack report lengths of low-newness attacks is larger than that of high-newness and medium-newness attacks. These suggest that there is no significant differences in the attack report length of varying attack newness, namely that the attack report length does not reflect attack newness. Figure 2b is the boxplot of the attack report length vs. attack sophistication. We observe that attack report lengths of low-sophistication attacks are substantially smaller than that of high-sophistication attacks. This suggests that attack report length indeed reflects the level of attack sophistication.

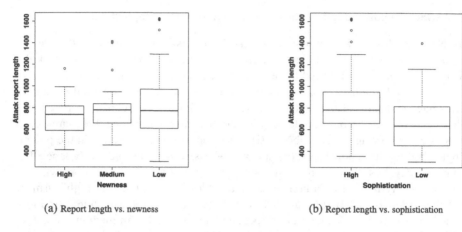

(a) Report length vs. newness

(b) Report length vs. sophistication

**Fig. 2.** Boxplots of attack report length vs. newness and report length vs. sophistication.

To formally confirm the intuitive findings mentioned above, we perform the following two statistical tests: ANOVA, which deals with the mean value of distributions [14], and Kruskal-Wallis, which is a non-parametric method dealing with distributions [29]. First, we use the ANOVA test to determine whether or not the mean attack report length is statistically the same in the three categories (i.e., high, medium, and low). For attack newness, the F statistic is 0.67 with a $p$-value 0.514, meaning that there is no statistical difference between the mean attack report length in the three categories of attack newness. For attack sophistication, the F statistic is 3.271 with a $p$-value 0.0736, where the small $p$-value indicates that there is a statistical difference between the mean attack report length in the two categories of attack sophistication. Second, we use the Kruskal-Wallis test to determine whether or not the attack report lengths in the two categories have the same distribution. For attack newness, the Kruskal-Wallis test statistic is 1.0182 with a $p$-value 0.601, which is consistent with the ANOVA test showing no statistical difference between the attack report length in the three categories of attack newness. For attack sophistication, the Kruskal-Wallis test statistic is 4.5789 with a $p$-value 0.0324, which confirms that there is a statistical difference between the attack report length of different attack sophistication. In summary, we draw:

**Insight 2.** *Attack report length reflects attack sophistication (i.e., less sophisticated attacks lead to shorter attack reports), but does not reflect attack newness (because a new attack is not necessarily sophisticated).*

# 3   Characterizing Time Series $X_t$, $Y_t$, $Z_t$

## 3.1   Basic Characteristics

From Fig. 1 we observe that $Y_t$ for $t < 100$ can be large while the corresponding $X_t$ is small, suggesting there are sophisticated attacks when $t < 100$. We observe that $Z_t$ for $t \geq 300$ is large, which is caused by large $X_t$ although the corresponding $Y_t$ is not large. When $Y_t$ is large, $Z_t$ can still be small (e.g., for $t < 100$); when $Y_t$ is not large, $Z_t$ can still be large (e.g., for $x > 300$). This discrepancy between $Y_t$ and $Z_t$ justifies the importance of analyzing both $Y_t$ and $Z_t$. In summary, we draw:

**Insight 3.** *The average attack report length and the total attack report length can exhibit different characteristics and can have different cybersecurity implications.*

From Fig. 1 we also observe that there exist large volatilities in all of the three time series. This phenomenon is confirmed by the basic statistics reported in Table 1, which shows that the variances are much larger than the corresponding mean values. This prompts that the classic Poisson process is not suitable for modeling these time series, and that we should investigate two statistical properties:

– Long-Range Dependence (LRD): This property, reviewed above, is important to characterize because it can guide the design of models to fit and forecast time series.
– Burstiness: This property, which is also reviewed above, is important because more analysts need to be allocated to cope with the bursts in attack events.

It is worth mentioning that *extreme values* [67] would be another property of interest, but the dataset contains too few data points to warrant an extreme value analysis.

**Table 1.** Basic statistics of $X_t$, $Y_t$, and $Z_t$ sample values.

| Time series | Min | Mean | Median | Var | Max |
|---|---|---|---|---|---|
| $X_t$ | 0 | 25.4153 | 21 | 394.0408 | 126 |
| $Y_t$ | 0 | 533.7213 | 509 | 44617.46 | 1700 |
| $Z_t$ | 0 | 14114.33 | 10884 | 170066803 | 81648 |

## 3.2   LRD Analysis

Figure 3a-3c plot the AutoCorrelation Function (ACF) of the three time series. We observe that in each case the ACF decays slowly, suggesting the presence of LRD. However, nonstationarity can also cause LRD [45,51], meaning that we need to study whether or not the slow decaying is indeed caused by the nonstationarity of the time series. For this purpose, we adopt the following widely-used hypothesis tests [37,47,51]:

– Unit root test: We use the Phillips-Perron test [46] to test if the slow decaying is caused by the *unit root* because it is nonparametric and robust against unspecified autocorrelation and heteroscedasticity. For time series $R_t$, we have:

$$H_0 : R_t \text{ has a unit root.} \qquad H_a : R_t \text{ is a stationary time series.}$$

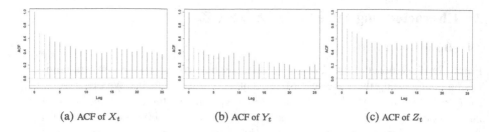

(a) ACF of $X_t$         (b) ACF of $Y_t$         (c) ACF of $Z_t$

**Fig. 3.** Plots of the ACF (AutoCorrelation Functions) of time series $X_t, Y_t, Z_t$.

– Nonstationary test: We use the Kwiatkowski-Phillips-Schmidt-Shin (KPSS) test [33] to test if the slow decaying is caused by *non-stationarity*.

$$H_0 : R_t \text{ is stationary.} \qquad H_a : R_t \text{ is nonstationary.}$$

The $p$-values of the Phillips-Perron test for $X_t$, $Y_t$, and $Z_t$ are all small ($< 0.01$), suggesting no evidence of unit root in any of the three time series. The $p$-values of the KPSS test are all small ($< 0.01$), suggesting that the three time series are nonstationary. Therefore, the LRD observed in the time series is indeed caused by nonstationarity.

**Insight 4.** *Cyber attack processes $X_t$, $Y_t$ and $Z_t$ are not Poisson but exhibit nonstationarity, which should be leveraged to design gray-box forecasting models.*

### 3.3 Burstiness Analysis

Burstiness is characterized via inter-event time. Since $X_t$, $Y_t$ and $Z_t$ are regular time series (i.e., their inter-event times are fixed), we pre-process them to obtain irregular time series of larger values, namely creating new time series by cutting off the small values of $X_t$, $Y_t$ and $Z_t$ to respectively obtain time series $X_t'$, $Y_t'$ and $Z_t'$. We here focus on obtaining $X_t'$ from $X_t$, while noting that $Y_t'$ and $Z_t'$ are obtained in the same fashion. Consider $X_t$ for $1 \leq t \leq 366$, we first sort them as $X_{t_1} \leq X_{t_2} \leq \ldots \leq X_{t_{366}}$. Then, we select a threshold $\zeta$ and omit any $X_t \leq X_{t_\zeta}$ because we only consider the larger values. That is, we will analyze the time series $X_t'$ for $t = 1, \ldots, 366$, where

$$X_t' = \begin{cases} X_t - X_{t_\zeta} & \text{if } X_t > X_{t_\zeta} \\ 0 & \text{otherwise.} \end{cases}$$

There are two guiding principles for selection threshold $\zeta$: (i) $X_t'$, $Y_t'$ and $Z_t'$ should have sufficiently many non-zero values for building statistically significant models; and (ii) $X_t'$, $Y_t'$ and $Z_t'$ should have roughly the same number of non-zero values, which assures that they are equally significant in a statistical sense. Based on $X_t'$, we can define *inter-arrival time* between two consecutive, non-zero weekly-binned number of attacks as follows. We say two non-zero values $X_t'$ and $X_{t'}'$ are *consecutive* if there does not exist $t^*$ such that $t < t^* < t'$ and $X_{t^*}' > 0$. Then, we define the *inter-arrival time* between the two consecutive, non-zero values $X_t'$ and $X_{t'}'$ as $t' - t$.

By examining the dataset, we set $\zeta = 257$ for $X_t$ to obtain $366 - \zeta = 109$ non-zero values for $X'_t$, where $X_{t_{257}} = X_{319} = 32$; we set $\zeta = 256$ for $Y_t$ to obtain 110 non-zero values for $Y'_t$, where $Y_{t_{256}} = Y_{297} = 610$; we set $\zeta = 256$ for $Z_t$ to obtain 110 non-zero values for $Z'_t$, where $Z_{t_{256}} = Z_{173} = 16,524$.

**Burstiness Analysis of $X'_t$.** Figure 4a plots time series $X'_t$. We observe that more attacks are detected in the later weeks perhaps because the networks grow over time. Figure 4b plots the inter-arrival time between two consecutive, non-zero values. We observe many small inter-arrival times (i.e., 1 week), indicating that many attacks are waged during consecutive weeks and that bursts are exhibited. Corresponding to Fig. 4b and according to Eq. (4), the burstiness measure is $B = 0.424$, meaning that attacks are bursty.

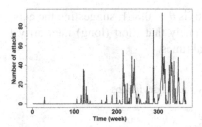
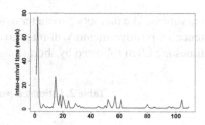

(a) Large number of attacks $X'_t$ ($y$-axis) vs. time $t$ ($x$-axis)

(b) Inter-arrival times in $X'_t$ ($y$-axis) vs. sequence # of intervals ($x$-axis)

**Fig. 4.** Plots of $X'_t$ and inter-arrival times between two consecutive, non-zero weekly-binned number of attacks.

To compute the more delicate burstiness measure $\delta$ given in Eq. (5), we need to fit the distribution of the inter-arrival times. Figure 5a plots the empirical density of inter-arrival times. We observe that the density is asymmetric with a long tail, meaning that distributions like normal, exponential, and weibull cannot fit inter-arrival times and that we should fit with a mixed distribution. We propose modeling the tail via GPD given in Eq. (7), and the other part by a lognormal distribution. The mixture density function is

$$f(x) = \begin{cases} \frac{1}{x\sigma_1\sqrt{2\pi}}e^{-\frac{(\ln x - \mu_1)^2}{2\sigma^2}}, & \text{if } x \leq u, \\ \frac{1}{k}\left(1 + \xi z\right)^{-\frac{1}{\xi}-1}, & \text{if } x \geq u, \end{cases} \tag{1}$$

where $\sigma_1$ and $\mu_1$ are respectively the standard deviation and the mean of lognormal distribution, $\xi$ is the shape parameter of GPD, $z = (x - u)/\sigma$, and $k = 1/(\sigma \cdot \xi)$ is the scale parameter. Figure 5b shows the qq-plot of the inter-arrival times. We observe that the proposed mixed distribution fits the inter-arrival times well except for one data point, while all of the data points are within the simulated 95% confidence interval.

Table 2 describes the estimated parameters of the fitted distribution and their standard deviations. We observe that threshold $u$ for the mixture distribution is 2.001, meaning that the tail proportion for the GPD fitting is around 21% and that the proposed mixture distribution fits the data well. Based on the fitted mixture distribution, we use Eq. (5) to derive the burstiness parameter $\delta = 0.556$, which suggests the presence of

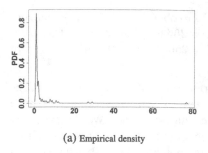

(a) Empirical density

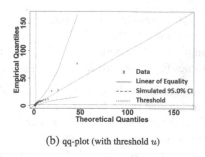

(b) qq-plot (with threshold $u$)

**Fig. 5.** Empirical density and qq-plot of inter-arrival time (CI stands for "confidence interval")

burstiness. The memory parameter defined in Eq. (6) is $\theta = 0.381$, suggesting the existence of positive memory in inter-arrival times, namely that short (long) inter-arrival times are often followed by short (long) inter-arrival times.

**Table 2.** Estimated parameters and standard deviations

|  | $\mu_1$ | $\sigma_1$ | $\xi$ | $k$ | $u$ |
|---|---|---|---|---|---|
| Parameter | 0.301 | 0.409 | 0.387 | 5.670 | 2.001 |
| Standard deviation | 0.045 | 0.045 | 0.379 | 4.481 | 0.001 |

**Burstiness Analysis of $Y_t'$.** Figure 6a plots time series of large, weekly-binned average attack report lengths. We observe that larger average lengths are mainly exhibited when $t < 100$ and when $t > 300$. This further confirms the non-uniformality in the average attack report length and therefore attack sophistication. Figure 6b plots the inter-arrival times between two consecutive, non-zero $Y_t'$ values, which are similarly defined as in the case of $X_t'$. We observe that most inter-arrival times between two large, average attack report lengths are very small (i.e., 1 week). Corresponding to Fig. 6b and according to Eq. (4), the burstiness measure $B$ is $B = 0.462$, which suggests that the inter-arrival times between large, average attack report lengths exhibit the burstiness property.

To obtain the more delicate burstiness measure $\delta$ given by Eq. (5), we fit the distribution of inter-arrival times. Figure 7a plots the empirical density of inter-arrival times of large average attack report lengths, which also shows asymmetry with a long tail. This also suggests us to use the mixture distribution in Eq. (1) to fit inter-arrival times in $Y_t'$. Figure 7b shows the qq-plot, indicating the mixed distribution fits inter-arrival times very well and all of the data points are within the simulated 95% confidence interval.

Table 3 summarizes the estimated parameters of the fitted distributions and their standard deviations. We observe that the threshold $u$ for the mixture distribution is 4.999, which means that the tail proportion for the GPD fitting is around 11%. Having fitted the distribution, we compute the burstiness parameter $\delta$ according to Eq. (5) to

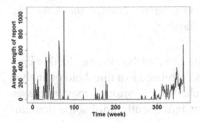
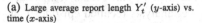

(a) Large average report length $Y_t'$ ($y$-axis) vs. time ($x$-axis)

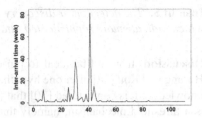

(b) Inter-arrival times in $Y_t'$ ($y$-axis) vs. sequence # of intervals ($x$-axis)

**Fig. 6.** Plots of $Y_t'$ and inter-arrival time between two consecutive, non-zero attack report lengths.

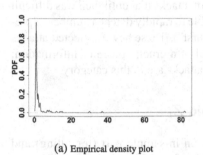

(a) Empirical density plot

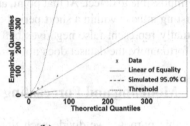

(b) qq-plot of inter-arrival times

**Fig. 7.** Empirical density and qq-plot of inter-arrival time (CI stands for confidence interval).

**Table 3.** Estimated parameters and standard deviations

|  | $\mu_1$ | $\sigma_1$ | $\xi$ | $k$ | $u$ |
|---|---|---|---|---|---|
| Parameter | 0.211 | 0.385 | 0.594 | 7.049 | 4.999 |
| Standard deviation | 0.040 | 0.031 | 0.487 | 3.938 | 0.001 |

have $\delta = 0.862$, which suggests the presence of strong burstiness. The memory parameter $\theta$ as defined in Eq. (6) is $\theta = 0.125$, which suggests positive memory, namely that short (long) inter-arrival times are often followed by short (long) inter-arrival times.

**Burstiness Analysis of $Z_t'$.** According to Eq.(4), burstiness of $Z_t'$ is $B = 0.425$. When we proceed to derive burstiness measure $\delta$, which requires to fitting the distribution of inter-arrival times of two consecutive, non-zero total report length in $Z_t'$, we did *not* find any accurate fitting of the distribution, despite we tried the normal, weibull, gamma, GPD, and several mixed models. This means burstiness measure $\delta$ as defined in Eq. (5) is not always practical because fitting distributions may not be feasible especially when the distribution involves many parameters. Nevertheless, the other burstiness measure defined in Eq. (6) is $\theta = 0.392$, implying positive memory in inter-arrival times, namely that long (short) inter-arrival times are often followed by short (long) inter-arrival times.

**Insight 5.** *The detection or discovery of cyber attacks is bursty. Longer attack reports, which indicate more sophisticated attacks, are also bursty.*

**Discussion**. It would be ideal to pin down the root cause of burstiness. To this end, Harang and Kott [28] offer one hypothesis: burstiness is related to a threshold of analyst knowledge. It is conjectured [30] that the common element of various bursty processes is a threshold mechanism, namely that events occur infrequently until some domain-specific quantity accumulates to a threshold value, at which point events "burst out" at a high frequency. It is interesting to note that cyber analysts recalled episodes that multiple attacks are detected after the arrival of a crucial piece of new information about a previously unknown attack behavior or characteristic [28]. This new information enables analysts to recognize a particular type of attacks that until then was difficult or impossible to detect. At that point, analysts are able to rapidly detect a number of pre-existing attacks within a short period of time (a "burst"). These newly detected attacks actually represent false negatives before the arrival of a crucial piece of information; unfortunately, the dataset does not describe which attacks are in this category.

## 4   Modeling and Forecasting $X_t$, $Y_t$, and $Z_t$

For this purpose, we divide each time series into an in-sample part (for fitting) and an out-of-sample part (for forecasting). We use the first 266 samples (e.g., $X_t$ for $1 \leq t \leq 266$) for in-sample fitting, and use the rest 100 samples (i.e., $X_t$ for $267 \leq t \leq 366$) for out-of-sample forecasting. We first need to test if there is correlation between $X_t$ and $Y_t$, while noting that there is correlation between $X_t$ and $Z_t$ and between $Y_t$ and $Z_t$ because $Z_t = X_t \times Y_t$. This is important because correlation, if existing, can be leveraged to achieve more accurate fitting and/or forecasting, as shown in other settings [44,55]. Examining Fig. 1a and 1b leads to the following *Hypothesis: there is a negative relation between the number of attacks and the length of reports.* One possible mechanism underpinning this hypothesis is that with many attacks arriving, there is less time to write long reports but, conversely, with few attacks arriving, there is more time to write long reports. Since we have showed the existence of temporal correlation within each individual time series $X_t$ and $Y_t$, we need to eliminate this temporal correlation so as not to interfere with the evaluation of the correlation between $X_t$ and $Y_t$. For this purpose, we measure the correlation between $X_t$ and $Y_t$ via the correlation between their respective standardized model residuals. By using the Pearson's correlation test to their residuals, we obtain a $p$-value of 0.117; by using the Kendall's rank correlation test, we obtain a $p$-value 0.5084. These $p$-values suggest no correlation between $X_t$ and $Y_t$.

### 4.1   Model Fitting $X_t$, $Y_t$ and $Z_t$ for $1 \leq t \leq 266$

Motivated by the presence of volatilities and burstiness in $X_t$ as shown above, we propose using the ARIMA+GARCH model reviewed above to fit $X_t$ because ARIMA can accommodate the mean part and GARCH can accommodate high volatilities or burstiness. To be flexible, we allow the orders of $p'$ and $q'$ in the mean part of

ARIMA$(p', d, q')$ to vary between 0 and 5, and we use the Akaike Information Criterion (AIC) criterion to select the orders. For the GARCH part, we fix the order as GARCH$(1, 1)$ because it is adequate to accommodate volatilities in the residuals, while recalling that higher-order GARCH models are not necessarily better than GARCH$(1, 1)$ [27].

Table 4 summarizes the selected model ARIMA$(0, 1, 1)$+IGARCH$(1, 1)$ with the skewed Student-T distribution innovations and estimated parameters. We observe that the estimated coefficients are all significant (i.e., statistical significance at level .05) except for $\omega$ which is the intercept of GARCH model.

**Table 4.** Estimated parameters (EST.) and their standard deviations (SD) of the ARIMA$(0, 1, 1)$+IGARCH$(1, 1)$ fitting of $X_t$.

|      | $\psi_1$ | $\omega$ | $\alpha_1$ | $\beta_1$ | $\xi$ | $\nu$ |
|------|------|------|------|------|------|------|
| EST. | -0.67 | 1.88 | 0.20 | 0.80 | 1.53 | 5.20 |
| SD   | 0.05 | 1.18 | 0.05 | – | 0.13 | 1.22 |

Figure 8a plots the ARIMA$(0, 1, 1)$+IGARCH$(1, 1)$ fitting of $X_t$, where red-colored crosses represent fitted values, black-colored empty circles represent the observed values, and blue-colored lines are the fitted 95% confidence intervals. We observe that the overall fitting is good except for a few very large data points, and that the fitted confidence intervals can accommodate the variability of $X_t$. Figure 8b plots the ACF of the model residuals. We observe that the residuals fall in between the two blue-colored dashed lines (indicating the 95% confidence limits) with probability 95%, meaning that none of the correlation is significant. Therefore, ARIMA$(0, 1, 1)$+IGARCH$(1, 1)$ can adequately fit the temporal dependence between the $X_t$'s with varying $t$. By using the Ljung-Box test on the standardized residuals, we obtain a $p$-value of 0.09, which implies that ARIMA$(0, 1, 1)$+IGARCH$(1, 1)$ can accurately fit $X_t$.

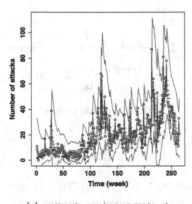

(a) ARIMA$(0, 1, 1)$+GARCH$(1, 1)$

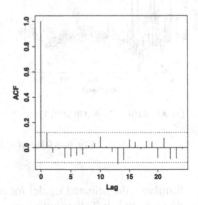

(b) Plot of ACF of model residuals

**Fig. 8.** ARIMA$(0, 1, 1)$+GARCH$(1, 1)$ fitting $X_t$ for $1 \leq t \leq 266$.

Using the same model fitting method as in fitting $X_t$, the selected model is also ARIMA$(0, 1, 1)$+IGARCH$(1, 1)$ with the skewed Student-T distribution innovations and estimated parameters summarized in Table 5. We again observe that all the coefficients for the are significant except for $\omega$.

Figure 9 plots the ARIMA$(0, 1, 1)$+IGARCH$(1, 1)$ fitting of $Y_t$, where red-colored crosses represent the fitted values, black-colored empty circles represent the observed values, and blue-colored lines represent the fitted 95% confidence intervals. We observe that the fitting is good and the fitted confidence intervals can accommodate the variability in $Y_t$. Figure 9b plots the ACF of the model residuals. We observe that none of the correlation is significant, meaning the selected ARIMA$(0, 1, 1)$+IGARCH$(1, 1)$ model can adequately fit the temporal dependence in $Y_t$. By applying the weighted Ljung-Box test to the standardized residuals, we obtain a $p$-value of $0.91$, which implies that ARIMA$(1, 0, 1)$+IGARCH$(1, 1)$ can accurately fit $Y_t$.

**Table 5.** Estimated parameters (EST.) and their standard deviations (SD) of the ARIMA$(0, 1, 1)$+IGARCH$(1, 1)$ fitting of $Y_t$.

|      | $\psi_1$ | $\omega$ | $\alpha_1$ | $\beta_1$ | $\xi$ | $\nu$ |
|------|------|--------|------|------|------|------|
| EST. | -0.76 | 237.00 | 0.20 | 0.80 | 1.42 | 5.81 |
| SD   | 0.37 | 176.21 | 0.05 | – | 0.13 | 1.72 |

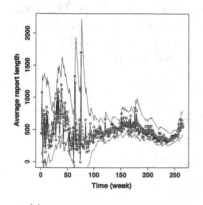

(a) ARIMA$(0, 1, 1)$+IGARCH$(1, 1)$

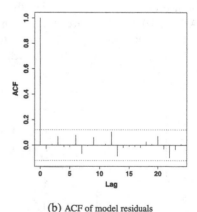

(b) ACF of model residuals

**Fig. 9.** ARIMA$(0, 1, 1)$+IGARCH$(1, 1)$ fitting $Y_t$.

Similarly, the selected model for fitting $Z_t$ is ARIMA$(1, 1, 1)$+GARCH$(1, 1)$ with the skewed Student-T distribution innovations and estimated parameters summarized in Table 6. The coefficient for AR, $\phi_1$, is significant at the $0.1$ level. The coefficients for GARCH$(1, 1)$ are significant, and $\alpha_1 + \beta_1 = 0.99$, which explains the burstiness.

**Table 6.** Estimated parameters (EST.) and standard deviations (SD) for the ARIMA$(1,1,1)$+GARCH$(1,1)$ fitting of $Z_t$

|       | $\phi_1$ | $\psi_1$ | $\omega$ | $\alpha_1$ | $\beta_1$ | $\xi$ | $\nu$ |
|-------|------|-------|--------|------|------|------|------|
| EST.  | 0.27 | -0.90 | 31343  | 0.08 | 0.91 | 1.59 | 6.14 |
| SD    | 0.07 | 0.05  | 187540 | 0.04 | 0.06 | 0.23 | 1.86 |

Figure 10 plots the ARIMA$(1,1,1)$+GARCH$(1,1)$ fitting result, where red-colored crosses represent the fitted values, black-colored empty circles represent the observed values, and blue-colored lines represent the fitted 95% confidence intervals. We observe that the overall fitting is good except for a few large samples, and the fitted confidence intervals can accommodate the variability in $Z_t$. Figure 10b plots the ACF of the model residuals, and shows that there is no significant correlation because they are within the blue-colored lines (i.e., the 95% confidence intervals). That it, the selected ARIMA$(1,1,1)$+GARCH$(1,1)$ model can adequately fit $Z_t$. By using the weighted Ljung-Box test to the standardized residuals, we obtain a $p$-value of 0.38, meaning ARIMA$(1,1,1)$+GARCH$(1,1)$ can accurately fit $Z_t$.

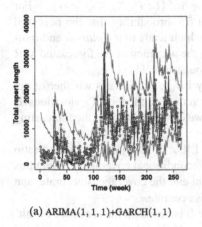

(a) ARIMA$(1,1,1)$+GARCH$(1,1)$

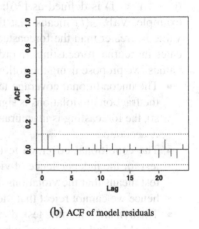

(b) ACF of model residuals

**Fig. 10.** ARIMA$(1,1,1)$+GARCH$(1,1)$ fitting $Z_t$.

**Insight 6.** *The three time series $X_t$, $Y_t$, and $Z_t$ can all be accurately fitted using the ARIMA+GARCH model because they exhibit the burstiness property (which can be accommodated by the GARCH part).*

## 4.2 Forecasting $X_t$, $Y_t$ and $Z_t$ for $267 \leq t \leq 366$

Now we use the fitted models of $X_t$, $Y_t$ and $Z_t$ for $1 \leq t \leq 266$ to forecast $X_t$, $Y_t$ and $Z_t$ for $267 \leq t \leq 366$, respectively. To provide more information, we propose forecasting distributions of $X_t$, $Y_t$ and $Z_t$ for $267 \leq t \leq 366$. To highlight ideas, we focus on forecasting $X_t$, but the idea is equally applicable to $Y_t$ and $Z_t$. Recall the fitted model $\Psi_{t-1} = \{X_{t-1}, X_{t-2}, \ldots, X_1\}$. Let $\{f_t(X|\Psi_{t-1})\}_{t=1}^{\infty}$ be a sequence of one-step ahead (i.e., one-week ahead in this paper) density forecasting conditioned on the information available at time $t-1$, The cumulative density forecasts are given by

$$F_t(X_t) = \int_{-\infty}^{X_t} f_t(u|\Psi_{t-1})du. \tag{2}$$

If the model is accurate, the sequence of probability integral transforms $\{F_t(X_t)|t = 1, 2, \ldots\}$ are independent and identically random variables $U(0,1)$ [1]. That is, the sequence of transforms being $U(0,1)$ means we cannot reject the model being accurate. To evaluating accuracy of the forecasted probability density, we use two metrics:

- Density accuracy: Berkowitz [4] developed a formal test for evaluating the performance of density forecasting. The basic idea is to transform $\{F_t(X_t)|t = 1, 2, \ldots\}$ to the standard normal distribution $N(0,1)$ by using the normal quantile function, and then test the normality of transformed data via the Lagrange Multiplier test [52]. Passing the test means the forecasted density is accurate.
- VaR violation: For a random variable $X_t$, the Value-at-Risk (VaR) at level $\alpha$ ($0 < \alpha < 1$) is defined as [36]: $\text{VaR}_\alpha(t) = \inf\{l : P(X_t \leq l) \geq \alpha\}$. For example, $\text{VaR}_{.95}(t)$ means that there is only a 5% probability that the observed value is greater than the forecasted $\text{VaR}_{.95}(t)$, which leads to a *violation* and indicates inaccurate forecasting. In order to evaluate the accuracy of the forecasted VaR values, we propose using the following three popular tests [11, 17]:
  - The unconditional coverage test, denoted by $\text{LR}_{uc}$: It evaluates whether or not the fraction of violations is significantly different from the model's violations. If so, the forecasting is inaccurate; otherwise, we cannot reject that the forecasting is accurate.
  - The conditional coverage test, denoted by $\text{LR}_{cc}$: It is a joint likelihood ratio test for the independence of violations and unconditional coverage. Passing this test means that the violations are independent and the coverage is accurate, and hence we cannot reject that the forecasting is accurate.
  - The dynamic quantile test, denoted by (DQ): It is based on the sequence of 'hit' variables and it measures whether the present VaR violations are uncorrelated to the past violations or not. Passing this test means they are uncorrelated, and hence we cannot reject that the forecasting is accurate.

**Forecasting Algorithm.** Algorithm 1 describes the *rolling* algorithm for forecasting $X_t$, and can be adapted to forecast $Y_t$ and $Z_t$ by replacing $X_t$ with $Y_t$ or $Z_t$ and by replacing $\hat{X}_t$ with $\hat{Y}_t$ or $\hat{Z}_t$, respectively.

**Forecasting Results.** Figure 11 plots the forecasting results of $X_t$, $Y_t$, and $Z_t$ for $267 \leq t \leq 366$. We observe that the forecasted values are close to the observed ones, and that the forecasted $\text{VaR}_{.95}$ can describe the violations well. We use rigorous statistical tests to validate the observation mentioned above. Table 7 summarizes the results. For $X_t$, the $p$-value of the Berkowitz test is $0.1925$, indicating the forecasted density functions

**Algorithm 1.** Algorithm for forecasting $X_t$

Input: Historical time series dataset $\{(t, X_t)|t = 1, \ldots, 366\}$
Output: The predicted $\{(t, \hat{X}_t)|t = 267, \ldots, 366\}$
    **for** $t = 266, \ldots, 365$ **do**
        Fit the historical data $\{(s, X_s)|s = 1, \ldots, t\}$ with the selected
        ARIMA$(p', d, q')$+GARCH$(1, 1)$ model
        Use the fitted model to forecast $X_{t+1}$, denoted the forecasted value by $\hat{X}_{t+1}$
    **end for**
    Return $\{(t, \hat{X}_t)|t = 267, \ldots, 366\}$

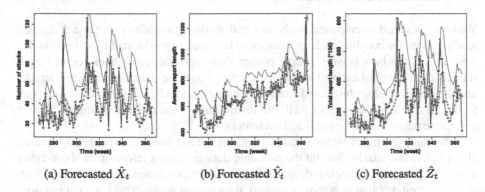

(a) Forecasted $\hat{X}_t$        (b) Forecasted $\hat{Y}_t$        (c) Forecasted $\hat{Z}_t$

**Fig. 11.** Plots of forecasting results based on ARIMA$(p', d, q')$+GARCH$(1, 1)$ models, where the red-colored $+$'s represent forecasted values, black-colored empty circles are the observed values, and blue-colored lines are the forecasted VaR$_{.95}$ (Color figure online).

are accurate. For the VaR violation tests at the $\alpha = .95$ level, all of the $p$-values are large. This also suggests that the ARIMA$(p', d, q')$+GARCH$(1, 1)$ models have a good accuracy performance. For $Y_t$, the $p$-value of the Berkowitz test is 0.5657, indicating the forecasted density functions are accurate. The VaR violation tests at the $\alpha = .95$ level all lead to large $p$-values, indicating the ARIMA$(p', d, q')$+GARCH$(1, 1)$ models have a good forecasting accuracy. For $Z_t$, the $p$-value of the Berkowitz test is 0.1550, meaning the forecasted density functions are accurate. The VaR violation tests at the $\alpha = .95$ level all lead to large $p$-values, meaning the ARIMA$(p', d, q')$+GARCH$(1, 1)$ models have a good forecasting accuracy. Summarizing the discussion, we draw:

**Table 7.** Statistical tests for forecast accuracy of ARIMA$(p', d, q')$+GARCH$(1, 1)$ models.

|  | Density accuracy | VaR violation metrics | | |
|---|---|---|---|---|
|  | Berkowitz | LR$_{uc}$ | LR$_{cc}$ | DQ |
| $X_t$ | 0.1925 | 0.3855 | 0.4027 | 0.2725 |
| $Y_t$ | 0.5657 | 0.6560 | 0.6147 | 1 |
| $Z_t$ | 0.1550 | 1 | 0.7664 | 0.999 |

**Insight 7.** *The distribution of the number of attacks, of the average report length, and of the total report length can be accurately forecasted, by using models that can accommodate the statistical properties exhibited by the data (i.e., burstiness in this case).*

Insight 7 is valuable because being able to predict report length, which reflects the sophistication of incoming attacks, provides a means to proactively allocate defense resources (e.g., human experts) to achieve more effective defense.

## 5   Conclusion and Discussion

We have presented an empirical study on a real-world high quality cyber attack dataset, leading to useful insights, such as: burstiness is commonly exhibited by cyber attacks; new attacks can be relatively simple (rather than sophisticated); attack report length reflects attack sophistication (but not attack newness); the detection of cyber attacks and the detection of sophisticated attacks are both bursty; ARIMA+GARCH model can fit the number of cyber attacks well; the distribution of the number of attacks against enterprise networks can be predicted accurately.

The study is limited by the dataset. First, the dataset does not provide information about individual attacks. Should the available dataset contain information about cyber attacks (e.g., types of attacks), deeper analysis can be conducted. Second, it is an outstanding open problem to precisely identify the cause of bursts. Third, the dataset does not provide any false negative information. Fourth, Insights 1 and 2 are drawn based on 100 random samples rather than all the 9,302 samples because we are only given the privilege to request for 100 random samples.

**Acknowledgment.** This work was supported in part by ARL Grant #W911NF-17-2-0127, NSF Grants #2122631 and #2115134, and Colorado State Bill 18-086.

## A   Statistical Preliminaries

**Long Range Dependence (LRD).** Intuitively, LRD means a stochastic process exhibits persistent temporal correlations, namely its autocorrelation decays slowly. Formally, a stationary time series $\{X_i, i \geq 1\}$ is said to possess LRD [50,54] if its autocorrelation function $\rho(h)$, which is defined below, has the following property:

$$\rho(h) = \text{Cor}(X_i, X_{i+h}) \sim h^{-\beta} L(h), \quad h \to \infty, \tag{3}$$

for $0 < \beta < 1$, where $\text{Cor}(\cdot, \cdot)$ is the correlation function and $L(\cdot)$ is a *slowly varying* function satisfying $\lim_{x \to \infty} \frac{L(tx)}{L(x)} = 1$ for $t > 0$ [16]. The degree of LRD is expressed by the Hurst parameter (H), which is related to the parameter $\beta$ in Eq. (3) as $\beta = 2 - 2H$. For LRD, we have $1/2 < H < 1$ and the degree of LRD increases as $H \to 1$.

**Burstiness.** Intuitively, burstiness indicates an abnormally large number of events within a short period of time when compared with the Poisson and regular stochastic processes. Unlike LRD, burstiness has no universally accepted definition. The simplest definition of burstiness is perhaps the *coefficient of variation*, namely $r = \sigma/\mu$, where $\sigma$ and $\mu$ are respectively the standard deviation and the mean of inter-event times [24,31]. Since $r$ can be an arbitrary number, a refined burstiness definition is [31]:

$$B = \frac{\sigma - \mu}{\sigma + \mu} = \frac{r - 1}{r + 1}, \tag{4}$$

where $B \in [-1, 1]$, $B = -1$ corresponds to the regular time series (i.e., $r = 0$), $B = 0$ corresponds to the Poisson process, and $B \to 1$ indicates bursty time series [31].

By observing that burstiness can be rooted in two deviations from the Poisson process (i.e., the distribution and memory of the inter-event time), burstiness has also been defined as a vector $(\delta, \theta)$, where $\delta, \theta \in [-1, 1]$ [24]. The first element $\delta$ reflects the distribution of inter-event time $\tau$ and is defined as

$$\delta = \frac{\text{sign}(\sigma - \mu)}{2} \int_0^\infty |\Pr(\tau) - \Pr_p(\tau)| d\tau, \tag{5}$$

where "sign" is the *sign* function, $\sigma$ is the standard deviation, $\mu$ is the mean of inter-event time $\tau$, $\Pr(\cdot)$ is the density function of $\tau$, and $\Pr_p(\cdot)$ is the exponential distribution. Note that $\delta$ measures the difference between $\Pr(\cdot)$ and $\Pr_p(\cdot)$, $\delta = -1$ indicates the regular time series, $\delta = 0$ indicates the Poisson process, and $\delta \to 1$ indicates bursty time series. The second element $\theta$ reflects the memory of inter-event times, namely the correlation coefficient of consecutive inter-event times

$$\theta = \frac{1}{N} \sum_{i=1}^{N-1} \frac{(T_i - \mu_1)(T_{i+1} - \mu_2)}{\sigma_1 \times \sigma_2}, \tag{6}$$

where $\mu_1$ ($\mu_2$) and $\sigma_1$ ($\sigma_2$) are respectively the sample mean and the standard deviation of inter-event times $T_1, \ldots, T_{N-1}$ ($T_2, \ldots, T_N$), and $N$ is the sample size. Note that $\theta$ is the memory coefficient describing the correlation of consecutive inter-event times, where $\theta > 0$ means positive memory, namely short (long) inter-event times are often followed by short (long) ones, and $\theta < 0$ means negative memory, namely short (long) inter-event times are often followed by long (short) ones.

**Generalized Pareto Distribution (GPD).** To characterize burstiness in inter-arrival times, we can first transform a regular time series into an irregular time series as follows. Given a sequence of independent and identically distributed (iid) observations $X_1, \ldots, X_n$, the excesses $X_i - \ell$ of some suitably large threshold $\ell$ can be modeled by, under certain mild conditions, the *generalized Pareto distribution* (GPD) [16,48]. For characterizing burstiness, we choose the $\ell$ such that 30% of the excess values will be investigated, which is in contrast to the study of extreme values that would only consider the largest 1% samples. The survival function of the GPD is

$$\bar{G}_{\xi,\sigma_1,\ell}(x) = 1 - G_{\xi,\sigma_1,\ell} = \begin{cases} \left(1 + \xi \frac{x-\ell}{\sigma_1}\right)^{-1/\xi}, & \xi \neq 0, \\ \exp\left\{-\frac{x-\ell}{\sigma_1}\right\}, & \xi = 0, \end{cases} \tag{7}$$

where $x \geq \ell$ if $\xi \in \mathbb{R}^+$, $x \in [\ell, \ell - \sigma_1/\xi]$ if $\xi \in \mathbb{R}^-$, and $\xi$ and $\sigma_1$ are respectively called the *shape* and *scale* parameters.

**ARIMA and GARCH Models**. ARIMA (AutoRegressive Integrated Moving Average) and GARCH (Generalized AutoRegressive Conditional Heteroskedasticity) are widely-used time series models [13]. Intuitively, ARIMA can model the mean of a time series, and GARCH can model the high volatility of a time series. Formally, let $\phi(x) = 1 - \sum_{j=1}^{p'} \phi_j x^j$, $\psi(x) = 1 + \sum_{j=1}^{q'} \psi_j x^j$, and $\epsilon_t$ be independent and identical normal random variables with mean 0 and variance $\sigma_\epsilon^2$. A time series $\{X_t\}$ is said to be a ARIMA$(p', d, q')$ process if $\phi(B)(1 - B)^d(X_t - \mu) = \psi(B)\epsilon_t$, where $d$ is the number of difference, $B$ is the back shift operator, and $\mu$ is the mean. A time series $\{Y_t\}$ is called a GARCH process [5] if $Y_t = \sigma_t \epsilon_t$, where $\epsilon_t$ (also called *innovation*) is the standard white noise. For the standard GARCH model, we have $\sigma_t^2 = w + \sum_{j=1}^{q'} \alpha_j \epsilon_{t-j}^2 + \sum_{j=1}^{p'} \beta_j \sigma_{t-j}^2$. A restricted version of the GARCH model is called the integrated GARCH (IGARCH) by requiring $\sum_{j=1}^{q'} \alpha_j + \sum_{j=1}^{p'} \beta_j = 1$. To accommodate more general classes of noise, we will use the skewed Student-T distribution, whose density can be written as [21]:

$$g(z; \vartheta_j) = \frac{2}{\xi + \xi^{-1}} \left[ t_\nu(\xi z) I(z < 0) + t_\nu v\left(\xi^{-1} z\right) I(z \geq 0) \right],$$

where $I(\cdot)$ is the indicator function, $\vartheta_j = (\xi_j, \nu_j)$, $\xi > 0$ is the skewness parameter, $\nu > 0$ is the shape parameter, and

$$t_\nu(z) = \frac{\Gamma((\nu+1)/2)}{\sqrt{\nu\pi}\Gamma(\nu/2)} \left[1 + z^2/\nu\right]^{-(\nu+1)/2}.$$

**Akaike's Information Criterion (AIC)**. When fitting time series, we need criteria for model section. AIC is a widely used criterion [13,36,45] for balancing the goodness-of-fit of a model and its complexity such that a smaller AIC value indicates a better model. Formally, AIC $= -2\log(\text{MLE}) + 2k$, where MLE measures the goodness-of-fit of a model and $k$ is the number of model parameters (indicating model complexity).

## References

1. Andersen, T.G., Bollerslev, T., Diebold, F.X., Labys, P.: Modeling and forecasting realized volatility. Econometrica **71**(2), 579–625 (2003)
2. Bakdash, J., et al.: Dataset associated with 'malware in the future? forecasting analyst detection of cyber events' (2019). https://osf.io/hjffm/
3. Bakdash, J.Z., et al.: Malware in the future? Forecasting of analyst detection of cyber events. J. Cybersecurity **4**(1) (2018)
4. Berkowitz, J.: Testing density forecasts, with applications to risk management. J. Bus. Econ. Stat. **19**(4), 465–474 (2001)
5. Bollerslev, T., Russell, J., Watson, M.W.: Volatility and Time Series Econometrics: Essays in Honor of Robert Engle. Oxford University Press, Oxford (2010)
6. Charlton, J., Du, P., Xu, S.: A new method for inferring ground-truth labels and malware detector effectiveness metrics. In: Lu, W., Sun, K., Yung, M., Liu, F. (eds.) SciSec 2021. LNCS, vol. 13005, pp. 77–92. Springer, Cham (2021). https://doi.org/10.1007/978-3-030-89137-4_6

7. Chen, H., Cho, J., Xu, S.: Quantifying the security effectiveness of firewalls and DMZs. In: Proceedings of the HoTSoS 2018, pp. 9:1–9:11 (2018)
8. Chen, H., Cam, H., Xu, S.: Quantifying cybersecurity effectiveness of dynamic network diversity. IEEE Trans. Dependable Secur. Comput. (2021). https://doi.org/10.1109/TDSC.2021.3107514
9. Chen, Y., Huang, Z., Xu, S., Lai, Y.: Spatiotemporal patterns and predictability of cyberattacks. PLoS One 10(5), e0124472 (2015)
10. Cho, J.H., Xu, S., Hurley, P.M., Mackay, M., Benjamin, T., Beaumont, M.: STRAM: measuring the trustworthiness of computer-based systems. ACM Comput. Surv. 51(6), 128:1–128:47 (2019)
11. Christoffersen, P.F.: Evaluating interval forecasts. International Economic Review, pp. 841–862 (1998)
12. Condon, E., He, A., Cukier, M.: Analysis of computer security incident data using time series models. In: International Symposium on Software Reliability Engineering, pp. 77–86 (2008)
13. Cryer, J.D., Chan, K.S.: Time Series Analysis With Applications in R. Springer, New York (2008). https://doi.org/10.1007/978-0-387-75959-3
14. Devore, J.L., Berk, K.N., Carlton, M.A.: Modern Mathematical Statistics with Applications. STS, Springer, Cham (2021). https://doi.org/10.1007/978-3-030-55156-8
15. Du, P., Sun, Z., Chen, H., Cho, J.H., Xu, S.: Statistical estimation of malware detection metrics in the absence of ground truth. IEEE T-IFS 13(12), 2965–2980 (2018)
16. Embrechts, P., Klüppelberg, C., Mikosch, T.: Modelling Extremal Events. AM, vol. 33. Springer, Heidelberg (1997). https://doi.org/10.1007/978-3-642-33483-2
17. Engle, R.F., Manganelli, S.: CAViaR: conditional autoregressive value at risk by regression quantiles. J. Bus. Econ. Stat. 22(4), 367–381 (2004)
18. Fachkha, C., Bou-Harb, E., Debbabi, M.: Towards a forecasting model for distributed denial of service activities. In: 2013 IEEE 12th International Symposium on Network Computing and Applications, pp. 110–117 (2013)
19. Fang, X., Xu, M., Xu, S., Zhao, P.: A deep learning framework for predicting cyber attacks rates. EURASIP J. Inf. Secur. 2019, 5 (2019)
20. Fang, Z., Xu, M., Xu, S., Hu, T.: A framework for predicting data breach risk: leveraging dependence to cope with sparsity. IEEE T-IFS 16, 2186–2201 (2021)
21. Fernandez, C., Steel, M.F.J.: On Bayesian modeling of fat tails and skewness. J. Am. Stat. Assoc. 93(441), 359–371 (1998)
22. Ganesan, R., Jajodia, S., Cam, H.: Optimal scheduling of cybersecurity analysts for minimizing risk. ACM Trans. Intell. Syst. Technol. 8(4), 52:1–52:32 (2017)
23. Garcia-Lebron, R., Myers, D.J., Xu, S., Sun, J.: Node diversification in complex networks by decentralized colouring. J. Complex Netw. 7(4), 554–563 (2019)
24. Goh, K.I., Barabási, A.L.: Burstiness and memory in complex systems. EPL (Europhys. Lett.) 81(4), 48002 (2008)
25. Han, Y., Lu, W., Xu, S.: Characterizing the power of moving target defense via cyber epidemic dynamics. In: HotSoS, pp. 1–12 (2014)
26. Han, Y., Lu, W., Xu, S.: Preventive and reactive cyber defense dynamics with ergodic time-dependent parameters is globally attractive. IEEE TNSE 8(3), 2517–2532 (2021)
27. Hansen, P.R., Lunde, A.: A forecast comparison of volatility models: does anything beat a GARCH (1, 1)? J. Appl. Economet. 20(7), 873–889 (2005)
28. Harang, R., Kott, A.: Burstiness of intrusion detection process: empirical evidence and a modeling approach. IEEE Trans. Inf. Forensics Secur. 12(10), 2348–2359 (2017)
29. Hollander, M., Wolfe, D.A., Chicken, E.: Nonparametric Statistical Methods. vol. 751. Wiley, Hoboken (2013)
30. Karsai, M., Kaski, K., Barabási, A.L., Kertész, J.: Universal features of correlated bursty behaviour. Sci. Rep. 2, 1–7 (2012)

31. Kim, E.K., Jo, H.H.: Measuring burstiness for finite event sequences. Phys. Rev. E **94**(3), 032311 (2016)
32. Kott, A., Arnold, C.: The promises and challenges of continuous monitoring and risk scoring. IEEE Secur. Priv. **11**(1), 90–93 (2013)
33. Kwiatkowski, D., Phillips, P.C., Schmidt, P., Shin, Y., et al.: Testing the null hypothesis of stationarity against the alternative of a unit root. J. Econometrics **54**(1–3), 159–178 (1992)
34. Li, X., Parker, P., Xu, S.: A stochastic model for quantitative security analyses of networked systems. IEEE TDSC **8**(1), 28–43 (2011)
35. Lin, Z., Lu, W., Xu, S.: Unified preventive and reactive cyber defense dynamics is still globally convergent. IEEE/ACM ToN **27**(3), 1098–1111 (2019)
36. McNeil, A.J., Frey, R., Embrechts, P.: Quantitative Risk Management: Concepts, Techniques, and Tools. Princeton University Press, Princeton (2010)
37. Mikosch, T., Starica, C.: Nonstationarities in financial time series, the long-range dependence, and the IGARCH effects. Rev. Econ. Stat. **86**(1), 378–390 (2004)
38. Mireles, J.D., Ficke, E., Cho, J., Hurley, P., Xu, S.: Metrics towards measuring cyber agility. IEEE Trans. Inf. Forensics Secur. **14**(12), 3217–3232 (2019)
39. Montañez Rodriguez, R., Longtchi, T., Gwartney, K., Ear, E., Azari, D.P., Kelley, C.P., Xu, S.: Quantifying psychological sophistication of malicious emails. In: Yung, M., et al. (eds.) SciSec 2023, LNCS, vol. 14299, pp. 319–331. Springer, Cham (2023)
40. Montañez, R., Atyabi, A., Xu, S.: Book chapter in "cybersecurity and cognitive science", chap. social engineering attacks and defenses in the physical world vs. cyberspace: a contrast study. Elsevier, pp. 3–41 (2022)
41. Montañez, R., Golob, E., Xu, S.: Human cognition through the lens of social engineering cyberattacks. Front. Psychol. **11**, 1755 (2020)
42. Pendleton, M., Garcia-Lebron, R., Cho, J.H., Xu, S.: A survey on systems security metrics. ACM Comput. Surv. **49**(4), 62:1–62:35 (2016)
43. Peng, C., Xu, M., Xu, S., Hu, T.: Modeling and predicting extreme cyber attack rates via marked point processes. J. Appl. Stat. **44**(14), 2534–2563 (2017)
44. Peng, C., Xu, M., Xu, S., Hu, T.: Modeling multivariate cybersecurity risks. J. Appl. Stat. **45**(15), 2718–2740 (2018)
45. Peter, B., Richard, D.: Introduction to Time Series and Forecasting. Springer, New York (2002). https://doi.org/10.1007/b97391
46. Phillips, P.C., Perron, P.: Testing for a unit root in time series regression. Biometrika **75**(2), 335–346 (1988)
47. Qu, Z.: A test against spurious long memory. J. Bus. Econ. Stat. **29**(3), 423–438 (2011)
48. Resnick, S.: Heavy-Tail Phenomena: Probabilistic and Statistical Modeling. Springer, New York (2007). https://doi.org/10.1007/978-0-387-45024-7
49. Rodriguez, R.M., Xu, S.: Cyber social engineering kill chain. In: Proceedings of International Conference on Science of Cyber Security (SciSec 2022), pp. 487–504 (2022)
50. Samorodnitsky, G.: Long range dependence. Founda. Trends Stoch. Syst. **1**(3), 163–257 (2006)
51. Shao, X.: A simple test of changes in mean in the possible presence of long-range dependence. J. Time Ser. Anal. **32**(6), 598–606 (2011)
52. Silvey, S.D.: The Lagrangian multiplier test. Ann. Math. Stat. **30**(2), 389–407 (1959)
53. Trieu-Do, V., Garcia-Lebron, R., Xu, M., Xu, S., Feng, Y.: Characterizing and leveraging granger causality in cybersecurity: framework and case study. EAI Endorsed Trans. Secur. Safety **7**(25), e4 (2020)
54. Willinger, W., Taqqu, M.S., Leland, W.E., Wilson, V.: Self-similarity in high-speed packet traffic: analysis and modeling of ethernet traffic measurements. Stat. Sci. **10**, 67–85 (1995)
55. Xu, M., Hua, L., Xu, S.: A vine copula model for predicting the effectiveness of cyber defense early-warning. Technometrics **59**(4), 508–520 (2017)

56. Xu, M., Schweitzer, K.M., Bateman, R.M., Xu, S.: Modeling and predicting cyber hacking breaches. IEEE T-IFS **13**(11), 2856–2871 (2018)
57. Xu, M., Xu, S.: An extended stochastic model for quantitative security analysis of networked systems. Internet Math. **8**(3), 288–320 (2012)
58. Xu, S.: Emergent behavior in cybersecurity. In: Proceedings of the HotSoS 2014, pp. 13:1–13:2 (2014)
59. Xu, S.: The cybersecurity dynamics way of thinking and landscape (invited paper). In: ACM Workshop on Moving Target Defense (2020)
60. Xu, S., Lu, W., Zhan, Z.: A stochastic model of multivirus dynamics. IEEE Trans. Dependable Secur. Comput. **9**(1), 30–45 (2012)
61. Xu, S.: Cybersecurity dynamics. In: Proceedings of the Symposium on the Science of Security (HotSoS 14), pp. 14:1–14:2 (2014)
62. Xu, S.: Cybersecurity dynamics: a foundation for the science of cybersecurity. In: Wang, C., Lu, Z. (eds.) Proactive and Dynamic Network Defense. Advances in Information Security, vol. 74, pp. 1–31. Springer, Cham (2019). https://doi.org/10.1007/978-3-030-10597-6_1
63. Xu, S.: SARR: a cybersecurity metrics and quantification framework (keynote). In: Lu, W., Sun, K., Yung, M., Liu, F. (eds.) SciSec 2021. LNCS, vol. 13005, pp. 3–17. Springer, Cham (2021). https://doi.org/10.1007/978-3-030-89137-4_1
64. Xu, S., Yung, M., Wang, J.: Seeking foundations for the science of cyber security. Inf. Syst. Front. **23**(2), 263–267 (2021)
65. Zhan, Z., Xu, M., Xu, S.: A characterization of cybersecurity posture from network telescope data. In: Proceedings of the InTrust, pp. 105–126 (2014)
66. Zhan, Z., Xu, M., Xu, S.: Characterizing honeypot-captured cyber attacks: Statistical framework and case study. IEEE T-IFS **8**(11), 1775–1789 (2013)
67. Zhan, Z., Xu, M., Xu, S.: Predicting cyber attack rates with extreme values. IEEE Trans. Inf. Forensics Secur. **10**(8), 1666–1677 (2015)
68. Zheng, R., Lu, W., Xu, S.: Preventive and reactive cyber defense dynamics is globally stable. IEEE TNSE **5**(2), 156–170 (2018)

# Cryptography and Authentication

Cryptography and Embedded ...

# MCVDSSE: Secure Multi-client Verifiable Dynamic Symmetric Searchable Encryption

Jinyang Li[1], Zhenfu Cao[1,2($\boxtimes$)], Jiachen Shen[1($\boxtimes$)], and Xiaolei Dong[1,2($\boxtimes$)]

[1] Shanghai Key Laboratory of Trustworthy Computing East China Normal
University, Shanghai 200062, China
{zfcao,jcshen,dongxiaolei}@sei.ecnu.edu.cn
[2] Research Center for Basic Theories of Intelligent Computing,
Research Institute of Basic Theories, Zhejiang Lab, Hangzhou 311121, China

**Abstract.** Existing multi-user dynamic symmetric searchable encryptions (DSSE) schemes unreasonably require private key sharing or require data owners to stay online. And existing verifiable DSSE is mostly based on public key primitives and confronts a large efficiency bottleneck. To address these issues, this paper proposes MCVDSSE: Secure Multi-Client Verifiable Dynamic Symmetric Searchable Encryption. In MCVDSSE, the data owner neither needs to stay online nor leak the critical private key to ensure secure data search and dynamic update. In addition, it further achieves forward and backward security, and is secure against replay attacks and collusion attacks. Finally, the security proof shows that the user has the ability to verify the integrity and timeliness.

**Keywords:** Secure ciphertext search · Verifiable · Multi-user

## 1 Introduction

Searchable symmetric encryption is an emerging research field under cloud storage technology. In order to save local storage overhead, data owners often choose to securely outsource data to cloud servers for storage. However, these ciphertext data still face the needs of sharing, updating and flexible retrieval, and most of the actual users and public cloud servers cannot be fully trusted [6]. Therefore, under the premise of ensuring data storage and search capabilities, the SSE field is very concerned about privacy protection, search efficiency, and correctness verification of search results [8].

Supporting multi-client ciphertext search is an important feature that meets the actual needs. In order to implement the single writer multiple readers (SWMR) model, the most common practice is that the users request a search token from the data owner (DO) before each search. The disadvantage is that DO needs to keep online. Another method is to share the private key owned by DO for generating search tokens to users, so that DO does not have to stay online, but leaking the private key will cause a large security loss.

In order to achieve user verifiability of search results, most of the existing schemes provide users with an additional verification algorithm, which allows the

© The Author(s), under exclusive license to Springer Nature Switzerland AG 2023
M. Yung et al. (Eds.): SciSec 2023, LNCS 14299, pp. 85–96, 2023.
https://doi.org/10.1007/978-3-031-45933-7_5

server to return a proof while returning a matching file set. Users apply their knowledge of keywords to this proof to generate a matching file set, or apply their knowledge to this matching file set to generate a comparison proof. If the two values are consistent, the server does not do evil in the search. The current research focuses more on correctness and completeness. Correctness means that the results are in line with expectations, and completeness means that the results are not omitted [13]. In addition, an anti-replay attack, namely timeliness, is considered, indicating that the search results should be the latest version rather than the historical version.

Our requirements for multi-user and verification mechanisms are as follows: DO neither need to keep online nor share critical private keys to users. Each user can only search specific keywords, and the server and insecure users cannot obtain search tokens. After obtaining the matching result set corresponding to the keyword, the user can independently verify its correctness and integrity. Our idea of implementing multi-user mechanism is to hand over the responsibility for passing tokens to the server, and ensure that the token will not be leaked to the server, but can only be decrypted when the authorized users request the token. In addition, the verification mechanism relies on the design of a common index structure. Our contributions are summarized as follows:

1. Design a SSE of multi-user model. Based on the idea of permission handover, the work of token distribution is given to the server, so that the data owner neither need to stay online nor need to share the most critical key with users. It has both search permission control and anti-collusion attack capabilities.
2. Add verification mechanism for multi-user DSSE. Resist data integrity attacks, replay attacks and read-write conflicts. The analysis shows the effectiveness of our verification method.
3. A search index that supports file dynamic update is used to make the scheme meet forward and backward security.

## 2  Related Work

### 2.1  Multi-user Searchable Searchable Encryption

Under the SWMR framework, [5] first proposed an searchable encryption (SE) scheme based on broadcast encryption (BE) technology, but the efficiency is low. The [9] scheme is based on [4], where users need to interact with DO to obtain a search token before each search. It requires DO to stay online, which is contrary to the requirements of outsourcing services. [10] and [16] are also multi-user schemes based on [4]. [10] avoids exposing the master key to the user by applying [15]'s trapdoor secret sharing scheme, but its method of generating tokens is impractical, because the generation of each token requires the participation of all key fragment holders. [16] improves the computational efficiency of the algorithm using set-constrainted pseudo-random functions(PRFs), and considers the collusion attack between clients, but DO still needs to generate and distribute the search-key for users, which does not really break the principle that the master key cannot be shared.

## 2.2   Verifiable Searchable Symmetric Encryption (VSSE)

[3,6] both use the Merkle Hash tree or its variants to construct a data structure that can be regarded as a proof to achieve verifiability, and are based on inverted index. [11] uses aggregate key to realize data sharing and result verification. Specifically, DO distributes aggregate keys to other users to achieve selective access control; users can search and verify with this aggregate key. [14] is the first VSSE scheme with public verifiability, that is, DO delegates the verification process to a third party. This scheme effectively reduces the computational burden of DO, but increases the communication overhead between users and the third party in the verification process. Non-interactive VSSE [3] is usually more effective than the scheme with interactive verification algorithm [14]. [7] implements a VSSE under the SWMR model, and considers the collusion attack between the client and the server. However, its verifiability is based on blockchain, which introduces additional complex module.

## 3   Preliminaries

This section lists the meaning of symbols (Table 1), and briefly introduces the relevant basic knowledge.

### 3.1   Broadcast Encryption Scheme

This paper adopts the broadcast encryption scheme in [2], roughly as follows:

$BE\_Setup(n) \rightarrow (PK, d_1, \ldots, d_n)$: Input $n$, the number of all authorized users, output the public key $PK$ and the private key $d_i$ of each user.

$BE\_Encrypt(S, PK) \rightarrow (Hdr, K)$: Input $PK$ and the authorization set $S$, output message encryption key $K$ and broadcast body $Hdr$. After encrypting the message $M$ to $C_M$ with $K$, the users in $S$ have decryption rights.

$BE\_Decrypt(S, i, d_i, Hdr, PK) \rightarrow (K)$: If a user $i$ belongs to $S$, it has the right to obtain $K$, thereby decrypting $C_M$ to obtain the broadcast plaintext $M$.

### 3.2   Merkle Patricia Tree

The Merkle Patricia Tree (MPT) is a deterministic tree structure that records key-value pairs. The path of the tree represents the key item, which is generally the basis of retrieval, such as the keyword information in SE. Each node in the path records one or more bits of the key item, thereby reducing the depth of the tree. The leaf node of the tree records the value item, which is generally expected to be retrieved, such as the index or proof corresponding to the keyword. Besides, MPT comes with a fast search and update algorithm. See [17] for details.

### 3.3   Incremental Hash

The incremental hash (IH) function can realize the addition and subtraction of two random strings without collision. For $IH(A)$, if a new element $a$ is added to the collection, there is no need to recalculate $IH(A \cup a)$, but calculate $IH(A) +_H IH(a)$. More detailed descriptions can be found in [1].

**Table 1.** Notations

| Notation | Meaning | Notation | Meaning |
|----------|---------|----------|---------|
| $\omega$ | Keyword | $C$ | The ciphertext set |
| $\tau_\omega$ | Search token | $C_{\omega_i}$ | The ciphertext set of keyword |
| $C_T$ | The ciphertext of token | $SPT$ | User search permissions table |
| $W$ | Keyword set | $I$ | Search index |
| $|W|$ | Total number of keywords | $V$ | Verify index |
| $n$ | Total number of users | $Proof_{\omega_i}$ | Proof of search results |
| $N_f$ | Total number of files | $U$ | Global update number |
| $D$ | The plaintext set | $R$ | Search result |
| $D_{\omega_i}$ | The plaintext set of keyword | $\delta$ | Update token |

## 4   Problem Statement

### 4.1   System Model

There are three kinds of roles in our system model, as shown in Fig. 1.

Data owner: Honestly manage files, authorize user groups, and their search permissions. DO needs to encrypt files and search tokens, create security indexes and send them to the server.

Authorized user group: Semi-honest user groups, with private keys, can obtain search tokens within the scope of authority to initiate searches, and can verify the integrity and timeliness of search results.

Cloud server: Untrusted role, providing storage and computing services, responsible for holding search tokens and performing search and update requests.

Encrypted Database · · · · · · · · · · · · · · · · · · · · · · · · · · ·   Search Result
Secure Index                                                              Search Token

Data Owner                    Cloud Server                    Data Users

**Fig. 1.** System architecture of MCVDSSE.

### 4.2   Security Definition

DO is honest, users are semi-honest, and the server is malicious. The authorized users may launch a collusion attack on the search token; the server will actively steal more information, and even tamper or forge search results. The generation of erroneous search results includes data integrity attacks and replay attacks. Subsequent discussions refer to the integrity and timeliness of search results.

**Definition 1 (Data integrity attack).** *Refers to a malicious server (or attacker) trying to tamper with search results or return empty results, resulting in authorized users unable to obtain complete and correct search results.*

**Definition 2 (Replay attack).** *Also known as data freshness attack, refers to the malicious server (or attacker) trying to return the past search results rather than the latest, in order to save computing power.*

# 5   The Proposed Scheme

This section will introduce the specific structure. The first module is *Preparation*, which is responsible for key generation and distribution, search token encryption, and security index construction. The second module is *Search*, including search token acquisition and search execution. The third module is *Verification* and the fourth module is *Update*, including file update and user group management.

$Setup(1^\lambda, n) \to K = \{K_{DO}, \{K_i\}, PK\}$: Enter the security parameters $\lambda$ and the number of authorized users $n$, randomly generates private key of the data owner $K_{DO} = \{k_f, k_g, k_{sym}\}$, the private key of the authorized user group $K_i = \{d_i, k_g, k_{sym}\}, i \in [1, n]$, and the public parameter $PK = \{PK_{BE}, U\}$.

In the above symbols, $k_f$ is the mastser private key held only by $DO$. The $k_f$ and $k_g$ are the private keys of pseudo-random functions $F$ and $G$ respectively, and $k_{sym}$ is the key of symmetric encryption scheme $SKE$. The $PRF$ values of $x$ are denoted by $F_{k_f}(x)$ and $G_{k_g}(x)$ respectively. In the specific scheme, $F_{k_f}$ is used to generate trapdoors, $G_{k_g}$ will be used to generate the value of $IH$. $d_i$ is the private key used by user $i$ for broadcast encryption, $\{d_i | i \in [1, n]\}$ and the broadcast encryption public key $PK_{BE}$ are generated by calling $BE\_Setup(n)$ algorithm (see 3.1). The global update number $U$ is initialized to 0.

$IndexCreate(K_{DO}, D) \to (C, I, V)$: A deterministic algorithm run by DO. See Algorithm 1 for details.

---

**Algorithm 1.** IndexCreate Algorithm

---

**Input:** $K_{DO}, D$
**Output:** $I, V$
1: Create a $|W| \times (N_f + 1)$ cross linked list $I$ and a MPT $V$.
2: **for** $i \in [1, |W|]$ **do**
3:     $I_{\omega_{i1}}.value = \tau_{\omega_i} = F_{k_f}(\omega_i \| U)$
4:     **for** $j \in [2, N_f + 1]$ **do**
5:         $\omega_{ij} =$ if $f_j$ contains $\omega_i, v_j = 0, I_{\omega_{ij}}.value = Enc_{k_{sym}}(\omega_{ij}, v_j)$
6:     **end for**
7:     $v_{\omega_i} = \sum_{f_j \in D_{\omega_i}} IH(G_{k_g}(f_j)) \| F_{k_f}(U)$
8:     $\lambda = \lambda.Insert(\tau_{\omega_i}, v_{\omega_i})$
9: **end for**

---

*Search Index Creation.* The search index $I$ is a cross linked list based on reverse index table. Each row corresponds to a keyword $\omega$, and each column corresponds to a file $f$(the first column is the head nodes). Each internal index node stores the inclusion relationship between the encrypted file and the keyword. Specifically, the head node is equivalent to the search node, storing the search token of the keyword $\tau_{\omega_i} = F_{k_f}(\omega_i \parallel U)$, where the global update number $U$ ensures the forward security. The $w_{ij}$ in internal index node represents whether the file $f_j$ contains the keyword $\omega_i$, 1 represents the inclusion, 0 is the opposite; $v_j$ is an update counter for file $j$, which prevents the server from inferring more information through the similarity of ciphertext $Enc_{k_{sym}}(w_{ij})$.

*Verify Index Creation.* The $<key, value>$ pairs stored in $V$ are respectively the keyword trapdoor $\tau_{\omega_i}$ and the incremental hash values $v_{\omega_i}$ of documents corresponding to $\omega_i$. Noted that the *key* is stored by the path and the *value* is stored by the leaf node. The incremental hash value in $v_{\omega_i}$ is completed by special addition $+_H$, and then added to the PRF value by binary string addition. $U$ is embedded to achieve anti-replay attacks. An example of some $<key, value>$ pairs in $V$ are shown in Table 2.

**Table 2.** An Example of key-value pairs in $V$

| Keyword | Token (Key) | Files | Value |
|---|---|---|---|
| $\omega_1$ | $'256ac'$ | $f_1, f_3, f_4$ | $(IH(f_1) +_H {}^1 IH(f_3) +_H IH(f_4)) \parallel F(U)$ |
| $\omega_2$ | $'93385'$ | $f_2, f_5$ | $(IH(f_2) +_H IH(f_5)) \parallel F(U)$ |

[1] $+_H$ represents incremental hashing addition.

**$SearchTokenEncrypt(SPT, PK_{BE}) \rightarrow C_T$:** DO encrypts the search tokens $\tau_{\omega_i}$ according to the search permission table $SPT$ to obtain the ciphertext $C_T$, and uploads it to the server together with $(C, I, V)$. Where $SPT = \{<\tau_{\omega_i}, S_i>\}$, represents the searchable user set $S_i$ corresponding to each keyword, such as $<\tau_{\omega_1}, \{u_1, u_2\}>$ and $<\tau_{\omega_2}, \{u_2\}>$.

Specifically, as in Algorithm 2, DO sequentially calls $BE\_Encrypt(S_i, PK_{BE})$ to obtain the symmetric key $K_i$, encrypts the first-round plaintext $M_{round_{1i}} = \tau_{\omega_i}$ to obtain the first-round ciphertext $C_{T_i}$ that can only be decrypted by the corresponding users. According to the above example, user 1 can decrypt $C_{T_1}$ to get $\tau_{\omega_1}$, and user 2 can decrypt $C_{T_1}$ and $C_{T_2}$ to get $\tau_{\omega_1}$ and $\tau_{\omega_2}$.

---

**Algorithm 2.** SearchTokenEncrypt Algorithm

---
**Input:** $SPT, PK_{BE}$
**Output:** $C_T$
1: $S_u = \{1, \ldots, n\}, M_{round_2} = U \parallel F_{k_f}(U)$
2: **for** $i \in [1, |W|]$ **do**
3:     $S_i = SPT[\omega_i], M_{round_{1i}} = \tau_{\omega_i}$
4:     $(Hdr_i, k_i) = BE\_Encrypt(S_i, PK_{BE})$
5:     $C_{T_i} = Enc_{k_i}(M_{round_{1i}}), M_{round_2} = M_{round_2} \parallel C_{T_i}$
6: **end for**
7: $(Hdr', k') = BE'\_Encrypt(S_u, PK'_{BE})$
8: $C_T = Enc_{k'}(M_{round_2})$

---

Then DO encrypts $M_{round_2} = C_{T_1} \parallel C_{T_2} \parallel \cdots \parallel C_{T_{|W|}} \parallel U_t \parallel F_{k_f}(U_t)$ with a new set of $BE'$ parameters to obtain the two-round ciphertext $C_T$. The authorization set $S_u$ entered by $BE'\_Encrypt$ is all authorized user groups $[1, n]$, ensuring that all users have the decryption ability of $C_T$ to get $M_{round_2}$ while the server does not. Here, splicing the current $U_t$ and $F_{k_f}(U_t)$ information is to achieve verifiability, in other words, to achieve anti-replay attacks.

$SearchTokenRequest(j, d_j) \rightarrow \{\tau_{\omega_i}\}$: Including two rounds of interaction between the user and server. $j$ corresponds to the user and $i$ corresponds to the keyword. User $j$ initiates a request to the server to obtain $C_T$. Then user j calls $BE'\_Decrypt(S_u, j, d_j, Hdr', PK'_{BE})$ to obtain the symmetric key $k'$ and decrypts $C_T$ to get $M_{round_2}$. $U_t$ and $F_{k_f}(U_t)$ are retained on the user for subsequent verification. Then, in the same way, user $j$ tries to call $BE\_Decrypt(S_i, j, d_j, Hdr_i, PK_{BE})$ for $i \in [1, n]$ to get $\tau_{\omega_i}$ within competence.

$Search(\tau_{\omega_i}, I, V) \rightarrow (R)$: The server search on $I$ and $V$ according to $\tau_{\omega_i}$ to obtain $R = (C_{\omega_i}, Proof_{\omega_i})$. Firstly, match the head node in $I$ and return the set of index nodes in $I_{w_i}$, which is noted as $C_{\omega_i} = \{Enc_{k_{sym}}(w_{ij}, v_j)\}$. Meanwhile, the $MPT'Search$ algorithm is called on $V$ to return the $value$ field $v_{\omega_i}$ of the leaf node on the matching path. The returned $value$ field is also denoted as $Proof_{\omega_i}$ for subsequent verification, such as $(IH(f_2) +_H IH(f_5)) \parallel F(U)$ in Table 2.

$Verify(K_i, \tau_{\omega_i}, R) \rightarrow (b)$: As in Algorithm 3, user $i$ accepts the search results only when timeliness and completeness are both satisfied. Timeliness verification includes anti readi-write conflict and anti replay attack. Firstly, verify whether $U_t$ is consistent with the latest $U$. If an inconsistency is found, it means a read-write conflict, that is, a sudden update occurs when the user searches. Once this occurs, the search should be terminated to maintain forward security. Then verify whether $F(U_t)$ is consistent with $F(U)$ in $Proof_{\omega_i}$, and inconsistency indicates a replay attack. In the integrity verification, the user recovers $Proof'_{\omega_i} = \sum_{f_i \in D_{\omega_i}} IH(G_{k_g}(f_i))$ based on decrypting the $C_{w_i}$ in $R$, and compares it with the $Proof_{\omega_i}$ in $R$. The integrity verification is passed if and only if the two $proofs$ are the same.

$FileUpdateRequest(K_{DO}, f_j, op) \rightarrow (\delta, CT')$: According to the update file $f_j$ and the operation type $op$, DO generates the update token $\delta = (updateI_j, updateW, updateMptValue)$ and the new search token ciphertext $C'_T$ (if $SPT$ changes), and sends them to the server. Specifically, $updateI_j$ is the new value of column $I_j$, $updateW$ is the set of related $\tau_\omega$, $updateMptValue = \pm IH(G_{k_g}(f_j)) \parallel F_{k_f}(U + 1)$ where '$\pm$' is determined by $op$ ('+' when $op = 'add'$). If $SPT$ changes, the new $M_{round_2}$ is set to $C_{T_1} \parallel \cdots \parallel C_{T_{|W|}} \parallel (U + 1) \parallel F_{k_f}(U + 1)$. If a new keyword appears, the new value in $I$ and $V$ should be calculated additionally.

$IndexUpdate(\delta) \rightarrow (I, V)$: The server performs updates on $I$ and $V$ based on $\delta$ send by DO. Specifically, update $updateI_j$ the $j_{th}$ column of $I$, updates $updateMptValue$ to the $value$ field of leaf nodes corresponding to $updateW$ in $V$, and store the new $C'_T$ field to replace the old $C_T$.

**Algorithm 3.** Verify Algorithm

---

**Input:** $K_i, \tau_{\omega_i}, R$
**Output:** $b$
1: $D = Dec_{k_{sym}(C)}, D_{\omega_i} = \emptyset, Proof'_{\omega_i} = F(U)$
2: **if** $U_t \neq U$ or $F(U_t) \neq F(U)$ **then**
3:     return 0
4: **else**
5:     **for** $j \in [2, N_f + 1]$ **do**
6:        **if** $(Dec_{k_{sym}}(C_{\omega_{ij}})).\acute{\omega}_{ij} = 1$ **then**
7:           $D_{\omega_i} = D_{\omega_i} \cup f_j, Proof'_{\omega_i} + = IH(G_{k_g}(f_j))$
8:        **end if**
9:     **end for**
10:    return whether $Proof'_{\omega_i} == Proof_{\omega_i}$
11: **end if**

---

$UserUpdate(n', SPT') \rightarrow (C'_T, PK'_{BE}, d_{n'})$: When the user joins, actively leaves or is revoked, the algorithm $BE\_Setup(n')$ is called to regenerate the $PK'_{BE}$ (and the $d_{n'}$ if user $n'$ joins). Then called the $SearchTokenEncrypt$ $(SPT', PK'_{BE})$ to regenerate $C'_T$ which will be updated to the server.

# 6 Security

This section will proof the security of MCVDSSE, including confidentiality, verifiability, forward and backward security, and anti-collusion attack.

**Definition 3 (Confidentiality).** *The server cannot learn any valid information about files and keywords through file ciphertexts, secure indexes, search and update tokens. Considering the following probability experiments (where $\mathcal{A}$ is a stateful adversary, $S$ is a stateful simulator, $\mathcal{L}$ is a stateful leakage function):*

*$\mathbf{Real}_{\mathcal{A}}(k)$: First, the challenger runs $Setup(1^\lambda)$. $\mathcal{A}$ selects a file set $D$ for the challenger, and the challenger runs $IndexCreate(K_{DO}, D)$ to generate $(I, V)$. Then $\mathcal{A}$ performs adaptive querys $q = \{\omega_i\}$ of a polynomial number. The challenger calls $SearchTokenRequest(i, d_i)$ and $Search(\tau_{\omega_i}, I, V)$ in turn, returns $C_T$ and $R$ for each query. Finally, $\mathcal{A}$ outputs a boolean bit $b$.*

*$\mathbf{Ideal}_{\mathcal{A}, S}(k)$: $\mathcal{A}$ chooses a file set $D$. $S$ has $\mathcal{L}(D)$, generates $(I, V)$ to $\mathcal{A}$. $\mathcal{A}$ performs $q = \{\omega_i\}$, and $S$ returns $(C_T, R)$ to $\mathcal{A}$ for each $q$. Finally, $\mathcal{A}$ outputs a boolean bit $b$.*

*If for all probabilistic polynomial time opponents $\mathcal{A}$, there is always a probabilistic polynomial time simulator $S$ satisfying the following conditions, then MCVDSSE is secure under $\mathcal{L}$.*

$$|Pr[\mathbf{Real}_{\mathcal{A}}(k) = 1] - Pr[\mathbf{Ideal}_{\mathcal{A}, S}(k) = 1]| \leq negl(k)$$

**Theorem 1.** *If the broadcast encryption scheme $BE$ is semantically secure and $F$ and $G$ are pseudo-random functions, then our MCVDSSE scheme is $\mathcal{L}$-secure.*

*Proof.* We prove that for all probabilistic polynomial time adversary $\mathcal{A}$, there always exists a probabilistic polynomial time simulator $S$ such that the outputs of $\mathbf{Real}_{\mathcal{A}}(k)$ and $\mathbf{Ideal}_{\mathcal{A},S}(k)$ are indistinguishable. In the token acquisition phase, the challenger in $\mathbf{Real}_{\mathcal{A}}(k)$ calls $SearchTokenRequest$ to get $C_T$, and $S$ in $\mathbf{Ideal}_{\mathcal{A},S}(k)$ generates $C'_T$ based on $\mathcal{L}(D)$. $C_T$ is calculated by $BE$, and here $C'_T$ is equivalent to a random number. Therefore, the indistinguishability of two games for $C_T$ can be attributed to the semantic security of $BE$.

Next, in the keyword search and verification phase, the challenger in $\mathbf{Real}_{\mathcal{A}}(k)$ calls $Search$ to get the $proof$, while $S$ generates $proof'$ according to $\mathcal{L}(D)$ in $\mathbf{Ideal}_{\mathcal{A},S}(k)$. $proof$ contains the value generated by PRF $F$ and $IH$, which takes the output of PRF $G$ as input. So the indistinguishability of two games for $proof$ can be attributed to the pseudo-randomness of $F$ and $G$. In summary, the output of $\mathbf{Real}_{\mathcal{A}}(k)$ and $\mathbf{Ideal}_{\mathcal{A},S}(k)$ is indistinguishable.

**Definition 4 (Verifiability).** *Verify the integrity and timeliness means preventing the integrity attack(Definition 1) and the replay attack (Definition 2). Consider the following probabilistic experiments:*

**Theorem 2.** *If the broadcast encryption scheme $BE$ is semantically secure, and $IH$ is collision-resistant, and $F$ and $G$ are PRFs, then MCVDSSE is verifiable.*

*Proof.* The integrity is judged by $proof_{\omega_i}$ which is generated by $IH$, $F$ and $G$. Assuming that the error search results can be verified by the user, it indicates that there is a collision in $v_{\omega_i} = \sum_{f_i \in D_{\omega_i}} IH(G_{k_g}(f_i)) \parallel F_{k_f}(U)$, so the key to integrity checking can be attributed to the collision of $IH, F$ and $G$.

The timeliness is judged by $U$ and $F_{k_f}(U)$. They are encrypted by $BE$, and transmitted as part of $M_{round_2}$ which can only be decrypted by legitimate users. If the malicious server has the ability to tamper with $U$ and $F_{k_f}(U)$, he also has the ability to destroy the semantic security of $BE$. Therefore, the timeliness can be attributed to the semantic security of the broadcast encryption scheme.

**Dynamic Security.** Forward security means that users can't use old tokens to search for new files, and backward security means that users can't search for deleted files with a new token. In MCVDSSE, $\tau_{\omega_i} = F_{k_f}(\omega_i \parallel U)$ is based on $U$, so the search token updates when each file is updated. This ensures forward security. Next, after deleting one file, the $\omega_{ij}$ in $I$ will be set to 0 which obviously ensures that the deleted file will not be placed in the result set.

**Anti-collusion Attack.** Only focus on the stage where the user obtains the search token, even if multiple semi-honest users collude, there is no way to obtain more information outside the scope of authority. The security is realized by the characteristics of broadcast encryption itself. See citation [2] for specific proof.

## 7    Efficiency Analysis

In this section, the functional and efficiency comparison are shown in Table 3 and Table 4 respectively, and our efficiency is summarized in Table 5.

In the Preparative algorithm, there are $3n$ power exponential operation on group $\mathbb{G}$, PRF $F$ operation for $|W| + 1$ times, PRF $G$ operation for $N_f$ times, symmetric encryption $Enc$ operation for $N_f \times |W| + N_f + 3|W| + 2$ times, $BE\_Encrypt$ operation for $|W| + 1$ times, and $IH$ operation for $\sum_{i \in [1,|W|]} |D_{\omega_i}|$ times. Among them, the time complexity of $Setup$ is $\mathcal{O}(n)$. The complexity of creating $I$ is $\mathcal{O}(|W| \times N_f)$, while $V$ costs $\mathcal{O}(\sum_{\omega_i \in W} h_{leaf_{\omega_i}}) \leq O(|W| \times h_{MPT})$, where $h_{leaf_{\omega_i}}$ is the height of corresponding leaf node in MPT, and $h_{MPT}$ is the height of the MPT. The complexity of token encryption is $\mathcal{O}(|W|)$.

In the Search algorithm, the user needs to interact with the server for four rounds. The first two rounds are used to securely obtain the search token, and the last two rounds are used to complete the search. A user $j$ needs to perform $BE\_Decrypt$ operation for $|W| + 1$ times and symmetric decryption $Dec$ operation for $|S_j| + 1$ times. Among them, the time complexity of the user to obtain the token is $\mathcal{O}(|W| + |S_j|)$. The time complexity of the server executing the search on $I$ is $\mathcal{O}(N_f)$, while searching on $V$ is $\mathcal{O}(h_{leaf_{\omega_i}}) \leq O(h_{MPT})$.

In the Verify algorithm, it is necessary to perform $N_f$ symmetric decryption $Dec$ operation, and PRF $G$ and $IH$ operation for $\alpha = |D_{\omega_i}|$ times. Among them, the time complexity of timeliness verification is only $\mathcal{O}(1)$, while $\mathcal{O}(\alpha)$ for regenerating the proof for integrity verification.

**Table 3.** Functional Comparison

| Reference | Dynamism | Multiuser | Timeliness | Integrity | Verify Efficiency |
|---|---|---|---|---|---|
| OK16 [12] | ✗ | ✗ | ✗ | ✓ | $\mathcal{O}(\alpha)$[a] |
| Bost16 [3] | ✓ | ✗ | ✓ | ✓ | $\mathcal{O}(\alpha)$ |
| GSSE18 [17] | ✓ | ✓ | ✓ | ✓ | $\mathcal{O}(\log(|W|))$ |
| MCVDSSE | ✓ | ✓ | ✓ | ✓ | $\mathcal{O}(\alpha)$ |

[a] $\alpha = |DB[\omega_i]|$, represents the number of entries matching the keyword.

**Table 4.** Efficiency Comparison

| Reference | Multiuser | Search[b,c] | Verify | FileUpdate |
|---|---|---|---|---|
| Bost16 [3] | ✗ | $\mathcal{O}(\alpha + \log^2 m)$ | $\mathcal{O}(MAC.Verify)$[d] | $\mathcal{O}(\log^2 m)$ |
| Sun20 [16] | ✓ | $\mathcal{O}(\alpha_{sterm})$ | ✗ | ✗ |
| MCVDSSE | ✓ | $\mathcal{O}(N_f + h_{leaf_{\omega_i}})$ | $\mathcal{O}(N_f)$ | $\mathcal{O}(|W| + \sum_{\omega_i \in W_{f_j}} h_{leaf_{\omega_i}})$ |

[b] $m = |DB|$ denotes the total number of (keyword, document) pairs.
[c] $sterm$ represents the least frequent keyword in the join query.
[d] Its complexity depends on the verification efficiency of the MAC used.

In the Update algorithm, there is a round of interaction between DO and the server. It is necessary to perform $|W|$ symmetric encryption $Enc$ operation, $|W_{f_j}| + 2$ PRF $F$ operation, where $|W_{f_j}|$ refers to the number of keywords contained in the file $f_j$. Also need to perform PRF $G$ and $IH$ operation for 1 time. The computational time complexity of DO is $\mathcal{O}(|W|)$, while the server-side update complexity is $\mathcal{O}(|W| + \sum_{\omega_i \in W_{f_j}} h_{leaf_{\omega_i}})$.

**Table 5.** Our Time Complexity

| Algorithm | Time complexity | Algorithm | Time complexity |
|---|---|---|---|
| Setup | $\mathcal{O}(n)$ | Search | $\mathcal{O}(N_f + h_{leaf_{\omega_i}})$ |
| TokenEncrypt | $\mathcal{O}(|W|)$ | TimelinessVerify | $\mathcal{O}(1)$ |
| TokenRequest | $\mathcal{O}(|W| + |S_j|)$ | IntegrityVerify | $\mathcal{O}(\alpha)$ |
| SearchIndexCreate | $\mathcal{O}(|W| \times N_f)$ | UpdateRequest | $\mathcal{O}(|W|)$ |
| VerifyIndexCreate | $\mathcal{O}(\sum_{\omega_i \in W} h_{leaf_{\omega_i}})$ | IndexUpdate | $\mathcal{O}(|W| + \sum_{\omega_i \in W_{f_j}} h_{leaf_{\omega_i}})$ |

# 8 Conclusion

This paper discusses a way to implement a multi-client verifiable dynamic symmetric searchable encryption scheme. In order to achieve anti-integrity attack and anti-replay attack, we combine incremental hashing technology and global version number to design a verification index based on Merkle Patricia Tree. In order to achieve effective data search and update, the cross linked list structure is used to design the search index. In addition, multi-user search access control against collusion attack is realized based on the idea of permission handover and broadcast encryption technology. Security analysis and efficiency analysis show that the scheme is quite safe and effective.

**Acknowledgement.** This work was supported in part by the National Key Research and Development Program of China (Grant No. 2020YFA0712300), in part by the National Natural Science Foundation of China (Grant No. 62132005, 62172162), in part by Shanghai Trusted Industry Internet Software Collaborative Innovation Center. Zhenfu Cao, Jiachen Shen, and Xiaolei Dong are the corresponding authors.

# References

1. Bellare, M., Goldreich, O., Goldwasser, S.: Incremental cryptography: the case of hashing and signing. In: Desmedt, Y.G. (ed.) CRYPTO 1994. LNCS, vol. 839, pp. 216–233. Springer, Heidelberg (1994). https://doi.org/10.1007/3-540-48658-5_22
2. Boneh, D., Gentry, C., Waters, B.: Collusion resistant broadcast encryption with short ciphertexts and private keys. In: Shoup, V. (ed.) CRYPTO 2005. LNCS, vol. 3621, pp. 258–275. Springer, Heidelberg (2005). https://doi.org/10.1007/11535218_16
3. Bost, R., Fouque, P.A., Pointcheval, D.: Verifiable dynamic symmetric searchable encryption: optimality and forward security. Cryptology ePrint Archive (2016)
4. Cash, D., Jarecki, S., Jutla, C., Krawczyk, H., Roşu, M.-C., Steiner, M.: Highly-scalable searchable symmetric encryption with support for Boolean queries. In: Canetti, R., Garay, J.A. (eds.) CRYPTO 2013, Part I. LNCS, vol. 8042, pp. 353–373. Springer, Heidelberg (2013). https://doi.org/10.1007/978-3-642-40041-4_20
5. Curtmola, R., Garay, J., Kamara, S., Ostrovsky, R.: Searchable symmetric encryption: improved definitions and efficient constructions. In: Proceedings of the 13th ACM Conference on Computer and Communications Security, pp. 79–88 (2006)

6. Ge, X., et al.: Towards achieving keyword search over dynamic encrypted cloud data with symmetric-key based verification. IEEE Trans. Dependable Secure Comput. **18**(1), 490–504 (2019)
7. Gharehchamani, J., Wang, Y., Papadopoulos, D., Zhang, M., Jalili, R.: Multi-user dynamic searchable symmetric encryption with corrupted participants. IEEE Trans. Dependable Secure Comput. (2021)
8. Huang, C., Liu, D., Yang, A., Lu, R., Shen, X.: Multi-client secure and efficient DPF-based keyword search for cloud storage. IEEE Trans. Dependable Secure Comput. (2023)
9. Jarecki, S., Jutla, C., Krawczyk, H., Rosu, M., Steiner, M.: Outsourced symmetric private information retrieval. In: Proceedings of the 2013 ACM SIGSAC Conference on Computer & Communications Security, pp. 875–888 (2013)
10. Kermanshahi, S.K., et al.: Multi-client cloud-based symmetric searchable encryption. IEEE Trans. Dependable Secure Comput. **18**(5), 2419–2437 (2019)
11. Liu, Z., Li, T., Li, P., Jia, C., Li, J.: Verifiable searchable encryption with aggregate keys for data sharing system. Futur. Gener. Comput. Syst. **78**, 778–788 (2018)
12. Ogata, W., Kurosawa, K.: Efficient no-dictionary verifiable searchable symmetric encryption. In: Kiayias, A. (ed.) FC 2017. LNCS, vol. 10322, pp. 498–516. Springer, Cham (2017). https://doi.org/10.1007/978-3-319-70972-7_28
13. Sharma, D.: Searchable encryption: a survey. Inf. Secur. J.: Glob. Perspect. **32**(2), 76–119 (2023)
14. Soleimanian, A., Khazaei, S.: Publicly verifiable searchable symmetric encryption based on efficient cryptographic components. Des. Codes Crypt. **87**(1), 123–147 (2019)
15. Sun, S.-F., Liu, J.K., Sakzad, A., Steinfeld, R., Yuen, T.H.: An efficient non-interactive multi-client searchable encryption with support for Boolean queries. In: Askoxylakis, I., Ioannidis, S., Katsikas, S., Meadows, C. (eds.) ESORICS 2016, Part I. LNCS, vol. 9878, pp. 154–172. Springer, Cham (2016). https://doi.org/10.1007/978-3-319-45744-4_8
16. Sun, S.F., et al.: Non-interactive multi-client searchable encryption: realization and implementation. IEEE Trans. Dependable Secure Comput. **19**(1), 452–467 (2020)
17. Zhu, J., Li, Q., Wang, C., Yuan, X., Wang, Q., Ren, K.: Enabling generic, verifiable, and secure data search in cloud services. IEEE Trans. Parallel Distrib. Syst. **29**(8), 1721–1735 (2018)

# A Graphical Password Scheme Based on Rounded Image Selection

Xinyuan Qin and Wenjuan Li$^{(\boxtimes)}$

Department of Electronic and Information Engineering,
The Hong Kong Polytechnic University, Hong Kong, China
`wenjuan.li@polyu.edu.hk`

**Abstract.** Graphical password is considered as an alternative to traditional textual password, but it also faces many threats such as shoulder-surfing attack. To design and build a more secure and robust graphical password system with the resistance to multiple attacks modalities, especially brute force attack, guessing attack and shoulder-surfing attack, it is important to avoid the credentials being captured in just one step, e.g., by adding several rounds of input. For example, with respect to shoulder-surfing attack resistance, the input design ought to incorporate a certain degree of fault tolerance, with the specific value determined based on the acceptable tolerance range. By integrating this fault tolerance characteristic, the system can effectively withstand shoulder-surfing attacks while preserving the integrity of the authentication procedure. In this work, we learn from the current literature and design a graphical password scheme based on rounded image selection (e.g., three rounds). We provide a detailed scheme design and perform a performance analysis via a user study. Our results indicate that our proposed scheme is viable and gets credit from the participants.

**Keywords:** Graphical Password · Password Security · Usability · Shoulder-surfing Attack · User Authentication · Image Selection

## 1 Introduction

Textual password-based authentication dominates the current user authentication, which requires an individual to input a correct string set of characters for authentication. However, the system based on textual passwords is often suffered from some typical limitations in the aspects of *security*, e.g., recording attack [4] and charging attack [31,33], and *usability*, e.g., the multiple password inference [29] and the limitation of long term memory [24,45].

To address this problem, graphical passwords have been widely researched in the literature, which require an individual to create credentials based on images [16]. The previous studies have proven that humans are better at remembering images than textual information [34,37]. One typical application of graphical password is Android Unlock Pattern, where an individual has to draw a pattern with 4 dots the minimum and 9 dots the maximum on a $3 \times 3$ grid [29].

© The Author(s), under exclusive license to Springer Nature Switzerland AG 2023
M. Yung et al. (Eds.): SciSec 2023, LNCS 14299, pp. 97–114, 2023.
https://doi.org/10.1007/978-3-031-45933-7_6

Though graphical password can enhance the usability of an authentication system, it is still vulnerable to the same security threats as textual passwords. For example, Android Unlock Pattern is vulnerable to brute force attack since its password space (possible patterns) is only 389,112 [1]. In addition, shoulder-surfing attack [47] is big threat to graphical password, which means that the credential input will be visible to an attacker. Hence there is a need to enhance the security of current graphical password schemes.

**Contributions.** To defeat some strong attacks such as shoulder-surfing attack, it is important to avoid the credentials being captured in one time. That is, it is a promising direction to design a more secure graphical password scheme with multiple rounds of input. The main goal is to create a password system that is both secure and easy-to-use, thus promoting a wide scheme adoption and reducing the risk of various attacks. Motivated by these, in this work, we design a graphical password with three rounds of image selection. The contributions of this work can be summarized as follows.

- We introduce a graphical password scheme that has three rounds of image selection, in order to achieve shoulder-surfing attack resistance. To facilitate the authentication, our scheme adopts a certain degree of fault tolerance (e.g., matching rate over 75%), with the specific value determined based on the acceptable tolerance range.
- Considering the resistance to both brute-force and shoulder-surfing attacks, we analyze the variable settings and perform probability calculation. These should be carefully considered with user experience when selecting the most suitable model and variables. Our scheme aims to ensure a balance between security and usability, optimizing the overall effectiveness of authentication.

**Road Map.** The reminder of this paper is structured as follows. Section 2 introduces related work on the design of graphical passwords. Section 3 describes our graphical password scheme in detail, including the scheme design, variable setting, and probability calculation. Section 4 provides scheme implementation and a small-scale user study. We conclude the work in Sect. 5.

## 2   Related Work

Graphical password requires an individual to create a credential by interacting with image(s). The current graphical password schemes can be categorized into four groups [30,40].

- *Recognition-based scheme.* Users have to create credentials by choosing one or multiple images from an image pool.
- *Pure recall-based scheme.* Users have to create credentials by leaving some secrets on an image.
- *Cued recall-based scheme.* Users have to create credentials by choosing an object or an area on an image.

– *Hybrid scheme.* The scheme can combine one or multiple design approaches above, e.g., recognition, pure recall, cured recall.

There are a plenty of graphical password schemes have been proposed in the literature. For instance, Wiedenbeck *et al.* [44] introduced *PassPoints*, where users have to click five points in an order on a selected image. Then Chiasson *et al.* [3] improved *PassPoints* by asking users to select one area on an image with a total of three images. One commercialized graphical password scheme was based on face selection, called *PassFaces* [36], in which a user has to pick up the correct face from 9 face images. Jermyn *et al.* [9] introduced a graphical password scheme based on drawing, called 'draw-a-secret', which requires users to draw a pattern on an image. Currently, Android unlock pattern is the most widely used graphical password scheme. Actually, it is a revised version of Pass-Go [42] scheme, where an individual has to create a pattern by selecting 4–9 dots on a grid with $3 \times 3$ points.

In addition, many graphical password schemes are hybrid. As an example, *CD-GPS* scheme [17] is a typical scheme with two main steps: 1) image selection and 2) secret drawing. First, an individual has to select several images out of an image pool, and then chooses one of them to draw a pattern via click-draw action. Several relevant research can refer to former studies [18–20,28]. Map-based graphical passwords are another type of hybrid schemes, such as *PassMap* [41] and *RouteMap* [22], demanding an individual to choose locations or create patterns on a world map. For example, *PassMap* requires an individual to choose two spots on a world map, while *RouteMap* requires an individual to generate a route on a world map.

However, there are many security threats on graphical passwords especially Android unlock pattern, such as recording attacks [35] and charging attacks [26,27,31]. All the attacks can capture the pattern input on the mobile devices. To enhance its security, there are many multimodel methods proposed in the research society, by combining biometric with unlock pattern [21]. For example, *TMGuard* [25] is a touch movement-based unlock scheme by integrating with touch dynamics, which aims to verify both touch behaviors and unlock pattern. There are also various studies trying to combine certain touch actions to enhance the security of Android unlock pattern, such as Pinch-to-Zoom [15], swipe action [10,11], double-click action [13,14], and handwritten [43]. Some more relevant studies on graphical passwords could be referred but not limited to [5,7,8,12,23,32,39,46].

## 3   Our Approach

When designing a graphical authentication scheme, it is imperative to secure it against various attacks. First, with respect to shoulder-surfing attack resistance, the input design ought to incorporate a certain degree of fault tolerance, with the specific value determined based on the acceptable tolerance range. By integrating this fault tolerance, the system can effectively withstand shoulder-surfing attacks while preserving the authentication integrity.

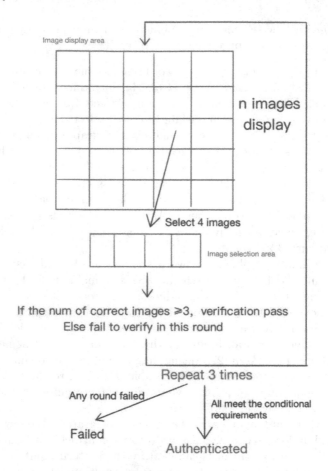

**Fig. 1.** Flow chart of our design.

Then when concerning the resistance to brute-force and guessing attacks, it is crucial to employ suitable variable settings and probability calculations. These should be carefully combined with prior information such as user experience, in order to select the most suitable model and variables.

## 3.1    Scheme Design

By considering the above points, we aim to design a graphical password with several rounds of image selection. This is because a) image selection is the most necessary and basic step in graphical password, and b) using multiple rounds of image selection can avoid the credentials being captured by attackers in just one go. Figure 1 depicts the high-level flow chart (workflow) of our designed graphical password scheme.

In our scheme, the images will be randomly generated by a program, similar to Dhamija and Perrig's design [6]. In this case, though the images randomly

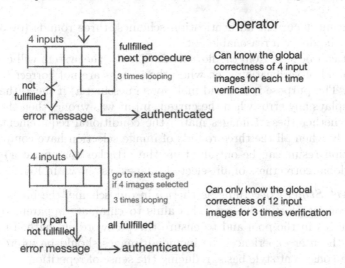

**Fig. 2.** Authentication method comparison.

generated are diversiform, the scheme is still not resistant to shoulder-surfing attack, because the password can be compromised once the attacker remembers all images a user selected. To address this issue, we devise a new mechanism: in the authentication process, users are required to select 4 different images from the pool. If the number of images matching the pre-selected image is equal or greater than 3, then this round of verification will be successful. Then, the user has to repeat the image selection for 3 times (rounds), if and only if all attempts meet the requirement, the authentication can be formally successful.

*Matching Rate.* The rationale behind the current matching rate at $\geq 75\%$ (3/4) for the selected images is to attain a "semi-anti shoulder-surfing" effect. In a scenario where attackers record all 4 chosen images, it is still not easy for them to ascertain whether all 4 images constitute the correct password. In such case, there will be five possible conditions: **1)** all 4 images are correct, or **2)**–**5)** one of the 4 images is not correct.

The reason why we call it "semi-anti" is that even if the attackers cannot compromise the password directly, they may still guess it through multiple attempts and speculation. One round of verification process with 4 selected images can provide 5 possible combinations, which means that a total of 5–125 ($5^3$) combinations exist during the whole authentication process, and the combination of passwords can be predicted by considering the frequency of selected images. As a result, if more selected image series are recorded, the password can still be compromised in the end.

*Three Rounds of Verification.* There are two main reasons for us to adopt three rounds of verification. Firstly, it is accessible that the more cumbersome the authentication is, the more secure the system will be. Considering the error tolerance in most current authentication schemes, three rounds (or 3-time) of verification should be a reasonable option.

Then, it can be seen from the flow chart (Fig. 1): the system will continue to the next round of verification even when the images are not correct in the current round. The purpose is to avoid malicious guessing: a) if the authentication stops or displays any error when the current input was wrong, then the attacker can know whether these 4 images match the conditional requirement immediately; b) only when all the three rounds of image selection have completed, the authentication result can be output; thus the attacker (or operator) can only know the global correctness of all selected images, as shown in Fig. 2.

***Similar Art Style.*** Concerning aesthetics in our scheme, the image pool can consist of consistent style images. This aims to eliminate apparent differences between images in the pool and to ensure the password strength. In addition, concerning the user experience, the included images should be interesting and playful on a consistent style basis, reducing the sense of repetition and boredom for users [38]. Backgrounds and sound effects can also be considered.

### 3.2   Variable Setting and Probability Calculation

To test the intensity of passwords, we first choose the $4 \times 4$ display area and $1 \times 4$ image selection area, which is more concise and user-friendly than the $5 \times 5$ display area, proved by a pre-held questionnaire survey. It is found that with the "$\geq 75\%$ matching rate", the number of pre-selected images that can appear in each round with $4 \times 4 = 16$ display cells can be: 3, 4, 5, 6, 7,..., 16, where the corresponding probability of a single round being cracked by random input can be computed as follows:

```
the number: 3   probobility: 0.007142857142857143
the number: 4   probobility: 0.026923076923076925
the number: 5   probobility: 0.06318681318681318
the number: 6   probobility: 0.11813186813186813
the number: 7   probobility: 0.19230769230769232
the number: 8   probobility: 0.2846153846153846
the number: 9   probobility: 0.3923076923076923
the number: 10  probobility: 0.510989010989011
the number: 11  probobility: 0.6346153846153846
the number: 12  probobility: 0.7554945054945055
the number: 13  probobility: 0.8642857142857143
the number: 14  probobility: 0.95
the number: 15  probobility: 1.0
the number: 16  probobility: 1.0
```

The probabilities of different numbers of pre-selected images displayed in each round lead to two basic recursive questions: (i) Why it is necessary to group different numbers of pre-selected images? (ii) How to determine the patterns of different numbers of pre-selected images and the proportion of their distribution?

In order to explain these two problems more succinctly and easily, we raise an assumption that the image pool has 100 images, and 20 of these images are pre-selected. In the case of a brute-force cracking, we assume that a large number of login interface display areas for each round are captured and recorded.

***The Necessity of Different Numbers of Pre-selected Images Displayed in Each Round.*** If the number of correct images (pre-selected images) for each round is set to a fixed value (e.g., 3, 4, or 5), as shown in Fig. 3, then it can follow a one-time binomial distribution. The attacker can conditionally do a filtration based on this fixed pre-condition, which means that there exists a specific set of 20 images from the 100 images, satisfying the condition that having only 3 (or 4, 5) images in the set for each display area with $4 \times 4 = 16$ images. In this case, the 20 pre-selected images can be identified by an exhaustive method.

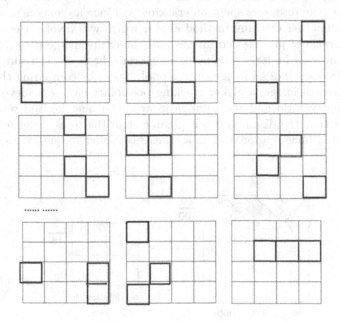

**Fig. 3.** The number of correct images for each round can be set to a fixed value of 3.

Another observation relating to the probability distribution is: given multiple rounds with the same expectation, a discrete probability distribution will have more functional space to achieve a more muscular password strength compared with a fixed probability. For example, the average expectation of being cracked in each round is 3%. Hence we can have two situations: a) S1: fixed probability and b) S2: uniform distributed probability.

If situation S2 to be with all three rounds of all conditions occurring once, then the expectation values for the three rounds can be calculated for these two situations as: S1) $3\% \times 3\% \times 3\% = 2.7 \times 10^{-5}$, and S2) $1\% \times 3\% \times *5\% = 1.5 \times 10^{-5}$. By comparing the expectation of a multi-round of fixed probability and a distributed probability when a single round has the same mean expectation, we can find that under certain conditions (e.g., a fixed distribution over a range), the distribution probability will have an average expectation smaller than the fixed probability over multiple rounds.

To summarize, under the same circumstances, the more patterns employed, the more negligible probability of being cracked and the more margin for further scheme enhancement.

***To Determine the Distribution of Different Numbers of Pre-selected Images Patterns.*** The next questions is how should the probability of each pattern be assigned? If the authentication system uses a random distribution with an average of 4/16, the approach mentioned above may become challenging to implement (the resources spent on cracking will increase exponentially).

However, another cracking method exists, which will involve the inequality between the ratio of pre-selected images and the entire image pool. As long as the attack can record a large number of login trials, he or she can simply record the display frequency of all images in the pool. Then, by comparing the display frequency for each image, a higher cracking possibility can be achieved. In this case, the display ratio and the total ratio are: 4/16 and 20/100 respectively, which means that the display ratio is greater than the total ratio, resulting in the pre-selected images appearing in a relatively higher frequency, as shown in Fig. 4:

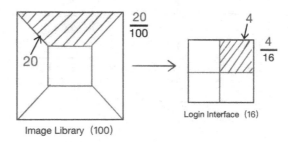

**Fig. 4.** The inequality between the ratio of pre-selected images and the entire image pool.

The appearance probability of a single image can be computed as below:

- Pre-selected: $(1/20) \times (4/16) = 1.25\%$
- Non Pre-selected: $(1/80) \times (12/16) = 0.9375\%$

Hence the probability of a single image to be the pre-selected images is 1.25%, while it is 0.9375% for non-pre-selected images, both can be significantly deviated by 1% (1/100 – the denominator is the total number of images in the pool). Attackers can record and analyze the images that appear more frequently, treating them as "hot-spot" images. This method can be considered as a low-cost but highly efficient brute-force attack.

Thus to avoid being cracked by this method, the average proportion of pre-selected images to the total images in the login process as well as the image pool should be configured equally. Assume the displayed image number is 16 per round and the total image number is 100, let variable $x$ and $n$ denote: the average number of pre-selected images displayed per round and the total number of pre-selected images, respectively. Then this leads to two corresponding ratios, $x/16$ and $n/100$, which have to be equal in mathematics, as shown in Fig. 5.

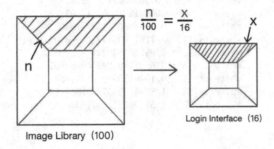

Image Library (100)

**Fig. 5.** The schematic diagram of the ratios $n/100$ and $x/16$.

On the premise that $x/16 = n/100$, the value of $x/16$ and $n/100$ is correlated with each other, meaning that if one is determined, the other can be derived. Therefore, we can decide the average ratio of pre-selected images in the login process to the displayed images per round ($x/16$) and then calculate the probability distribution. As the probability will be considered relatively high when 6 pre-selected images or more are included in the total displayed 16 images, patterns with 3–6 pre-selected images should be used to reduce the occurrence of small probability events (being successfully guessed to crack).

Defining that the probability distribution of 3–6 pre-selected images appearing in each round is $pi$, and the probability of one round being cracked at random is $P(c)$. We can compute $x$ and $P(c)$ as below:

$$x = 3 \times p3 + 4 \times p4 + 5 \times p5 + 6 \times p6 = 16 \times (n/100),$$
$$p3 + p4 + p5 + p6 = 1,$$
$$P(c) = 0.00714 \times p3 + 0.02692 \times p4 + 0.063187 \times p5 + 0.11813 \times p6$$

where p3, p4, p5, and p6 can be calculated by *sampling without the replacement* method[1]. Assuming that randomly selecting 16 images from the image pool of 100 images and recording the 16 pre-selected images, then the probabilities of 3–6 pre-selected images picked can be fitted with the total probability of 1 to regain the values of p3, p4, p5, and p6.

For example, when the number of pre-selected images is 25 from the image pool of 100 images (n = 25), we can have the below:

```
white is 0  = 0.006352847680254038
white is 1  = 0.042352317868360245
white is 2  = 0.12497405272630893
white is 3  = 0.21635292998855632
white is 4  = 0.2455434046695521
white is 5  = 0.19336543117727226
white is 6  = 0.10907793553589715
white is 7  = 0.04485889123770662
white is 8  = 0.013558097724829243
white is 9  = 0.0030129106055176093
white is 10 = 0.0004890521562579308
white is 11 = 5.7161940341836065e-05
white is 12 = 4.696403549211883e-06
white is 13 = 2.6091130828954906e-07
white is 14 = 9.190613599240476e-09
white is 15 = 1.8215630557053196e-10
white is 16 = 1.5179692130877663e-12
```

Fitting:

$$p3 = 0.21635/(0.21635 + 0.24554 + 0.19337 + 0.10908) = 0.28305$$
$$p4 = 0.24554/(0.21635 + 0.24554 + 0.19337 + 0.10908) = 0.32124$$
$$p5 = 0.19337/(0.21635 + 0.24554 + 0.19337 + 0.10908) = 0.25299$$
$$p6 = 0.10908/(0.21635 + 0.24554 + 0.19337 + 0.10908) = 0.14271$$
$$x0 = 3 \times 0.28305 + 4 \times 0.32124 + 5 \times 0.25299 + 6 \times 0.14271 = 4.25532$$
$$\neq 16 \times (25/100) = 4$$

Slightly refitting to match 25/100:

$$p3 = 0.34505, p4 = 0.38124, p5 = 0.19299, p6 = 0.08071,$$
$$0.34505 + 0.38124 + 0.19299 + 0.08071 = 0.99999 \approx 1,$$
$$x1 = 3 \times 0.34505 + 4 \times 0.38124 + 5 \times 0.19299 + 6 \times 0.08071 = 4.0093 \approx 4,$$
$$P(c) = 0.00714 \times 0.34505 + 0.02692 \times 0.38124 + 0.063187 \times 0.19299 + 0.11813$$
$$\times 0.08071 = 0.034455,$$
$$P(c)^{(3)} = 4.09045e - 05$$

Therefore, under the 25/100 model, the average probability of the entire login process being randomly cracked is $4.09045 \times 10^{-5}$, where the password strength

---

[1] It means a subset of the observations are selected randomly, and once an observation is selected it cannot be selected again.

is more muscular than the four-digit password $(10^{-4})$ but is weaker than the five-digit password $(10^{-5})$. For the standard password strength, it is still relatively weak, hence a smaller value of $n$ can be considered.

When the number of pre-selected images is 20 in the image pool of 100 images $(n = 20)$, we can have the below.

```
white is 0 = 0.020030470130923787
white is 1 = 0.09861154525993249
white is 2 = 0.21291129090212693
white is 3 = 0.26693355874296515
white is 4 = 0.21688351647865914
white is 5 = 0.12070039177942772
white is 6 = 0.047418011056203746
white is 7 = 0.01335718621301514
white is 8 = 0.0027131784495187
white is 9 = 0.00039644616613971874
white is 10 = 4.1251830801024786e-05
white is 11 = 3.0001331491654392e-06
white is 12 = 1.480328856496105e-07
white is 13 = 4.732320020766769e-09
white is 14 = 9.100615424551479e-11
white is 15 = 9.215813088153397e-13
white is 16 = 3.5999269875599205e-15
```

Fitting:

$$p3 = 0.26693/(0.26693 + 0.21688 + 0.12070 + 0.04742) = 0.40944$$
$$p4 = 0.21688/(0.26693 + 0.21688 + 0.12070 + 0.04742) = 0.33267$$
$$p5 = 0.12070/(0.26693 + 0.21688 + 0.12070 + 0.04742) = 0.18514$$
$$p6 = 0.04742/(0.26693 + 0.21688 + 0.12070 + 0.04742) = 0.07273$$
$$x0 = 3 \times 0.40944 + 4 \times 0.33267 + 5 \times 0.18514 + 6 \times 0.07273 = 3.9211$$
$$\neq 16 \times (20/100) = 3.2$$

Slightly refitting to match 20/100:

$$p3 = 0.84, p4 = 0.13, p5 = 0.02, p6 = 0.01,$$
$$0.84 + 0.13 + 0.02 + 0.01 = 1,$$
$$x1 = 3 \times 0.84 + 4 \times 0.13 + 5 \times 0.02 + 6 \times 0.01 = 3.2,$$
$$P(c) = 0.00714 \times 0.84 + 0.02692 \times 0.13 + 0.063187 \times 0.02 + 0.11813 \times 0.01$$
$$= 0.011942,$$
$$P(c)^3 = 1.703065e - 06$$

The probability of being randomly cracked when $n = 20$ is $1.70317 \times 10^{-6}$, where password strength is more muscular than the five-digit password $(10^{-5})$ and is close to the six-digit password $(10^{-6})$, which is much better than $n = 25$ in the same situation. However, under the situation with $n = 20$, the appearance probability of $n = 3$ can reach 84%, meaning that the user will need to frequently locate all three pre-selected images out of the 16 displayed images, which might be unfavorable to the user experience.

Table 1 shows the possibilities with different values of $n$.

**Table 1.** Possibility $P(c)$ with different values of $n$.

| n (/100) | 20 | 21 | 22 | 25 |
|---|---|---|---|---|
| $P(c)$ | 0.011942 | 0.0164479 | 0.021140 | 0.034455 |
| $P(c)^3$ | 1.703065e−06 | 4.449698e−06 | 9.447457e−06 | 4.09045e−05 |

## 4    Implementation and Evaluation

### 4.1    System Deployment

There are three main components in a graphical password system: authentication interface, image pool, and user database. The interface is a bridge for calling and manipulating the image pool and user database, which ultimately coordinates the authentication operations. The image pool in our implementation contains 100 image files (PNG format) numbered from 1 to 100, and the user database stores serialized user accounts and their corresponding passwords.

**Fig. 6.** The interface of our proposed scheme.

Figure 6 shows the interface of our graphical password scheme. Compared with traditional login systems, our main interface only requires to input a user name. If the user is new, then he or she has to register first and create the credentials. Clicking on the "Register" button can automatically close the main interface and display the registration window, as shown in Fig. 7.

For the registration, 20 images (5 × 4) are displayed per round. Users are supposed to click and select 5 images, after which the system automatically refreshes the images for a new round. There are totally 4 rounds with 80 different images presented, where users have to choose 20 images in total as their graphical password (4 rounds × 5 images per round). Upon completion of this step, users can proceed to input their username and register their account.

In the login process, a user has to input the username first and then click the "Login" button (can refer to Fig. 6). Then, the login interface will display 16 images for one-time verification, as shown in Fig. 8. Users are required to select 4 images per verification round, and the images will be refreshed automatically

**Fig. 7.** The interface of system registration.

**Fig. 8.** The interface of system login.

**Fig. 9.** Several images from the image pool.

each round. With a total of 3 rounds verification, users have to select 12 images as their credentials (3 rounds × 4 images per round). Once this step is completed, users can click on the login button to verify whether the input is correct.

***Images Design for the System.*** As mentioned earlier, the similar art style of images may help increase the password strength and enhance user experience. In our implementation, the used images were obtained from the mobile game – "3 Tiles"[2] and possessed a dimension of approximately 150 × 150 pixels, as shown in Fig. 9. Further, the predominant artistic image style is characterized by the simplicity and charm.

---

[2] https://www.appsyoulove.com/3tiles..

**Fig. 10.** Examples of similar distracters in the image pool.

In practice, the selection of images is based on common things in people's daily life, such as fruits, desserts and plants. The purpose is to leverage the existing knowledge of users whenever possible rather than requiring them to memorize entirely new and/or random information [2]. While the image pool should contain similar distracters that share common characteristics, such as strawberries and drinks, as shown in Fig. 10. For users who have interacted with the system after a period of time, these distracters are easy to differentiate, but it could be difficult for attackers (e.g., shoulder surfing) to distinguish and remember in the beginning rounds.

## 4.2 Initial User Study

With the implementation of our scheme, we performed a small-scale user study with 20 participants (students with 8 males and 12 femals) for around one week, and collected their feedback about the scheme performance.

It is found that, during the first 1–3 days, a majority of participants were not easy at memorizing the credentials, but towards the end of the user study, most participants were able to log into the system proficiently and quickly, with an average login time of approximately 18 s and an average login success rate of 80%. In the study, we also collected user feedback from a questionnaire survey regarding our system. The key questions are summarized as follows:

- Do you think this system is feasible?
- Would you be willing to use this system?
- What is the most attractive feature of this system to you?
- What do you think are the shortcomings of this system?

For the response to the first question, we found that all participants expressed their positive attitude towards the usage of our system. For the second question, all participants indicated their willingness to use our system in practice.

Regarding the third question, participants' feedback focused primarily on the aesthetic appeal of images. Most participants consider these images as "cute" and "attractive". Participants also mentioned that they believed the system has good operability and can defeat shoulder-surfing attack. In addition, one participant highlighted that our scheme imposed fewer restrictions than textual passwords, such as not requiring upper case and special characters, which was an advantage for practical deployment. For the fourth question, we aim to explore the limitations of our current implementation. Most participants indicated that they need some time to get familiar with the system. Also, one participant stated that the login time could be further decreased.

On the whole, most participants' feedback and responses were primarily positive, giving support to the system usability. However, the results can be further verified as the participants in this study were mainly college students. In future, we plan to include more participants with diverse background.

### 4.3  Discussion

In the user study, the observations indicated the feasibility and usability of our proposed scheme, but there are still many aspects that can be explored and enhanced in our future work.

- *Time consumption.* In the current implementation, the time consumption for login can be improved. For graphical password, it is an open challenge on how to make a balance between security and usability. To ensure the security level against shoulder-surfing attack, it is usually required to decrease the usability to some degree.
- *Diverse background.* In the current user study, we involved 20 participants, most of which are students. This is an initial study to explore the feasibility of our scheme. In our next step, we plan to validate the scheme performance by involving more participants (e.g., 100 users or more) with diverse background (e.g., students, business people, staff).
- *Comparison with similar schemes.* As mentioned in related work, there are many similar schemes that use image selection as the key step. In our future work, we plan to perform a systematic comparison between our proposed scheme and the existing state-of-the-art.

## 5  Conclusion

The design of graphical passwords aims to complement the traditional textual password, as users can remember images better than textual information. In this work, we develop a more secure graphical password scheme with three rounds of image selection, which can increase the cracking difficulty for various attacks, e.g., shoulder-surfing attack. This is because multiple rounds of image selection can avoid the credentials being captured at one time. In addition, our scheme adopts a method of matching rate, where a successful authentication requires a matching rate over 75%. In our initial user study, most participants expressed a positive response regarding the scheme performance, and most of them would like to use our scheme in practice. Future work can include performing a larger user study with more diverse participants and making a comparison between our scheme and the state-of-the-art.

**Acknowledgments.** We would like to thank the participants during the user study. This work was partially supported by the Startup Fund from The Hong Kong Polytechnic University.

# References

1. Aviv, A.J., Gibson, K., Mossop, E., Blaze, M., Smith, J.M.: Smudge attacks on smartphone touch screens. In: Proceedings of the 4th USENIX Conference on Offensive Technologies, pp. 1–7, USENIX Association (2010)
2. Biddle, R., Chiasson, S., Van Oorschot, P.: Graphical passwords: learning from the first twelve years. ACM Comput. Surv. **44**(4), 1–41 (2012)
3. Chiasson, S., Stobert, E., Forget, A., Biddle, R.: Persuasive cued click-points: design, implementation, and evaluation of a knowledge-based authentication mechanism. IEEE Trans. Dependable Secure Comput. **9**(2), 222–235 (2012)
4. Chakraborty, N., Anand, S.V., Mondal, S.: Towards identifying and preventing behavioral side channel attack on recording attack resilient unaided authentication services. Comput. Secur. **84**, 193–205 (2019)
5. Dirik, A.E., Memon, N., Birget, J.C.: Modeling user choice in the passpoints graphical password scheme. In: Proceedings of the 3rd Symposium on Usable Privacy and Security (SOUPS), New York, NY, USA, pp. 20–28. ACM (2007)
6. Dhamija, R., Perrig, A.: Deja Vu: a user study using images for authentication. In: Proceedings of the 9th USENIX Security Symposium (2000)
7. Dunphy, P., Yan, J.: Do background images improve "draw a secret" graphical passwords? In: Proceedings of the 14th ACM Conference on Computer and Communications Security (CCS), pp. 36–47 (2007)
8. Gołofit, K.: Click passwords under investigation. In: Biskup, J., López, J. (eds.) ESORICS 2007. LNCS, vol. 4734, pp. 343–358. Springer, Heidelberg (2007). https://doi.org/10.1007/978-3-540-74835-9_23
9. Jermyn, I., Mayer, A., Monrose, F., Reiter, M.K., Rubin, A.D.: The design and analysis of graphical passwords. In: Proceedings of the 8th Conference on USENIX Security Symposium, pp. 1–14. USENIX Association, Berkeley (1999)
10. Li, W., Tan, J., Meng, W., Wang, Yu., Li, J.: SwipeVLock: a supervised unlocking mechanism based on swipe behavior on smartphones. In: Chen, X., Huang, X., Zhang, J. (eds.) ML4CS 2019. LNCS, vol. 11806, pp. 140–153. Springer, Cham (2019). https://doi.org/10.1007/978-3-030-30619-9_11
11. Li, W., Tan, J., Meng, W., Wang, Y.: A swipe-based unlocking mechanism with supervised learning on smartphones: design and evaluation. J. Netw. Comput. Appl. **165**, 102687 (2020)
12. Li, W., Meng, W., Furnell, S.: Exploring touch-based behavioral authentication on smartphone email applications in IoT-enabled smart cities. Pattern Recognit. Lett. **144**, 35–41 (2021)
13. Li, W., Wang, Y., Tan, J., Zhu, N.: DCUS: evaluating double-click-based unlocking scheme on smartphones. Mob. Netw. Appl. **27**(1), 382–391 (2022)
14. Li, W., Tan, J., Zhu, N.: Double-X: towards double-cross-based unlock mechanism on smartphones. In: Meng, W., Fischer-Hübner, S., Jensen, C.D. (eds.) SEC 2022. IFIP Advances in Information and Communication Technology, vol. 648, pp. 412-C428. Springer, Cham (2022). https://doi.org/10.1007/978-3-031-06975-8_24
15. Li, W., Gleerup, T., Tan, J., Wang, Y.: A security enhanced android unlock scheme based on pinch-to-zoom for smart devices. IEEE Trans. Consum. Electron. (2023)

16. Meng, W.: Graphical authentication. In: Jajodia, S., Samarati, P., Yung, M. (eds.) Encyclopedia of Cryptography, Security and Privacy, pp. 1–4. Springer, Heidelberg (2021). https://doi.org/10.1007/978-3-642-27739-9_1581-1
17. Meng, Y.: Designing click-draw based graphical password scheme for better authentication. In: Proceedings of the 7th IEEE International Conference on Networking, Architecture, and Storage (NAS), pp. 39–48 (2012)
18. Meng, Y., Li, W.: Evaluating the effect of tolerance on click-draw based graphical password scheme. In: Chim, T.W., Yuen, T.H. (eds.) ICICS 2012. LNCS, vol. 7618, pp. 349–356. Springer, Heidelberg (2012). https://doi.org/10.1007/978-3-642-34129-8_32
19. Meng, Y., Li, W.: Evaluating the effect of user guidelines on creating click-draw based graphical passwords. In: Proceedings of the 2012 ACM Research in Applied Computation Symposium (RACS), pp. 322–327 (2012)
20. Meng, Y., Li, W., Kwok, L.-F.: Enhancing click-draw based graphical passwords using multi-touch on mobile phones. In: Janczewski, L.J., Wolfe, H.B., Shenoi, S. (eds.) SEC 2013. IAICT, vol. 405, pp. 55–68. Springer, Heidelberg (2013). https://doi.org/10.1007/978-3-642-39218-4_5
21. Meng, W., Wong, D.S., Furnell, S., Zhou, J.: Surveying the development of biometric user authentication on mobile phones. IEEE Commun. Surv. Tutor. **17**(3), 1268–1293 (2015)
22. Meng, W.: RouteMap: a route and map based graphical password scheme for better multiple password memory. In: NSS 2015. LNCS, vol. 9408, pp. 147–161. Springer, Cham (2015). https://doi.org/10.1007/978-3-319-25645-0_10
23. Meng, W.: Evaluating the effect of multi-touch behaviours on android unlock patterns. Inf. Comput. Secur. **24**(3), 277–287 (2016)
24. Meng, W., Li, W., Jiang, L., Meng, L.: On multiple password interference of touch screen patterns and text passwords. In: ACM Conference on Human Factors in Computing Systems (CHI 2016), pp. 4818–4822 (2016)
25. Meng, W., Li, W., Wong, D.S., Zhou, J.: TMGuard: a touch movement-based security mechanism for screen unlock patterns on smartphones. In: Manulis, M., Sadeghi, A.-R., Schneider, S. (eds.) ACNS 2016. LNCS, vol. 9696, pp. 629–647. Springer, Cham (2016). https://doi.org/10.1007/978-3-319-39555-5_34
26. Meng, W., Lee, W.H., Liu, Z., Su, C., Li, Y.: Evaluating the impact of juice filming charging attack in practical environments. In: Kim, H., Kim, D.-C. (eds.) ICISC 2017. LNCS, vol. 10779, pp. 327–338. Springer, Cham (2018). https://doi.org/10.1007/978-3-319-78556-1_18
27. Meng, W., Fei, F., Li, W., Au, M.H.: Harvesting smartphone privacy through enhanced juice filming charging attacks. In: Nguyen, P., Zhou, J. (eds.) ISC 2017. LNCS, vol. 10599, pp. 291–308. Springer, Cham (2017). https://doi.org/10.1007/978-3-319-69659-1_16
28. Meng, W., Li, W., Kwok, L.-F., Choo, K.-K.R.: Towards enhancing click-draw based graphical passwords using multi-touch behaviours on smartphones. Comput. Secur. **65**, 213–229 (2017)
29. Meng, W., Li, W., Lee, W.H., Jiang, L., Zhou, J.: A pilot study of multiple password interference between text and map-based passwords. In: Gollmann, D., Miyaji, A., Kikuchi, H. (eds.) ACNS 2017. LNCS, vol. 10355, pp. 145–162. Springer, Cham (2017). https://doi.org/10.1007/978-3-319-61204-1_8
30. Meng, W., Lee, W.H., Au, M.H., Liu, Z.: Exploring effect of location number on map-based graphical password authentication. In: Pieprzyk, J., Suriadi, S. (eds.) ACISP 2017. LNCS, vol. 10343, pp. 301–313. Springer, Cham (2017). https://doi.org/10.1007/978-3-319-59870-3_17

31. Meng, W., Jiang, L., Wang, Y., Li, J., Zhang, J., Xiang, Y.: JFCGuard: detecting juice filming charging attack via processor usage analysis on smartphones. Comput. Secur. **76**, 252–264 (2018)
32. Meng, W., Zhu, L., Li, W., Han, J., Li, Y.: Enhancing the security of fintech applications with map-based graphical password authentication. Futur. Gener. Comput. Syst. **101**, 1018–1027 (2019)
33. Meng, W., Jiang, L., Choo, K.K.R., Wang, Y., Jiang, C.: Towards detection of juice filming charging attacks via supervised CPU usage analysis on smartphones. Comput. Electr. Eng. **78**, 230–241 (2019)
34. Nelson, D.L., Reed, V.S., Walling, J.R.: Pictorial superiority effect. J. Exp. Psychol.: Hum. Learn. Mem. **2**(5), 523–528 (1976)
35. Nyang, D., et al.: Two-thumbs-up: physical protection for PIN entry secure against recording attacks. Comput. Secur. **78**, 1–15 (2018)
36. Passfaces. http://www.realuser.com/
37. Shepard, R.N.: Recognition memory for words, sentences, and pictures. J. Verbal Learn. Verbal Behav. **6**(1), 156–163 (1967)
38. Setchi, R., Asikhia, O.K.: Exploring user experience with image schemas, sentiments, and semantics. IEEE Trans. Affect. Comput. **10**(2), 182–195 (2019)
39. Sun, Y., Meng, W., Li, W.: Designing in-air hand gesture-based user authentication system via convex hull. In: Proceedings of The 19th Annual International Conference on Privacy, Security and Trust (PST), pp. 1–5. IEEE (2022)
40. Suo, X., Zhu, Y., Owen, G.S.: Graphical passwords: a survey. In: Proceedings of the 21st Annual Computer Security Applications Conference (ACSAC), pp. 463–472. IEEE Computer Society, USA (2005)
41. Sun, H., Chen, Y., Fang, C., Chang, S.: PassMap: a map based graphical-password authentication system. In: Proceedings of AsiaCCS, pp. 99–100 (2012)
42. Tao, H., Adams, C.: Pass-go: a proposal to improve the usability of graphical passwords. Int. J. Netw. Secur. **2**(7), 273–292 (2008)
43. Wang, L., Meng, W., Li, W.: Towards DTW-based unlock scheme using handwritten graphics on smartphones. In: The 17th International Conference on Mobility, Sensing and Networking (IEEE MSN), pp. 486–493 (2021)
44. Wiedenbeck, S., Waters, J., Birget, J.-C., Brodskiy, A., Memon, N.: Passpoints: design and longitudinal evaluation of a graphical password system. Int. J. Hum Comput Stud. **63**(1–2), 102–127 (2005)
45. Yan, J., Blackwell, A., Anderson, R., Grant, A.: Password memorability and security: empirical results. IEEE Secur. Priv. **2**, 25–31 (2004)
46. Yu, X., Wang, Z., Li, Y., Li, L., Zhu, W.T., Song, L.: EvoPass: evolvable graphical password against shoulder-surfing attacks. Comput. Secur. **70**, 179–198 (2017)
47. Zhou, T., Liu, L., Wang, H., Li, W., Jiang, C.: PassGrid: towards graph-supplemented textual shoulder surfing resistant authentication. In: Meng, W., Furnell, S. (eds.) SocialSec 2019. CCIS, vol. 1095, pp. 251–263. Springer, Singapore (2019). https://doi.org/10.1007/978-981-15-0758-8_19

# Implementation of the Elliptic Curve Method

Xiuxiu Li[1,2,3], Wei Yu[1,2,3](✉), and Kunpeng Wang[1,3]

[1] State Key Laboratory of Information Security, Institute of Information
Engineering, CAS, Beijing, China
yuwei@iie.ac.cn, yuwei_1_yw@163.com
[2] State Key Laboratory of Cryptology, P. O. Box 5159, Beijing 100878, China
[3] School of Cyberspace Security, University of Chinese Academy of Sciences,
Beijing, China

**Abstract.** The Elliptic Curve Method (ECM) is a versatile algorithm
for determining prime factors of medium-sized composite integers. It is
commonly used in the sieving phase of the Number Field Sieve to accelerate factorization. In this paper, we propose new approaches to improve
the efficiency of ECM. Specifically, we utilize a combination of twisted
Hessian curves and Montgomery curves, along with a new combination
of prime factors, to achieve improved performance. We present the scalar
multiplication costs, which constitute the most computationally demanding aspect of ECM. Experimental results demonstrate that our approach
achieves comparable efficiency to state-of-the-art ECM implementations.

**Keywords:** Elliptic curve method · scalar multiplication · twisted
Hessian curve · Montgomery curve · twisted Edwards curve ·
double-base representation

## 1 Introduction

The Elliptic Curve Method (ECM) [16], proposed by Lenstra in 1987, is a powerful algorithm for finding prime factors of plenty of medium size integers. ECM is
a generalization of Pollard's $P - 1$ method [21], using addition groups of elliptic
curves instead of multiplication groups. Although it is not the fastest universal
integer factorization method, ECM can speed up the factorization in Number
Field Sieve (NFS), for example, used in the cofactorization phase because this
phase needs to detect whether plenty of integers are $B_1$-powersmooth. An integer is $B_1$-powersmooth if all prime powers dividing this integer are not greater
than $B_1$, where $B_1$ is a smooth bound.

Various factorization techniques are available to test the smoothness of integers, such as the Pollard $P - 1$ [21], quadratic sieve [22], and ECM. However,
ECM has some advantages over these methods, including its versatility and ability to speed up the factorization in NFS. ECM is used to test the smoothness

Supported by the National Natural Science Foundation of China (No. 62272453,
U1936209, 61872442, and 61502487).

© The Author(s), under exclusive license to Springer Nature Switzerland AG 2023
M. Yung et al. (Eds.): SciSec 2023, LNCS 14299, pp. 115–126, 2023.
https://doi.org/10.1007/978-3-031-45933-7_7

of integers in the cofactorization phase when NFS decomposes large integers as shown in the following two examples: in [9], the cofactorization procedure of a 200-digit Rivest-Shamir-Adleman (RSA) modulus takes about 20% of the sieving phase of NFS; when factoring the 768-bit RSA number [15], the cofactorization takes approximately 33% of the sieving time. ECM has also been used to test the smoothness of large integers in discrete logarithm problems (DLP-240), retaining only the most promising integers that go through the ECM. For instance, when factoring 795-bit integers, ECM can extract all prime factors up to 75 bits with a high probability [6].

In practice, the efficiency of ECM depends heavily on the speed of scalar multiplication, which plays a significant role in the algorithm. Many efforts have been made to improve ECM by optimizing scalar multiplication. Montgomery and Brent contributed to the most recent ECM improvements at the end of 1985 [8,19]. Earlier, Montgomery curves [19] and some techniques [24] were used to implement ECM, with their approach incorporated into GMP-ECM [25], an advanced ECM implementation. Later, based on the work of Edwards [13], Bernstein et al. [2] applied Edwards curves to ECM to reduce the time spent on ECM scalar multiplication. Bos and Kleinjung [5] further optimize Edwards ECM by generating and combining addition-subtracting chains. Bouvier et al. [7] extended these ideas of [5] to process the scalar by batches of primes and discuss the mixing of twisted Edwards and Montgomery representations. Despite these advancements, the computation of scalar multiplication remains time-consuming. Therefore, the goal of this article is to reduce scalar multiplication's time by combining different curves and finding optimal combinations of prime numbers.

In this paper, we present novel techniques for factorizing medium-size integers using the ECM. Firstly, we analyze the use of twisted Edwards curves with $a = -1$ and show that their scalar multiplication costs are comparable to previous works. Specifically, our approach yields a cost of 2786M, which is on par with [7]. Secondly, we leverage the birational equivalence between twisted Hessian curves and Montgomery curves to combine the two types of curves. We provide transformation formulas from one model to another for different values of $q$ modulo 3. This mixed approach results in a cost of 3020M. To optimize elliptic operations on different curve types, we generate double-base expansions for twisted Edwards curves and twisted Hessian curves, and Lucas chains for Montgomery curves. Our experiments show that the arithmetic costs per bit can be as low as 7.6M in the best case.

The rest of this paper is organized as follows. In Sect. 2, we give the general ECM algorithm and the basic concepts used in this work. Section 3 describes the scalar multiplication algorithm and its development. Section 4 presents our method for reducing the costs of scalar multiplication in the ECM implementation. Section 5 concludes the paper.

## 2    Preliminaries

This paper aims to compare the computational cost of various point operations and scalar multiplications in the context of ECM. For the purpose of generality and to facilitate comparison with prior work, we use $\mathbf{M}$ and $\mathbf{S}$ to denote modular multiplication and squaring, and consider two cases: $1\mathbf{S} = 1\mathbf{M}$ and $1\mathbf{S} = 0.8\mathbf{M}$.

## 2.1 Twisted Edwards Curves

The Edwards ECM method with $a = -1$ has been found to be fast in practice, making it a popular choice for scalar multiplication in ECM. As a result, we will focus exclusively on Edwards curves in first case.

**Definition 1 (Twisted Edwards Curve).** *A twisted Edwards curve over a field $K$ denoted by $E_{A,B}^E$ is given by the equation*

$$ax^2 + y^2 = 1 + dx^2y^2 \tag{1}$$

*where $a, d \in K$ such that $ad(a - d) \neq 0$.*

Our analysis will leverage well-known point doubling and point addition algorithms for twisted Edwards curves, which were introduced in [14]. In addition, we will make use of an efficient point tripling algorithm for these curves, which is described in [4]. This is one of the main reasons why we chose to focus on twisted Edwards curves for our analysis.

## 2.2 Twisted Hessian Curves

The authors [3] proposed a new standard type of elliptic curve, on which fast elliptic curve operations are our concern.

**Definition 2 (Twisted Hessian Curve).** *An affine twisted Hessian curve over a field $K$, denoted by $E_{a,d}^H$, is a curve of the form*

$$ax^3 + y^3 + 1 = dxy \tag{2}$$

*where $a, d \in K$ and $a(27a - d^3) \neq 0$.*

In the following, we will utilize the formula from [3] to calculate point addition, point doubling, and point tripling on twisted Hessian curves. Table 1 summarizes the input and output formats and the point operations cost from other related works that we will use in this paper. The "ADD$_\text{mixed}$" denotes mixed addition on the elliptic curves. The symbols "affi" and "proj" denote affine coordinate and projective coordinate, respectively.

**Table 1.** The cost of point operations on twisted Hessian curves with $1\mathbf{S} = 1\mathbf{M}$

| Point Operation | Notation | Input | $\rightarrow$ | Output | cost |
|---|---|---|---|---|---|
| Addition | ADD | proj. | $\rightarrow$ | proj. | 11M |
|  | ADD$_\text{mixed}$ | affi. or proj. | $\rightarrow$ | proj. | 9M |
| Doubling | DBL | proj. | $\rightarrow$ | proj. | $6\mathbf{M} + 2\mathbf{S} = 8\mathbf{M}$ |
| Tripling | TPL | proj. | $\rightarrow$ | proj. | $6\mathbf{M} + 6\mathbf{S} = 12\mathbf{M}$ |

## 2.3  Montgomery Curves

Montgomery curves have become increasingly popular in elliptic curve cryptography due to their efficient point operations [10], which can lead to significant speedup in scalar multiplication.

**Definition 3 (Montgomery Curve).** *A Montgomery curve over a field $K$ denoted by $E^M_{A,B}$ is given by the equation*

$$By^2 = x^3 + Ax^2 + x \tag{3}$$

*where $A, B \in K$ such that $B(A^2 - 4) \neq 0$.*

Table 2 shows the cost of two common point operations on Montgomery curves using $XZ$ coordinates, namely point doubling and differential addition. Point doubling, denoted as dDBL, requires three multiplications and two squarings. Differential addition, denoted as dADD, involves four multiplications and two squarings.

It should be noted that although these point operations are fast, the use of differential addition requires specific scalar multiplication algorithms. Nonetheless, the efficiency of Montgomery curves makes them an attractive option for applications where speed is critical.

**Table 2.** The cost of point operations on Montgomery curves with $1M = 1S$ from [7]

| Point Operation | Notation | Coordinates | Cost |
|---|---|---|---|
| Doubling | dDBL | $XZ$ | $3M + 2S = 5M$ |
| Differential Addition | dADD | $XZ$ | $4M + 2S = 6M$ |

## 2.4  Double-Base Number System

The double-base number system (DBNS) is a way to represent a positive integer $n$ as a sum of terms of the form $\pm 2^{b_i} 3^{t_i}$, where $b_i$ and $t_i$ are non-negative integers. The DBNS representation is not unique, but for any given $n$, there is a unique representation that has no repeated pairs $(b_i, t_i)$.

DBNS was first proposed for use in elliptic curve cryptography [11], where it can be used to speed up scalar multiplication [17,18]. Scalar multiplication is computing $kP$, where $P$ is a point on the curve and $k$ is a scalar. DBNS can be used to speed up the computation of $kP$ by representing $k$ in DBNS form, and then using the properties of the double-base representation to perform the scalar multiplication more efficiently.

A double-base chain (DBC) is a special type of DBNS representation, where the exponents $b_i$ and $t_i$ are chosen such that $b_m \geq b_{m-1} \geq \cdots \geq b_0$ and $t_m \geq t_{m-1} \geq \cdots \geq t_0$. The optimal DBC [23] for a given integer $n$ is the DBC that requires the fewest modular multiplications when computing scalar multiplication. In general, finding the optimal DBC is a difficult problem, but there are algorithms that can be used to find good approximations to the optimal DBC.

To further reduce redundancy in the DBC representation, it is common to choose exponents such that there are no common factors between the terms of the DBC. This can be achieved using algorithms such as the Duursma-Lee algorithm or the Aoki-Joichi algorithm.

## 2.5    Lucas Chains

When computing scalar multiplication on a Montgomery curve, a Lucas chain is typically employed [20]. A Lucas chain is a sequence of integers generated using a recurrence relation of the form $U_n(P,Q) = P^n + Q^n$, where $P$ and $Q$ are fixed integers. The Lucas chain is then formed by repeatedly applying this recurrence relation, starting with $U_0 = 2$ and $U_1 = P + Q$. Each subsequent element in the chain is the sum of the two previous elements, and the corresponding difference must also have appeared in the chain.

To compute $[k]P$ on a Montgomery curve, a near-optimal Lucas chain must first be selected. For example, if we wish to compute $[1009]P$, we may choose to use the last element of the chain as $1009/\phi \approx 624$, where $\phi = (1 + \sqrt{5})/2$ is the golden ratio. It is important to note that the term $1009 - 624 = 385$ must also be included in the chain. A portion of the Lucas chain for this example is $1009 \rightarrow 624 \rightarrow 385 \rightarrow 239 \rightarrow 146 \rightarrow 93 \rightarrow 53 \rightarrow 40 \rightarrow 13$. At this moment, we have to iteratively reduce the pair $(d, e)$ using another transform instead of using the transform $(d, e) \rightarrow (d, d - e)$. The alternative transform uses from 1 to 4 point duplicates or additions, eventually reaching $d = 1$.

## 2.6    The Elliptic Curve Method

The Elliptic Curve Method (ECM) is a factoring algorithm that leverages the properties of elliptic curves. It is a generalization of the Pollard $p - 1$ method and is particularly useful for factoring integers with small factors. The algorithm involves selecting a random elliptic curve $E$ defined over the field of integers modulo $N$ ($\mathbb{Z}/N\mathbb{Z}$), where $N$ is the integer to be factored. A random point $P = (X : Y : Z)$ on $E$ is also chosen. The next step is to compute the scalar multiplication $Q = [k]P$, where $k$ is a smooth number with respect to a given factor base, and $B_1$ is the smooth bound that controls the size of the factor base. Specifically, $k$ is selected as the least common multiple of the integers $\{1, 2, \ldots, B_1\}$.

If $p$ is a prime factor of $N$, and the order of $E$ over the finite field $\mathbb{F}_p$ (i.e., the number of points on $E$ modulo $p$) is $B_1$-powersmooth, then $k$ is divisible by the order of $E$ modulo $p$, denoted by $\#E(\mathbb{F}_p)$. As a result, when the point $Q = [k]P$ is reduced modulo $p$, it is the point at infinity on $E(\mathbb{F}_p)$. If the greatest common divisor of $N$ and the $Z$-coordinate of $Q$ is not equal to 1, then a non-trivial factor of $N$ can be obtained, as the $Z$-coordinate of $Q$ is a multiple of $N$ with high probability.

## 3    A Overview of Scalar Multiplication

Scalar multiplication is multiplying a point $P$ on an elliptic curve by a scalar $k$, resulting in another point on the same curve, denoted by

$$k = \mathrm{lcm}(1, 2, \cdots, B_1) = \prod_{p \text{ prime } \leq B_1} p^{\lfloor \log_p(B_1) \rfloor}.$$

The scalar $k$ is typically a large integer and may be generated using a smoothness bound $B_1$ based on the prime factorization of the numbers 1 to $B_1$. Our goal is to speed up the computation of scalar multiplication $[k]P$ in the ECM context.

Several algorithms have been proposed to improve the efficiency of ECM. Dixon and Lenstra [12] proposed an algorithm dedicated to factorizing medium-sized integers that processes the scalars of ECM by batches of primes, rather than considering each prime individually. This approach takes advantage of the fact that the sum of the Hamming weights of individual primes $\sum_i \omega(p_i)$ may be greater than the Hamming weight of their product $\omega(\prod_i p_i)$. Bouvier and Imbert [7] introduced another method that improves arithmetic cost by mixing the twisted Edwards and Montgomery models. They choose integers with low Hamming weights in the double-base chains and the double-base expansion and test their smoothness. Then, among those integers that are $B_1$-powersmooth, a more exhaustive approach is used to find a suitable partition of prime factors of $k$ that minimizes the cost in the final computation. The critical issue is to divide all the prime factors of $k$ reasonably. For $B_1 = 256$, the optimal chains found so far lead to a scalar multiplication algorithm that requires 292 doublings, 42 triplings, and 29 additions. However, the main disadvantage of this method is the significant amount of time required to obtain near-optimal double-base chains for integers.

In this work, we use three representations for integers that are analogous to those used by Dixon and Lenstra, Bos and Kleinjung, and Bouvier and Imbert. Specifically, we define the term "block" in Tables 3 and 4 to denote the product of several primes, and we use $\mathcal{B}$ to represent the set of all blocks derived from double base expansions, double base chains, and Lucas chains. For each block $b \in \mathcal{B}$, $n(b)$ denotes the integer allied to $b$, and the prime factors of $n(b)$ compose the multiset $\mathcal{M}_b$. We use $\mathrm{cost}(b)$ to denote the cost of computing scalar multiplication by $n(b)$, and $\mathrm{acpb}(b)$ to denote the arithmetic cost per bit, which is computed as $\mathrm{cost}(b)/\mathrm{log2}(n(b))$. Dixon and Lenstra used an addition chain to represent each block, while Bos and Kleinjung denote each block by an addition–subtraction chain under the action of NAF expansions. Bouvier and Imbert considered a more general decomposition, proposing to represent integers by the double-base number system and Lucas chains, which significantly reduces the cost of scalar multiplication.

## 4    Our Implementation of ECM

In this section, we present our implementation of the ECM for reducing the cost of scalar multiplication. We first discuss our approach to combining various chains. Next, we describe the performance of our combinatorial algorithm in two different cases using different curves.

## 4.1  Combination of Blocks for ECM

A complete exhaustive search for the ECM is impossible, even for relatively small values of $B_1$. However, we can employ a relatively exhaustive but potentially optimal strategy based on the algorithm proposed by Bouvier and Imbert [7]. The core of algorithm is to find a block subset $S$ of $B$ that satisfies $k = \prod_{b \in S} n(b)$ and $\text{cost}(S)$ is minimized.

To compute $S$, we employ the algorithm proposed in [5] and [7]. First, we obtain an upper bound cost $C$ of the set $S$ using Bos and Kleinjung's algorithms [5]. We add blocks to build our solution set in increasing order of their arithmetic cost per bit (acpb). We start with $M = M_{B_1}$ and let $S_0 \subseteq B$ be the initial set that satisfies $\bigcup_{b \in S_0} M_b \subsetneq M_{B_1}$. When $M$ is not an empty set, we select the block $b$ from the set $B$ such that $M_b \subseteq M$. The cost per bit cannot exceed 9M. Next, we add $b$ to $S_0$ and remove $M_b$ from $M$. When the loop terminates, our algorithm returns a solution set $S$ that contains $S_0$ and satisfies $\text{cost}(S) < C$. The algorithm starts by sorting the blocks according to their acpb value.

In this section, we first present the algorithm cost of previous related works when using the above algorithm. The cost of our work will be detailed in the next two sections. When $\mathbf{S} = \mathbf{M}$, the arithmetic costs of scalar multiplication in [5] are 2844M, and the arithmetic costs per bit are 7.8M. The arithmetic costs of scalar multiplication in [7] are 2748M, and the arithmetic costs per bit are 7.5M. When $\mathbf{S} - 0.8\mathbf{M}$, the arithmetic costs of scalar multiplication in [5] are 2560M, and the arithmetic costs per bit are 7.0M. The arithmetic costs of scalar multiplication in [7] are 2496M, and the arithmetic costs per bit are 6.8M.

## 4.2  Individual Representation

Selecting appropriate curves that require a low cost of elliptic operations is a crucial aspect of improving the probability of ECM successfully factoring integers. In this regard, we implemented ECM using only the twisted Edwards curve with $a = -1$, which has been found to have low arithmetic costs [1]. In practical implementations of scalar multiplication on twisted Edwards curves, projective or extended coordinates are usually employed to represent the input and output points. We report our implementation results for the twisted Edwards curve in Table 3. Our experiments demonstrate that when $\mathbf{S} = \mathbf{M}$, the arithmetic costs of scalar multiplication using the twisted Edwards curve are 2786M, and the arithmetic costs per bit are 7.6M. Furthermore, when $\mathbf{S}=0.8\mathbf{M}$, the arithmetic costs of scalar multiplication using the twisted Edwards curve are 2525.2M, and the arithmetic costs per bit are 6.9M.

**Table 3.** The set of blocks from our algorithm using the twisted Edwards curve with $a = -1$ for $B_1 = 256$. Notation e denotes double-base expansions, notation c denotes double-base chains.

| Blocks | Type | | Cost |
|---|---|---|---|
| $151 \cdot 31 \cdot 7$ | c | $2^{15} - 1$ | 114M |
| $167 \cdot 149 \cdot 5$ | c | $2^9 3^5 - 1$ | 132M |
| $227 \cdot 73 \cdot 67 \cdot 17$ | c | $2^{21} 3^2 + 1$ | 180M |
| $193 \cdot 127 \cdot 109 \cdot 107 \cdot 61 \cdot 13 \cdot 7$ | c | $2^{12} 3^{18} - 1$ | 309M |
| $251 \cdot 43 \cdot 41$ | c | $2^{14} 3^3 + 2^4 3^2 + 1$ | 151M |
| $241 \cdot 229 \cdot 19$ | c | $2^{20} + 2^4 - 1$ | 157M |
| $211 \cdot 139 \cdot 13 \cdot 11$ | c | $2^{22} - 2^8 - 1$ | 171M |
| $233 \cdot 191 \cdot 173 \cdot 157$ | c | $2^{27} 3^2 + 2^{18} 3 - 1$ | 230M |
| $223 \cdot 137 \cdot 103 \cdot 83 \cdot 37$ | c | $2^{30} 3^2 + 2^{11} - 1$ | 251M |
| $179 \cdot 101 \cdot 97 \cdot 47 \cdot 29 \cdot 23 \cdot 5$ | c | $2^{38} - 2^3 - 1$ | 283M |
| $181 \cdot 131 \cdot 89 \cdot 59 \cdot 11$ | c | $2^{24} 3^4 + 2^{17} 3^4 - 2^8 - 1$ | 241M |
| $239 \cdot 199 \cdot 197 \cdot 163 \cdot 113 \cdot 79 \cdot 71 \cdot 53$ | e | $2^{46} 3^6 + 2^{42} + 2^{14} + 3^3$ | 421M |
| $5 \cdot 3 \cdot 3 \cdot 3 \cdot 3 \cdot 3$ | c | $2^7 3^2 + 2^6 - 1$ | 90M |
| $2 \cdot 2 \cdot 2 \cdot 2 \cdot 2 \cdot 2 \cdot 2 \cdot 2$ | c | $2^8$ | 56M |
| Total | | | 2786M |

## 4.3   Mixed Representation

In this section, we present a mixed representation of blocks using twisted Hessian curves and Montgomery curves. Specifically, we provide the birationally equivalent relation of the Montgomery curve and twisted Edwards curve for different values of $q$ modulo 3. By combining the transformation from the twisted Hessian curve to the Weierstrass curve and the transformation from the Weierstrass curve to the Montgomery curve, we can obtain the birational equivalence between the twisted Hessian curve and the Montgomery curve.

We consider the twisted Hessian curve $E^H_{a,d}$ and the Montgomery curve $E^M_{A,B}$ over field $K$, where $\mathrm{char}(K) > 3$, $a, d \in K$ with $d^3 \neq a$ and $a \neq 0$. The twisted Hessian curve is defined by

$$E^H_{a,d} : aU^3 + V^3 + W^3 = 3dUVW. \tag{4}$$

and the Montgomery curve is defined by

$$E^M_{A,B} : BY^2 Z = X^3 + AX^2 Z + XZ^2. \tag{5}$$

If $q \equiv 1 \bmod 3$, then there exists an $\varepsilon \in K$ such that $\varepsilon^2 + \varepsilon + 1 = 0$. In projective coordinates, we can obtain a birational equivalence between the twisted Hessian curve and the Montgomery curve with $s = d/2, t = (d^3 - a)/54$ via the substitution of variables

$$U = 3B'X + A'B'Z,$$
$$V = -3s\varepsilon B'X - 3B'^2(\varepsilon + 2)Y - (s\varepsilon A'B' + 9t\varepsilon)Z, \tag{6}$$
$$W = 9B'(1 + \varepsilon)X + 3B'^2(\varepsilon - 1)Y + (3A'B' + 9t)(\varepsilon + 1)Z.$$

where $A', B'$ are numbers given by $a$, $d$.

If $q \equiv 2 \bmod 3$, each Hessian curve $au^3 + v^3 + 1 = 3duv$ can be converted into a more general Hessian curve $u^3 + v^3 + 1 = 3d'uv$ for some $d'$. The equivalency between the Hessian curve and the Montgomery curve is given by the substitution of variables

$$U = \frac{9d'B'}{8(d'^3 - 1)}X + \frac{9B'^2}{8(d'^3 - 1)}Y + \frac{3d'A'B' + 12(d'^2 + d' + 1)}{8(d'^3 - 1)}Z,$$
$$V = \frac{9d'B'}{8(d'^3 - 1)}X + \frac{-9B'^2}{8(d'^3 - 1)}Y + \frac{3d'A'B' + 12(d'^2 + d' + 1)}{8(d'^3 - 1)}Z, \tag{7}$$
$$W = -\frac{9B'}{4(d'^3 - 1)}X - \frac{3A'B' + 12(d'^2 + d' + 1)}{4(d'^3 - 1)}Z.$$

where $A', B', d'$ are numbers given by $a$, $d$.

The projective twisted Hessian curve can be mapped to the projective Montgomery curve using maps (6) and (7). For all prime factors of the integer $N$, if the point at infinity modulo $q$ on the twisted Hessian curve is obtained, then the corresponding Montgomery curve must have a point with $Z \equiv 0 \ (mod \ q)$. Applying (6) and (7), the point $(0 : 1 : 0)$ at infinity on $E_{a,d}^H$ maps to the point $(0 : : 0)$ on $E_{A,B}^M$. The $XZ$ coordinates of the Montgomery curve can be represented as $(X : : Z)$, with no additional computational cost. However, it should be noted that once converted to the Montgomery curve, it is not possible to switch back to the Hessian curve.

Table 4 shows the implementation results when mixing the twisted Hessian curve and the Montgomery curve. The experiment shows that when $\mathbf{S} = \mathbf{M}$, the arithmetic costs of scalar multiplication using the mix of the Montgomery curve and twisted Hessian curve are 3020M, and the arithmetic costs per bit are 8.3M. When $\mathbf{S} = 0.8\mathbf{M}$, the arithmetic costs of scalar multiplication using the mix of the Montgomery curve and twisted Hessian curve are 2827.6M, and the arithmetic costs per bit are 7.6M.

Both in theory and practice, individual representations are better than mixed representations. This is because the point operations cost on the Edwards curve is less than that on the Hessian curve. Even the combination of the Montgomery curve and Hessian curve is not as efficient as the Edwards curve. Although the experimental results of the mixed representation are not as good as those of the individual representation, the mixed representation provides theoretical value for future research on ECM.

This paper explores two methods used to improve the efficiency of scalar multiplication, and both have their own merits in practice and theory. The implemented ECM uses C++ programs compiled by Microsoft Visual Studio 2019, and the code is publicly available at https://github.com/18801268593/ECM-Implementation.

**Table 4.** The set of blocks from our algorithm for $B_1 = 256$ mixes the twisted Hessian curve and Montgomery curve. Notation e denotes double-base expansions, notation c denotes double-base chains on the Hessian curve, and notation m denotes blocks on the Montgomery curve.

| Blocks | Type | | Cost |
|---|---|---|---|
| $227 \cdot 167 \cdot 149 \cdot 101 \cdot 59 \cdot 47$ | c | $2^{12}3^{18} - 2^{10}3^{14} - 2^{10}3^6 + 1$ | 343M |
| $191 \cdot 173 \cdot 151 \cdot 103 \cdot 89 \cdot 83$ | c | $2^{18}3^{15} + 2^{16}3^{12} + 2^{10}3^7 + 1$ | 355M |
| $241 \cdot 223 \cdot 157 \cdot 131 \cdot 79 \cdot 23$ | e | $2^6 3^{22} + 2^4 3^{10} - 2^6 3^4 + 1$ | 343M |
| $193 \cdot 127 \cdot 109 \cdot 107 \cdot 61 \cdot 41 \cdot 31 \cdot 13 \cdot 11 \cdot 7 \cdot 5 \cdot 5 \cdot 3$ | e | $2^{32}3^{18} - 2^{12}3^{18} - 2^{20} + 1$ | 503M |
| $239 \cdot 233 \cdot 211 \cdot 73 \cdot 67 \cdot 53 \cdot 19 \cdot 5$ | e | $2^{10}3^{24} + 2^7 3^{19} - 3^5 - 2^5$ | 401M |
| $229 \cdot 199 \cdot 197 \cdot 97 \cdot 37 \cdot 17 \cdot 11$ | e | $2^6 3^{23} + 3^{15} - 2^2 3^{10} + 2^1$ | 357M |
| $113 \cdot 13 \cdot 3$ | m | | 101M |
| $181 \cdot 7$ | m | | 86M |
| $43 \cdot 29$ | m | | 86M |
| $179$ | m | | 64M |
| $251 \cdot 3$ | m | | 80M |
| $139 \cdot 3$ | m | | 73M |
| $163$ | m | | 63M |
| $137 \cdot 3$ | m | | 73M |
| $71$ | m | | 52M |
| $2 \cdot 2 \cdot 2 \cdot 2 \cdot 2 \cdot 2 \cdot 2 \cdot 2$ | m | $2^8$ | 40M |
| Total | | | 3020M |

## 5    Conclusion

The ECM is a widely used algorithm for factoring medium-sized integers that arise in the NFS sieving phase. To speed up ECM, we propose using twisted Edwards curves with $a = -1$ as well as a combination of twisted Hessian curves and Montgomery curves. Given two points on the Hessian curve, to achieve a birational equivalence conversion between the twisted Hessian curve and the Montgomery curve, we employ the partial addition-and-switch operation to compute the addition of two points on the Montgomery curve. We generate a large number of Lucas chains, double-base chains, or expansions, following the works of Bos and Kleinjung and Bouvier and Imbert, and combine these representations using an asymptotic exhaustive approach. In the case of $\mathbf{S} = \mathbf{M}$, our arithmetic costs are 2786M when using twisted Edwards curves with $a = -1$, and 3020M when using a mix of twisted Hessian curves and Montgomery curves. In the best case, our arithmetic costs per bit are about 7.6M. Our experimental results demonstrate that our individual representation is superior to the work of [5] and comparable to that of Bouvier et al. [7]. Although the improvements we achieve are not drastic, our work provides a novel approach to implementing ECM. To further enhance ECM efficiency in the future, it is crucial to accelerate scalar multiplication, that is, obtaining the $n$-th multiple of a point at a low cost.

# References

1. Barbulescu, R., Bos, J.W., Bouvier, C., Kleinjung, T., Montgomery, P.: Finding ECM-friendly curves through a study of Galois properties. In: ANTS X: Proceedings of the Tenth Algorithmic Number Theory Symposium. Open Book Series, vol. 1, pp. 63–86 (2013). https://doi.org/10.2140/obs.2013.1.63
2. Bernstein, D.J., Birkner, P., Lange, T., Peters, C.: ECM using Edwards curves. Math. Comput. **82**, 1139–1179 (2013)
3. Bernstein, D.J., Chuengsatiansup, C., Kohel, D., Lange, T.: Twisted hessian curves. In: Lauter, K., Rodríguez-Henríquez, F. (eds.) LATINCRYPT 2015. LNCS, vol. 9230, pp. 269–294. Springer, Cham (2015). https://doi.org/10.1007/978-3-319-22174-8_15
4. Bernstein, D.J., Chuengsatiansup, C., Lange, T.: Double-base scalar multiplication revisited. Cryptology ePrint Archive, Report 2017/037 (2017). https://eprint.iacr.org/2017/037
5. Bos, J.W., Kleinjung, T.: ECM at work. In: Wang, X., Sako, K. (eds.) ASIACRYPT 2012. LNCS, vol. 7658, pp. 467–484. Springer, Heidelberg (2012). https://doi.org/10.1007/978-3-642-34961-4_29
6. Boudot, F., Gaudry, P., Guillevic, A., Heninger, N., Thomé, E., Zimmermann, P.: Comparing the difficulty of factorization and discrete logarithm: a 240-digit experiment. In: Micciancio, D., Ristenpart, T. (eds.) CRYPTO 2020. LNCS, vol. 12171, pp. 62–91. Springer, Cham (2020). https://doi.org/10.1007/978-3-030-56880-1_3
7. Bouvier, C., Imbert, L.: Faster cofactorization with ECM using mixed representations. In: Kiayias, A., Kohlweiss, M., Wallden, P., Zikas, V. (eds.) PKC 2020. LNCS, vol. 12111, pp. 483–504. Springer, Cham (2020). https://doi.org/10.1007/978-3-030-45388-6_17
8. Brent, R.P.: Some integer factorization algorithms using elliptic curves. Austral. Comput. Sci. Comm. **8**, 149–163 (1986)
9. The CADO-NFS Development Team: CADO-NFS, An Implementation of the Number Field Sieve Algorithm (2017). http://cado-nfs.gforge.inria.fr/. Release 2.3.0
10. Costello, C., Smith, B.: Montgomery curves and their arithmetic: the case of large characteristic fields. J. Cryptogr. Eng. **8**(3), 227–240 (2017)
11. Dimitrov, V., Imbert, L., Mishra, P.K.: Efficient and secure elliptic curve point multiplication using double-base chains. In: Roy, B. (ed.) ASIACRYPT 2005. LNCS, vol. 3788, pp. 59–78. Springer, Heidelberg (2005). https://doi.org/10.1007/11593447_4
12. Dixon, B., Lenstra, A.K.: Massively parallel elliptic curve factoring. In: Rueppel, R.A. (ed.) EUROCRYPT 1992. LNCS, vol. 658, pp. 183–193. Springer, Heidelberg (1993). https://doi.org/10.1007/3-540-47555-9_16
13. Edwards, H.M.: A normal form for elliptic curves. Bull. Am. Math. Soc. **44**, 393–422 (2007)
14. Hisil, H., Wong, K.K.-H., Carter, G., Dawson, E.: Twisted Edwards curves revisited. In: Pieprzyk, J. (ed.) ASIACRYPT 2008. LNCS, vol. 5350, pp. 326–343. Springer, Heidelberg (2008). https://doi.org/10.1007/978-3-540-89255-7_20
15. Kleinjung, T., et al.: Factorization of a 768-bit RSA modulus. In: Rabin, T. (ed.) CRYPTO 2010. LNCS, vol. 6223, pp. 333–350. Springer, Heidelberg (2010). https://doi.org/10.1007/978-3-642-14623-7_18
16. Lenstra, H.W.: Factoring integers with elliptic curves. Ann. Math. **126**(3), 649–673 (1987)

126     X. Li et al.

17. Meloni, N., Hasan, M.: Efficient double bases for scalar multiplication. IEEE Trans. Comput. **64**(8), 2204–2212 (2015)
18. Méloni, N., Hasan, M.A.: Elliptic curve scalar multiplication combining Yao's algorithm and double bases. In: Clavier, C., Gaj, K. (eds.) CHES 2009. LNCS, vol. 5747, pp. 304–316. Springer, Heidelberg (2009). https://doi.org/10.1007/978-3-642-04138-9_22
19. Montgomery, P.L.: Speeding the Pollard and elliptic curve methods of factorization. Math. Comput. **48**(177), 243–264 (1987)
20. Montgomery, P.L.: Evaluating recurrences of form $Xm + n = f(Xm, Xn, Xm - n)$ via Lucas chains. (1983). Unpublished
21. Pollard, J.M.: Theorems on factorization and primality testing. Proc. Camb. Philos. Soc. **76**, 521–528 (1974)
22. Pomerance, C.: The quadratic sieve factoring algorithm. In: Beth, T., Cot, N., Ingemarsson, I. (eds.) EUROCRYPT 1984. LNCS, vol. 209, pp. 169–182. Springer, Heidelberg (1985). https://doi.org/10.1007/3-540-39757-4_17
23. Yu, W., Musa, S.A., Li, B.: Double-base chains for scalar multiplications on elliptic curves. In: Canteaut, A., Ishai, Y. (eds.) EUROCRYPT 2020. LNCS, vol. 12107, pp. 538–565. Springer, Cham (2020). https://doi.org/10.1007/978-3-030-45727-3_18
24. Zimmermann, P., Dodson, B.: 20 years of ECM. In: Hess, F., Pauli, S., Pohst, M. (eds.) ANTS 2006. LNCS, vol. 4076, pp. 525–542. Springer, Heidelberg (2006). https://doi.org/10.1007/11792086_37
25. Zimmermann, P., et al.: GMP-ECM (elliptic curve method for integer factorization) (2012). https://gforge.inria.fr/projects/ecm/

# Almost Injective and Invertible Encodings for Jacobi Quartic Curves

Xiuxiu Li[1,2,3], Wei Yu[1,2,3]([✉]), Kunpeng Wang[1,3], and Luying Li[1,3]

[1] State Key Laboratory of Information Security, Institute of Information Engineering, CAS, Beijing, China
yuwei@iie.ac.cn, yuwei_1_yw@163.com
[2] State Key Laboratory of Cryptology, P. O. Box 5159, Beijing 100878, China
[3] School of Cyberspace Security, University of Chinese Academy of Sciences, Beijing, China

**Abstract.** This paper introduces a novel encoding scheme for hashing values from the finite field $\mathbb{F}_p$ to points on Jacobi quartic curves. These curves possess efficient group law and are immune to timing attacks. The proposed encoding scheme achieves almost injective and invertible mappings of the input values into Jacobi quartic curves. When $p \equiv 3 \mod 4$, our encoding saves $2\mathbf{I} + \mathbf{D} - 8\mathbf{M} - 4\mathbf{S}$ compared to existing methods. This improvement amounts to approximately 50% on average when compared to existing methods. The encoding scheme can be used in a variety of cryptographic applications that rely on elliptic curves, such as identity-based encryption schemes and private set intersection protocols.

**Keywords:** Jacobi quartic curves · Injective and invertible encoding · Inverse map · $B$-well-distributed

## 1 Introduction

Since the introduction of elliptic curves into cryptography by Miller [25] and Koblitz [23], elliptic curve cryptography has become a major branch in the field of cryptography. The group structure of elliptic curves has become a focus of research under the impetus of cryptography. All elliptic curves are considered to have the Weierstrass form, which is parametrized by a cubic equation $y^2 = x^3 + ax + b$. In the real domain, the addition law on a Weierstrass curve can be described by three points where the line intersects the curve, with the unit element point being the infinity point.

To achieve better efficiency in various protocols, many different forms of elliptic curves have been studied in elliptic curve cryptography, including Edwards form, Montgomery form, and Jacobi model. The Jacobi quartic is one of the two Jacobi models. Compared with the Montgomery form and Edwards forms, the extended Jacobi quartic form includes more curves. Billet and Joye [4] showed

Supported by the National Natural Science Foundation of China (No. 62272453, U1936209, 61872442, and 61502487).

© The Author(s), under exclusive license to Springer Nature Switzerland AG 2023
M. Yung et al. (Eds.): SciSec 2023, LNCS 14299, pp. 127–138, 2023.
https://doi.org/10.1007/978-3-031-45933-7_8

that every elliptic curve containing a point of order two could be written as a curve in Jacobi quartic form and provided the birational map between Weierstrass elliptic curves with a point of order two and the Jacobi quartic curves.

In [22], Hisil, Wong, Carter, and Dawson provided doubling formulae on Jacobi quartic curves that involves two field multiplications and five field squarings. According to Bernstein and Lange Explicit-Formulas Database [3], it is one of the fastest doubling formulae without loss of information. Meanwhile, [22] shows that the Jacobi quartic curves are competitive with twisted Edwards curve in variable-singlepoint-variable-single-scalar multiplication. Jacobi quartic curves can also be employed in pairing calculations [13,14,33].

Hashing into elliptic curves is an important procedure that encodes arbitrary values into points on elliptic curves over a finite field. This process is wildly employed in elliptic curve cryptosystems, including password-authenticated key exchanges [7], identity based encryption [5], and Boneh-Lynn-Shacham signatures [6,30].

The "try and increment" method, also referred to as "sample and reject" and "hint and pick", is the first map in hashing into elliptic curves. This encoding involes repeatedly sampling the value of $x$ and testing whether $x$ could be the $x$-coordinates of a point on elliptic curves until a satisfactory $x$ has been found.

Shallue and Woestijne employed Skalba's equality [27] proposed Shallue-van de Woestijne map [26]. Three candidates of the $x$-coordinates were proposed, and at least one of them could be the $x$-coordinates of a point. Ulas [29] and Brier et al. [8] subsequently simplified the Shallue-van de Woestijne map. Their simplifications are referred to as SWU encoding and brief/simplified SWU encoding. Recently Wahby and Boneh further sped up this model and constructed the mapping for the BLS-12 381 curve in CHES2019 [30].

In order to avoid censorship, Bernstein, Hamburg, Krasnova, and Lange proposed Elligator 1 and Elligator 2 encodings [2]. Both Elligator 1 and Elligator 2 are almost injective and invertible maps, with Elligator 2 can be employed on more curves. The injectivity and invertibility of Elligator 2 allow it to make the points indistinguishable from random strings at less cost. The IETF [18] prefers the Elligator 2 encodings over other encodings and has speed up the Elligator 2 encoding on Montgomery curves and Edwards curves.

Boneh and Franklin put forwarded a deterministic mapping for a specific type of supersingular curve over a finite field $\mathbb{F}_p$ with $p = 2 \mod 3$. Icart generalized Boneh and Franklin's method for Weierstrass curves over a finite field $\mathbb{F}_p$ with $p \equiv 2 \mod 3$. The SWU encodings, Elligator encodings, and Icart encoding have been extended and adapted into many other forms in the literature [12,17, 20,21,31,32]. Additionally, there has been significant research on the security of hashing into elliptic curves [1,8,11,16,19,24,28].

The current Jacobi quartic curves encoding and decoding process is inefficient due to the computational expense of the Elligator 2 algorithm used to deterministically encode arbitrary values into Jacobi quartic curves from $\mathbb{F}_p$ with $p \equiv 3 \mod 4$. Additionally, the resulting curve points are not uniformly distributed. To address these issues, this paper proposes a new encoding method that constructs almost-injective and invertible encodings for Jacobi quartic curves. The proposed encoding method is based on Elligator 2 and employs projective coordinates to reduce the number of inversions required in our mapping. An inverse map is provided to ensure that the resulting points are indistinguishable from uniform random strings. In theory, our new encoding method achieved a 50% reduction in time compared to previous square root encoding. This paper provides a solution to the inefficiency of the current Jacobi quartic curve encoding and decoding process, and demonstrates the effectiveness of our proposed method through experimentation.

The paper is organized as follows: Sect. 2 provides necessary background information for encoding, Sect. 3 presents the theorems and proofs about the map and inverse map, Sect. 4 introduces our injective and invertible encoding for Jacobi quartic curves, Sect. 5 compares our encoding's time complexity to previous works, and Sect. 6 concludes this paper.

## 2    Background

Let $K$ be a field of characteristic not equal to 2. The Jacobi quartic curves are elliptic curves of the form:

$$y^2 = (1 - x^2)(1 - k^2 x^2), \qquad k \neq \pm 1.$$

where $k$ is a nonzero field element. Chudnovsky and Chudnovsky [10] introduced a variant of the Jacobi quartic curve in the form

$$y^2 = x^4 + ax^2 + b,$$

which they used to construct inversion-free addition formulas. Billet and Joye further extended the Jacobi quartic form to

$$y^2 = dx^4 + 2ax^2 + 1$$

where $a, d \in K$, $a, d \neq 0$, and $\Delta = 256d(a^2 - d)^2 \neq 0$.

Any elliptic curve in Weierstrass form $E : y^2 = x^3 + ax + b$, that has a point $(\theta, 0)$ of order two, is birational equivalent to the curve $E_{a',d'} : y^2 = d'x^4 + 2a'x^2 + 1$ in Jacobi quartic form, where $d = -(3\theta^2 + 4a)/16$ and $a' = 3\theta/4$. The birational map from $E$ to $E_{a'd'}$ is given by

$$\phi : E \to E_{a',d'}$$

$$(x, y) \mapsto (\frac{x - \theta}{y}, \frac{(2x + \theta)(x - \theta)^2 - y^2}{y^2}), \tag{1}$$

where $(x, y) \neq \mathcal{O}, (\theta, 0)$ and $O$ denotes the point at infinity. $\phi(\mathcal{O}) = (0, 1)$, $\phi(\theta, 0) = (0, -1)$.

**Group Law.** On Jacobi quartic curve, the $(0,1)$ is the identity point, and $(0,-1)$ is a point of order two. The negative of a point $(x,y)$ is $(-x,y)$. Given two points $(x_1,y_1)$ and $(x_2,y_2)$ on the curve $E_{a,d}$, their sum is the point $(x_3,y_3)$ with

$$x_3 = \frac{x_1y_2 + y_1x_2}{1 - dx_1^2x_2^2},$$

$$y_3 = \frac{(y_1y_2 + 2ax_1x_2)(1 + dx_1^2x_2^2) + 2dx_1x_2(x_1^2 + x_2^2)}{1 - dx_1^2x_2^2},$$

where $a$ and $d$ are parameters of the curve. Hisil et al. have also shown that $y_3$ can be alternatively represented as followings:

$$y_3 = \frac{(y_1y_2 - 2ax_1x_2)(x_1^2 + x_2^2) - 2x_1x_2(1 + dx_1^2x_2^2)}{(x_1y_2 - y_1x_2)^2},$$

$$y_3 = \frac{(y_1y_2 + 2ax_1x_2)(x_1y_2 - y_1x_2) + 2(x_2y_2 - x_1y_1)}{(x_1y_2 - y_1x_2)(1 - dx_1^2x_2^2)},$$

and

$$y_3 = \frac{x_1y_1(2 + 2ax_1^2 - y_1^2) - x_2y_2(2 + 2ax_2^2 - y_2^2)}{(x_1y_2 - y_1x_2)(1 - dx_1^2x_2^2)}.$$

## 3    The Map and the Inverse Map

First we employ the Elligator 2 to construct the deterministic encoding from $\mathbb{F}_p$ with $p \equiv 3 \mod 4$ to Jacobi quartic curves. Inspired by [30] and [18], we adopt the projective form to improve the efficiency of the encoding. The encoding process first maps the values to the curve $Y'^2Z' = X'(X'^2 - 4aX'Z' + (4a^2 - 4d)Z'^2)$ over $\mathbb{F}_p$, and then maps them to the Jacobi quartic curve $Y^2Z^2 = dX^4 + 2aX^2Z^2 + Z^4$.

Let $u$ be an element in $\mathbb{F}_p$ for the encoding. Generally, we select $u$ from set

$$S = \{u \in \mathbb{F}_p \mid u \neq 0, \pm 1, 16a^2u^2 \neq 4(a^2 - 4d)(1 + u^2)^2\}.$$

Here $S$ corresponds to the set $R$ given by Theorem 5 in [2]. With this selection, we can derive the following theorem.

**Theorem 1.** *Let $p$ be a odd prime power satisfies $p \equiv 3 \mod 4$. Let $u \in S$, where $S$ is defined above. Let*

$$D = 1 - u^2$$
$$U = 64a^3 - 64a^3D + 16a(a^2 - 4d)D^2$$
$$V = D^3$$
$$R = (UV)(UV^3)^{\frac{p-3}{4}}$$

*If $VR^2 = U$, let*

$$(X',Y',Z') = (-4a, R'D, D),$$

*where $R'$ is the even one in $\{R, -R\}$. Else let*

$$(X', Y', Z') = (-4a(1 - D), R'D, D),$$

*where $R'$ is the odd one in $\{uR, -uR\}$. Then the followings can be obtained:*

*(1) $DUVRX'Y'Z' \neq 0$, $D \neq 0$, $R \neq 0$, $V \neq 0$, $U \neq 0$, $Y' \neq 0$, $Z' \neq 0$, $X' \neq 0$,.*
*(2) $(X' : Y' : Z')$ is a point on curve $Y'^2 Z' = X'(X'^2 - 4aX'Z' + (4a^2 - 4d)Z'^2)$.*
*(3) $(X : Y : Z) = (2Y'Z' : X'^2 - 4(a^2 - d)Z'^2 : X'^2 - 4aX'Z' + (4a^2 - 4d)Z'^2)$*
    *is a point on Jacobi quartic curve $Y^2 Z^2 = dX^4 + 2aX^2 Z^2 + Z^4$.*
*(4) Let $g(x) = x(x^2 - 4ax + 4(a^2 - d))$ and denote $\sqrt{\cdot}$ as the principle square root,*
    *i.e., the even one in the two square roots, then if $X' = 4a$, $R' = \sqrt{g(X'/Z')}$.*
    *If $X' = -4a(1 - D)$, $R' = -\sqrt{g(X'/Z')}$.*

*Proof.* Let $A = -4a$, $B = 4(a^2 - 4d)$. Then

$$4a/D = -A/(1 - u^2) \triangleq v,$$

$$U/V = v^3 + Av^2 + Bv,$$

and

$$\sqrt{U/V} = R.$$

Inserting these expressions in Theorem 5 in [2], choose the principle square root as the even one, then (1) and (2) can be obtained. (3) is derived from (2) and the birational map in [22] §2.3.2. (4) is obvious.

**Theorem 2.** *Let $\varphi = \psi \circ \tau$ be map provided in Theorem 1, where $\tau$ is the map from $S$ to points on curve $Y'^2 Z' = X'(X'^2 - 4aX'Z' + (4a^2 - 4d)Z'^2)$ and $\psi$ is the map from curve $Y'^2 Z' = X'(X'^2 - 4aX'Z' + (4a^2 - 4d)Z'^2)$ to Jacobi quartic curve $Y^2 Z^2 = dX^4 + 2aX^2 Z^2 + Z^4$. Then*

*(1) For any $u \in S$, if $\varphi(u)$ and $\varphi(u')$ denote the same projective point, then $u = \pm u'$.*
*(2) If $(X : Y : Z) \in \varphi(S)$ then the following element $\bar{u}$ of $S$ is defined and $\varphi(\bar{u}) = (X : Y : Z)$:*

$$
\bar{u} = 
\begin{cases}
\sqrt{\dfrac{Z^2 - aX^2 + YZ}{Z^2 + aX^2 + YZ}}, & \text{if } \dfrac{4YZ^2 + 4Z^3 + 4aX^2Z}{X^3} \text{ is even,} \\[2ex]
\sqrt{\dfrac{Z^2 + aX^2 + YZ}{Z^2 - aX^2 + YZ}}, & \text{if } \dfrac{4YZ^2 + 4Z^3 + 4aX^2Z}{X^3} \text{ is odd.}
\end{cases}
\tag{2}
$$

*Proof.* Let $u$, $X'$, $Y'$, $Z'$, $X$, $Y$, and $Z$ be defined as in Theorem 1. (1) By the birational map $\psi$ and its inverse map $\psi'$ given in [22] §2.3.2, $\varphi(u) = \varphi(u')$ follows that $\tau(u) = \tau(u')$. According to Theorem 7 in [2], $u' = \pm u$. (Note that when $u \in S$, $X', Y'$, $Z'$, $X$, $Z$ are not zero.) (2) can be derived by the birational map $\psi'$ and Theorem 7 in [2].

# 4    Hash into Jacobi Quartic Curves

## 4.1    $B$-Well-Distributed Property

Recall the definition of $B$-well-distributed in [15].

**Definition 1** ([15]). *Let $X$ be a smooth projective curve over a finite field $\mathbb{F}_p$, $J$ its Jacobian, $f$ a function $\mathbb{F}_p \to X(\mathbb{F}_p)$ and $B$ a positive constant. We say that $f$ is $B$-well-distributed if for any nontrivial character $\chi$ of $J(\mathbb{F}_p)$, the character sum $S_f(\chi)$ satisfies the following equation:*

$$|S_f(\chi)| \leq B\sqrt{p}.$$

In the following, we introduce the basic theorem for the $B$-well-distributed property.

**Theorem 3 (Theorem 7 in [15]).** *Let $h : \tilde{X} \to X$ be a nonconstant morphism of curves, and $\chi$ be any nontrivial character of $J(\mathbb{F}_p)$, where $J$ is the Jacobian of $X$. Assume that $h$ does not factor through a nontrivial unramified morphism $Z \to X$. Then*

$$\left| \sum_{P \in \tilde{X}(\mathbb{F}_p)} \chi(X(P)) \right| \leq (2\tilde{g} - 2)\sqrt{p}$$

*where $\tilde{g}$ is the genus of $\tilde{X}$. Furthermore, if $p$ is odd and $\varphi$ is a nonconstant rational function on $\tilde{X}$, then*

$$\left| \sum_{P \in \tilde{X}(\mathbb{F}_p)} \chi(X(P)) \left( \frac{\varphi(P)}{p} \right) \right| \leq (2\tilde{g} - 2 + 2 \deg \varphi)\sqrt{p},$$

*where $\left( \frac{\cdot}{\cdot} \right)$ denotes the Legendre symbol.*

**Theorem 4.** *Let $\varphi$ be the encoding defined in Theorem 1, $p \equiv 3 \mod 4$. For any nontrivial character $\chi$ of $E(\mathbb{F}_p)$, the character sum $S_\varphi(\chi)$ satisfies:*

$$|S_\varphi(\chi)| \leq 16\sqrt{p} + 43.$$

*Proof.* Let $S$, $R'$, $D$ be defined as in Sect. 3. Let $\bar{S} = \mathbb{F}_p \backslash S$. Then for any $u \in S$, the following equivalents are established:

$$X' = -4a \Leftrightarrow u^2 - \omega = 0$$

$$X' = -4a(1 - D) \leftrightarrow u^2 - \frac{1}{\omega} = 0$$

where $\omega = (1 + ax^2 - y)/(1 - ax^2 + y)$. The coordinates $x = X/Z$ and $y = Y/Z$ are from equation (2). Let two coverings $h_j : C_j \to E$, $j = 1, 2$ be the smooth projective curves whose function field are the extensions of $\mathbb{F}_p(x, y)$ defined by $u^2 - \omega = 0$ and $u^2 - 1/\omega = 0$. Then the parameter $u$ is a rational function on each of the $C_j$ giving rise to morphisms $g_j : C_j \to \mathbb{P}^1$, such that any point in $\mathbb{A}^1(S)$ has exactly two preimages in $C_j(\mathbb{F}_p)$ for one of $j = 1, 2$, and none in the other. It follows that $h_j$ is ramified if and only if $u = 0$ or $u = \infty$. Hence by Riemann-Hurwitz formula,

$$2g_{C_j} - 2 = 0 + 1 + 1 = 2.$$

Hence curves $C_j$ are of genus 2. Denote the map from $C_j$ to $\mathbb{F}_p$ that maps $P = (u, x, y)$ to $R'$ by $\bar{\varphi}$. We have $\deg \bar{\varphi} = 6$. Let $S_j = h_j^{-1}(\bar{S} \cup \{\infty\})$

$$\left| \sum_{u \in S} \chi(\varphi(u)) \right| = \left| \sum_{\substack{P \in C_0(\mathbb{F}_p) \backslash S_0, \\ \left(\frac{R'}{p}\right) = 1}} \chi(h_0(P)) + \sum_{\substack{P \in C_1(\mathbb{F}_p) \backslash S_1, \\ \left(\frac{R'}{p}\right) = -1}} \chi(h_1(P)) \right|$$

$$\leq \left| \sum_{\substack{P \in C_0(\mathbb{F}_p), \\ \left(\frac{R'}{p}\right) = 1}} (h_0^* \chi)(P) \right| + \left| \sum_{\substack{P \in C_1(\mathbb{F}_p), \\ \left(\frac{R'}{p}\right) = -1}} (h_1^* \chi)(P) \right| + \#S_0 + \#S_1$$

And we have

$$2 \left| \sum_{\substack{P \in C_0(\mathbb{F}_p), \\ \left(\frac{R'}{p}\right) = 1}} (h_0^* \chi)(P) \right| = \left| \sum_{P \in C_0(\mathbb{F}_p)} (h_0^* \chi)(P) + \sum_{P \in C_0(\mathbb{F}_p)} (h_0^* \chi)(P) \cdot \left(\frac{R'}{p}\right) \right.$$

$$\left. - \sum_{\substack{P \in C_0(\mathbb{F}_p), \\ \left(\frac{R'}{p}\right) = 0}} (h_2^* \chi)(P) \right|$$

$$\leq \left| \sum_{P \in C_0(\mathbb{F}_p)} (h_0^* \chi)(P) \right| + \left| \sum_{P \in C_0(\mathbb{F}_p)} (h_0^* \chi)(P) \cdot \left(\frac{R'}{p}\right) \right|$$

$$+ \#\{u \mid R' = 0\}$$

By Theorem 3, we have

$$\left| \sum_{P \in C_j(\mathbb{F}_p)} (h_j^* \chi)(P) \right| \leq (2g_{C_j} - 2)\sqrt{p} = 2\sqrt{p}$$

and

$$\left| \sum_{P \in C_j(\mathbb{F}_p)} (h_j^* \chi)(P) \cdot \left(\frac{R'}{p}\right) \right| \leq (2g_{C_j} - 2 + 2 \deg \bar{\varphi})\sqrt{p} = 14\sqrt{p}$$

Since for all $u \in S$, $R' \neq 0$, and $\#\bar{S} \leq 1+7 = 8$. We have $\#S_j \leq 2(\#\bar{S}+1) \leq 18$. It follows that

$$|S_\varphi(\chi)| = \left| \sum_{u \in S} \chi(\varphi(u)) \right| \leq 16\sqrt{p} + 43$$

## 4.2  Indifferentiable from Random Oracle

According to Theorem 3 in [15], our encoding $\varphi$ described in the previous paragraph is a well-distributed encoding. Furthermore, Corollary 2 in [15] states that if $h_1$ and $h_2$ are two independent random oracle hash functions, then the following construction:

$$H(m) = \varphi(h_1(m)) + \varphi(h_2(m))$$

is indifferentiable from a random oracle.

## 4.3  Points Indistinguishable from Uniform Random Strings

Since our encoding is almost injective and invertible, points on Jacobi quartic curves can be encoded as strings to avoid censorship by the inverse map given in Theorem 2. Based on the $B$-well-bounded property of our encoding, it is easy to prove that our encoding is $(d, B)$-well-bounded. Therefore, the Elligator Square method can be applied to our encoding to make the resulting points indistinguishable from uniform random strings. However, it should be noted that the Elligator Square method is time-consuming. For further details on the $(d, B)$-well-bounded property and the Elligator Square method, please refer to [28].

## 5  Time Complexity

In the rest of this work, we use **I** denotes field inversion, **E** denotes field exponentiation, **M** denotes field multiplication, and **S** denotes field squarings for simplification. Then the cost of our almost-injective and invertible encoding can be summarized as follows:

1. Compute $u^2$ require one **S**, and it is enough for $D$.
2. Compute $D^2$ and $V = D^3$ require $\mathbf{M} + \mathbf{S}$.
3. 2**M** are required in the computation of $U$ since $64a^3$ and $16a(a^2 - 4d)$ can be pre-computed.
4. Computing $R$ as $R = (U \cdot V) \cdot ((UV) \cdot V^2)^{(q-3)/4}$ costs $\mathbf{E} + 3\mathbf{M} + \mathbf{S}$
5. Checking whether $VR^2 = U$ costs $\mathbf{M} + \mathbf{S}$
6. $(X', Y', Z')$ can be computed within 3**M**.
7. Computing $X'^2, Z'^2, 2X'Z' = (X' + Z')^2 - X'^2 - Z'^2$. And then compute $(X, Y, Z)$ by $Y', Z', X'^2, Z'^2, 2X'Z'$ and pre-computed values $4a$ and $4(a^2 - d)$. This procedure costs $3\mathbf{M} + 3\mathbf{S}$ in total to obtain $X, Y$ and $Z$.

To sum up, $\varphi$ costs $\mathbf{E} + \mathbf{13M} + \mathbf{7S}$. And the inverse map $\varphi^{-1}$ can be computed as follows:

1. Computing $Z^2$, $X^2$, $X^3$, $aX^2$ and $YZ$ in $\mathbf{3M} + \mathbf{2S}$.
2. Employing Montgomery's technique compute the inversion $s = 1/(X^3(Z^2 + aX^2 + YZ)(Z^2 - aX^2 + YZ))$ by $\mathbf{I} + \mathbf{2M}$.
3. Using $\mathbf{3M} + \mathbf{S}$ to check the parity of $4YZ^2 + 4aX^2Z + 4Z^3/X^3 = 4Z(Z^2 + aX^2 + YZ)^2(Z^2 - aX^2 + YZ)s$.
4. If the parity is even, let $(U, V) = (Z^2 + aX^2 + YZ, Z^2 - aX^2 + YZ)$, and else let $(U, V) = (Z^2 - aX^2 + YZ, Z^2 + aX^2 + YZ)$. This step needs no cost.
5. $\bar{u} = \sqrt{U/V} = (UV)(UV^3)^{(p-3)/4}$ is obtained in $\mathbf{E} + \mathbf{3M} + \mathbf{S}$.

Let $f_A$ denote the encoding proposed by Alasha [1], $f_{YS}$ and $f_{YI}$ denotes the encoding proposed by Yu et al. [31], which are based on brief SWU encoding and Icart encoding respectively. Table 1 shows the theoretical time complexity of these encodings compared with ours. Specifically, when the finite field $\mathbb{F}_p$ satisfying $p \equiv 3 \mod 4$, our encoding $\varphi$ saves $2\mathbf{I} + \mathbf{D} - \mathbf{8M} - \mathbf{4S}$ compared to $f_{YS}$. According to Bernstein and Lange Explicit-Formulas Database [3], if the ratio $\mathbf{I/M} = 100$, our encoding on Jacobi quartic curves is more than 50% faster than $f_{YS}$ when $p \equiv 3 \mod 4$.

**Table 1.** Theoretical time cost of different encodings on Jacobi quartic curves

| Encodings | Field condition | Costs |
|---|---|---|
| $f_A$ | $p \equiv 2 \mod 3$ | $\mathbf{E} + \mathbf{2I} + \mathbf{8M} + \mathbf{3S}$ |
| $f_{YI}$ | $p \equiv 2 \mod 3$ | $\mathbf{E} + \mathbf{I} + \mathbf{9M} + \mathbf{5S}$ |
| $f_{YS}$ | $p \equiv 3 \mod 4$ | $\mathbf{E} + \mathbf{2I} + \mathbf{5M} + \mathbf{3S} + \mathbf{D}$ |
| $\varphi$ | $p \equiv 3 \mod 4$ | $\mathbf{E} + \mathbf{13M} + \mathbf{7S}$ |

To compare the efficiency of our encoding and $f_{YS}$, both running on the finite field $\mathbb{F}_p$ with $p \equiv 3 \mod 4$, we conducted experiments using SageMath for big number arithmetic. The experiments were performed on a 12th Gen Intel(R) Core(TM) i7-12700H 2.30 GHz processor, with each encoding running 1,000,000 times, where $u$ was randomly chosen on $\mathbb{F}_{P256}$ and $\mathbb{F}_{P384}$. The primes $P256$ and $P384$ were selected as the NIST primes [9]. The experiments results are presented in Table 2.

**Table 2.** Time cost ($\mu$s) comparison on $\mathbb{F}_{P256}$ and $\mathbb{F}_{P384}$

| Encodings | P256 | P384 |
|---|---|---|
| $f_{YS}$ | 87 | 152 |
| $\varphi$(ours) | 45 | 75 |

According to above experimental results, our encoding is 48.3% faster than $f_{YS}$ on field $\mathbb{F}_{P256}$ and 50.7% faster on field $\mathbb{F}_{P384}$. The experimental results are consistent with the previous theoretical results.

## 6   Conclusion

This paper presents an almost-injective and invertible encoding scheme for Jacobi quartic curves using Elligator 2 encoding and projective coordinates. The proposed encoding reduces the number of inversions required for mapping, resulting in a faster algorithm compared to previous square root encoding techniques. The inverse map is also provided to ensure that the encoded points are indistinguishable from uniform random strings. Our results show that the proposed encoding technique outperforms previous methods by reducing computation time by approximately 50%. Additionally, the decoding of points on elliptic curves into finite fields is also addressed in this paper.

## References

1. Alasha, T.: Constant-time encoding points on elliptic curve of different forms over finite fields (2012)
2. Bernstein, D., Hamburg, M., Krasnova, A., Lange, T.: Elligator: Elliptic-curve points indistinguishable from uniform random strings, pp. 967–980 (2013). https://doi.org/10.1145/2508859.2516734
3. Bernstein, D., Lange, T.: Explicit-formulas database (2020). http://hyperelliptic.org/EFD/
4. Billet, O., Joye, M.: The Jacobi model of an elliptic curve and side-channel analysis. In: Fossorier, M., Høholdt, T., Poli, A. (eds.) AAECC 2003. LNCS, vol. 2643, pp. 34–42. Springer, Heidelberg (2003). https://doi.org/10.1007/3-540-44828-4_5
5. Boneh, D., Franklin, M.: Identity-based encryption from the weil pairing. In: Kilian, J. (ed.) CRYPTO 2001. LNCS, vol. 2139, pp. 213–229. Springer, Heidelberg (2001). https://doi.org/10.1007/3-540-44647-8_13
6. Boneh, D., Lynn, B., Shacham, H.: Short signatures from the weil pairing. In: Boyd, C. (ed.) ASIACRYPT 2001. LNCS, vol. 2248, pp. 514–532. Springer, Heidelberg (2001). https://doi.org/10.1007/3-540-45682-1_30
7. Boyko, V., MacKenzie, P., Patel, S.: Provably secure password-authenticated key exchange using Diffie-Hellman. In: Preneel, B. (ed.) EUROCRYPT 2000. LNCS, vol. 1807, pp. 156–171. Springer, Heidelberg (2000). https://doi.org/10.1007/3-540-45539-6_12
8. Brier, E., Coron, J.-S., Icart, T., Madore, D., Randriam, H., Tibouchi, M.: Efficient indifferentiable hashing into ordinary elliptic curves. In: Rabin, T. (ed.) CRYPTO 2010. LNCS, vol. 6223, pp. 237–254. Springer, Heidelberg (2010). https://doi.org/10.1007/978-3-642-14623-7_13
9. Chen, L., Moody, D., Regenscheid, A., Randall, K.: Draft nist special publication 800-186 recommendations for discrete logarithm-based cryptography: elliptic curve domain parameters. Technical report, National Institute of Standards and Technology (2019)

10. Chudnovsky, D., Chudnovsky, G.: Sequences of numbers generated by addition in formal groups and new primality and factorization tests. Adv. Appl. Math. **7**(4), 385–434 (1986). https://doi.org/10.1016/0196-8858(86)90023-0
11. Chávez-Saab, J., Rodríguez-Henrquez, F., Tibouchi, M.: SwiftEC: Shallue-van de Woestijne indifferentiable function to elliptic curves (2022). https://eprint.iacr.org/2022/759
12. Diarra, N., Sow, D., Khlil, A.Y.O.C.: On indifferentiable deterministic hashing into elliptic curves. Eur. J. Pure Appl. Math. **10**, 363–391 (2017)
13. Doss, S., Kaondera-Shava, R.: An optimal Tate pairing computation using Jacobi quartic elliptic curves. J. Comb. Optim. **35**(4), 1086–1103 (2018). https://doi.org/10.1007/s10878-018-0257-y
14. Duquesne, S., Fouotsa, E.: Tate pairing computation on Jacobi's elliptic curves. In: Abdalla, M., Lange, T. (eds.) Pairing 2012. LNCS, vol. 7708, pp. 254–269. Springer, Heidelberg (2013). https://doi.org/10.1007/978-3-642-36334-4_17
15. Farashahi, R., Fouque, P.A., Shparlinski, I., Tibouchi, M., Voloch, J.: Indifferentiable deterministic hashing to elliptic and hyperelliptic curves. IACR Cryptol. ePrint Arch. **2010**, 539 (2010). https://doi.org/10.1090/S0025-5718-2012-02606-8
16. Farashahi, R.R., Shparlinski, I.E., Voloch, J.F.: On hashing into elliptic curves. J. Math. Cryptol. **3**(4), 353–360 (2009)
17. Farashahi, R.R.: Hashing into hessian curves. In: Nitaj, A., Pointcheval, D. (eds.) AFRICACRYPT 2011. LNCS, vol. 6737, pp. 278–289. Springer, Heidelberg (2011). https://doi.org/10.1007/978-3-642-21969-6_17
18. Faz-Hernández, A., Scott, S., Sullivan, N., Wahby, R.S., Wood, C.A.: Hashing to elliptic curves. Internet-Draft draft-irtf-cfrg-hash-to-curve-13, Internet Engineering Task Force (2021). https://datatracker.ietf.org/doc/html/draft-irtf-cfrg-hash-to-curve-13
19. Fouque, P.-A., Tibouchi, M.: Estimating the size of the image of deterministic hash functions to elliptic curves. In: Abdalla, M., Barreto, P.S.L.M. (eds.) LATIN-CRYPT 2010. LNCS, vol. 6212, pp. 81–91. Springer, Heidelberg (2010). https://doi.org/10.1007/978-3-642-14712-8_5
20. Fouque, P.-A., Tibouchi, M.: Indifferentiable hashing to Barreto–Naehrig curves. In: Hevia, A., Neven, G. (eds.) LATINCRYPT 2012. LNCS, vol. 7533, pp. 1–17. Springer, Heidelberg (2012). https://doi.org/10.1007/978-3-642-33481-8_1
21. He, X., Yu, W., Wang, K.: Hashing into generalized huff curves. In: Lin, D., Wang, X.F., Yung, M. (eds.) Inscrypt 2015. LNCS, vol. 9589, pp. 22–44. Springer, Cham (2016). https://doi.org/10.1007/978-3-319-38898-4_2
22. Hisil, H., Wong, K.K.-H., Carter, G., Dawson, E.: Jacobi quartic curves revisited. In: Boyd, C., González Nieto, J. (eds.) ACISP 2009. LNCS, vol. 5594, pp. 452–468. Springer, Heidelberg (2009). https://doi.org/10.1007/978-3-642-02620-1_31
23. Koblitz, N.: Elliptic curve cryptosystems. Math. Comput. **48**, 203–209 (1987)
24. Koshelev, D.: Indifferentiable hashing to ordinary elliptic $F\_q$-curves of $j = 0$ with the cost of one exponentiation in $F\_q$. Designs Codes Cryptogr. **90** (2022). https://doi.org/10.1007/s10623-022-01012-8
25. Miller, V.S.: Use of elliptic curves in cryptography. In: Williams, H.C. (ed.) CRYPTO 1985. LNCS, vol. 218, pp. 417–426. Springer, Heidelberg (1986). https://doi.org/10.1007/3-540-39799-X_31
26. Shallue, A., van de Woestijne, C.E.: Construction of rational points on elliptic curves over finite fields. In: Hess, F., Pauli, S., Pohst, M. (eds.) ANTS 2006. LNCS, vol. 4076, pp. 510–524. Springer, Heidelberg (2006). https://doi.org/10.1007/11792086_36

27. Skałba, M.: Points on elliptic curves over finite fields. Acta Arith. **117**(3), 293–301 (2005)
28. Tibouchi, M.: Elligator squared: uniform points on elliptic curves of prime order as uniform random strings. In: Christin, N., Safavi-Naini, R. (eds.) FC 2014. LNCS, vol. 8437, pp. 139–156. Springer, Heidelberg (2014). https://doi.org/10.1007/978-3-662-45472-5_10
29. Ulas, M.: Rational points on certain hyperelliptic curves over finite fields. arXiv Number Theory (2007)
30. Wahby, R.S., Boneh, D.: Fast and simple constant-time hashing to the BLS12-381 elliptic curve. IACR Trans. Cryptogr. Hardw. Embed. Syst. **2019**, 154–179 (2019)
31. Yu, W., Wang, K., Li, B., He, X., Tian, S.: Hashing into Jacobi quartic curves. In: Lopez, J., Mitchell, C.J. (eds.) ISC 2015. LNCS, vol. 9290, pp. 355–375. Springer, Cham (2015). https://doi.org/10.1007/978-3-319-23318-5_20
32. Yu, W., Wang, K., Li, B., He, X., Tian, S.: Deterministic encoding into twisted Edwards curves. In: Liu, J.K., Steinfeld, R. (eds.) ACISP 2016. LNCS, vol. 9723, pp. 285–297. Springer, Cham (2016). https://doi.org/10.1007/978-3-319-40367-0_18
33. Zhang, F., Li, L., Wu, H.: Faster pairing computation on Jacobi quartic curves with high-degree twists. In: Yung, M., Zhu, L., Yang, Y. (eds.) INTRUST 2014. LNCS, vol. 9473, pp. 310–327. Springer, Cham (2015). https://doi.org/10.1007/978-3-319-27998-5_20

# SeeStar: An Efficient Starlink Asset Detection Framework

Linkang Zhang[1,2,3], Yunyang Qin[1,2,3], Yujia Zhu[1,2,3(✉)], Yifei Cheng[1,2,3],
Zhen Jie[1,2], and Qingyun Liu[1,2,3]

[1] Institute of Information Engineering, Chinese Academy of Sciences, Beijing, China
zhuyujia@iie.ac.cn
[2] National Engineering Research Center of Information Security, Beijing, China
[3] School of Cyber Security, University of Chinese Academy of Sciences,
Beijing, China

**Abstract.** Starlink is a new communication network architecture that uses thousands of low-orbiting satellites to provide high-speed, low-latency Internet services. However, there is still much information about Starlink that has not been disclosed to the public. The details of Starlink network architecture, and key nodes which are important to deeply understand and evaluate the performance, security, and impact of Starlink, etc. are still not known. In this paper, we propose an efficient Starlink asset detection framework based on active detection, passive detection, and non-intrusive search engine-based detection methods for the effective discovery and identification of Starlink assets. Based on this framework, this paper implements SeeStar, a Starlink asset mapping system, and provides a detailed analysis of Starlink ground stations and key nodes, exploring their roles and characteristics in the network. Finally, this paper provides an aggregated analysis of Starlink assets in terms of device and service dimensions, and attempts to evaluate their security. The work in this paper provides a powerful methodology and system to unravel the mystery of Starlink network.

**Keywords:** Starlink · Satellite Internet · Network Asset · LEO · Detection · Mapping

## 1 Introduction

Starlink [1] is a LEO (Low Earth Orbit) satellite Internet communication system built by SpaceX (Space Exploration Technologies Corp.) in recent years, aiming to provide high-speed, low-latency, and highly stable Internet services through the deployment of low-orbit broadband satellites with global coverage. As of March 2023, Starlink has more than 3,800 satellites in orbit [2], providing Internet services to 50 countries and regions, and currently has more than 1 million subscribers [3,4], with plans to deploy 12,000 satellites and eventually expand to 42,000 [5].

As a project to provide Internet services using space technology, Starlink has attracted global attention since its launch in 2015. Starlink uses an innovative

© The Author(s), under exclusive license to Springer Nature Switzerland AG 2023
M. Yung et al. (Eds.): SciSec 2023, LNCS 14299, pp. 139–156, 2023.
https://doi.org/10.1007/978-3-031-45933-7_9

network system that assigns dedicated IP addresses and protocols to satellites and hardware devices. In addition, Starlink has installed laser cross-link technology on the satellites, changing the traditional satellite communication model. This technology allows satellites to transmit and forward data to each other, reducing reliance on ground stations. Starlink also uses a new P2P network protocol and end-to-end hardware encryption technology that outperforms conventional Internet technologies in terms of security and prevents data theft or cracking.

Although Starlink has launched a large number of satellites and provided services to some countries and regions, the status of its assets is not well known. Starlink assets are familiarly known as satellite operations, and known TLE files [6] and websites publish the status of space segment assets. However, as an Internet service provider, little is known about its ground segment network assets. In order to deeply analyze and understand the characteristics and advantages of the Starlink network, its ground segment network assets need to be effectively detected and identified. At present, some cyberspace search engines [7–11] have started to include Starlink network assets, but because their search scope is too broad and lacks targeting, and there is no unified Starlink network asset identification standard and dynamic update mechanism, the data they provide often have poor timeliness, much noise, and low accuracy. To address these problems, this paper proposes a Starlink asset detection method that integrates active detection, passive detection, and non-intrusive search engine-based detection methods, aiming to improve the efficiency and accuracy of Starlink asset detection.

Specifically, the main contributions of this paper are as follows:

1. Propose **the first efficient and targeted Starlink asset detection framework** based on the Starlink network architecture and open data sources, and integrate a heuristic algorithm for Starlink asset detection.
2. Deploy **Starlink Asset Mapping System called SeeStar**, which enables continuous dynamic detection, Starlink ground station discovery and critical node classification and mapping of Starlink IPv4 and some IPv6 assets.
3. Obtain Starlink asset data and perform aggregation analysis to **refine the characteristics of Starlink assets**, and can **infer Starlink key assets** based on Starlink asset attributes and **analyze its security** based on the characteristics of assets.

The rest of the paper is organized as follows: Sect. 2 reviews related work. Section 3 introduces the starlink network architecture and definition of starlink asset. Section 4 describes in detail the starlink asset detection framework. Evaluations are presented in Sect. 5. Finally, we make final remarks in Sect. 6 and Sect. 7.

## 2    Related Works

Satellite Internet mainly relies on the space satellite constellation to achieve seamless global Internet connection and provide broadband Internet access to users anytime and anywhere, which is the new generation of Internet infrastructure and is the inevitable trend of future network infrastructure development. The academic and industrial communities are also increasingly interested in studying the satellite Internet represented by the Starlink network.

In the study of Starlink, Michel et al. [12] summarized the analysis of Starlink performance conducted by researchers using active measurement methods, as well as their evaluation of performance under load and packet loss using QUIC. M. M. Kassem et al. [13] utilized a web browser extension to measure Starlink connectivity performance, aiming to answer questions such as how Starlink connectivity compares to other ISPs in the same geographic region, whether connection quality changes over time, and whether weather affects performance. Stock et al. [14] discusses the use of distributed on-demand routing for LEO mega-constellations, using the Starlink case study as an example. Ma et al. [15] uses experiments and observations to study the network characteristics and performance of Starlink, the largest LSN constellation, with a focus on end-to-end user experiences.

Another research aspect related to our work is asset detection, Feng et al. [16] proposes a scalable framework for profiling physical devices on the Internet, using network reconnaissance and banner grabbing to extract device information. Meidan et al. [17] describes the development of a multi-stage meta classifier that utilizes machine learning algorithms to accurately identify and classify IoT devices based on network traffic data, which was collected and labeled from a heterogeneous set of devices. Leonard et al. [18] summarizes the authors' use of the IRLscanner tool to perform 21 Internet-wide experiments for service discovery, and analyzes feedback generated while suggesting novel approaches for reducing blowback.

Currently, mainstream search engines such as ZoomEye [7] have scanned and collected some Starlink asset data, but their exploration is too broad to be comprehensive or dynamic. Due to Starlink's continuous development and changing IP allocation and assets, identifying its assets is inconsistent, leading to suboptimal data accuracy and timeliness. We propose a combination of exploration methods based on active detection, passive detection and search engines, along with continuous exploration using publicly available data sources, for comprehensive and timely identification of Starlink assets with high accuracy.

## 3   Starlink Architecture and Assets Definition

As a satellite constellation system, Starlink uses a constellation of LEO satellites designed to provide high-speed Internet service to rural and remote areas where Internet connectivity is unreliable or non-existent, ultimately achieving global Internet coverage. The Starlink system is divided into three main parts: the user segment, the space segment, and the ground segment. The Starlink system architecture is shown in Fig. 1.

The user segment belongs to the user intranet. This segment contains mainly user devices and Starlink terminals. The wireless router is used to provide Wi-Fi

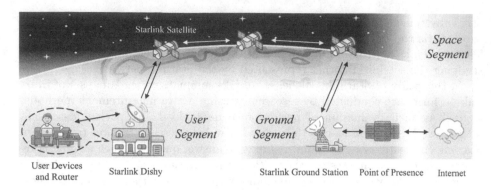

**Fig. 1.** Starlink System Architecture

signals for user devices to access the local LAN empowered by Starlink. The satellite receiver with phased-array antennas is used to track Starlink satellite signals in near-Earth orbit.

The space segment is part of the inter-satellite network. In this segment, the components communicate and relay via a proprietary, new hardware encryption technology that has not been officially disclosed by Starlink, rather than the traditional TCP/IP protocol stack. Because of the extremely limited public data available for this segment, the lack of clarity on its specific technical details, and the insufficient work at the protocol level, it is currently not possible to probe this segment through existing mapping techniques, and electromagnetic or signal analysis may serve as an entry point for subsequent research.

The ground segment contains the ground gateway station, the point of presence, and the data center network. The ground station is used to communicate with the satellite and send data read from the satellite to the point of presence. The data center is where Internet service providers aggregate their networks and share bandwidth. In the Starlink network, the ground gateway station does not directly access the Internet but instead accesses a nearby point of presence via fiber optics to reach the data center, which provides Internet access.

In short, in the Starlink network, user devices send data packets to Starlink user terminals through the local LAN, and the user terminals encode the packets and send the data to the LEO satellite through the uplink. After the inter-satellite transmission, the LEO satellite sends the data to a specific ground gateway station through the downlink according to the pre-designed routing algorithm, and the ground gateway station further processes the data and transmits the data to the point of presence through the optical fiber. Finally, the data enter the Internet through the data center.

In this paper, according to the Starlink network architecture, Starlink assets are divided into three parts according to function and location: user segment assets, space segment assets and ground segment assets. User segment assets mainly include Starlink terminals and other devices used by users. Space segment assets mainly include Starlink satellite network, sensors and other devices, and

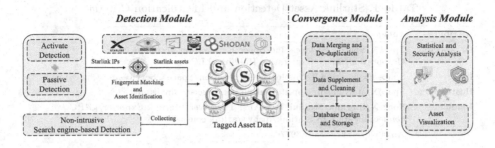

**Fig. 2.** Starlink Asset Detection System Architecture

ground segment assets mainly include Starlink ground stations and PoP(point of presence) and other key nodes. This paper argues that because the Starlink network uses a new communication protocol and data transmission method different from the traditional TCP/IP protocol stack, as well as its fewer public data sources, it is difficult for user segment and space segment assets to be effectively detected and mapped. Therefore, the research in this paper **focuses on the detection of assets in ground segment**. The public information of Starlink's key nodes and the locations and parameters of ground stations are obtained by collecting and collating public data sources. Then, active detection, passive detection, and non-intrusive search engine-based detection are carried out on top of this to obtain more information about Starlink assets. Finally, this paper constructs the Starlink asset detection framework and deploys the Starlink asset mapping system to achieve the detection and mapping of Starlink ground segment assets.

## 4  Starlink Asset Detection Framework

In this paper, we propose an efficient Starlink asset detection framework. The framework is divided into three modules: detection, convergence and analysis. The architecture of Starlink asset detection framework is shown in Fig. 2.

### 4.1  Deteciton

**Detection Criteria.** At present, there exists a huge amount of network assets on the Internet, and it is a very significant challenge to accurately obtain the IPs associated with Starlink and identify their assets from the large-scale network assets. In order to improve the accuracy and efficiency of Starlink asset identification, we develop the following Starlink asset identification criteria to achieve accurate identification by using Starlink official data [1], Whois database [19], BGP database [20] and other diversified intelligence. For the collected network assets, we verify whether the relevant attributes in the whois database satisfy the rules shown in Table 1, and any one of them will be considered as a candidate Starlink-associated asset.

**Table 1.** Starlink Asset Detection and Identification Criteria

| Attribute | Value |
| --- | --- |
| Organization | SpaceX/SpaceX Services, Inc/Space Exploration Technologies Corporation/Space Exploration Holdings LLC |
| ISP | Space Exploration Technologies Corporation/SpaceX Services,Inc/Spacex.com |
| Hostname | SpaceX/starlinkisp/.pop.starlinkisp.net./.mc.starlinkisp.net |
| Banner | SpaceX/Starlink/Starlink GW AP FTP server/Starlink Exporter/gRPC connection state to Starlink/dishy.starlink.com |
| AS | AS14593/AS27277/AS397763 |

**Detection Methods.** This section focuses on the introduction of Starlink asset detection methods. The detection section mainly utilizes active detection, passive detection and non-intrusive search engine based detection methods, aiming to enhance the ability of Starlink asset data acquisition and improve the validity and accuracy of Starlink asset data by using multi-source data and multi-dimensional methods.

**Active Detection.** Active Detection enables the acquisition of Starlink asset information by performing alive scanning and port scanning to the destination host. The input of the active detection is a list of network IP addresses and the output is the IP addresses associated with Starlink. The main function of this module is to extract the IP addresses associated with Starlink from the Internet IP address space for subsequent acquisition of Starlink-associated asset data. Specifically, the network IP addresses are probed in an active way using a combination of ICMP, TCP, and other types of probing packets, combined with Zmap [21] and Nmap [22]. The Whois information corresponding to the IP addresses contains information about the segment name, the organization to which it belongs, etc. Therefore, the network IP addresses are scanned to collect the data. By scanning the IP addresses of the whole network and collecting the surviving IP addresses, combined with the information in the Whois database, the IP addresses associated with Starlink can be mined based on the asset identification criteria above accurately and efficiently. Then the Starlink-associated IP addresses are labeled to build the Starlink asset-associated IP dataset. Finally, active detection identifiers are added to each piece of data. In addition, as a supplement, combining DNS data, autonomous system information, and public information, it is also possible to mine Starlink-associated IP addresses with certain characteristics among them. The rDNS [23] can be obtained by scanning the global IP address space for rDNS probes. If the domain name text in the detection result contains Starlink-associated keywords such as starlinkisp, the corresponding IP is considered a candidate Starlink-associated IP address. Since the current global IP address space is unusually large and inefficient scanning cannot accomplish the goal of fast scanning, a distributed scanning method is used to efficiently scan global IP addresses based on global scanning nodes and using a distributed architecture.

**Passive Detection.** Passive Detection uses network sniffing tools to get the traffic, analyze the IP and other data in it, and combine the IP corresponding

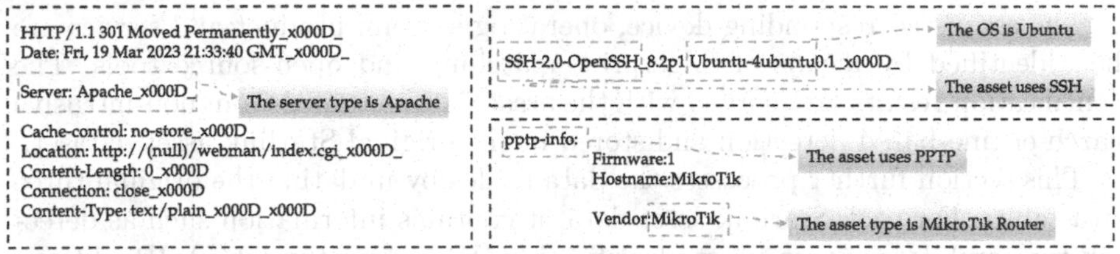

HTTP/1.1 301 Moved Permanently_x000D_
Date: Fri, 19 Mar 2023 21:33:40 GMT_x000D_

Server: Apache_x000D_    ··· The server type is Apache

Cache-control: no-store_x000D_
Location: http://(null)/webman/index.cgi_x000D_
Content-Length: 0_x000D_
Connection: close_x000D_
Content-Type: text/plain_x000D_x000D_D

······ The OS is Ubuntu
SSH-2.0-OpenSSH_8.2p1 Ubuntu-4ubuntu0.1_x000D_

····· The asset uses SSH

pptp-info:
    Firmware:1 ··········· The asset uses PPTP
    Hostname:MikroTik

    Vendor:MikroTik ····· The asset type is MikroTik Router

**Fig. 3.** Identifying assets based on banner

domain name information and Whois information to get Starlink assets. The input of the Passive Detection is the traffic on the Internet (which may involve Starlink-associated IP addresses) obtained by using network sniffing tools, and the output is the Starlink-associated IP addresses. For the IPs in the collected traffic, we first query their Whois information to determine whether their domain names, organization names, etc. match the characteristics of Starlink assets, such as the domain name is customer.*.starlinkisp.net or the organization name is SpaceX, etc. The Starlink-associated IP addresses are also labeled and incorporated into the Starlink asset-associated IP dataset, and a passive probe identifier is added to each piece of data.

**Non-intrusive Search Engine-Based Detection.** At present, the mainstream cyberspace search engines have included a portion of Starlink asset data, however, due to the detection cycle and other factors, these data may have poor timeliness and other problems. In order to get a more comprehensive overview of Starlink assets, a non-intrusive detection method based on cyberspace search engines is used to obtain Starlink asset data to expand the data obtained by active and passive detection. The input of the search engine-based non-intrusive detection module is a search statement designed based on asset identification criteria, and the output is a list of Starlink-associated assets. Relying on the current relatively mature cyberspace search engines Zoomeye [7], Shodan [8], Quake [9], fofa [10], and Censys [11], such cyberspace search engines are earlier developed and more mature in technology, and usually provide information such as IP, port number, geographic location information, service type, device type, and product type. Therefore, the Starlink asset detection and identification criteria developed above are used to construct search statements that conform to the syntax of each search engine. The Starlink-associated asset data is retrieved by the APIs provided by each search engine, and each piece of data is included in the Starlink asset-associated IP data set, adding a non-intrusive search engine-based detection marker and tagging its data source.

**Asset Identification Methods.** For the list of Starlink-associated IP addresses obtained, a list of candidate scanning ports is designed based on the search engine results, combined with experience. We use TCP protocol to connect to the specified port of the target IP and print out the information returned from the port to get the banner shown in Fig. 3.

The asset's corresponding device, operating system, product and service are then identified by means of fingerprint matching and open-source tools. The identification results are merged with the asset data obtained from non-intrusive search engine-based detection and stored in the original Starlink asset dataset.

This section further processes the data tables by updating the original three data tables. For each Starlink asset data, it contains information such as detection type and data source, asset identification (IP and port number), IP address, port number, hostname, autonomous system number, service, device, operating system, product, latitude, longitude, country and banner and other raw information.

---

**Algorithm 1:**

---

**Input:** $I, T, R, P$
**Output:** Starlink Assets Table $A$

1  IPs← Randomly Rearrange(IPs);
2  **for** $IP$ in $I$ **do**
3  |   **if** IsAlive($IP$) **then**
4  |   |   **for** $r$ in $R$ **do**
5  |   |   |   **if** RuleTest($IP, r$) **then**
6  |   |   |   |   Store((Active, $IP$), IP$_{SL}$);
7  |   |   |   **end**
8  |   |   **end**
9  |   **end**
10 **end**
11 **for** $t$ in $T$ **do**
12 |   $IP$←ExtractIP($t$) **for** $r$ in $R$ **do**
13 |   |   **if** RuleTest($IP, r$) **then**
14 |   |   |   Store((Passive, $IP$), IP$_{SL}$);
15 |   |   **end**
16 |   **end**
17 **end**
18 **for** $IP$ in $IP_{SL}$ **do**
19 |   **for** $p$ in $P$ **do**
20 |   |   $b \leftarrow$ ExtractBanner($IP, p$);
21 |   |   **if** FingerprintMatch($b$) **then**
22 |   |   |   Store((Type, $IP, p$, Details), $A$);
23 |   |   **else**
24 |   |   |   run Open Source Tools and store the results into $A$;
25 |   |   **end**
26 |   **end**
27 **end**
28 **for** $r$ in $R$ **do**
29 |   $Query$←GenerateQuery($r$);
30 |   $Res$←SearchUsingEngine($Query$);
31 |   Store($Res, A$);
32 **end**

---

**Heuristic Algorithm.** Based on the description of the Starlink asset detection framework, we have summarized a related heuristic algorithm shown in Algorithm 1.

The algorithm is mainly used for Starlink asset detection. The input of this algorithm is the list of IP addresses to be probed $I$ (because the IPv6 address space is too large, the list of IP addresses to be probed in this algorithm description is a subset of the IPv4 address space), the passive traffic data $T$, the Starlink asset discrimination rules $R$ and the list of port numbers to be scanned $P$. The output is Starlink asset data table $A$.

In order to avoid intensive scanning of the same network segment, which may cause the defense mechanism of the target segment and thus affect the detection process, the list of IPs to be scanned is first randomly rearranged. If the IP is alive, the IP is identified as a Starlink-associated IP based on a predefined Starlink asset identification criteria (e.g., domain name information, autonomous system information, organization information, and other keywords), and if it passes the identification policy, the IP address is stored in the $IPSL$ list with an active detection identifier. If the identification policy is passed, the IP address is stored in the $IPSL$ list and the active detection mark is attached.

For the collected passive traffic data $T$, extract the IP address from each piece of information and determine whether it is a Starlink-associated IP as described above, and if it passes the identification policy, store the IP address in the $IPSL$ list and attach a passive detection mark. The type of service/request of Starlink-associated IPs and the port number and protocol they use may be present in the passive traffic, and this part of data is saved to help in the subsequent acquisition of asset data.

For each IP in $IPSL$, each port in the port list $P$ to be scanned is scanned, the banner returned by the port is extracted and fingerprinted for matching, and if there is matching information, the asset data is stored in the asset data table A in the form of (probe type (active/passive), IP address, port number, asset details).

Next, Starlink asset discrimination rules are used to generate a series of cyberspace search engine query statements and retrieve them, and store each piece of information in the returned results in the form of (probe type (search engine), IP address, port number, asset details) in the asset data table $A$.

## 4.2   Convergence

This section focuses on merging and de-duplicating Starlink asset data obtained through active detection, passive detection, and non-intrusive search engine-based detection methods. The data is extracted, verified, and cleaned according to Starlink asset characteristics. And then we implement database design and data storage. Data extraction and verification make the same object data from different sources conform to a unified form and provide guarantees for the fusion of multiple heterogeneous data. Data cleaning can improve the credibility and validity of the subsequent data fusion analysis results.

**Data Merging and De-duplication.** Starlink assets obtained through active and passive detection and non-intrusive search engine-based detection methods may have duplicate data, so this part of the assets needs to be merged and de-duplicated to form the final Starlink asset dataset.

**Data Supplementation.** For Starlink asset data from search engines, there may be missing asset data, or there may be cases where the original alive assets no longer exist. Therefore, this data needs to be supplemented. Firstly, we use alive scanning to screen the alive assets, and then we use a combination of Nmap and traditional fingerprint matching methods to identify the services and operating systems of assets, etc.

**Data Cleaning.** Preliminary analysis of the acquired asset data reveals that there are cases of different descriptions of the same operating system, service, product, and device information in Starlink asset information due to user-defined or vendor differences. In addition, there are cases of inconsistent country names (e.g. using country Chinese names, country English names, country codes, etc.) and inconsistent default values, so the data cleaning rules were developed to facilitate statistics and analysis.

– Attribute naming unification (eg. Ubuntu/ubuntu/UBUNTU)
– Empty value replacement (eg. convert unknown/UNKNOWN to null value)
– Attribute case conversion (eg. convert UBUNTU/Ubuntu to ubuntu)
– Conversion of country names (eg. convert US to United States)
– IP asset expansion (eg. unroll the list of IP addresses)
– Data correction (eg. modify the incorrect data such as ASN)
– Same asset data conflict resolution(eg. Assets change over time, and different conflicting methods may yield different results depending on the time of data collection).

### 4.3  Analysis

This section focuses on the in-depth analysis of Starlink asset data, which mainly involves the following three aspects: screening of Starlink asset features, aggregation and security assessment of asset data, and visual display of asset data.

**Screening of Starlink Asset Features.** This aspect mainly involves extracting Starlink asset data from the Starlink asset database according to certain criteria such as operating system, product, device type, and so on, and extracting Starlink asset data that meet the conditions. For example, if you need to analyze the data of all assets whose product type is camera, the system will return a table containing the data of all assets whose product type is camera for subsequent analysis.

**Asset Data Aggregation and Security Assessment.** This aspect is mainly to aggregate and analyze the asset data obtained according to specific asset characteristics in different dimensions such as open ports, operating systems, device types, etc., and generate corresponding statistical charts. At the same

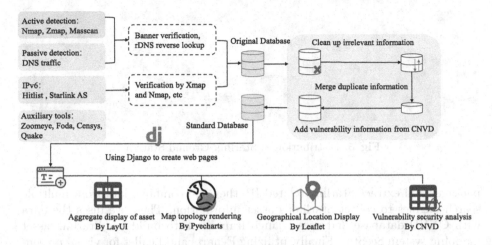

**Fig. 4.** Starlink System Implementation

time, the vulnerability knowledge base such as CNVD (National Information Security Vulnerability Sharing Platform) [24] is used for aggregation analysis, and the historical vulnerabilities and current unpatched vulnerabilities of asset data are queried according to their operating systems and product types to achieve security assessment of asset data.

**Visualization Display of Asset Data.** This aspect mainly utilizes some visualization libraries such as Echarts and Leaflet to present the results of convergence analysis and security assessment in a more intuitive and visual way, such as through charts and maps to show the overall situation and distribution characteristics of Starlink's asset data.

## 5   Evaluation

### 5.1   Implementation

This paper proposes a Starlink asset detection framework and implements a mapping system SeeStar that can automatically detect, identify and analyze Starlink assets based on this framework. The system mainly consists of detection, aggregation and analysis from the bottom up.

As mentioned above, the overall system operation is divided into three parts, namely detection, aggregation and analysis: in the detection stage, it is mainly active and passive detection to obtain asset information and use tools for verification; in the aggregation stage, it is mainly data processing and integration; in the analysis stage, it is mainly statistics and presentation of assets from various dimensions. The main technologies and the overall architecture flow used in each segment are shown in Fig. 4. In this paper, we adopt an active detection method, based on public data sources and open source tools, to send detection

**Fig. 5.** Distribution of Starlink Ground Station

packages and extract Starlink-related IP. then, we combine data from multiple search engines to collect Starlink's asset information. Then, we fuses the data with CNVD database and uses Django framework to build our Starlink asset mapping system SeeStar. Finally, utilizing Echarts and Leaflet for visual presentation of the data.

### 5.2   Application

**Ground Station Discovery and Critical Node Classification.** Based on the public data from FCC [25], Google Map [26], etc., a total of 211 ground stations are collected and analyzed, which are distributed in 27 countries. We mapped the geographic locations of the ground stations onto the map and finally obtained their global distribution which is shown in Fig. 5.

We conducted an in-depth analysis of the IP addresses of the Starlink network, revealing the structure and characteristics of the Starlink network in terms of autonomous systems, host names and network architecture. In this paper, we first count the autonomous systems to which Starlink-associated IP addresses belong, and find that these IP addresses are mainly distributed in three autonomous systems, AS14593, AS27277 and AS397763 [27]. Then, this paper performs reverse domain name resolution on these IP addresses to obtain their host names, and classifies and identifies the nodes of Starlink network according to the naming rules of host names. This paper finds that there are mainly the following types of nodes in the Starlink network:

- **PoP node:** This is an important node in the Starlink network, which is located near the ground station, communicates with the satellite, and connects to the Internet core to provide Internet access services for Starlink users. the PoP node can also interconnect with other PoP nodes or MC nodes to achieve interconnection and redundancy of the network.
- **MC node:** This is another important node in the Starlink network, which is responsible for coordinating data transmission between satellites, earth stations and user terminals, as well as assigning IP addresses and subnet masks, etc. The MC node is the core component of the Starlink user access network.

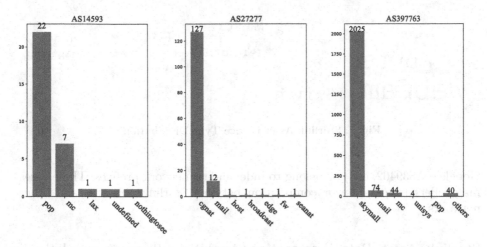

**Fig. 6.** Distribution of Hostname in AS14593, AS27277, AS397763

- **CGNAT node:** This is a node in the Starlink network used to solve the IPv4 address exhaustion problem and protect the security of user terminals. CGNAT node can realize the conversion between private IP addresses and public IP addresses, thus supporting two-way communication.

In this paper, we analyzed the types of host names in the three autonomous systems, as shown in Fig. 6, and came to the following conclusions:

- **AS14593** is the most important autonomous system in the Starlink network, which contains a large number of PoP nodes and MC nodes, which are key components of Starlink in providing satellite Internet access services and managing network devices. Among these nodes, the number of PoP nodes is significantly more than the number of MC nodes. The highest percentage of assets is also found in AS14593 in the detection results.
- **AS27277** consists mainly of CGNAT nodes, which are used to solve the IPv4 address exhaustion problem and to protect user terminal security. In addition, this autonomous system contains a small number of mail services and special nodes, which may be related to other functions of Starlink.
- **AS397763** has a large number of hostnames that are not obviously associated with Starlink, such as hostnames containing keywords like skymall, mail, unisys, etc. These hostnames may be historical legacies of this autonomous system or domain names planned for future use. In addition, there are a certain number of MC nodes and a small number of PoP nodes in the autonomous system, so it is presumed that the autonomous system has not been fully utilized by Starlink and may be in the construction or testing stage.

Finally, this paper also found that, in addition to AS14593, AS27277 and AS397763 provided by public data sources, there are cases that Starlink associated IPs are attached to other autonomous systems (e.g., AT&T's AS7018,

**Fig. 7.** Starlink Asset Device Type Distribution

Google's AS39462, etc.) or belong to independent network prefixes. These cases may reflect cooperation or competition between Starlink networks and other networks.

**Starlink Asset Data Aggregation Analytics.** Based on our deployed SeeStar, we have accumulated 10,188 Starlink-associated IPs and 23,132 asset data. We analyzed Starlink asset data specifically and found that all asset data involved 2073 open ports, 219 open services, 22 operating systems. Then we provide an in-depth analysis of the device types in Starlink assets and finds that Starlink assets contain 40 different device types, with firewalls and webcams as the main ones. These device types reflect the current trend of network security and IoT, and also reveal the main uses and functions of Starlink assets. Figure 7 shows the specific distribution of device types.

This paper identifies and counts the device products in Starlink assets, and finds that Starlink assets mainly use products from some well-known companies, such as Amerest, Hikvision, foscam, Dahua and other webcam products, and Sonic WALL, Fortinet, pfSense, and FortiGate and other firewall products. The features and performance of these products can help us further understand the management and configuration strategies of Starlink assets, as well as their potential attack surfaces and vulnerabilities. In order to understand the vulnerability of Starlink assets, this paper integrates CNVD based on the device and product data of Starlink assets to obtain the vulnerabilities that have existed in Starlink assets and those that have not been fixed.

In addition, this paper also counts the number of open ports in the IP addresses of Starlink assets and finds that there are large differences in the number of open ports in the IP addresses of Starlink assets, which may be related to their roles and importance in the Starlink network architecture. We believe that the higher the number of open ports, the more functional and utilized the IP address is, so the importance of Starlink IP addresses can be assessed based on the number of open ports.

Taking an IP shown in Fig. 8 with the highest number of open ports as an example, the IP 129.222.251.*** is located in the United States and its host name is customer.*.pop.starlinkisp.net, so we judge that this host exists as a PoP node by collecting data coming from Internet satellites from ground stations and importing it into the terrestrial Internet. The high number of open ports

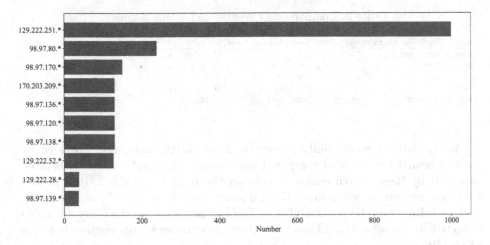

**Fig. 8.** The top 10 IPs of the detected port openings

indicates that it has more types of services, reflecting that this IP is in a more important position in Starlink's network architecture.

**System Accuracy and Proof of Advantage.** The purpose of this section is to evaluate the advantages of the system in terms of both accuracy and breadth of data and timeliness of data.

In terms of data breadth and accuracy, this paper adopts three methods: active detection, passive detection and non-intrusive detection based on search engines, and conducts in-depth analysis for Starlink IP allocation, which significantly improves the data breadth through multi-source data fusion. At the same time, this paper uses the Whois database and other Starlink asset features to conduct a comprehensive detection and identification of the entire network IP address space, which also improves the accuracy of the data. In order to verify the detection effect of this paper, the detection results and data volume are compared with the Starlink asset data included in major search engines in this paper, as shown in the table. From the table, we can see that the asset detection framework in this paper can obtain more Starlink asset data more effectively and accurately.

In terms of data timeliness, SeeStar built in this paper achieves real-time updates of Starlink assets across the network by continuously detecting and using Starlink IP to assign public data sources, Whois information and other Starlink asset features. In contrast, the conventional cyberspace search engine is a broad detection of all kinds of assets on the whole network and does not focus on Starlink assets, so it is more lenient in the development of detection criteria and will miss a large number of Starlink assets, and because the search scope of the cyberspace search engine is larger and the operation cycle is longer, the timeliness of the assets it obtains is lower compared with this system, and the system includes the proportion of surviving IPs is higher.

**Table 2.** Comparison of SeeStar and search engines

| System | IPs | Surviving IPs | Survivability | Assets | Surviving Assets | Survivability | Service | Device | OS | Vulnerabilities | Whois |
|---|---|---|---|---|---|---|---|---|---|---|---|
| ZoomEye | 7388 | 442 | 5.98% | 13777 | 924 | 6.70% | • | • | • | • | • |
| FOFA | 3279 | 455 | 13.8% | 8045 | 1325 | 16.4% | • | • | • | ○ | ○ |
| Quake | 3255 | 388 | 11.92% | 9172 | 782 | 8.52% | • | • | • | ○ | ○ |
| CenSys | 1696 | 354 | 20.87% | 3506 | 680 | 19.39% | • | • | • | ○ | • |
| **SeeStar** | **10188** | **3015** | **29.59%** | **23132** | **7317** | **31.63%** | • | • | • | • | • |

We performed survivability probes on Starlink IPs and assets provided by various search engines and compared our system (SeeStar) results with those provided by these search engines to obtain the results in Table 2. It is proved that our system provides more IP and asset data in terms of data accuracy and breadth. Both IP survivability (29.59%) and asset survivability (31.63%) detected by our system are higher compared to existing search engines in terms of timeliness.

## 6    Discussion

This paper proposes an efficient Starlink asset detection framework that combines active detection, passive detection and non-intrusive search engine based detection techniques, aiming to acquire Starlink asset data efficiently and accurately.

In this paper, a Starlink asset mapping system SeeStar is implemented using this framework to map Starlink ground stations and key nodes, achieve ground station discovery and key node classification, and perform aggregation and analysis of Starlink asset data. The detection scheme in this paper has high accuracy because it combines publicly available data sources assigned by Starlink IP, uses multiple detection methods, and uses a combination of fingerprint matching and open source tools.

The work in this paper also has some limitations. On the one hand, this paper has a relatively limited understanding of Starlink IP assignments. Since there is a time lag between the release of Starlink public data sources and the allocation and use of IPs, and the Starlink IP allocation is more complex, there are some IPs attached to other autonomous systems or independent network prefixes in addition to the three publicly available autonomous systems. On the other hand, due to the oversized IPv6 address space, this paper does not explore IPv6 assets in depth, although the public data sources and existing studies indicate the existence of IPv6 assets in Starlink.

## 7    Conclusion

In this paper, we propose a Starlink asset detection framework that combines active detection, passive detection, and non-intrusive search engine based detection methods, and implement a Starlink asset mapping system SeeStar based on this framework.

The system implements ground station discovery and critical node classification. In terms of ground stations, the study finds that the distribution of ground stations matches the countries and regions where Starlink has officially released its services. By analyzing Starlink IP, we found that Starlink IP distribution is not only limited to AS14593, AS27277 and AS397763, but also a small portion is attached to other autonomous systems or affiliated with independent network prefixes. In terms of key nodes, the study found that Starlink key nodes are mainly classified into three categories: PoP, MC, and CGNAT, in addition to other special nodes and unknown nodes. In addition, we specifically analyzed Starlink asset data, aggregating and analyzing Starlink assets from different dimensions including service, device, operating system, and product. Finally, we compare the performance difference between the proposed system and existing search engines and demonstrate that the system outperforms them in terms of accuracy, breadth and effectiveness. In the future, we will further fuse data from multiple sources, expand Starlink IP allocation intelligence to increase the amount of Starlink asset data, and combine IPv6 address space prediction algorithms to more completely probe Starlink IPv6 assets. We will also further optimize the detection strategy by adjusting the detection range and detection period to achieve more effective and efficient detection.

**Acknowledgments.** This work is supported by the Scaling Program of Institute of Information Engineering, CAS (Grant No. E3Z0191101) and the Strategic Priority Research Program of the Chinese Academy of Sciences with No. XDC02030400.

# References

1. Starlink. https://www.starlink.com/
2. McDowell, J.: Starlink Launch Statistics. Planet4589 (2022). https://planet4589.org/space/con/star/stats.html. Accessed 18 Dec 2022
3. SpaceX [@SpaceX]. Starlink now has more than 1,000,000 active subscribers (Tweet) (2022). https://twitter.com/SpaceX/status/1604872936976154624. Accessed 13 Mar 2023
4. Starlink Internet Review 2023: Plans, Pricing, and Speeds. https://www.satelliteinternet.com/providers/starlink/
5. Launches. https://www.spacex.com/launches/
6. CelesTrak: NORAD Two-Line Element Set Format. https://celestrak.org/NORAD/documentation/tle-fmt.php
7. Zoomeye. https://www.zoomeye.org/
8. What is Shodan? - Shodan Help Center. Shodan. https://help.shodan.io/the-basics/what-is-shodan. Accessed 11 Nov 2021
9. Quake. https://quake.360.net/quake/#/index
10. FOFA. https://fofa.info/
11. Censys. https://search.censys.io/
12. Michel, F., Trevisan, M., Giordano, D., Bonaventure, O.: A first look at starlink performance. In: 22nd ACM Internet Measurement Conference (IMC 2022), pp. 130–136. Association for Computing Machinery, New York (2022)

13. Kassem, M.M., Raman, A., Perino, D.: A browser-side view of starlink connectivity. In: 22nd ACM Internet Measurement Conference (IMC 2022), pp. 151–158. Association for Computing Machinery, New York (2022)
14. Stock, G., Fraire, J.A., Hermanns, H.: Distributed on-demand routing for LEO mega-constellations: a starlink case study. In: 2022 11th Advanced Satellite Multimedia Systems Conference and the 17th Signal Processing for Space Communications Workshop (ASMS/SPSC), Graz, Austria, pp. 1–8. IEEE (2022). https://doi.org/10.1109/ASMS/SPSC55670.2022.9914716
15. Ma, S., Chou, Y.C., Zhao, H., Chen, L., Ma, X., Liu, J.: Network characteristics of LEO satellite constellations: a starlink-based measurement from end users. arXiv (2022). http://arxiv.org/abs/2212.13697. Accessed 22 Apr 2023
16. Feng, X., et al.: Active profiling of physical devices at internet scale. In: 2016 25th International Conference on Computer Communication and Networks (ICCCN), Waikoloa, HI, USA, pp. 1–9. IEEE (2016). https://doi.org/10.1109/ICCCN.2016.7568486
17. Meidan, Y., et al.: ProfilIoT: a machine learning approach for IoT device identification based on network traffic analysis. In: Proceedings of the Symposium on Applied Computing, Marrakech Morocco, pp. 506–509. ACM (2017) https://doi.org/10.1145/3019612.3019878
18. Leonard, D., Loguinov, D.: Demystifying internet-wide service discovery. IEEE/ACM Trans. Netw. **21**(6), 1760–1773 (2013). https://doi.org/10.1109/TNET.2012.2231434
19. ASN/IP Whois Query–IPIP.NET. https://whois.ipip.net/
20. BGP.Tools. https://bgp.tools/
21. Durumeric, Z., Wustrow, E., Halderman, J.A.: ZMap: fast internet-wide scanning and its security applications (2013)
22. Lyon, G.F.: NMap Network Scanning: The Official NMap Project Guide to Network Discovery and Security Scanning (2009)
23. rDNS. https://en.wikipedia.org/wiki/Reverse_DNS_lookup
24. National Information Security Vulnerability Sharing Platform. https://www.cnvd.org.cn/
25. Satellite Earth Station: License. https://fcc.report/IBFS/Filing-List/SES-LIC
26. Starlink Global Gateways & PoPs. https://www.google.com/maps/d/viewer?mid=1805q6rlePY4WZd8QMOaNe2BqAgFkYBY&hl=en_US&ll=47.6144489%2C-122.33867770000002&z=8
27. Starlink AS. https://whois.ipip.net/search/SPACEX

# Privacy-Enhanced Anonymous and Deniable Post-quantum X3DH

Kaiming Chen, Atsuko Miyaji$^{(\boxtimes)}$, and Yuntao Wang

Osaka University, 2-1 Yamadaoka, Suita, Osaka, Japan
{kaiming,miyaji,wang}@cy2sec.comm.eng.osaka-u.ac.jp

**Abstract.** End-to-end encryption (E2EE) is widely used in instant messaging applications to protect data privacy. Forward secrecy (FS) and post-compromised security (PCS) are two essential features that aim to protect security when the session keys are compromised. Among E2EE applications, Signal is known for being the first one that guarantees FS and PCS concurrently by implementing the extended triple Diffie-Hellman (X3DH) protocol and double ratchet protocol. However, the original X3DH and double ratchet protocols cannot resist quantum attacks and require post-quantum implementation. While a post-quantum double ratchet protocol has been proposed, the issues of post-quantum X3DH protocols persist. Some post-quantum X3DH protocols are claimed to be anonymous and deniable. However, their anonymity only protects the communication content, not the identity key that can be distinguished. Additionally, their identity certificates must be delivered through a trusted channel during authentication. If these certificates are considered evidence, their deniability will be broken. To address these problems, we propose a solution that leverages ephemeral keys to hide the identity keys for enhancing anonymity. The identity is automatically authenticated without the trusted channel to exclude evidence for deniability.

**Keywords:** Anonymity · Deniability · End-to-end Encryption · Authenticated Key Exchange · Ring Learning with Errors

## 1 Introduction

### 1.1 Signal Protocol

Authenticated key exchange (AKE) is a cornerstone of the internet. It allows both parties in network communication to confirm each other's identity and share the same key to ensure security. Instant messaging is an essential application scenario for AKE. The number of daily active users is estimated to exceed one billion [3]. Together, these users produce an immeasurable amount of data, and leakage of these data can have disastrous consequences. To solve this problem, end-to-end encryption (E2EE) is proposed [17], wherein with the exception of the communicating parties, no other third parties can obtain the communication content or identity information. Signal is a well-known E2EE application [18]. Its core components, i.e., the extended triple Diffie-Hellman (X3DH) and double ratchet protocols, are widely used in various applications, such as WhatsApp, Facebook Messenger, and Signal itself, as well as research on group protocols [8, 10].

© The Author(s), under exclusive license to Springer Nature Switzerland AG 2023
M. Yung et al. (Eds.): SciSec 2023, LNCS 14299, pp. 157–177, 2023.
https://doi.org/10.1007/978-3-031-45933-7_10

In addition, because the current X3DH is always run on mobile phones, a hardware security module (HSM) can be applied to protect the long-term keys. An HSM is a machine that stores the long-term private key and makes it difficult to steal. It can only be used to execute specific algorithms on the input parameters.

However, with the emergence of quantum computers and quantum algorithms, AKE based on DH and integer factorization assumptions will no longer be secure. It is necessary to design post-quantum X3DH and double ratchet schemes because their original versions are vulnerable to quantum attacks. At present, a post-quantum double ratchet construction can resist quantum attacks [1]. Nonetheless, some features of current post-quantum X3DH (PQ-X3DH) protocols [5,6,13,15], such as anonymity and deniability, require improvement.

## 1.2 Privacy-Enhanced Anonymity and Deniability

Anonymity is important for preventing identity information from being leaked. Deniability allows users to deny that they used to send messages or participate in a conversation, but they can still authenticate each other [20]. The anonymity of the current PQ-X3DH protocols is weak, as only the communication data is anonymous while that of data containing long-term public keys is not protected. Attackers can verify the identity data to determine the owner of the anonymous data and break the anonymity. Additionally, the deniability of the current PQ-X3DH protocols is easy to break. Because the authentication of these PQ-X3DH protocols should be performed through a trusted channel, e.g., scanning the QR codes (certificates) face-to-face, authenticated QR codes provide irrefutable evidence that one has communicated with another offline, which compromises the deniability.

Here, we take X3DH and Signal as examples. The user account in Signal is related to the phone number. The authentication of two users is performed through QR codes. Before generating this QR code, each user should send at least one message to the other user's account. Then, the users can see the QR codes from each other. Their phone numbers are correct (authenticated) if they have the same QR code. Assuming that the message server is malicious, it can distinguish the users' identities according to user accounts, which breaks the anonymity. If two users have the same QR code that is reported to a judge, the judge can decide that they have exchanged messages, which reduces the deniability. Thus, we propose a PQ-X3DH protocol with enhanced anonymity and deniability for privacy protection.

## 1.3 Reconciliation

At present, two methods are adopted for quantum-resistant key exchange. The first one is based on the key encapsulation mechanism (KEM), which is considered a black box [5,15]. The second one involves specific assumptions like supersingular isogeny Diffie-Hellman (SIDH) [13], learning with errors (LWE) [12], and ring learning with errors (RLWE) [11]. Because protocols based on LWE and RLWE can embed all secrets into the shared values, Zhang et al. [21] proposed a scheme based on RLWE that achieves the idea of the HMQV protocol [16], where no signature is necessary. A protocol without signatures can claim that it is deniable because the judge cannot determine whether a message is from the sender, owing to the lack of evidence. Therefore, we follow the idea of Zhang et al. [21] to construct a PQ-X3DH protocol.

Most RLWE protocols have reconciliation structures [11,12] that eliminate the effect of small errors for the shared secrets. Consider the following scenario: let $\mathbf{a}$ be a uniformly random public polynomial; $\mathbf{x}$ and $\mathbf{e}$ are small polynomials sampled from an error distribution, $A$ is the initiator Alice, and $B$ is the responder Bob. Given $\mathbf{b}_A := \mathbf{a}\mathbf{x}_A + \mathbf{e}_B$ and $\mathbf{b}_B := \mathbf{a}\mathbf{x}_B + \mathbf{e}_B$, let $sk_{AB} := \mathbf{b}_A\mathbf{x}_B$ and $sk_{BA} := \mathbf{b}_B\mathbf{x}_A$. It can be observed that $sk_{AB} \neq sk_{BA}$ because of the small error polynomials. To obtain the correct shared secrets, we need a function to hint at $sk_{AB}$ and $sk_B A$ and a reconciliation function to derive these secrets. The output of the hint function is computationally independent of the hint function input [12].

## 1.4  Related Work

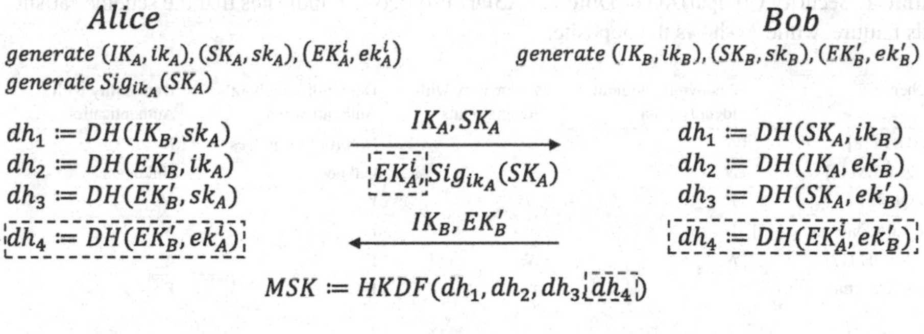

Alice                                                                                      Bob

$generate\ (IK_A, ik_A), (SK_A, sk_A), (EK_A^i, ek_A^i)$        $generate\ (IK_B, ik_B), (SK_B, sk_B), (EK_B', ek_B')$
$generate\ Sig_{ik_A}(SK_A)$

$dh_1 := DH(IK_B, sk_A)$             $IK_A, SK_A$         $dh_1 := DH(SK_A, ik_B)$
$dh_2 := DH(EK_B', ik_A)$    $\overline{EK_A^i, Sig_{ik_A}(SK_A)}$    $dh_2 := DH(IK_A, ek_B')$
$dh_3 := DH(EK_B', sk_A)$                      $dh_3 := DH(SK_A, ek_B')$
$\overline{dh_4 := DH(EK_B', ek_A')}$     $IK_B, EK_B'$      $\overline{dh_4 := DH(EK_A^i, ek_B')}$

$$MSK := HKDF(dh_1, dh_2, dh_3, \overline{dh_4})$$

**Fig. 1.** X3DH Protocol. The dotted box means to be optional. This protocol generates at least three DH values, and thus it is called the extended triple Diffie-Hellman key exchange, X3DH. HKDF is a hash key-derived function. $MSK$ is the shared secret of Alice and Bob.

**X3DH.** The flows of X3DH are shown in Fig. 1. During the execution, each user should generate $(pk, sk)$ pairs, where $pk$ represents the public key and $sk$ represents the private key, including a long-term key $(IK, ik)$ that represents the user identity and cannot be changed, a signed pre-key $(SK, sk)$ that can be used several times, and a set of optional ephemeral keys $(EK, ek)$. The intention of $SK$ and $EK$ is to ensure perfect forward secrecy (PFS) [9], which means that a long-term private key $ik$ being compromised will not lead to the disclosure of the previously shared secret $MSK$ and session keys. $SK$ aims to ensure that if there is no $EK$, the communication is still available. The signature of $SK$ is to avoid the impersonation attack. If the $sk$ and $ik$ of one user are all compromised, PFS will hold if there is $ek$. X3DH is responder obliviousness and asynchronous. From Fig. 1, we can see that Alice does not need to know Bob's identity in the first flow. This feature is called responder obliviousness. The asynchrony indicates that Bob can use the shared key directly without waiting for a response from Alice.

**KEM-Based Protocols.** Post-quantum AKEs based on KEM [4] work as follows: Alice forms a public and private key pair $(pk, sk)$ and sends $pk$ to Bob. Bob runs the encapsulation function $Enc$ on Alice's $pk$ to obtain ciphertext and a related secret $k$.

He then sends the ciphertext to Alice. Alice decrypts the ciphertext using $sk$ to obtain the secret $k$. To achieve the PFS, the Signal-Conforming [15] and SPQR [5] protocols require two or three KEMs whose ciphertexts and $pks$ should be signed. The shared secrets are the derived KEM keys, which will be used to generate the session keys. Thus, if one KEM secret is compromised, the attackers cannot obtain the session keys.

SI-X3DH [13] is an X3DH-like protocol based on SIDH [14]. It is almost identical to the original X3DH but adds an FO transform and zero-knowledge proof of long-term keys to resist key-reuse attacks so that if the long-term key is fixed, the attackers cannot recover the private key. However, its security against an attack on SIDH needs to be reconsidered [7].

## 1.5  Our Contribution

**Table 1.** Security Comparison of Different X3DH Protocols. $Y$ indicates that the scheme satisfies this feature, while $N$ shows the opposite.

| Scheme | Anonymity without Identity Data | Anonymity with Identity Data | Deniability without Authentication | Deniability with Authentication |
|---|---|---|---|---|
| X3DH [18] | N | N | No proof for judges | N |
| Split KEM [6] | N | N | Unknown | Unknown |
| SPQR [5] | Y | N | Y | N |
| Signal-Conforming [15] | Y | N | Y | N |
| SI-X3DH [13] | N | N | Y | N |
| B-X3DH (this work) | Y | Y | Y | Y |

In this paper, we propose a PQ-X3DH protocol Blind-X3DH (B-X3DH) based on RLWE assumptions. A comparison of the X3DH and PQ-X3DH protocols is presented in Table 1.

*HSM Security and PFS*: To catch the scenario that the private keys are protected in the HSM, we set up a new model based on the one proposed in [15] by using oracles to access the HSM. In our scheme, the long-term private key is stored in the HSM and can only be accessed through reconciliation functions. The difference between HSM security and PFS is that PFS allows the adversary to reveal the private keys, whereas HSM offers only the reconciliation oracle and thus less information. PCS is beyond the scope of this study because it is available in post-quantum double ratchet protocols.

*Anonymity*: We redefine a notion of anonymity, in which attackers are allowed to observe and control all transcripts, including the identity data (the long-term public key), and should attempt to distinguish the transcripts from different users. In our anonymity model, the protocols [5,6,13,15] are not secure, because they do not hide their identity data and can be recognized.

*Deniability*: We follow the definition of deniability presented in [5], except that the adversary can query the authentication data. In our deniability model, a simulator attempts to generate the transcripts with knowledge of the long-term private keys. The adversary should distinguish the transcripts with the authentication data from the simulator and honest users. In our deniability model, the protocols [5,6,13,15] are not secure, because their authentication data are the evidence for breaking deniability.

*B-X3DH*: In our scheme, there are three different public key and private key pairs: the long-term keys whose private key is fixed in an HSM, the blind keys, and the ephemeral keys. The long-term public key is encrypted by the blind public key to be the blinded ciphertext. The blinded ciphertext, the ephemeral public key of one user, and all private keys of another are the input of the reconciliation function in HSM to generate the shared secrets. These secrets are used to encrypt authenticated data and other messages. We prove that B-X3DH is correct, anonymous, deniable, and has HSM security and PFS.

**Outline of Our Paper**: In Sect. 2, we introduce the preliminaries. In Sect. 3, we set up the models of HSM security, PFS, anonymity, and deniability. Section 4 presents the construction of B-X3DH. In Sect. 5, we prove the correctness and security of B-X3DH. In Sect. 6, we conclude the paper and discuss the limitations of the study, along with future research.

## 2 Preliminaries

### 2.1 Notations

Let $a = b$ be that $a$ is equal to $b$, and $a := b$ assign $b$ to $a$. $[n]$ is the set $\{1, 2, 3, ..., n-1, n\}$, $\mathcal{M}$ is the metadata space and $\mathcal{K}$ is the key space. Let $\mathbb{Z}$ be the field of all integers, $\mathbb{Z}_q := \{i \bmod q | i \in \mathbb{Z}\}$ the clinic group where $q$ is a prime number, and $\mathbb{Z}^n$ the field of all n dimension integer vectors module $q$, we shift the $\mathbb{Z}_q$ to be $\{-\frac{q-1}{2}, ..., \frac{q-1}{2}\}$. $\mathbb{R}_q := \mathbb{Z}_q[x]/f_m(x)$ is the ring of integer polynomials in which $f_m(x) := x^n + 1$ is the $m$-th cyclotomic polynomial and $n := m/2$. Let $\mathbf{a} := a_0 + a_1 x + a_2 x^2 + ... + a_{n-1} x^{n-1} \in \mathbb{R}_q$ be the polynomial that is presented by its coefficients vector $\mathbf{a} := (a_0, a_1, a_2, ..., a_{n-1})$, we respectively define the addition and multiplication of $\mathbf{a}, \mathbf{b} \in \mathbb{R}_q$ as $\mathbf{a} + \mathbf{b} \bmod (q, x^n + 1)$ and $\mathbf{a} \cdot \mathbf{b} \bmod (q, x^n + 1)$ and omit the $\bmod (q, x^n + 1)$ part. For any real number vector $\mathbf{r}$ and a positive number $\sigma$, $\rho_{\sigma,\mathbf{r}}(\mathbf{x}) := e^{-\frac{\pi \|\mathbf{x}-\mathbf{r}\|^2}{\sigma^2}}$ is the Gaussian function. Let $S$ be the subset of $\mathbb{Z}^n$, $\rho_{\sigma,\mathbf{r}} := \sum_{x \in S} \rho_{\sigma,\mathbf{r}}(x)$ the discrete integral of $\rho_{\sigma,\mathbf{r}}$ over $S$, and $\mathcal{D}_{S,\sigma,\mathbf{r}}$ the discrete Gaussian distribution over $S$. For any $\mathbf{x} \in S$, there is $\mathcal{D}_{S,\sigma,\mathbf{r}}(\mathbf{x}) := \frac{\rho_{\sigma,\mathbf{r}}(\mathbf{x})}{\rho_{\sigma,\mathbf{r}}(S)}$.

Here, $S$ is fixed to $\mathbb{Z}^n$ and $\mathbf{r}$ is set to be $\mathbf{0}$. For short, $\mathcal{D}_{\mathbb{Z}^n,\sigma} := \mathcal{D}_{\mathbb{Z}^n,\sigma,\mathbf{0}}$. Let $\xleftarrow{\$} \mathcal{D}$ be the sampling from a distribution $\mathcal{D}$, $\|\cdot\|_\infty$ the $l_\infty$ norm, and $\|\cdot\|_2$ the $l_2$ norm. The following two lemmas are required:

**Lemma 1.** *[19]. For $\sigma > 0, r \geq \frac{1}{\sqrt{2\pi}}$, if $x \xleftarrow{\$} \mathcal{D}_{\mathbb{Z}^n,\sigma}$, then, $\|x\|_2 \leq r\sigma\sqrt{n}$ except a negligible rate $\varepsilon_{rlwe} := \Pr(\|x\|_2 \geq r\sigma\sqrt{n}; x \xleftarrow{\$} \mathcal{D}_{\mathbb{Z}^n,\sigma}) \leq (\sqrt{2\pi e r^2} e^{-\pi r^2})^n$.*

**Lemma 2.** *[11]. For $a, b \in \mathbb{R}_q$, $\|a \cdot b\|_\infty \leq \|a\|_2 \cdot \|b\|_2$.*

### 2.2 Ring Learning with Errors

All adversaries $\mathcal{A}$ in this paper are quantum probabilistic polynomial time (QPT). Let $q$ be a prime, $\mathcal{D}_{\mathbb{Z}^n,\sigma}$ be the discrete Gaussian distribution over $\mathbb{Z}^n$, $\mathbf{a} \xleftarrow{\$} \mathbb{R}_q$ be a random public parameter. Let $x \in R_q$, $e \xleftarrow{\$} \mathcal{D}_{\mathbb{Z}^n, \sigma}$ where $x$ is a secret and $e$ is an error polynomial. The following assumptions hold for any QPT adversaries:

- **Definition 1.** *Standard Search RLWE: Given a list of* $(a, b := a \cdot x + e \in R_q)$, *the adversary reveals* $x$ *only in a negligible rate.*
- **Definition 2.** *Standard Decision RLWE: Given a list of* $(a, b_b)$, $b \in \{0, 1\}$ *where* $b_0 \xleftarrow{\$} R_q$, $b_1 \leftarrow a \cdot x + e_i \in R_q$, *the adversary cannot distinguish* $b_0$ *and* $b_1$ *except a negligible rate.*

| $Game^{\mathcal{A}}_{sRLWE}(pp)$ | $Game^{\mathcal{A}}_{dRLWE}(pp)$ | $Oracle\ O_1$ |
|---|---|---|
| 1: $a \xleftarrow{\$} R_q$ | 1: $b \xleftarrow{\$} \{0, 1\}$ | 1: $e \xleftarrow{\$} \mathcal{D}_{\mathbb{Z}^n, \sigma}$ |
| 2: $x \xleftarrow{\$} \mathcal{D}_{\mathbb{Z}^n, \sigma}$ | 2: $a \xleftarrow{\$} R_q$ | 2: **return** $b \leftarrow a \cdot x + 2e$ |
| 3: $x' \leftarrow \mathcal{A}^{O_1}(pp, a)$ | 3: $x \xleftarrow{\$} \mathcal{D}_{\mathbb{Z}^n, \sigma}$ | $Oracle\ O_2$ |
| 4: **return** $x = x'$ | 4: $b' \leftarrow \mathcal{A}^{O_2}(pp, a)$ | 1: **if** $b = 0$ : **return** $b_0 \xleftarrow{\$} R_q$ |
| | 5: **return** $b = b'$ | 2: **otherwise** : **return** $O_1$ |

**Fig. 2.** Attack games of sRLWE and dRLWE.

**Definition 3.** *HNF-RLWE [2]: Let* $pp$ *be the public parameters* $(n, q, \sigma)$. *Given an integer* $1 < t < q$ *where* $t$ *and* $q$ *are co-prime, if* $x \xleftarrow{\$} \mathcal{D}_{\mathbb{Z}^n, \sigma}$, *the following Search RLWE and Decision RLWE assumptions are still hard (t is fixed to 2 in this paper):*

- **Search RLWE under HNF-RLWE (sRLWE):** *Given a list of* $(a, b := a \cdot x + te \in R_q)$, *the adversary reveals* $x$ *only in a negligible rate. The model of sRLWE is shown in Fig. 2. The probability that adversary wins* $Game^{\mathcal{A}}_{sRLWE}(pp)$ *is* $\varepsilon_{srlwe} := \Pr(x = x')$.
- **Decision RLWE under HNF-RLWE (dRLWE):** *Given a list of* $(a, b_b)$, $b \in \{0, 1\}$ *where* $b_0 \xleftarrow{\$} R_q$, $b_1 \leftarrow a \cdot x + te \in R_q$, *the adversary cannot distinguish* $b_0$ *and* $b_1$ *except a negligible advantage* $\varepsilon_{drlwe}$. *The model of dRLWE is shown in Fig. 2. The probability that adversary wins* $Game^{\mathcal{A}}_{dRLWE}(pp)$ *is* $\varepsilon_{drlwe} := \left| \Pr(b = b') - \frac{1}{2} \right|$.

### 2.3 Hint Function and Extractor

**Hint Function.** Because the products of $b_0 \cdot x_1$ and $b_1 \cdot x_0$ where $b_i := ax_i + 2e_i$ and $i \in \{0, 1\}$ are not equal, it requires a hint function to indicate a shared secret. Let $\lfloor x \rceil$ be the least integer close to $x$, $\mathbb{Z}_{\frac{q}{2}} := [\lfloor -\frac{q}{4} \rfloor, \lfloor \frac{q}{4} \rfloor]$, $\mathbb{Z}_{\frac{q}{2}+1} := [\lfloor -\frac{q}{4} + 1 \rfloor, \lfloor \frac{q}{4} + 1 \rfloor]$. In this paper, the following hint functions are used:

$$ht_0(x) := \begin{cases} 0, & x \in \mathbb{Z}_{\frac{q}{2}} \\ 1, & otherwise \end{cases} \qquad ht_1(x) := \begin{cases} 0, & x \in \mathbb{Z}_{\frac{q}{2}+1} \\ 1, & otherwise \end{cases}.$$

**Assistant Extractor.** We define an assistant extractor function to derive hint values over $\{0, 1\}^n$ where $b$ is fixed to be 0 or 1:

$$helpExt(\mathbf{a}) := \{ht_b(a_0), ht_b(a_1), ..., ht_b(a_{n-1})\} \in \{0, 1\}^n$$

**Extractor.** An extractor is defined as:

$$Ext(\mathbf{x}, \mathbf{t}) := \{(x_i + t_i \frac{q-1}{2} \bmod q) \bmod 2 | \mathbf{x} \in \mathbb{R}_q, \mathbf{t} := helpExt(\mathbf{x})\}$$

The extractor $Ext(\mathbf{x}, \mathbf{t})$ is robust if Lemma 3 holds.

**Lemma 3.** *[12]. Let $q > 8$ be an odd prime, $Ext(\mathbf{a}, \mathbf{t}) = Ext(\mathbf{b}, \mathbf{t})$ if the error tolerance of $Ext$ is at most $\frac{q}{4} - 2$, that is, $\|\mathbf{x} - \mathbf{y}\|_\infty \leq \frac{q}{4} - 2$, and the adversary cannot distinguish the distributions of $Ext(\mathbf{a}, \mathbf{t})$ whenever the hint function is $ht_0$ or $ht_1$.*

## 2.4 Other Definitions

**Pseudo-Random Function.** A pseudo-random function (PRF) is a set of functions that is indistinguishable from all real random functions. Let $Fc[M, \mathcal{Y}] : M \to \mathcal{Y}$ be the set of all functions that map $M$ to $\mathcal{Y}$, $F : \mathcal{K} \times M \to \mathcal{Y}$ is a function family, $\mathcal{Y}$ is the output space. The definition of secure PRF is as follows.

**Definition 4.** *Secure Pseudorandom Function: A pseudo-random function F is secure if it is indistinguishable from a real random function $Fc[M, \mathcal{Y}]$. The model of the secure PRF is shown in Fig. 3. Adversary can win the $Game^{\mathcal{A}}_{PRF}(F, Fc)$ with a negligible rate $|\Pr(b = b') - \frac{1}{2}| = \varepsilon_{PRF}$.*

| $Game^{\mathcal{A}}_{PRF}(F, Fc)$ | $Game^{\mathcal{A}}_{Sig}(pp_{sig})$ |
|---|---|
| 1: $b \xleftarrow{\$} \{0, 1\}, k \xleftarrow{\$} \mathcal{K}$ | 1: $pk, sk \leftarrow SKgen(pp_{sig})$ |
| 2: $f_0 \leftarrow F(k, \cdot)$ | 2: $(m', sig') \leftarrow \mathcal{A}^{Sign(m_i, sk)}(pk, pp_{sig})$ |
| 3: $f_1 \xleftarrow{\$} Fc[M, \mathcal{Y}]$ | 3: **if** $(m', sig') \neq (m_i, sig_i)$ : |
| 4: $b' \leftarrow \mathcal{A}^{y_i \leftarrow f_b(m_i)}(F, Fc)$ | 4: $b' \leftarrow Verify(pk, m', sig')$ |
| 5: **return** $b = b'$ | 5: **return** $b'$ |
| | 6: **otherwise: return** $\bot$ |

Fig. 3. Attack games of PRF and signature.

**Signature.** A signature scheme includes the following algorithms with the public parameter $pp_{sig}$:

- $pk, sk \leftarrow SKgen(pp_{sig})$. It is to generate a signing key $sk$ and a public verification key $pk$.
- $sig \leftarrow Sign(m, sk)$. It is to sign a message by signing key $sk$.
- $b \leftarrow Verify(pk, m, sig)$. It is to verify a signature by $pk$. If $sig$ is generated from $Sign(m, sk)$, $b := 1$ otherwise $b := 0$.

**Definition 5.** *Let $\varepsilon_{sig}$ be a negligible rate, a signature scheme is correct and existential unforgeable under chosen message attack (EUF-CMA) if the advantage of $\mathcal{A}$ to win $Game^{\mathcal{A}}_{Sig}(pp_{sig})$ shown in Fig. 3 is $\varepsilon_{sig} := \Pr(Verify(pk, m', sig') = 1)$.*

**Symmetric Encryption.** Let $k \xleftarrow{\$} \mathcal{K}, m_1, m_2 \leftarrow \mathcal{M}$, an encryption scheme $\mathcal{E} := (E_S, D_S)$ is secure if given $E_S(k, m_0), E_S(k, m_1)$, adversary cannot distinguish these two ciphertexts.

# 3   Security Model

In this section, we introduce our security model, which is mainly based on the model of Signal-Conforming [15]. To catch the use of the HSM and blind key, we additionally provide **accessHSM** and **RevRnd** oracles.

## 3.1   Execution Environment

There are at most $\mu$ parties s.t. $P := \{P_i | i \in [\mu]\}$. Each party $P_i$ contains $\gamma$ oracles s.t. $\pi_i := \{\pi_i^s | s \in [\gamma]\}$. Each oracle $\pi_i^s$ can access the long-term public and private keys $(lpk_i, lsk_i)$ and the following local variables:

- $id_i$ : The identity of user $P_i$
- $sid_i^s$ : The session identifier of oracle $\pi_i^s$
- $pid_i^s$ : The partner identity of $P_i$ in session $sid_i^s$
- $\mathfrak{R}_i^s \in \{\mathcal{I}, \mathcal{R}\}$ : The role of $P_i$ in session $\pi_i^s$: initiator or responder
- $sk_i^s$ : The session key generated in $\pi_i^s$
- $rnd_i^s$ : The random value generated by $P_i$ in session $\pi_i^s$
- $esk_i^s$ : The ephemeral key generated by $\pi_i^s$
- $\Phi_i^s \in \{\perp, \textbf{accept}, \textbf{reject}\}$ : The state of $\pi_i^s$: not started, finished successfully, and session rejected

Each session $\pi_i^s$ initiates its variables and long-term keys as $\perp$. An AKE protocol has two partner oracles, and the one who sends the first message is called the initiator $\mathcal{I}$. The one who receives the first message is called the responder $\mathcal{R}$. $\pi_i^s$ is finished successfully if the session key $sk_i^s$ is computed, i.e., $sk_i^s \neq \perp \iff \Phi_i^s = \textbf{accept}$. We now define the partner oracle.

**Partner Oracles and Correctness.** To define the correctness, it is required that two partner oracles faithfully run the AKE protocol and reach the **accept** stage. For any $(i, j, s, t) \in [\mu]^2 \times [\gamma]^2$ with $i \neq j$, $\pi_i^s$ and $\pi_j^t$ are partners if: (I) $pid_i^s = id_j$ and $pid_j^t = id_i$; (II) $\mathfrak{R}_i^s \neq \mathfrak{R}_j^t$; (III) $sid_i^s = sid_j^t$.

**Definition 6.** *Valid Partners: Two oracles are valid partners if and only if 1) they have the same session identifier and 2) they generate the different session keys with a negligible probability. Let $S := [\mu] \times [\gamma]$; then, the following properties hold:*

1. *$\mathcal{VP}$ is a set of valid partners if $\mathcal{VP} := \{((i, s), (j, t)) | (i, s), (j, t) \in S \wedge i < j\}$.*
2. *If $\pi_i^s$ has two valid partners $\pi_j^t$ and $\pi_h^r$, we have $(j, t) = (h, r)$.*

Thus, each oracle $\pi_i^s$ has a unique partner $\pi_j^t$ and is ordered in sequence.

**Definition 7.** *Correctness: An AKE protocol $\prod_{AKE}$ is correct for any $\mathcal{VP}$ except a negligible rate $\varepsilon$, if $\prod_{AKE}$ is executed between all $((i, s), (j, t)) \in \mathcal{VP}$ honestly where $\pi_i^s$ and $\pi_j^t$ are partners. It holds that*

$$Pr(\Phi_i^s = \Phi_j^t = accept \wedge sk_i^s = sk_j^t) \geq (1 - \varepsilon).$$

## 3.2 Security Definition

In this section, the HSM security and PFS of the AKE protocol $\prod_{AKE}$ are modeled as $Game_{HSM}^{\mathcal{A}}$ through the security game played between a challenger $C$ and an adversary $\mathcal{A}$, which is described in Sect. 3.3.

## 3.3 Security Model

**Setup:** $C$ performs the following operations:

1. Initiate a trusted server **CA**.
2. Generate $\mu$ long-term key pairs $\{(lpk_i, lsk_i) | i \in [\mu]\}$.
3. Register the users to **CA**.
4. Initiate all sessions $\{\pi_i^s | (i, s) \in \mathcal{S}\}$.
5. Publish the public parameters, long-term public keys, and certificates.
6. Run the adversary $\mathcal{A}$.

**Stage 1:** $\mathcal{A}$ can access the following oracles at any time:

- **Send**$(i, s, m)$: This oracle allows $\mathcal{A}$ to send any message to $\pi_i^s$. $\pi_i^s$ responds to this message according to its current state. Message $m$ should be *"init, j, role"* to start a new session. $C$ will set $\mathfrak{R}_i^s := role$, $pid_i^s := id_j$ if the partner oracles share some secrets. If $\pi_i^s$ is the initiator, $C$ returns the first message of the protocol. Otherwise, let $mt$ be the metadata selected by the protocol shared in this session, and $lpk_i^s$ and $lpk_j^t$ be the public keys related to the long-term keys, $C$ sets the session identifier as $sid_i^s := mt \| lpk_i^s \| lpk_j^t$. The transcripts of this new session will be the response to $\mathcal{A}$.
- **Register**$(i)$: This oracle allows $\mathcal{A}$ to register a user if $i := \mu + 1$. We simply consider that the order of the new user is one plus that of the latest user. The **CA** sets $\mu := \mu + 1$ s.t. $[\mu] := [\mu - 1] \cup \{i\}$, generates its $lpk_i, lsk_i$, and returns them to $\mathcal{A}$. Here, $lsk_i$ is *revealed*.
- **RevIK**$(i)$: $\mathcal{A}$ can obtain the long-term private key $lsk_i$ through this oracle if $i \in [\mu]$. $lsk_i$ is *revealed*.
- **RevRnd**$(i, s)$: When this oracle is queried, $\mathcal{A}$ can obtain the random values used in $\pi_i^s$.
- **RevEsk**$(i, s)$: This query returns the ephemeral private key of $P_i$ in $\pi_i^s$.
- **RevSkey**$(i, s)$: When querying this oracle, $\mathcal{A}$ can receive the session key $sk_i^s$ that is *revealed*.
- **accessHSM**$(i, x)$: This oracle, instead of returning the long-term secret, responds with the outputs of $f(lsk_i, x)$ that is a function only executed in the HSM.

**Test:** When **Stage 1** is over, $\mathcal{A}$ queries the **Test** oracle as follows:

- **Test**$(i, s)$: $C$ sets $b \xleftarrow{\$} \{0, 1\}$. If $(i, s) \notin \mathcal{S}$ or $\Phi_i^s \neq$ **accept**, $C$ returns $\perp$. Otherwise, $C$ sets $k_0 \xleftarrow{\$} \mathcal{K}$ and $k_1 \leftarrow sk_i^s$ and then responds with $k_b$ to $\mathcal{A}$.

**Stage 2:** $\mathcal{A}$ accesses the oracles as in **Stage 1** except that it cannot query the **RevSkey** of the tested oracle.

**Chall:** $\mathcal{A}$ outputs $b' \in \{0, 1\}$ after terminating **Stage 2**. If the tested oracle is not *fresh* in **Test** stage, $C$ outputs a random $b'$ instead. A session $\pi_i^s$ with $pid_i^s = id_j$ is *fresh* s.t.

1. **Register**($i$) is not queried.
2. **RevSkey**($i, s$) is not issued .
3. If $\pi_i^s$ has a valid partner oracle $\pi_j^t$, **RevSkey**($j, t$) is not issued.
4. $lsk_i$, $rnd_i^s$ and $esk_i^s$ are not revealed .
5. If $\pi_i^s$ has a valid partner oracle $\pi_j^t$, $lsk_j$, $rnd_j^t$ and $esk_j^t$ are not revealed.
6. If $\pi_i^s$ does not have a partner, $lsk_j$ can be revealed only after $\pi_i^s$ successfully finishes the protocol execution.
7. **accessHSM**($i, x$) is not queried.
8. If $\pi_i^s$ has a valid partner oracle $\pi_j^t$, **accessHSM**($j, x$) is not queried.
9. If $\pi_i^s$ does not have a partner, **accessHSM**($j, x$) can be accessed after $\pi_i^s$ successfully finishes the protocol execution .

The adversary can query at most two oracles of each partner among **RevIK/accessHSM**, **RevRnd**, and **RevEsk**. $\mathcal{A}$ wins $Game_{HSM}^{\mathcal{A}}$ if $b = b'$ with the advantage:

$$Adv_{\Pi_{AKE}}^{HSM}(\mathcal{A}) := \left| \Pr(b = b') - \frac{1}{2} \right|.$$

**Definition 8.** *Security of AKE protocol: An AKE protocol is secure under the HSM if* $Adv_{\Pi_{AKE}}^{HSM}(\mathcal{A})$ *is negligible for any QPT adversary* $\mathcal{A}$.

**Security Properties.** In this section, we describe the security properties of our model.

1. **Key Independence.** The revealed session keys are excluded from the *fresh* partners in items 2 and 3. Thus, the session keys cannot be computed because of the revealed keys. The reveal of other session keys gives no information about the tested *fresh* session keys, which indicates the key independence.
2. **PFS.** Items 4 to 6 allow $\mathcal{A}$ to reveal any secret information, including the long-term private keys $lsk$, the generated random values $rnd$, and the ephemeral keys $esk$ of both partner oracles. $\mathcal{A}$ cannot reveal all the secrets in one session. For item 6, it defines the PFS: after some sessions are finished and the session keys are generated, $\mathcal{A}$ can corrupt both parties and test the target session. Because a valid partner can only be corrupted after the *fresh* oracle is finished, our model can catch the impersonation attack owing to the leakage of keys.
3. **HSM.** Items 7 to 9 allow $\mathcal{A}$ to access the **accessHSM**($i, x$) instead of revealing the long-term private keys. In the HSM, the secrets are secure, but the adversary can access the HSM oracle to do the operations involving private keys. The HSM security is at least as strong as PFS.

### 3.4 Anonymity

Some X3DH-like protocols [5, 15] are claimed to be anonymous because their message flows are separated from the identity keys, but the flows of the identity keys are known to the public and will reveal the identity of each partner oracle if there is only one session. Thus, we provide a definition to catch this case.

Assuming that the long-term public key and the certificate are considered as the user identity, we define anonymity to ensure that even if the adversary can reveal the long-term private keys and certificates, he/she cannot link the transcripts to this user. The anonymity model $Game^{\mathcal{A}}_{AMS}$ is similar to $Game^{\mathcal{A}}_{HSM}$. However, in the **Test** stage, the adversary should run two different *fresh* sessions and obtain the transcripts. In **Stage** 2, **RevRnd** cannot be queried for each tested partner oracle. In the **Chall**, the adversary should determine the owner of the transcripts. The attack model of anonymity $Game^{\mathcal{A}}_{AMS}$ is shown in Fig. 4.

**Definition 9.** *Anonymity: Assume that $\Pi_{AKE}$ is the AKE executed between users, and let $TSP_{j_0}$ and $TSP_{j_1}$ be the sets of all transcripts in sessions $\pi_{j_0}$ and $\pi_{j_1}$ with $\pi_i^s$, respectively, where the roles of $P_{j_0}$ and $P_{j_1}$ are the same. $\pi_i^s$ is fresh and does not have a partner. The advantage of $\mathcal{A}$ for wining $Game^{\mathcal{A}}_{AMS}$ is*

$$Adv^{AMS}_{\Pi_{AKE}}(\mathcal{A}) := \left| \Pr(b = b') - \frac{1}{2} \right|.$$

*$\Pi_{AKE}$ is anonymous if $Adv^{AMS}_{\Pi_{AKE}}(\mathcal{A})$ is negligible.*

| $Game^{\mathcal{A}}_{AMS}(pp)$ | $Oracle\ Test_b(j_0, j_1, i, s, role)$ |
|---|---|
| 1 :  $b \xleftarrow{\$} \{0,1\}$ | 1 :  $m_b \leftarrow$ "*init*", $j_b, role$ |
| 2 :  $b' \leftarrow \mathcal{A}^{Test_b(j_0, j_1, i, s, role)}(pp)$ | 2 :  $TSP_{j_b} \leftarrow Send(i, s, m_b)$ |
| 3 :  **return** $b = b'$ | 3 :  **return** $TSP_{j_b}$ |

**Fig. 4.** The model of anonymity.

### 3.5 Deniability

In this section, we give the definition of deniability. Informally, the deniability is the property that when two parties $P_i, P_j$ are engaged in a protocol, and $P_i$ provides some evidence generated by the protocol indicating that $P_j$ is communicating with $P_i$, $P_i$ could generate these evidence without having any knowledge of $P_j$. That is, $P_j$ can deny he/she is communicating with $P_i$ because the evidence may be generated by $P_i$. The attack model of deniability is close to $Game^{\mathcal{A}}_{HSM}$. But in **Test** stage, the challenger will run a simulator and randomly send the transcripts of the simulator or that of the honest users to the adversary.

Our attack model of deniability is similar to $Game^{\mathcal{A}}_{HSM}$. However, in the **Test** stage, the challenger runs a simulator and randomly sends the transcripts of the simulator or that of the honest users to the adversary. Meanwhile, the adversary can query the certificates and authentication results.

**Definition 10.** *Deniability: Let $P_i$ and $P_j$ be the parties who execute a protocol $\prod_{AKE}$ with a key generation function $lsk, lpk \leftarrow KeyGen(pp)$. Let $TSP_{ORG}$ be the original transcripts and authenticated data generated by $P_i$ and $P_j$ in $\prod_{AKE}$, and let $TSP_{SIM}$ be the transcripts and authenticated data of the simulator $SIM$ who attempts to simulate the transcripts of $P_i$ and $P_j$ with the knowledge of the long-term private key $lsk_i$ of $P_i$ or $lsk_j$ of $P_j$. If $\prod_{AKE}$ is deniable, the probability that $\mathcal{A}$ can distinguish $TSP_{ORG}$ and $TSP_{SIM}$ is negligible.*

## 4   Our Constructions

### 4.1   Our Concept of B-X3DH

An overview of B-X3DH is presented in Fig. 5. It differs from the previous X3DH protocols [5, 13, 15] in the following three points:

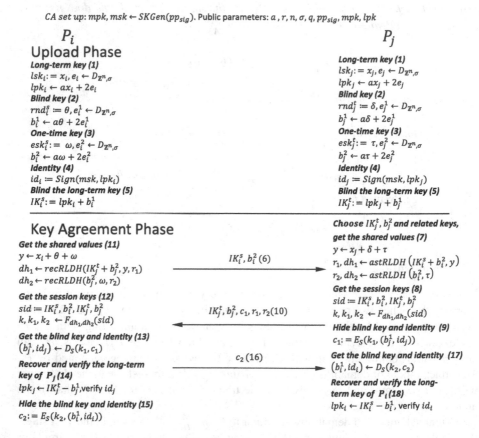

**Fig. 5.** Overview of B-X3DH Executed between $\pi_i^s$ and $\pi_j^t$.

- **Anonymity.** We blind the long-term public key with a random value, as indicated by lines (2,4,5) in Fig. 5. The result of the blinded public key is regarded as a ciphertext of a one-time pad because the random values are only used once. Thus, the adversary cannot reveal users' identities (long-term public keys). Even if the current random value is leaked, the anonymity of previous sessions will not be affected.
- **Authentication.** Instead of verifying the identity via a third channel or scanning QR codes face-to-face, B-X3DH introduces a signature scheme as a certificate system, as indicated by line (4) in Fig. 5.
- **Semi-asynchrony.** Asynchrony means that the responder can use the shared key without ensuring the correctness of the initiator. Compared with [5,13,15], the third flow is necessary to verify the identities of the initiator. Therefore, B-X3DH is not an asynchronous protocol. However, we can provide semi-asynchrony. After the first two flows, the responder can obtain the shared session key, although it is not verified. It can be proven that if the AKE is secure under HSM, the participants will share the same key. This is why the responder can directly use the unverified key. This feature is called semi-asynchrony.

## 4.2 B-X3DH Construction

| $astRLDH(\mathbf{b}, \mathbf{x})$ | $recRLDH(\mathbf{b}, \mathbf{x}, \mathbf{r})$ | $F_{dh_1, dh_2}(mt)$ |
|---|---|---|
| 1: $\mathbf{e} \xleftarrow{\$} \mathcal{D}_{\mathbb{Z}^n, \sigma}$ | 1: $\mathbf{e} \xleftarrow{\$} \mathcal{D}_{\mathbb{Z}^n, \sigma}$ | 1: $dh \leftarrow dh_1 \oplus dh_2$ |
| 2: $\mathbf{g} \leftarrow \mathbf{bx} + 2\mathbf{e}$ | 2: $\mathbf{g} \leftarrow \mathbf{bx} + 2\mathbf{e}$ | 2: $k_1 \leftarrow F(dh, (mt, "First"))$ |
| 3: $\mathbf{r} \leftarrow HelpExt(\mathbf{g})$ | 3: $dh \leftarrow Ext(\mathbf{g}, \mathbf{r})$ | 3: $k_2 \leftarrow F(dh, (mt, "Second"))$ |
| 4: $dh \leftarrow Ext(\mathbf{g}, \mathbf{r})$ | 4: **return** $dh$ | 4: $k \leftarrow F(dh, (mt, "Shared"))$ |
| 5: **return** $\mathbf{r}, dh$ | | 5: **return** $k, k_1, k_2$ |

**Fig. 6.** The HSM functions $astRLDH, recRLDH$ and the secure PRF

This section presents details regarding the construction of the B-X3DH protocol, i.e., $\prod_{B-X3DH}$, which requires the following building blocks.

- $KeyGen(pp_{rlwe})$: This is the key generation function, which is similar to $sk, pk \leftarrow KeyGen(pp_{rlwe})$.
- $astRLDH(\mathbf{b}, \mathbf{x})$: This is the assistant algorithm based on RLWE in the HSM, where the long-term key $lsk$ is a constant. For clarity, the additions related to $lsk$ are shown in the procedure. On input of a public key $\mathbf{b}$ and a secret parameter $\mathbf{x}$, it returns an assistant vector $\mathbf{r}$ and a shared secret $dh$. It is shown in Fig. 6.
- $recRLDH(\mathbf{b}, \mathbf{x}, \mathbf{r})$: This the reconciliation algorithm in the HSM. On input $\mathbf{b}, \mathbf{x}, \mathbf{r}$, it returns a shared secret $dh$, which is shown in Fig. 6.
- $k, k_1, k_2 \leftarrow F_{dh_1, dh_2}(mt)$: This is a secure PRF. It is shown in Fig. 6.
- $\prod_{Sign} = (Setup, SKGen, Sign, Verify)$: This is a EUF-CMA secure signature algorithm.
- $\prod_S = (E_S, D_S)$: This is a secure symmetric encryption scheme.

**CA Setup.** First, **CA** initializes the signature public parameters $pp_{sig}$. Then, it runs $msk, mpk \leftarrow SKgen(pp_{sig})$. Next, it selects $q, \sigma, n, \mathbf{a} \xleftarrow{\$} \mathbb{R}_q$ as $pp_{rlwe}$. Finally, it does the **User Register** as follows and publishes parameters $pp_{sig}, pp_{rlwe}, mpk$, and public identities $\{lpk_i, Cert_i\}_{i=1 \text{ to } n}$.

**User Register.** Each user $P_i$ generates $lsk_i := (x_i, e_i), lpk_i$, as indicated by line (1) in Fig. 5, and submits $lpk_i$ to **CA**. **CA** runs $Cert_i \leftarrow Sign(msk, lpk_i)$ and returns $Cert_i$. $P_i$ sets $id_i := Cert_i$. Finally, $P_i$ sets $lsk_i$ to its HSM.

**Construction.** $\prod_{B-X3DH}$ is run between $P_i$ and $P_j$ in some sessions, such as $\pi_i^s$ and $\pi_j^t$. Details are presented below:

- **First Flow:** The initiator $\pi_i^s$ sets $\mathfrak{R}_i^s := I$ and generates blind key pairs $(\theta, \mathbf{e}_i^1), \mathbf{b}_i^1 \leftarrow KeyGen(pp_{rlwe})$ and ephemeral key pairs $(\omega, \mathbf{e}_i^2), \mathbf{b}_i^2 \leftarrow KeyGen(pp_{rlwe})$. Then, it calculates the blinded public key $IK_i^s := lpk_i + \mathbf{b}_i^1 \in \mathbb{R}_q$. Next, it sets $rnd_i^s := (\theta, \mathbf{e}_i^1), esk_i^s := (\omega, \mathbf{e}_i^2)$. Finally, $IK_i^s, \mathbf{b}_i^2$ is sent to $\pi_j^t$.
- **Second Flow:** The responder $\pi_j$ sets $\mathfrak{R}_j^t := \mathcal{R}$ and generates the following.
  - Blind key pair: $(\delta, \mathbf{e}_j^1), \mathbf{b}_j^1 \leftarrow KeyGen(pp_{rlwe})$
  - Ephemeral key pair: $(\tau, \mathbf{e}_j^2), \mathbf{b}_j^2 \leftarrow KeyGen(pp_{rlwe})$

  It then calculates the blinded public key $IK_j^t := lpk_j + \mathbf{b}_j^1 \in \mathbb{R}_q$. Next, it sets $rnd_j^t := (\delta, \mathbf{e}_j^1), esk_j^t := (\tau, \mathbf{e}_j^2)$. Subsequently, it accesses its HSM and performs the following operations:
  - $\mathbf{y}_j := lsk_j.\mathbf{x}_j + \delta + \tau, \mathbf{X}_i := IK_i^s + \mathbf{b}_i^2$
  - $\mathbf{r}_j^1, dh_j^1 \leftarrow astRLDH(\mathbf{X}_i, \mathbf{y}_j)$
  - $\mathbf{r}_j^2, dh_j^2 \leftarrow astRLDH(\mathbf{b}_i^2, \tau)$
  - $sid_j^t := (IK_i^s, \mathbf{b}_i^2, IK_j^t, \mathbf{b}_j^2). k, k_1, k_2 \leftarrow F_{dh_j^1, dh_j^2}(sid_j^t)$
  - $c_j^t \leftarrow E_S(k_1, (\mathbf{b}_j^1, id_j))$

  Finally, it sets $sk_j^t := k$ and returns $IK_j^t, \mathbf{b}_j^2, \mathbf{r}_1, \mathbf{r}_2, c_j^t$ to $\pi_i^s$.
- **Third Flow of $I$:** After obtaining the response, $\pi_i^s$ accesses its HSM and performs the following operations:
  - $\mathbf{y}_i := lsk_i.\mathbf{x}_i + \theta + \omega, \mathbf{X}_j := IK_j^t + \mathbf{b}_j^2$
  - $dh_j^1 \leftarrow recRLDH(\mathbf{X}_j, \mathbf{y}_i, \mathbf{r}_j^1)$
  - $dh_i^2 \leftarrow recRLDH(\mathbf{b}_j^2, \omega, \mathbf{r}_j^2)$
  - $sid_i^s := (IK_i^s, \mathbf{b}_i^2, IK_j^t, \mathbf{b}_j^2). k, k_1, k_2 \leftarrow F_{dh_i^1, dh_i^2}(sid_i^s)$
  - $m \leftarrow D_S(k_1, c_j^t)$. If $m = \perp$, it terminates this session with $\Phi_i^s := $ **reject**. Otherwise,
  - $\mathbf{b}_j^{1*}, id_j^* \leftarrow m$
  - $lpk_j^* \leftarrow IK_j^t - \mathbf{b}_j^{1*} \in \mathbb{R}_q$
  - If $Verify(mpk, lpk_j^*, id_j^*) \neq 1$, it terminates this session with $\Phi_i^s := $ **reject**. Otherwise,
  - Reply $c_i^s \leftarrow E_S(k_2, (\mathbf{b}_i^1, id_i))$ to $\pi_j^t$

  Finally, it sets $pid_i^s := id_j^*, sk_i^s := k, \Phi_i^s := $ **accept**.
- **Third Flow of $\mathcal{R}$:** After receiving $c_i^s$, it performs the following operations:

- $m \leftarrow D_S(k_2, c_i^s)$. If $m = \perp$, it terminates this session with $\Phi_j^t := $ **reject**. Otherwise,
- $\mathbf{b}_i^{1*}, id_i^* \leftarrow m$
- $lpk_i^* \leftarrow IK_i^s - \mathbf{b}_i^{1*} \in \mathbb{R}_q$
- If $Verify(mpk, lpk_i^*, id_i^*) \neq 1$, it terminates this session with $\Phi_j^t := $ **reject**. Otherwise, $pid_j^t := id_i^*, \Phi_j^t := $ **accept**

## 5 Security Analysis

In this section, we prove the correctness, HSM security, anonymity, and deniability of B-X3DH. The correctness of B-X3DH is proven by Theorem 1. Lemma 4 proves that if two partner oracles run B-X3DH, they share the same secrets. Theorem 2 shows that our protocol has HSM security and PFS if the tested oracle is *fresh*. Theorem 3 explains the anonymity of B-X3DH. Theorem 4 explains the deniability.

**Lemma 4.** *If* $40(r\sigma\sqrt{n})^2 \leq \frac{q}{4} - 2$, *we have* $dh_i^1 = dh_j^1$ *and* $dh_i^2 = dh_j^2$.

*Proof.* In $astRLDH$ and $recRLDH$, let $\mathbf{X}_i := IK_i^s + b_i^2 = \mathbf{a}(\mathbf{x}_i + \theta + \omega) + 2(\mathbf{e}_i + \mathbf{e}_i^1 + \mathbf{e}_i^2)$,

$$\mathbf{g}_j^1 := \mathbf{X}_i\mathbf{y}_j + 2\mathbf{e}_j^3 = \mathbf{a}(\mathbf{x}_i + \theta + \omega)(\mathbf{x}_j + \delta + \tau) + 2(\mathbf{e}_i + \mathbf{e}_i^1 + \mathbf{e}_i^2)(\mathbf{x}_j + \delta + \tau) + 2\mathbf{e}_j^3 \quad (1)$$

$$\mathbf{g}_i^1 := \mathbf{X}_j\mathbf{y}_i + 2\mathbf{e}_i^3 = \mathbf{a}(\mathbf{x}_i + \theta + \omega)(\mathbf{x}_j + \delta + \tau) + 2(\mathbf{e}_j + \mathbf{e}_j^1 + \mathbf{e}_j^2)(\mathbf{x}_i + \theta + \omega) + 2\mathbf{e}_i^3 \quad (2)$$

$dh_i^1 := Ext(\mathbf{g}_i^1, \mathbf{r}_i^1)$ and $dh_j^1 := Ext(\mathbf{g}_j^1, \mathbf{r}_j^1)$. According to Lemmas 1 and 2,

$$
\begin{aligned}
\|\mathbf{g}_j^1 - \mathbf{g}_i^1\|_\infty &= \|2(\mathbf{e}_i + \mathbf{e}_i^1 + \mathbf{e}_i^2)(\mathbf{x}_j + \delta + \tau) - \\
&\quad 2(\mathbf{e}_j + \mathbf{e}_j^1 + \mathbf{e}_j^2)(\mathbf{x}_i + \theta + \omega) + 2\mathbf{e}_j^3 - 2\mathbf{e}_i^3\|_\infty \\
&\leq 2\|(\mathbf{e}_i + \mathbf{e}_i^1 + \mathbf{e}_i^2)(\mathbf{x}_j + \delta + \tau)\|_\infty + \\
&\quad 2\|(\mathbf{e}_j + \mathbf{e}_j^1 + \mathbf{e}_j^2)(\mathbf{x}_i + \theta + \omega)\|_\infty + 2\|\mathbf{e}_j^3\|_\infty + 2\|\mathbf{e}_i^3\|_\infty \\
&\leq 36\|\mathbf{xe}\|_\infty + 2\|\mathbf{ex}\|_\infty + 2\|\mathbf{ex}\|_\infty \\
&\leq 40\|\mathbf{xe}\|_\infty \leq 40\|\mathbf{x}\|_2\|\mathbf{e}\|_2 \leq 40(r\sigma\sqrt{n})^2
\end{aligned}
$$

If the error tolerance satisfies $40(r\sigma\sqrt{n})^2 \leq \frac{q}{4} - 2$, Lemma 3 can be achieved s.t. $Ext(\mathbf{g}_j^1, \mathbf{r}_j^1) = Ext(\mathbf{g}_i^1, \mathbf{r}_i^1)$. The proof of $dh_i^2 = dh_j^2$ with tolerance $20(r\sigma\sqrt{n})^2$ is identical.

**Theorem 1.** *If* $\prod_{Sign}$ *is a* $(1 - \varepsilon_{sig})$-*correct signature,* $F(\cdot)$ *is a secure PRF, and* $40(r\sigma\sqrt{n})^2 \leq \frac{q}{4} - 2$, $\prod_{B-X3DH}$ *is* $(1 - (\mu\gamma(\varepsilon_{sig} + 6\varepsilon_{rlwe}) + 2\mu\varepsilon_{rlwe} + 2\varepsilon_{srlwe}))$-*correct.*

*Proof.* Because $F(\cdot)$ is a secure PRF, both parties share a wrong secret only when a vector is too large with probability $\varepsilon_{rlwe}$, which leads to an input error of the PRF. The session will terminate the protocol only when an error signature verification occurs with probability $\varepsilon_{sig}$. There are at most $\mu$ parties and $\gamma$ sessions for each party. Each session samples six polynomials (four for blind and ephemeral keys and two for reconciliation functions), and each party samples its two long-term secrets. Thus, the error probability is $\mu\gamma(\varepsilon_{sig} + 6\varepsilon_{rlwe}) + 2\mu\varepsilon_{rlwe}$. As for the authentication, if the adversary cannot obtain the long-term private keys of users, it is impossible for him/her to encrypt or decrypt the ciphertext of the certificate, which implies that if the adversary cannot solve the sRLWE problem to obtain the long-term private keys with probability $\varepsilon_{srlwe}$, he/she cannot break the authentication of $\prod_{B-X3DH}$. Thus, $\prod_{B-X3DH}$ is correct, except that an error occurs with expectation $\mu\gamma(\varepsilon_{sig} + 6\varepsilon_{rlwe}) + 2\mu\varepsilon_{rlwe} + 2\varepsilon_{srlwe}$.

**Theorem 2.** *Let F be a secure PRF and Sign be a secure signature scheme. Assume that there are at most $\mu$ parties and that each party has at most $\gamma$ sessions. The QPT adversary $\mathcal{A}$ can break the HSM and PFS security of $\prod_{B-X3DH}$ with the following negligible advantage:*

$$Adv^{HSM}_{\prod_{B-X3DH}}(\mathcal{A}) \leq \mu\gamma(\varepsilon_{sig} + 6\varepsilon_{rlwe}) + 2\mu\varepsilon_{rlwe} + 2\varepsilon_{srlwe} + \varepsilon_{PRF} + \mu(\mu - 1)\gamma^2\varepsilon_{drlwe}.$$

*Proof Sketch.* Here, we give the simple proof of Theorem 1. Please refer to the appendix for the full proof.

Initially, the challenger runs the game $Game^{\mathcal{A}}_{HSM}$. The proof consists of three steps. First, we exclude the error situations that occur when the protocol runs honestly. Second, the challenger guesses the tested oracle selected by the adversary. Third, the adversary queries the oracles to reveal some private keys but not all, and the challenger replaces the unrevealed private keys in $Game^{\mathcal{A}}_{HSM}$ with random polynomials. The distinguishing game played between the $Game^{\mathcal{A}}_{HSM}$ and the modified game forms the games of dRLWE in Fig. 2 and secure PRF in Fig. 3. Thus, we can reduce the HSM security and PFS to the security of dRLWE and the secure PRF.

**Theorem 3.** *Let F be a secure PRF and Sign be a secure signature scheme. Assume that there are at most $\mu$ parties and that each party has at most $\gamma$ sessions. The QPT adversary $\mathcal{A}$ can break the anonymity of $\prod_{B-X3DH}$ with the following negligible advantage:*

$$Adv^{AMS}_{\prod_{B-X3DH}}(\mathcal{A}) \leq 2\varepsilon_{drlwe} + 2\varepsilon_{PRF}.$$

*Proof.* Let game $G_0, G_1$ be the $Game^{\mathcal{A}}_{AMS}$ where $Test$ returns $TSP_{j_0}$ and $TSP_{j_1}$, respectively. Then, $Game^{\mathcal{A}}_{AMS}$ can be regarded as the game to distinguish $G_0$ and $G_1$. Let $G_2$ be the same as $G_0$ except that it replaces the blind key $b^1_{j_0}$ with a random value, which an adversary can use to solve dRLWE. $G_3$ is the same as $G_2$ except that it replaces the blinded long-term key $IK_{j_0}$ with a random one. $G_4$ replaces the shared $dh_1$ with a random value. $G_5$ changes the PRF $F$ to a real random function, which an adversary can use to distinguish the secure PRF from random functions. Games from $G_6$ to $G_9$ perform the operations of $G_2$ to $G_5$, respectively, for $G_1$. From the viewpoint of $\mathcal{A}$, $G_9$ and $G_5$ are identical because their distributions are real random functions. Above all, the advantage for adversary $\mathcal{A}$ to break $Game^{\mathcal{A}}_{AMS}$ is

$$Adv^{AMS}_{\prod_{B-X3DH}}(\mathcal{A}) \leq 2\varepsilon_{drlwe} + 2\varepsilon_{PRF}.$$

**Theorem 4.** *Let F be a secure PRF and Sign be a secure signature scheme. Assume that there are at most $\mu$ parties and that each party has at most $\gamma$ sessions. The QPT adversary $\mathcal{A}$ can break the malicious deniability of $\prod_{B-X3DH}$ with the following negligible advantage:*

$$Adv^{DNY}_{\prod_{B-X3DH}}(\mathcal{A}) \leq 2\varepsilon_{drlwe} + 2\varepsilon_{PRF}.$$

*Proof.* $TSP_{ORG}$ are the transcripts generated by $\pi_i^s$ and $\pi_j^t$ and observed by $\mathcal{A}$:

1. First flow from $P_i$ to $P_j$: $IK_i^s := lpk_i + \mathbf{b}_i^1, \mathbf{b}_i^2$.
2. Second flow from $P_j$ to $P_i$: $IK_j^t := lpk_j + \mathbf{b}_j^1, \mathbf{b}_j^2, \mathbf{r}_1, \mathbf{r}_2, c_j^t$.
3. Third flow from $P_i$ to $P_j$: $c_i^s$.

Assume that the simulator SIM can obtain the long-term key of $P_i$ or $P_j$. SIM simulates as follows and obtains $TSP_{SIM}$:

1. First flow: Because $\mathbf{b}_i^1, \mathbf{b}_i^2$ are random and one-time, SIM can simulate them by running $KeyGen(pp_{rlwe})$ and calculating $IK_i^s$.
2. Second flow: SIM can calculate $IK_j^t := lpk_j + \mathbf{b}_j^1, \mathbf{b}_j^2$, similar to the first flow. If SIM has the long-term key of $P_i$ or $P_j$, it can calculate $\mathbf{g} := (lpk_j + \mathbf{b}_j^1 + \mathbf{b}_j^2)(lsk_i + rnd_i^s + esk_i^s) + 2\mathbf{e}$ or $\mathbf{g} := (lpk_i + \mathbf{b}_i^1 + \mathbf{b}_i^2)(lsk_j + rnd_j^t + esk_j^t) + 2\mathbf{e}$ and obtain the correct $r_1, r_2, dh_1, dh_2$ to generate the $k, k_1, k_2$. Thus, SIM can simulate the second flow if it has $lsk_i$ or $lsk_j$.
3. Third flow: If SIM obtains $k_2$, it can calculate $c_i^s$ easily.

Because the simulator can generate the shared secret, the authenticated data that are decrypted from $c_j^t$ and $c_i^s$ and the verification results will always be the same as these of honest users. If adversaries query the authentication data, the SIM returns the $c_j^t$ and $c_i^s$ with $k_1$ and $k_2$. If $\mathcal{A}$ can distinguish $TSP_{ORG}$ and $TSP_{SIM}$, there exists an adversary $\mathcal{B}$ that can use $\mathcal{A}$ to address dRLWE. The proof of this follows the proof of $Game_{AMS}^{\mathcal{A}}$. Thus,

$$Adv_{\Pi_{B-X3DH}}^{DNY}(\mathcal{A}) \leq 2\varepsilon_{drlwe} + 2\varepsilon_{PRF}.$$

# 6 Conclusion

In this paper, we propose a PQ-X3DH protocol called B-X3DH based on RLWE problems, which is inspired by the HMQV protocol. We set up a security model and slightly modified this model to catch the anonymity and deniability definitions of AKE. We show that the construction of B-X3DH without a signature can simultaneously provide HSM security, and PFS, together with privacy-enhanced anonymity and deniability, in contrast to other schemes. Because we can only achieve deniability that requires further knowledge of the long-term private keys, future work should be done to realize stronger deniability without knowledge of the secrets.

**Acknowledgement.** This work is partially supported by JSPS KAKENHI Grant Number JP21H03443 and JP21K11751, and SECOM Science and Technology Foundation.

# A  The Full Proof for the HSM Security and PFS of B-X3DH

This section gives the full proof of the HSM security and PFS of B-X3DH.

**Theorem 5.** *Let F be a secure PRF and Sign be a secure signature scheme. Assume that there are at most $\mu$ parties and that each party has at most $\gamma$ sessions. The QPT adversary $\mathcal{A}$ can break the HSM and PFS security of $\prod_{B-X3DH}$ with the following negligible advantage:*

$$Adv_{\prod_{B-X3DH}}^{HSM}(\mathcal{A}) \le \mu\gamma(\varepsilon_{sig} + 6\varepsilon_{rlwe}) + 2\mu\varepsilon_{rlwe} + 2\varepsilon_{srlwe} + \varepsilon_{PRF} + \mu(\mu-1)\gamma^2\varepsilon_{drlwe}.$$

*Proof.* Let $\mathcal{A}$ play the $Game_{HSM}^{\mathcal{A}}$ with challenger $C$. We now prove that the advantage $Adv_{\prod_{B-X3DH}}^{HSM}(\mathcal{A})$ is negligible. Let $W_i$ be the event that $b = b'$ in game $G_i$. Assume in each game, $\mathcal{A}$ always issues a **Test** oracle.

**Game $G_0$.** Let game $G_0$ be the original game $Game_{HSM}^{\mathcal{A}}$. Then there is:

$$Pr(W_0) = Adv_{\prod_{B-X3DH}}^{HSM}(\mathcal{A}).$$

**Game $G_1$.** Let game $G_1$ be like $G_0$ except that an event $\mathcal{E}_{fail}$ occurs. This event happens when there exist two partner oracles that don't agree on the same key or don't authenticate with each other. The aim of $G_1$ is to exclude the passive error situations when $\mathcal{A}$ actively queries oracles. $\mathcal{A}$ can distinguish these two games if and only if $\mathcal{E}_{fail}$ occurs. Recall Theorem 1, there is:

$$|Pr(W_0) - Pr(W_1)| \le Pr(\mathcal{E}_{fail})$$
$$\le \mu\gamma(\varepsilon_{sig} + 6\varepsilon_{rlwe}) + 2\mu\varepsilon_{rlwe} + 2\varepsilon_{srlwe}.$$

**Game $G_2$.** Let game $G_2$ be like $G_1$ except that $C$ tries to randomly select two oracles $\pi_{i*}^{s*}$ and $\pi_{j*}^{t*}$ to guess the tested session in **Test** oracle. Let $E_{TO}$ be the event that $\pi_{i*}^{s*}$ and $\pi_{j*}^{t*}$ are neither tested nor partner oracles. If $C$ checks that $E_{TO}$ occurs, it aborts the game. $C$ will successfully guess the tested oracles with the probability at least $\frac{1}{\mu\gamma} \cdot \frac{1}{(\mu-1)\gamma}$. Thus,

$$Pr(W_2) \ge \frac{1}{\mu(\mu-1)\gamma^2} Pr(W_1).$$

**Game $G_3$.** In this game, the two guessed partner oracles are $\pi_{i*}^{s*}$ and $\pi_{j*}^{t*}$. $C$ will modify the response of $\pi_{i*}^{s*}$ for the second flow. In $astRLDH$, instead of generating the $dh_{i*}^1, dh_{i*}^2$ from $Ext(\mathbf{g}_{i*}^1, \mathbf{r}_j^1), Ext(\mathbf{g}_{i*}^2, \mathbf{r}_j^2)$ respectively, it derives $dh_{i*}^1 = Ext(\mathbf{g}_{j*}^1, \mathbf{r}_j^1)$ and $dh_{i*}^2 = Ext(\mathbf{g}_{j*}^2, \mathbf{r}_j^2)$. This change makes sense. Otherwise, they will not be partners, and $E_{TO}$ will happen. From the view of $\mathcal{A}$, $G_3$ and $G_2$ are the same. Thus,

$$Pr(W_3) = Pr(W_2).$$

**Game $G_4^{(t,l)}$.** They are a set of games, in which assume $\mathcal{A}$ can access HSM to get $\mathbf{g}_{i*}^1$ and $\mathbf{g}_{j*}^1$ (We ignore the others because they have been included in the above condition). $C$ will replace $IK_{i*}^{s*}$ or $\mathbf{b}_{i*}^2$ with a random value $\mathbf{z}_0 \xleftarrow{\$} \mathbb{R}_q$ according to the choice of $\mathcal{A}$.

$(t, l) \in [3] \times [3]$ where $t$ stands for the not revealed keys of $P_{i*}$ and $l$ for the $P_{j*}$. 1 to 3 indicate the $lsk, rnd, esk$ by sequence. Let

$$\mathbf{g}_{j*}^1 := (IK_{i*}^{s*} + \mathbf{b}_{i*}^2)(lsk + rnd + esk) + 2\mathbf{e}$$

Because $\mathcal{A}$ cannot reveal all of the secrets $(lsk, rnd, esk)$ of one party, we just consider $(IK_{i*}^{s*} + \mathbf{b}_{i*}^2)$. If $t \in [2]$, $IK_{i*}^{s*}$ is replaced with $\mathbf{z}_0$, $\mathbf{b}_{i*}^2$ otherwise. Let the result be $\mathbf{g}'$. Then, the distinguishing between $G_4^{(t,l)}$ and $G_3$ is to distinguish $\mathbf{g}_{j*}^1$ and $\mathbf{g}'$, which forms the game of dRLWE assumption.

Assume $\mathcal{B}'$ is to distinguish $G_3$ and $G_4^{(t,l)}$, there exists an adversary $\mathcal{B}$ who can make use of the output of $\mathcal{B}'$ to break dRLWE. $\mathcal{B}$ receives $\mathbf{z}_0$ from the dRLWE challenger and does the replacement as mentioned before. If $\mathbf{z}_0$ is in the form of RLWE, the game played by $\mathcal{B}'$ is the same as $G_3$. According to the output of $\mathcal{B}'$, $\mathcal{B}$ can decide if $\mathbf{z}_0$ is random or not and break dRLWE. Then,

$$|\Pr(W_4) - \Pr(W_3)| \leq \varepsilon_{drlwe}.$$

**Game $G_5$.** It is the same as $G_4^{(t,l)}$, except that $\mathbf{g}'$ is replaced with a random $\mathbf{z}_1 \xleftarrow{\$} \mathbb{R}_q$. From the view of $\mathcal{A}$, $G_5$ and $G_4^{(t,l)}$ are the same. Thus,

$$\Pr(W_5) = \Pr(W_4).$$

**Game $G_6$.** It is the same as $G_5$, except that $C$ replace $dh_{i*}^1 = dh_{j*}^1 = Ext(\mathbf{z}_1, \mathbf{r}_1)$ with a random value $\mathbf{z}_2 \xleftarrow{\$} \{0, 1\}^n$. Because $Ext(\mathbf{z}_1, \mathbf{r}_1)$ is a deterministic algorithm, and $\mathbf{z}_1$ is random, it means that the output of $Ext(\mathbf{z}_1, \mathbf{r}_1)$ can be regarded as a random vector. Thus, from the point of view of $\mathcal{A}$:

$$\Pr(W_6) = \Pr(W_5).$$

**Game $G_7$.** It is the same as $G_6$, except that $C$ derives $k, k_1, k_2 \xleftarrow{\$} \mathcal{K}$ instead of $k, k_1, k_2 \leftarrow F_{dh_1, dh_2}(sid)$. Because the ciphertext is generated from semantic secure encryption $\prod_S$, the difference between $G_6$ and $G_7$ is to compare the secure PRF $F$ from random, which forms the $Game_{PRF}^{\mathcal{A}}$. Thus, from the point of view of $\mathcal{A}$:

$$|\Pr(W_7) - \Pr(W_6)| \leq \varepsilon_{PRF}.$$

Above all, we have:

$$Adv_{\prod_{B\text{-}X3DH}}^{HSM}(\mathcal{A}) \leq \mu\gamma(\varepsilon_{sig} + 6\varepsilon_{rlwe}) + 2\mu\varepsilon_{rlwe} + 2\varepsilon_{srlwe} + \varepsilon_{PRF} + \mu(\mu - 1)\gamma^2\varepsilon_{drlwe}.$$

# References

1. Alwen, J., Coretti, S., Dodis, Y.: The double ratchet: security notions, proofs, and modularization for the signal protocol. In: Ishai, Y., Rijmen, V. (eds.) EUROCRYPT 2019. LNCS, vol. 11476, pp. 129–158. Springer, Cham (2019). https://doi.org/10.1007/978-3-030-17653-2_5

2. Applebaum, B., Cash, D., Peikert, C., Sahai, A.: Fast cryptographic primitives and circular-secure encryption based on hard learning problems. In: Halevi, S. (ed.) CRYPTO 2009. LNCS, vol. 5677, pp. 595–618. Springer, Heidelberg (2009). https://doi.org/10.1007/978-3-642-03356-8_35
3. Batra, B.: News communication through whatsapp. Int. J. Inf. Futur. Res. 3(10), 3725–3733 (2016)
4. Bos, J., et al.: Crystals-kyber: a CCA-secure module-lattice-based KEM. In: 2018 IEEE European Symposium on Security and Privacy (EuroS&P), pp. 353–367. IEEE (2018)
5. Brendel, J., Fiedler, R., Günther, F., Janson, C., Stebila, D.: Post-quantum asynchronous deniable key exchange and the signal handshake. In: Hanaoka, G., Shikata, J., Watanabe, Y. (eds.) PKC 2022, pp. 3–34. Springer, Cham (2022). https://doi.org/10.1007/978-3-030-97131-1_1
6. Brendel, J., Fischlin, M., Günther, F., Janson, C., Stebila, D.: Towards post-quantum security for signal's X3DH handshake. In: Dunkelman, O., Jacobson, Jr., M.J., O'Flynn, C. (eds.) SAC 2020. LNCS, vol. 12804, pp. 404–430. Springer, Cham (2021). https://doi.org/10.1007/978-3-030-81652-0_16
7. Castryck, W., Decru, T.: An efficient key recovery attack on SIDH (preliminary version). Cryptology ePrint Archive (2022)
8. Chen, K., Chen, J.: Anonymous end to end encryption group messaging protocol based on asynchronous ratchet tree. In: Meng, W., Gollmann, D., Jensen, C.D., Zhou, J. (eds.) ICICS 2020. LNCS, vol. 12282, pp. 588–605. Springer, Cham (2020). https://doi.org/10.1007/978-3-030-61078-4_33
9. Cohn-Gordon, K., Cremers, C., Dowling, B., Garratt, L., Stebila, D.: A formal security analysis of the signal messaging protocol. J. Cryptol. 33(4), 1914–1983 (2020)
10. Cohn-Gordon, K., Cremers, C., Garratt, L., Millican, J., Milner, K.: On ends-to-ends encryption: asynchronous group messaging with strong security guarantees. In: Proceedings of the 2018 ACM SIGSAC Conference on Computer and Communications Security, pp. 1802–1819 (2018)
11. Ding, J., Gao, X., Takagi, T., Wang, Y.: One sample ring-LWE with rounding and its application to key exchange. In: Deng, R.H., Gauthier-Umaña, V., Ochoa, M., Yung, M. (eds.) ACNS 2019. LNCS, vol. 11464, pp. 323–343. Springer, Cham (2019). https://doi.org/10.1007/978-3-030-21568-2_16
12. Ding, J., Xie, X., Lin, X.: A simple provably secure key exchange scheme based on the learning with errors problem. Cryptology ePrint Archive (2012)
13. Dobson, S., Galbraith, S.D.: Post-quantum signal key agreement with SIDH. Cryptology ePrint Archive (2021)
14. Galbraith, S.D., Petit, C., Shani, B., Ti, Y.B.: On the security of supersingular isogeny cryptosystems. In: Cheon, J.H., Takagi, T. (eds.) ASIACRYPT 2016. LNCS, vol. 10031, pp. 63–91. Springer, Heidelberg (2016). https://doi.org/10.1007/978-3-662-53887-6_3
15. Hashimoto, K., Katsumata, S., Kwiatkowski, K., Prest, T.: An efficient and generic construction for signal's handshake (X3DH): post-quantum, state leakage secure, and deniable. J. Cryptol. 35(3), 1–78 (2022)
16. Krawczyk, H.: HMQV: a high-performance secure Diffie-Hellman protocol. In: Shoup, V. (ed.) CRYPTO 2005. LNCS, vol. 3621, pp. 546–566. Springer, Heidelberg (2005). https://doi.org/10.1007/11535218_33
17. Padlipsky, M.A., Snow, D.W., Karger, P.A.: Limitations of end-to-end encryption in secure computer networks. Technical report, Mitre Corp, Bedford, MA (1978)
18. Signal. Signal protocol: Technical documentation. https://whispersystems.org/docs/. Accessed 25 June 2022

19. Stephens-Davidowitz, N.: Discrete gaussian sampling reduces to CVP and SVP. In: Proceedings of the Twenty-Seventh Annual ACM-SIAM Symposium on Discrete Algorithms, pp. 1748–1764. SIAM (2016)
20. Unger, N., Goldberg, I.: Deniable key exchanges for secure messaging. In: Proceedings of the 22nd ACM SIGSAC Conference on Computer and Communications Security, pp. 1211–1223 (2015)
21. Zhang, J., Zhang, Z., Ding, J., Snook, M., Dagdelen, Ö.: Authenticated key exchange from ideal lattices. In: Oswald, E., Fischlin, M. (eds.) EUROCRYPT 2015. LNCS, vol. 9057, pp. 719–751. Springer, Heidelberg (2015). https://doi.org/10.1007/978-3-662-46803-6_24

# AI for Security

# Enhancing the Anti-steganalysis Ability of Image Steganography via Multiple Adversarial Networks

Bin Ma[1], Kun Li[1], Jian Xu[2]([⊠]), Chunpeng Wang[1], Jian Li[1], and Liwei Zhang[3]

[1] Qilu University of Technology (Shandong Academy of Sciences), Jinan 250300, China
[2] Shandong University of Finance and Economics, Jinan 250300, China
sdfixj@126.com
[3] Integrated Electronic Systems Lab Co., Ltd., Jinan 250300, China

**Abstract.** Existing steganographic methods based on adversarial images can only design adversarial images for a single steganalyzer and cannot resist detection from the latest steganalyzers using convolutional neural networks such as SRNet and Zhu-Net. To address this issue, this paper proposes a novel method for enhancing the security of image steganography using multiple adversarial networks and channel attention modules. The proposed method employs generative adversarial networks based on the U-Net structure to generate high-quality adversarial images and uses the self-learning properties of the adversarial networks to iteratively optimize the parameters of multiple adversarial steganographic networks. This process generates high-quality adversarial images capable of misleading multiple steganalyzers. Additionally, the proposed scheme adaptsively adjusts the distribution of adversarial noise in the original image using multiple lightweight channel attention modules in the generator, thus enhancing the anti-steganalysis ability of adversarial images. Furthermore, the proposed method utilizes multiple discrimination losses and MSE loss, dynamically combined to improve the quality of adversarial images and facilitate the network's rapid and stable convergence. Extensive experimental results demonstrate that the proposed algorithm can generate adversarial images with a PSNR of up to 42.8 dB, and the success rate of misleading the advanced steganalyzers is over 93%. The security and generalization of the algorithm we propose exceed those of the compared steganographic methods.

**Keywords:** Steganography · Steganalysis · Adversarial image · Channel attention

This work was partially supported by National Natural Science Foundation of China (62272255); National key research and development program of China (2021YFC3340602); Shandong Provincial Natural Science Foundation Innovation and Development Joint Fund (ZR202208310038); Ability Improvement Project of Science and Technology SMES in Shandong Province (2022TSGC2485); Project of Jinan Research Leader Studio (2020GXRC056); Project of Jinan Introducing Innovation Team (202228016); Project of Jinan City-School Integration Development (JNSX2021030); Youth Innovation Team of Colleges and Universities in Shandong Province (2022KJ124);The "Chunhui Plan" Cooperative Scientific Research Project of Ministry of Education (HZKY20220482);Shandong Provincial Natural Science Foundation (ZR2020MF054).

© The Author(s), under exclusive license to Springer Nature Switzerland AG 2023
M. Yung et al. (Eds.): SciSec 2023, LNCS 14299, pp. 181–192, 2023.
https://doi.org/10.1007/978-3-031-45933-7_11

# 1  Introduction

Image steganography has always attracted attention as a branch of information hiding, which aims to hide secret messages into covers in an imperceptible manner. According to different steganography mechanisms, the existing image steganography can be classified into original image embedded based steganography, coverless steganography, and adversarial attacks based steganography.

The original image embedded based steganography can be categorized into spatial domain steganography and transform domain steganography, based on the embedding domain. Spatial domain steganography is used to conceal secret messages by modifying the pixel values of the cover. Some commonly used spatial domain steganographic techniques include LSB [1], HUGO [2], WOW [3], and others. Embedding at different positions on the cover image can have different effects, therefore it is important to choose a suitable location for embedding. To address this, J. Fridrich et al. proposed the additive distortion Syndrome Trellis Codes (STC) [4], which can be combined with any additive distortion cost function to develop a steganographic method. Since then, researchers have focused on improving the security of the distortion function. Transform domain steganography, on the other hand, conceals secret messages by modifying the frequency domain coefficients of the cover image. Some typical transform domain steganographic methods include J-UNIWARD [5], UED [6], and its variant [7].

To enhances the security of image steganography, coverless image steganography was proposed. This technique does not require modification of the cover but instead hides secret messages by designing mapping rules between image features and secret data [8, 9] or by synthesizing the secret data into the image texture using specific algorithms [10, 11]. Despite the advantages of coverless steganography, there are still some challenges to be overcome, such as low embedding capacity, the need for large image databases, and unsatisfactory image quality. More recently, coverless methods based on deep learning [12, 13] have been developed to improve embedding capacity.

With the development of deep learning, image steganalysis based on convolutional neural networks (CNNs) [14–17] has achieved high performance, traditional image steganographic methods cannot resist the detection of these steganalyzers, so there is an urgent need to enhance the security of image steganography. On the one hand, benefiting from the introduction of zero-sum game theory in Generative Adversarial Networks (GANs) [18], a series of GAN-based steganographic methods [19, 20, 25] have emerged and performed well. They use the adversarial training between the generator and discriminator to enhance security. On the other hand, the introduction of adversarial attacks has brought new inspiration to image steganography. Adversarial attacks in steganography research mainly focus on two aspects. One is to add slight perturbations to the original image to form an adversarial image as a new cover image. The stego image generated by the adversarial image can make the steganalyzers output misclassification results with high confidence, these methods include [21, 22, 26]. The other is to first generate the original modification probability map using a distortion cost function, then select the appropriate modification direction based on the gradient sign and magnitude, and finally use adaptive steganographic methods for secret information embedding, representative methods are [27] and [28].

In this paper, we build a multi-adversarial network to train a generator that can quickly generate a large number of adversarial images for various steganalysis networks and design a new steganalysis loss function to improve performance by distinguishing between original and adversarial images. To enhance the visual quality and anti-steganalysis capability of the adversarial images, we add multiple channel attention modules [29] to the generator to dynamically adjust the distribution of adversarial noise in the original image. Our contribution can be summarized as follows: (1) we propose a framework to generate robust adversarial images for multiple steganalysis networks; (2) we improve the performance of the generator by designing a new loss function to enhance steganalysis performance; (3) we enhance the visual quality and anti-steganalysis capability of adversarial images by adding multiple channel attention modules to the generator.

## 2  Related Work

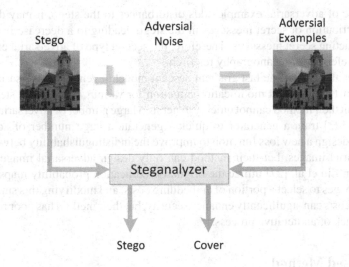

**Fig. 1.** Adversarial examples based on stego image.

Adversarial examples in image steganography can be classified into two groups according to the distinct stages of the steganographic procedure. The first type of adversarial examples is based on stegos, which are examples formed by adding carefully constructed and imperceptible subtle disturbances to the stegos. As shown in Fig. 1, these examples can cause neural network-based steganalyzers to output incorrect classification results with high confidence. The second type of adversarial examples is based on original images, which involve adding carefully constructed and imperceptible subtle disturbances to the original images. As illustrated in Fig. 2, using these examples as new covers to embed secret messages and generate stegos can cause the neural network-based steganalyzer to output incorrect classification results with high confidence. While

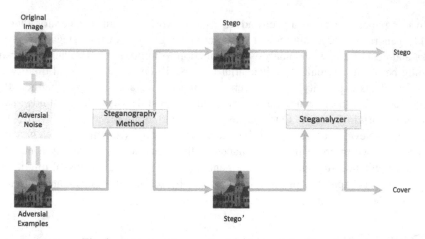

**Fig. 2.** Adversarial examples based on original image.

the first type of adversarial example adds disturbances to the stego, it may distort the original distribution of secret messages in the stego, leading to a decrease in the accuracy of extracting secret messages. Therefore, the second type of adversarial example is generally preferred in steganography research.

Zhang et al. [21] use the fast gradient descent model to generate adversarial images and can adjust the gradient modification direction for various CNN-based steganalysis networks, but their method cannot quickly generate a large number of adversarial images. Zhou et al. [22] train a generator to quickly generate a large number of adversarial images and design a new loss function to improve the indistinguishability between stego and adversarial images, but their method can only design adversarial images for one steganalyzer. Liu et al. [27] utilize the original modification probability maps of cover and stego images to select a portion of embedding costs, and modifying this small part of embedding costs can significantly enhance security, but their method has poor robustness due to the lack of an iterative process.

## 3   Proposed Method

### 3.1   Overview of the Proposed Model

As illustrated in Fig. 3, our proposed model comprises of three main components: (1) A generator $G$, which takes the original image $X$ as input, produces the adversarial noise $V$, and combines it with the original image to generate the adversarial image $X_V$. (2) A steganography network $SN$, which takes either the adversarial image $X_V$ or the original image $X$ as input and outputs the enhanced stego $X_{VS}$ or the stego $X_S$, respectively. (3) Multiple steganalysis networks $SD$, which take the original image $X$ and its corresponding adversarial image $X_V$ or the adversarial image $X_V$ and its corresponding enhanced stego $X_{VS}$ as inputs, and try to distinguish between the stego and the cover by assigning different scores to them. The steganalysis networks also include two sub-networks, the steganalysis optimization networks SON and the steganalysis adversarial networks SAN.

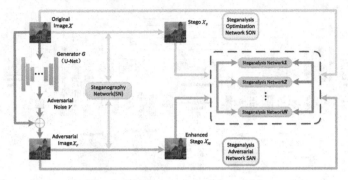

**Fig. 3.** The framework of the model.

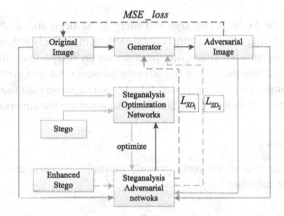

**Fig. 4.** Experimental flowchart.

We train our model based on Fig. 4, following the steps: (1) input the original image; (2) generate the adversarial image using the generator; (3) optimize the steganalysis networks by inputting the original image and the stego image into the steganalysis optimization networks, then build the steganalysis adversarial networks; (4) use the steganalysis adversarial networks to distinguish between the original image and the adversarial image; (5) use the steganalysis adversarial network to distinguish between the adversarial image and the enhanced stego image; (6) update the steganalysis optimization networks; (7) update the steganalysis adversarial networks; (8) update the generator network.

## 3.2  The Structure of Generator G

As depicted in Fig. 5, we utilize U-Net to construct the generator network structure. U-Net is an improved fully convolutional neural network (FCN) architecture that can extract two-dimensional image features. It solves the problem of FCN models losing many details during training and having difficulty in completely reconstructing the original image features. U-Net uses skip connections based on FCN to link the upper and lower

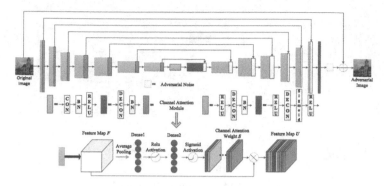

**Fig. 5.** The structure of the generator G.

sampling parts of the FCN. It overlays and merges the deep and shallow features of the original image during image feature reconstruction, effectively enhancing the ability of original image feature reconstruction to generate an image that closely resembles the original. By adjusting the size of the adversarial noise, we can control its intensity within a certain range. Finally, we inject the adversarial noise into the original image and constrain the pixel values to a specific range to generate the adversarial image.

### 3.3 Loss Functions

The steganalysis adversarial networks provide multiple discrimination loss functions $L_{SD1}$, which can be calculated as:

$$L_{SD_1} = -\sum_{i=1}^{2} x_i' \log(x_i) \tag{1}$$

where $x_1$ denotes the probability of original image $X$, $x_2$ denotes the probability of adversarial image $X_V$, $x_1', x_2'$ denote the labels corresponding to the original image $X$ and the adversarial image $X_V$.

The steganalysis adversarial networks provide multiple discrimination loss functions $L_{SD2}$, which can be calculated as:

$$L_{SD_2} = -\sum_{i=1}^{2} y_i' \log(y_i) \tag{2}$$

where $y_1$ denotes the probability of the adversarial image $X_V$ (steganalyzers treat it as the original image), $y_2$ denotes the probability of the enhanced stego $X_{VS}$, $y_1', y_2'$ denote the labels corresponding to the adversarial image $X_V$ and the enhanced stego $X_{VS}$.

In our proposed method, we need to ensure the visual indistinguishability between the adversarial image $X_V$ and the original image $X$, so we use the mean square error loss $L_{mse}$ to represent the loss of image distortion, which can be calculated as:

$$L_{mse} = MSE(X, X_V) = \frac{1}{C \times H \times W} \|X - X_V\|_2^2 \tag{3}$$

So the loss function of the generator $G$ can be calculated as:

$$L_G = k_1(-\alpha \cdot L_{SD_1} - \beta \cdot L_{SD_2}) + \ldots + k_n(-\alpha \cdot L_{SD_1} - \beta \cdot L_{SD_2}) + \lambda \cdot L_{mse} \qquad (4)$$

where $n$ denotes the number of steganalytic netwoks, $k$, $\alpha$, $\beta$ and $\lambda$ are all weight parameters.

## 4  Experiments Results

In our experiments, we use 10,000 grayscale images from BOSSBase [24]. Firstly, all images are resized to $256 \times 256$ using *"imresize"* in Matlab with default settings, and then randomly divided into two non-overlapping parts. The first part consists of 8,000 images used for training, while the remaining 2,000 images are used for testing. Two typical steganographic methods, namely ASDL-GAN [19] and UT-SCA-GAN [20], in the spatial domain are considered. Four CNN-based steganalytic methods, namely Xu-Net [14], Ye-Net [15], SRNet [16], Zhu-Net [17], and a traditional steganalyzer SRM [30] are also considered. Based on the experimental results described in their papers and our experimental verification, the detection abilities of these steganalytic methods for stego are ranked as follows: Zhu-Net > SRNet > Ye-Net > SRM > Xu-Net. In our experiments, we employ Adam (learning rate = 0.0001, beta_1 = 0.5, beta_2 = 0.99) as the optimizer for the generator. The steganalytic optimizers and their parameters used are given in the original papers.

There are four important parameters in our method, that is the number $n$ of steganalytic networks, the discriminant factor $\alpha$, $\beta$, and the image quality factor $\lambda$. We fixed $\alpha = 0.1$, $\beta = 0.9$, $\lambda = 0.2$ via comparing the experimental results with different parameters.

### 4.1  Steganalytic Networks Ablation Experiments

**Table 1.** The results of steganalytic networks ablation experiments.

| Different combinations of Steganalytic networks | Average PSNR | Average SSIM | Xu-Net Stego | Original | Ye-Net Stego | Original | SRNet Stego | Original | Zhu-Net Stego | Original | Trainning Time |
|---|---|---|---|---|---|---|---|---|---|---|---|
| Xu-Net+YeNet+SRNet+Zhu-Net | 34.8937 | 0.8736 | 50.1% | 84.9% | 50.2% | 86.6% | 49.6% | 89.1% | 50.5% | 89.4% | 1956 min |
| Ye-Net+SRNet+Zhu-Net | 37.1986 | 0.9101 | 50.1% | | 49.8% | | 50.4% | | 50.5% | | 1745 min |
| Ye-Net+SRNet | 38.2019 | 0.9379 | 49.6% | | 50.4% | | 50.6% | | 52.3% | | 1328 min |
| Ye-Net+Zhu-Net | 39.2194 | 0.9559 | 50.5% | | 49.7% | | 52.1% | | 50.8% | | 1289 min |
| SRNet+Zhu-Net | **39.9251** | **0.9601** | **50.4%** | | **50.6%** | | **49.4%** | | **50.6%** | | **1469 min** |
| SRNet | 43.3911 | 0.9792 | 49.6% | | 49.1% | | 50.8% | | 52.8% | | 996 min |
| Zhu-Net | 43.0458 | 0.9782 | 51.2% | | 51.1% | | 52.5% | | 50.9% | | 903 min |

In order to maintain the security of steganography and ensure the high quality of adversarial images, while reducing the number of network parameters and shortening the training time, we conducted steganalytic network ablation experiments. The steganographic method used in the experiment is UT-SCA-GAN with an embedded capacity of 0.4 bpp. Table 1 shows the experimental results, the optimal result should be 0.5, indicating that the steganalyzers are unable to distinguish whether the input image is the original image or the stego. The table data show that selecting SRNet+Zhu-Net as the combination of steganalytic networks achieves the best balance between steganography security and image quality. Therefore, we set the number of steganalytic networks $n$ to 2, $k_1 = k_2 = 5$, and choose this combination for subsequent experiments.

## 4.2 Channel Attention Modules Ablation Experiments

**Table 2.** The results of channel attention modules ablation experiments.

| Channel Attention Module Add Location | PSNR | Xu-Net | Ye-Net | SRNet | Zhu-Net |
|---|---|---|---|---|---|
| Encoding stage | **41.3556** | **50.2%** | **50.4%** | **49.6%** | **50.5%** |
| Decoding stage | 40.8628 | 49.5% | 50.5% | 50.6% | 50.7% |
| Encoding stage+Decoding stage | 40.2447 | 50.3% | 49.4% | 50.5% | 49.5% |
| Don't add | 39.9251 | 50.4% | 50.6% | 49.4% | 50.6% |

The use of channel attention mechanism in deep learning allows the network to learn to emphasize crucial features and disregard irrelevant ones. When generating adversarial images, the adversarial noise is combined with the original image features, which have different significance for the adversarial image. Therefore, using the channel attention mechanism may enhance the anti-steganalysis ability of adversarial images. Furthermore, adding adversarial noise to various channels affects image visual quality differently, thus using the channel attention mechanism can potentially enhance the quality of adversarial images. So we conducted ablation experiments to evaluate the effect of channel attention addition locations on our proposed model, we tested four variations: (1) add the channel attention modules in the encoding stage; (2) add the channel attention modules in the decoding stage; (3) add the channel attention modules in both the encoding and decoding stages; and (4) don't add the channel attention modules. The information embedding method is UT-SCA-GAN, the embedding capacity is 0.4 bpp, and the channel attention modules we use are SENet [29]. The experimental results are shown in Table 2, when choosing to add channel attention modules in the encoding stage, both the adversarial image quality and the anti-steganalysis ability have achieved the best results.

## 4.3 The Quality of Adversarial Images

In this section, we use the parameters determined in the previous experiment to train our generator, and then generate 2,000 adversarial images based on the test set. According to statistics, the average PSNR value is 41.3556. Figure 6 displays the differences between the original images and adversarial images. To clearly illustrate these differences, we also include their corresponding histograms. As shown in Fig. 6, the original and adversarial images are very similar and indistinguishable by the human visual system.

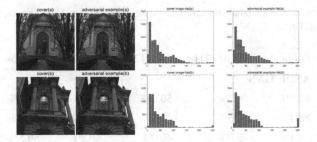

**Fig. 6.** Comparison between original images and adversarial images.

### 4.4 Comparative Experiments

Anti-steganalysis ability is the core index to evaluate the performance of adversarial image. Therefore, in this section, we compare the experimental results of our approach with three adversarial attacks-based methods. The steganographic methods are ASDL-GAN and UT-SCA-GAN, respectively, and the embedding capacity is 0.4 bpp. The experimental results are shown in Table 3. It shows whether adversarial images can deceive steganalyzers that has already been trained on the original images. The pre-trained steganalyzers used for detection have the same framework as steganalyzers being attacked. From Table 3, we can see that, when the steganalyzer used for detection is Xu-Net or Ye-Net, or SRM, compared with the original images without any modification, the adversarial images, whatever generated by the three methods or our method, both reduce the accuracy of steganalysis network to around 0.5. But when faced with the detection of SRNet or Zhu-Net, the other methods can not deceive these steganalysis networks. While our method has good results when faced with the detection of the five steganalyzers, it effectively improve the security of the image steganography.

Furthermore, we take the generated adversarial images and enhanced stego images as the training set, retrain five steganalyzers, and use the retrained steganalyzers to detect the anti-steganalysis ability of the adversarial images generated by the four methods. The experimental results are shown in Table 4. As can be seen, the anti-steganalysis ability of the four methods has varying degrees of decline. However, our method still achieves better performance than the other three methods. That is because Zhang's method fixes steganalyzers when training the generator, Zhou's method updates the steganalyzer only by discriminating whether secret information is embedded in adversarial images, Liu's method lacks an iterative process. While our method utilizes multi-adversarial training, designs novel loss functions, and introduces channel attention mechanism to enhance the robustness of generated adversarial images, thereby effectively improving the security of steganography.

**Table 3.** Accuracy of the steganalyzers trained on original images (Steganographic methods: ASDL-GAN and UT-SCA-GAN).

| Steganalyzer | Original | Zhang [21] | Zhou [22] | Liu [27] | Proposed method |
|---|---|---|---|---|---|
| **ASDL-GAN** | | | | | |
| Xu-Net | 86.8% | 50.3% | **50.2%** | 50.3% | **49.7%** |
| Ye-Net | 88.4% | 50.3% | 49.2% | 49.3% | **50.2%** |
| SRNet | 92.1% | 45.9% | 46.0% | 46.8% | **50.5%** |
| Zhu-Net | 92.6% | 46.2% | 46.4% | 47.1% | **49.4%** |
| SRM | 87.4% | 50.4% | 50.4% | 50.3% | **49.8%** |
| **UT-SCA-GAN** | | | | | |
| Xu-Net | 84.9% | 50.2% | 49.8% | 49.8% | **50.2%** |
| Ye-Net | 86.6% | 49.7% | 50.2% | 50.2% | **50.4%** |
| SRNet | 89.1% | 53.1% | 53.4% | 46.4% | **49.6%** |
| Zhu-Net | 89.4% | 53.8% | 46.7% | 46.9% | **50.5%** |
| SRM | 85.8% | 50.2% | 49.7% | 49.7% | **49.8%** |

**Table 4.** Accuracy of the steganalyzers retrained on adversarial images (Steganographic methods: ASDL-GAN and UT-SCA-GAN).

| Steganalyzer | Original | Zhang [21] | Zhou [22] | Liu [27] | Proposed method |
|---|---|---|---|---|---|
| **ASDL-GAN** | | | | | |
| Xu-Net | 86.8% | 61.4% | 58.4% | 58.6% | **53.5%** |
| Ye-Net | 88.4% | 63.8% | 59.6% | 59.6% | **53.8%** |
| SRNet | 92.1% | 70.3% | 67.6% | 65.3% | **59.4%** |
| Zhu-Net | 92.6% | 71.2% | 67.8% | 66.9% | **60.7%** |
| SRM | 90.6% | 61.4% | 58.4% | 59.2% | **53.5%** |
| **UT-SCA-GAN** | | | | | |
| Xu-Net | 84.9% | 60.2% | 57.8% | 57.4% | **52.1%** |
| Ye-Net | 86.6% | 60.9% | 58.4% | 58.2% | **52.7%** |
| SRNet | 89.1% | 69.6% | 66.3% | 64.8% | **58.6%** |
| Zhu-Net | 89.4% | 71.7% | 66.7% | 65.3% | **59.7%** |
| SRM | 90.2% | 60.2% | 57.8% | 58.0% | **52.1%** |

# 5 Conclusion

In this letter, we train a generator based on the U-Net architecture that can rapidly and efficiently produce robust adversarial images as new cover images by designing new loss functions, introducing channel attention mechanism, and multiple adversarial training. Experimental results demonstrate that the enhanced stego images generated by our method are effective in resisting detection by both deep learning-based and traditional steganalyzers. In our future work, we will focus on improve the generalization of the model.

# References

1. Mielikainen, J.: LSB matching revisited. IEEE Signal Process. Lett. **13**(5), 285–287 (2006)
2. Pevný, T., Filler, T., Bas, P.: Using high-dimensional image models to perform highly undetectable steganography. In: Böhme, R., Fong, P.W.L., Safavi-Naini, R. (eds.) Information Hiding, pp. 161–177. Springer, Heidelberg (2010). https://doi.org/10.1007/978-3-642-16435-4_13
3. Holub, V., Fridrich, J.: Designing steganographic distortion using directional filters. In: 2012 IEEE International Workshop on Information Forensics and Security (WIFS), pp. 234–239. IEEE (2012)
4. Filler, T., Judas, J., Fridrich, J.: Minimizing additive distortion in steganography using syndrome-trellis codes. IEEE Trans. Inf. Forensics Secur. **6**(3), 920–935 (2011)
5. Holub, V., Fridrich, J., Denemark, T.: Universal distortion function for steganography in an arbitrary domain. EURASIP J. Inf. Secur. **2014**(1), 1–13 (2014). https://doi.org/10.1186/1687-417X-2014-1
6. Guo, L., Ni, J., Shi, Y.Q.: An efficient JPEG steganographic scheme using uniform embedding. In: 2012 IEEE International Workshop on Information Forensics and Security (WIFS), pp. 169–174. IEEE (2012)
7. Guo, L., Ni, J., Shi, Y.Q.: Uniform embedding for efficient JPEG steganography. IEEE Trans. Inf. Forensics Secur. **9**(5), 814–825 (2014)
8. Zhang, X., Peng, F., Long, M.: Robust coverless image steganography based on DCT and LDA topic classification. IEEE Trans. Multimed. **20**(12), 3223–3238 (2018)
9. Luo, Y., Qin, J., Xiang, X., Tan, Y.: Coverless image steganography based on multi-object recognition. IEEE Trans. Circuits Syst. Video Technol. **31**(7), 2779–2791 (2021)
10. Wu, K., Wang, C.: Steganography using reversible texture synthesis. IEEE Trans. Image Process. **24**(1), 130–139 (2015)
11. Xu, J., Mao, X., Jin, X., et al.: Hidden message in a deformation-based texture. Vis. Comput. **31**, 1653–1669 (2015)
12. Chen, X., Zhang, Z., Qiu, A., Xia, Z., Xiong, N.N.: Novel coverless steganography method based on image selection and StarGAN. IEEE Trans. Netw. Sci. Eng. **9**(1), 219–230 (2022)
13. Peng, F., Chen, G., Long, M.: A robust coverless steganography based on generative adversarial networks and gradient descent approximation. IEEE Trans. Circuits Syst. Video Technol. **32**(9), 5817–5829 (2022)
14. Xu, G., Wu, H.-Z., Shi, Y.-Q.: Structural design of convolutional neural networks for steganalysis. IEEE Signal Process. Lett. **23**(5), 708–712 (2016)
15. Ye, J., Ni, J., Yi, Y.: Deep learning hierarchical representations for image steganalysis. IEEE Trans. Inf. Forensics Secur. **12**(11), 2545–2557 (2017)
16. Boroumand, M., Chen, M., Fridrich, J.: Deep residual network for steganalysis of digital images. IEEE Trans. Inf. Forensics Secur. **14**(5), 1181–1193 (2019)

17. Zhang, R., Zhu, F., Liu, J., Liu, G.: Depth-wise separable convolutions and multi-level pooling for an efficient spatial CNN-Based steganalysis. IEEE Trans. Inf. Forensics Secur. **15**, 1138–1150 (2020)
18. Goodfellow, I., et al.: Generative adversarial nets. In: Proceedings of the Advances in Neural Information Processing Systems, pp. 2672–2680 (2014)
19. Tang, W., Tan, S., Li, B., Huang, J.: Automatic steganographic distortion learning using a generative adversarial network. IEEE Signal Process. Lett. **24**(10), 1547–1551 (2017)
20. Yang, J.H., Liu, K., Kang, X.Q., et al.: Spatial image steganography based on generative adversarial network. arXiv: Multimedia, https://arxiv.org/abs/1804.07939v1 (2019)
21. Zhang, Y.W., Zhang, W.M., Chen, K.J., Liu, J.Y., Liu, Y.J., Yu, N.H.: Adversarial examples against deep neural network based steganalysis. In: Proceedings of the 6th ACM Workshop Information Hiding Multimedia Security, pp. 67–72. ACM (2018)
22. Zhou, L., Feng, G., Shen, L., Zhang, X.: On security enhancement of steganography via generative adversarial image. IEEE Signal Process. Lett. **27**, 166–170 (2020)
23. Ronneberger, O., Fischer, P., Brox, T.: U-Net: convolutional networks for biomedical image segmentation. In: Navab, N., Hornegger, J., Wells, W.M., Frangi, A.F. (eds.) MICCAI 2015. LNCS, vol. 9351, pp. 234–241. Springer, Cham (2015). https://doi.org/10.1007/978-3-319-24574-4_28
24. Bas, P., Filler, T., Pevný, T.: Break our steganographic system": the ins and outs of organizing BOSS. In: Filler, T., Pevný, T., Craver, S., Ker, A. (eds.) IH 2011. LNCS, vol. 6958, pp. 59–70. Springer, Heidelberg (2011). https://doi.org/10.1007/978-3-642-24178-9_5
25. Tan, J., Liao, X., Liu, J., Cao, Y., Jiang, H.: Channel attention image steganography with generative adversarial networks. IEEE Trans. Netw. Sci. Eng. **9**(2), 888–903 (2022)
26. Qin, X., Li, B., Tan, S., Tang, W., Huang, J.: Gradually enhanced adversarial perturbations on color pixel vectors for image steganography. IEEE Trans. Circuits Syst. Video Technol. **32**(8), 5110–5123 (2022)
27. Liu, M., Luo, W., Zheng, P., Huang, J.: A new adversarial embedding method for enhancing image steganography. IEEE Trans. Inf. Forensics Secur. **16**, 4621–4634 (2021)
28. Qin, X., Tan, S., Tang, W., Li, B., Huang, J.: Image steganography based on iterative adversarial perturbations onto a synchronized-directions sub-image. In: 2021 IEEE International Conference on Acoustics, Speech and Signal Processing (ICASSP), pp. 2705–2709. IEEE (2021)
29. Hu, J., Shen, L., Sun, G.: Squeeze-and-excitation networks. In: 2018 IEEE/CVF Conference on Computer Vision and Pattern Recognition, pp. 7132–7141. IEEE (2018)

# An Empirical Study of AI Model's Performance for Electricity Load Forecasting with Extreme Weather Conditions

Fusen Guo$^{(\boxtimes)}$, Jian-Zhang Wu, and Lei Pan

School of Information Technology, Deakin University,
Waurn Ponds, VIC 3216, Australia
s222159067@deakin.edu.au

**Abstract.** Electricity load forecast is critical to grid safety. Various artificial intelligence (AI) models have been proposed to forecast the short-term load, but little research has been conducted to investigate the effect of environmental factors like extreme weather. We aim to identify the most accurate AI models, discuss their implications for grid safety, and analyze the performance of the most accurate AI models under hot and cold wave conditions. Based on the experiment, we use Mean absolute percentage error (MAPE) and daily percentage error as measurement methods to evaluate the accuracy and stability of the model. We observe that the Long-Short-Term Memory (LSTM) model outperforms XgBoosting and Support vector machines (SVM) to forecast load at the cost of unstable daily percentage errors when extreme weather conditions occur. The grid operators who need to develop more accurate forecasting models and discuss the implications of short-term load forecasting for grid safety can benefit from this finding.

**Keywords:** Grid safety · load forecasting · AI models · extreme weather condition

## 1 Introduction

Demands for grid safety are becoming increasingly stringent, and excellent grid planning and dispatching are essential for power supply companies to ensure the safe and stable functioning of the power grid [19]. A model for short-term load forecasting (STLF) is used to facilitate the capacity of suppliers to offer continuous electricity to consumers and prevent operating and maintenance losses [1]. If a flawed model is chosen, significant fluctuations in power production will occur, thus affecting grid safety. Because the power station cannot efficiently store surplus electricity, the ability to create power must react to variable demand [6]. The grid dispatching system should take the necessary measures to preserve stability and limit the gap between peaks and troughs, which will maintain grid

© The Author(s), under exclusive license to Springer Nature Switzerland AG 2023
M. Yung et al. (Eds.): SciSec 2023, LNCS 14299, pp. 193–204, 2023.
https://doi.org/10.1007/978-3-031-45933-7_12

safety. Especially, weather conditions significantly impact power grid stability, with severe weather events posing major challenges to its safe operation [10]. Drawing upon the example of Shanghai, China, in the winter of 2020, persistent cold waves triggered an unprecedented peak load of 33.389 million kilowatts. This event marked the first time that winter peak load surpassed that of summer, creating a considerable shortfall in the electricity supply [3]. Therefore, the STLF model inside the dispatching system is proposed to help dispatchers regulate the grid system in real-time and make more accurate safety assessments ahead to prevent disasters [17].

The elements contributing to grid safety issues are highly complex and can happen quickly. During grid operation, dispatch centers employ centralized operational monitoring to guarantee the safe functioning and dispatch of the grid. Nevertheless, all data is transmitted and stored in the data center, making this method susceptible to data security threats and making it impossible to discern the pattern of load fluctuations in each area precisely [11]. But some model lakes simulate the electrical load of extreme weather and other special events [9]. Therefore, if some disasters, like hot and cold wave conditions, are not successfully evaluated, or controlled, which will affect the safety of people and property and cause system failure and blackouts, resulting in substantial economic losses for the nation. Regarding safety, when the power load is lower than the actual demand, it is easy to cause large-scale blackouts or even system disintegration because the stability is destroyed. However, when the load level is too high, it will shorten the life of the power supply equipment in the system and generate frequency collapse, thus seriously threatening the safe operation of the grid. Therefore, using STLF models to avoid and mitigate grid safety issues is a good approach [19].

The contributions of this paper can be summarized as follows:

1. We evaluate the performance of three AI models. It shows that the LSTM model has a better performance than XgBoosting and SVM.
2. We assess the forecasting ability of the most accurate AI model under cold and hot weather conditions.
3. We propose a measurement method to evaluate the performance of the STLF model on the safety assessment.

The rest of the paper is organized as follows. Section 2 summarizes the recently related works. Section 3 introduces our dataset, data set preparation, and methodology and illustrates the performance evaluation metrics. Based on our datasets, we discussed the performance of three AI models and simulation results under hot and cold wave conditions in Sect. 4. Section 5 concludes this work and proposes future work.

## 2    Related Works

Predictive load models are split into three categories: (1) statistical model, (2) AI model, and (3) hybrid models [16]. In the past few decades, research has been carried out to develop various models based on AI for STLF. These models, such as artificial neural networks (ANN), recurrent neural networks (RNN), SVM, LSTM, XgBoosting, and fuzzy logic (FG) have been utilized in some real-world cases and have produced good results [5]. Considering that the accuracy of forecasting directly affects the safety and quality of the power system, it is used to solve the problems of real-time control of the power grid and the safety of large-scale power outages [3]. As for a good model, it has high accuracy, and the daily forecast percentage error does not fluctuate significantly. Furthermore, the accurate model under extreme weather conditions aids power dispatch, improves safety in areas with weak grids and heavy loads, and lessens peak load impact on the power grid [4].

ANN has been widely used in STLF due to better self-learning and error tolerance. Nevertheless, ANN has limited generalization capacity and is easily trapped in the local minimum. As a result, applying ANN to STLF generated several errors and instabilities. To counter the shortcoming of ANN, specific models based on SVM have been developed to increase generalization capabilities and solve the problems of small sample size, nonlinearity, and local minimum, even though it has the difficulty of processing large samples slowly [22]. Furthermore, LSTM effectively improves RNN, which have been proven successful in load forecasting. The proposed LSTM-based model has been shown to predict building electricity consumption better than SVM, RF, and multi-layer perceptron neural network (MLP) models in nine out of twelve months [20].

Furthermore, XgBoosting is more accurate than the BP neural network in experiments because it can more accurately fit the dynamic trends of short-term power loads [21]. In summary, some AI models have proven effective in real cases, like SVM, XgBoosting, and LSTM. However, some previous research only considers a single variable: historical electric load, but many external circumstances simultaneously influence the load. Moreover, they only evaluate the accuracy of the model in general and need more research on the safety and stability of the power grid. Therefore, we evaluate the model's performance based on how to maintain the safety and stability of the grid.

Based on this finding, this paper evaluates the performance of three widely used AI models for high-accuracy short-term load forecasting: LSTM, XgBoosting, and SVM. Since there are no trials to evaluate the performance of these models under hot and cold waves. Therefore, the difference in their performance warrants further investigation, and the stability of those models is doubtful.

## 3   Dataset and Methodology

### 3.1   Data Exploration

Based on publicly accessible data sets, case studies were done to evaluate the model proposed before. This section describes the data set in detail. To verify the accuracy of the proposed model, two real data sets are employed. For the first case study, the data are taken on an open-access base from 1 January 2015 to 31 December 2019 in Victoria state, Australia [8]. The collected information contains a demand for the five years with 1826 samples and a sampling frequency of 1 day. We use a second dataset containing four years of data with the same features for Spain from Kaggle [7]. In addition, some meteorological characteristics are believed to impact the electric load: the daily load of electric load, price, min temperature, max temperature, and daily rainfall, are selected as features listed in Table 1.

**Table 1.** Selected Features

| Features | Description |
| --- | --- |
| Min-temperature | Minimum temperature during the day(°C) |
| Max-temperature | temperature during the day(°C) |
| Rainfall | Daily rainfall in mm |
| Daily price | The average price per MWh |
| Load | The total daily electricity demand in MWh |

Figure 1 shows the 5-year trend of Victoria's electricity load. It has fluctuations during the five years, with the load rising and then falling during the middle of the year but always peaking around the end of the year. It can be inferred that electricity load positively correlates with winter and summer. Furthermore, the range between peaks and troughs increases in the five years. It would be considered a data set affected by the season in this short-term loading study because it did not have multiple large fluctuations in the short term but with seasonal trends. It also shows the real power load trend in Spain for four years. The values of the daily electrical loads are tens of times larger than in data set 1, and the range is significantly more extensive. It shows that the daily data variation has more pronounced fluctuations, but the peaks and troughs are not concentrated at any specific time period. It has no apparent seasonal curve, but the maximum and minimum values have increased. It would be considered a data set influenced by the extreme value in this short-term loading study.

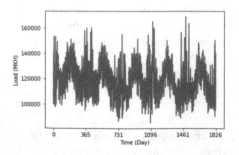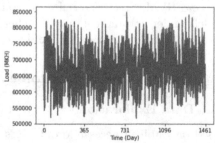

**Fig. 1.** The trend of datasets

## 3.2  Data Pre-processing

During the experiment, the dataset is divided into 70% for training and 30% for testing. In developing the models used in this paper, the pre-processing techniques for the input data are important because these models are concerned with the values of the variables rather than with the distribution of the variables and the conditional probabilities between the variables. Therefore, we apply the min-max normalization method to normalize each feature in the data set in the range [0, 1]. The definition of normalization function is as follows:

$$x_{new} = \frac{x_i - x_{min}}{x_{max} - x_{min}} \tag{1}$$

where $x_{new}$ represent the pre-processed value and $x_i$ is the initial value of data $i$. $x_{min}$ and $x_{max}$ represent the minimum and the maximum value of the sequence $\{x_1, x_2, \ldots, x_i\}$, respectively.

## 3.3  Methodology

LSTM is a specific recurrent neural network (RNN) architecture that overcomes the gradient explosion or gradient disappearance problem that occurs when RNNs process long-time sequences because the hidden layer of RNN has only one state [12]. The key components of the LSTM are the cell and three kinds of gates: input gate, output gate, and forger gate. Each cell acts as a memory location capable of retaining data for indefinite amounts of time, and three gates control the influx of information into the cell. These gates in the LSTM cell can store a persistent error that can be propagated backward through time and layers, enabling the recurrent network to continue learning across numerous time steps [14]. These gates collaborate to acquire and store long-term and short-term sequence-related data. It uses the sigmoid function to modulate the output between 0 and 1. When the gate value is 0, the signal is not let through. To achieve the best accuracy in the experiment, the LSTM-based model includes one input layer, two LSTM layers, a dropout layer added to each hidden layer, and one output layer. The Adam optimization algorithm is used in the network training.

XgBoosting model stands for Extreme Gradient Boosting, a machine learning system based on Gradient Boosting design with higher performance [2]. It mainly uses the gradient-boosting decision tree to improve speed in tree-based AI calculations and performance. It tries to correct the residuals of all previous weak learners by adding new weak learners because one learner may not be enough to get good results. When these learners are combined for the final prediction, the accuracy is higher than that of individual learners. The final prediction is the sum of the scores of each learner [13]. To increase the accuracy and to be able to customize the loss function, it performs a second-order Taylor expansion on the loss function. Considering the influence of the parameters on the performance, we set the parameter of Booster as Gbtree to use a tree structure to run data; the learning rate is set as 0.05 to control the step size of each iteration; max_depth is set the height of the tree as 5; n_estimators is set as 1000 to control the maximum number of iterations, and the rest of parameters are set to default value.

The SVM model is used to find the mapping relationship between the input vectors and output vectors. During the process of building short-term load forecasting model, it requires extracting feature vectors from historical load data and other influenced factors, which are the component of training sets and test sets of the model [22]. It is an intelligent prediction method based on statistical learning theory, and the principle of structural risk minimization [18]. And it can seek the best compromise between the complexity of the model and the learning capability based on the limited sample information to obtain the best generalization capability. Therefore, it is well suited for predicting short-term loads with stochastic, nonlinear characteristics. During the experiment, the kernel is set as rbf, $C$ is set as 8 to control the strength of the regularization, and $\epsilon$ is set as 0.01, which defines the tolerance of mis-classification.

### 3.4   Measurement Method

Performance analysis by error or evaluation metrics is an effective method to assess a model. FOr performance evaluation and comparison, the mean absolute percentage error (MAPE) is calculated for each model as a performance measurement. It provides a fundamental understanding of the disparity between the predicted and actual values. The measurement method is defined as follows:

$$MAPE = \frac{1}{n}\sum_{i=1}^{n}\left|\frac{x_{act(i)}-X_{pred(i)}}{x_{act}(i)}\right|. \tag{2}$$

The range of MAPE is $[0, +\infty]$, and the magnitude of MAPE is positively correlated with the magnitude of the error as its value increases, indicating that the error also increases.

Furthermore, in addition to the overall prediction accuracy of the model, we also looked at the daily predictions from a microscopic perspective. In this paper, we use the percentage error for evaluation to observe how much error there is in the daily load forecast. If there is an extreme value, this will greatly affect the safety and stability of the grid. The smaller the difference between the minimum and the lowest value, the more stable model. The percentage error is defined as follows:

$$Percentage\ Error = \left|\frac{V_{\text{Forecast}} - V_{\text{Actual}}}{V_{\text{Actual}}}\right|. \tag{3}$$

The range of percentage Error is $[0, 1]$, and it is negatively correlated with the accuracy of the model. As its value increases, the accuracy of prediction decrease.

# 4    Evaluation

## 4.1    Evaluation Results

Both data sets were used in this experiment to evaluate the accuracy of the prediction capacity of the three models. The results of an analysis using the MAPE metric are shown in Table 2, which compares the performance of the LSTM, XgBoosting, and SVM models. The min-max scaling approach is used to pre-process the data that all models require. Based on the findings, we can observe that the LSTM model consistently produces excellent results across both data sets, but the SVM model has the poorest performance on dataset 1, and the XgBoosting model was the model with the lowest performance on dataset 2. Regarding dataset 1, the difference in MAPE achieved via SVM and LSTM models may reach 3.3%. With regard to dataset 2, the MAPE curve does not demonstrate a significant decline; instead, it reveals just roughly a 2% change. It is acceptable for there to be a performance variance of around one to two percent between the use of the same model on various datasets. Unfortunately, because the MAPE is more than 5%, none of the aforementioned models performed well. It shows that models' performance is affected by the impacted factor.

**Table 2.** The Performance Results of Models

| Model | MAPE(Dataset 1) | MAPE(Dataset 2) |
|---|---|---|
| LSTM | 7.123 | 7.873 |
| XgBoosting | 7.874 | 9.868 |
| SVM | 10.489 | 8.126 |

Each model performs differently in terms of the different values of influenced factors on each dataset. As for the LSTM, it identifies the trend of dataset 1 well but has a poor ability to predict the minima, as shown in Fig. 2. The difference between the predicted and actual values can reach about 30,000. As for the XgBoosting model, it shows the same trend as the LSTM model, but the gap between the predictive value and the actual widened further when the minimum occurred. This is what causes the MAPE of the XgBoosting model to be larger than the LSTM model. The forecast curve of SVM shows an opposite trend, and its single-day forecast does not show a significant difference. As shown in Fig. 2, it cannot accurately predict the change in power load, but only the average power load during this period. This is a very bad outcome. If we follow the forecasts for safety assessment and generation planning, it is likely to lead to power outages, which can greatly threaten our power safety.

However, when the data set fluctuates a lot, as shown in data set 2, the predictive power of the LSTM model deteriorates significantly. The line in Fig. 3 clearly shows the big gap between actual load and prediction. Therefore, we infer that with more influencing factors and more fluctuations in load, the LSTM model

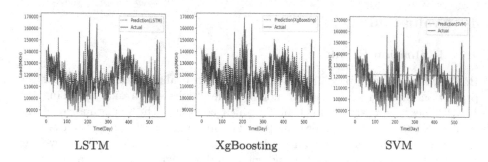

**Fig. 2.** The actual load and forecasting load on dataset 1.

does not work as well as other researchers have claimed. As for the XgBoosting model, compared with the LSTM model, it can identify the trend of large fluctuations in load in the short term, but the prediction results are still not satisfactory. Its prediction result is larger than the actual result, which results in inaccurate prediction. It means that planning power generation according to the forecast results will easily lead to power wastage. When observing the prediction line of the SVM model in Fig. 3, it has the same problem as we find in dataset 1. It fell into the problem of under-fitting, and it cannot predict large fluctuations in power load.

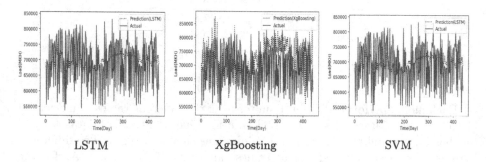

**Fig. 3.** The actual load and forecasting load on dataset 2.

The difference in daily percentage error also needs to maintain stability which directly affects power generation planning and grid safety. We expect a small percentage error gap between maximum and minimum because a 1% increase in STLF error could lead to a $10 million increase in annual operational expenses based on estimation [15]. When the maximum prediction error occurs, the maximum economic loss is expected to be as high as $370 million. In addition, when the difference in the percentage error between two adjacent days is too large, there may be a surplus of electricity on one day and a shortage on the other, which can seriously affect power safety and lives of people. Based on Table 3, shows that the LSTM model has the smallest max error and min error on data set 1 compared with others. Therefore, it will be regarded as the best model for this dataset. However, the gap between max error and min error achieves around 30%, which is very high. If we adopt this model in real situations, grid safety will still be compromised due to unstable forecasting. As for the dispatcher, they prefer to choose a model with stable forecasting even though the accuracy is not the best, even though the model has performance fluctuates between high and low accuracy. Because the large daily fluctuations in prediction errors are a big challenge for equipment, damaging the safety of the grid. As for dataset 2, the SVM model will be regarded as the best model but has the same problem. The maximum error is thousands of times the minimum error, even though it achieves very high accuracy in one day.

**Table 3.** The min-max daily percentage error of different models

| Model | Max Error(1) | Min Error(1) | Max Error(2) | Min Error(2) |
|-------|-------------|-------------|-------------|-------------|
| LSTM | 0.29 | 0.0001 | 0.35 | 0.0002 |
| XgBoosting | 0.37 | 0.0001 | 0.37 | 0.0002 |
| SVM | 0.36 | 0.0003 | 0.28 | 0.0001 |

The empirical study and analysis lead us to conclude that the LSTM model performs better than the SVM model. Although the aforementioned prediction models are accurate when just the electric load factor is examined, including other influencing variables at this time greatly reduces their accuracy. We need to improve the models more to cope with the influence of the influencing elements, which results from the complexity of the actual environment, the cumulative effect of the number of consumers, and the occurrence of severe events. The electricity system will become unstable once short-term predictions vary substantially since all models have a large difference between the maximum and minimum percentage errors. All of the repercussions of this are disastrous. Overloading equipment, for instance, increases the risk of malfunction and may cause blackouts. That is why it is important to remember that adding new uncertainty may significantly impact how well the model predicts. In other words, the aforementioned model does not safeguard the grid if we use it to schedule generation and anticipate severe situations. If we want to keep the prediction steady and prevent threats to safety, we need to reduce the MAPE and daily percentage error of the model.

## 4.2  Discussion

Based on the experiment result, the LSTM model has the best performance, which can learn the advantages of long-range time-series dependence, identify the changing pattern of the load itself from the time dimension, and identify the nonlinear effects of weather, power price, and other factors on load from the influence factor dimension. However, when the load fluctuates too much, the prediction performance will be degraded, especially when the load is lower than the average load, and there will be a larger error. If we adopt this model in the real case, the safety of the grid will be affected seriously because the stability will be destroyed.

**Table 4.** The forecasting on different conditions for LSTM models

| Dataset | Error | Max-temp | Min-temp |
|---|---|---|---|
| 1(High error) | 0.29 | 18 | 10 |
| 1(High error) | 0.27 | 16 | 9 |
| 1(Heat wave) | 0.04 | 44 | 14 |
| 1(Hear wave) | 0.04 | 43 | 21 |
| 1(Cold wave) | 0.1 | 13 | 0.8 |
| 1(Cold wave) | 0.09 | 11 | 1.9 |
| 2(High error) | 0.35 | 15 | 3 |
| 2(High error) | 0.28 | 15 | 0 |
| 2(Heat wave) | 0.07 | 37 | 25 |
| 2(Hear wave) | 0.17 | 36 | 22 |
| 2(Cold wave) | 0.07 | −0.10 | 16 |
| 2(Cold wave) | 0.07 | 0 | 12 |

We discuss the performance of the LSTM model when extreme conditions occur. It helps us to know if it can predict heat and cold waves. When we observe the influence of introduced factors, Table 4 shows a good prediction of the heat and cold waves in dataset 1, which has a percentage error below 1% and remains stable. However, the errors are large when the min-temperate fluctuates around 10 °C, and the max-temperature fluctuates around 17 °C on the same day. As for the performance on dataset 2, LSTM also shows a terrible ability to predict load when the max-temperature fluctuates around 15 °C, and the min-temperature fluctuates around 2. Besides, it shows a better ability to forecast cold waves than heat waves. In summary, LSTM performs well when extreme weather conditions occur, but daily percentage error is unstable with a sudden weather change.

In a real-world grid system, LSTM is an accurate AI model for predicting load demand. It assists operators in strategically scheduling maintenance or escalating generation capacity to ensure seamless operations under hot and cold wave conditions. In terms of benefits to operators, it contributes to greater operational efficiency and stability. By having a reliable estimation of future load, operators can make informed decisions on resource allocation, reducing the likelihood of grid failures or costly over-provisioning. Moreover, it may assist operators in predicting and managing peak load periods, which are often associated with higher operational costs and grid safety. However, the benefits will only be fully realized when the limitation of the model is addressed, such as high MAPE. We recommend further research and improvements on the model to better cater to the practical needs of grid operators.

## 5  Conclusion and Future Work

In this study, we evaluate the performance of three AI models considered to be the most accurate in terms of power load forecasting with the introduction of more influencing factors, discuss the implications for forecasting models for grid safety, and suggest that we also need to observe the daily error percentage as the most important evaluation criterion in terms of grid safety. For the model analysis, we believe that the LSTM model generally has a more accurate prediction capability and can maintain it during hot and cold waves. However, our experimental results still confirm that all three prediction models are still risky regarding grid safety, as the daily error ratio fluctuates significantly, affecting stability.

We believe that it is a good direction for future research to develop hybrid models that combine two or more algorithms to better handle the impact of multiple influences on the electric load to predict the load more accurately and thus maintain grid safety. Meanwhile, our next work plan will focus on optimizing model parameters on the AI model to obtain better performance and lower time complexity. Our future research will address prediction accuracy, training efficiency, and algorithm improvement in different scenarios.

## References

1. Azeem, A., Ismail, I., Jameel, S.M., Harindran, V.R.: Electrical load forecasting models for different generation modalities: a review. IEEE Access **9**, 142239–142263 (2021). https://doi.org/10.1109/ACCESS.2021.3120731
2. Chen, T., Guestrin, C.: XGBoost: a scalable tree boosting system. In: Proceedings of the 22nd ACM SIGKDD International Conference on Knowledge Discovery and Data Mining, KDD 2016, pp. 785–794. Association for Computing Machinery, New York (2016). https://doi.org/10.1145/2939672.2939785
3. Deng, X., et al.: Bagging-XGBoost algorithm based extreme weather identification and short-term load forecasting model. Energy Rep. **8**, 8661–8674 (2022). https://doi.org/10.1016/j.egyr.2022.06.072. https://www.sciencedirect.com/science/article/pii/S2352484722012124
4. Fan, C., Xiao, F., Wang, S.: Development of prediction models for next-day building energy consumption and peak power demand using data mining techniques. Appl. Energy **127**, 1–10 (2014)
5. Ganguly, P., Kalam, A., Zayegh, A.: Short term load forecasting using fuzzy logic. In: Proceedings of the International Conference on Research in Education and Science, pp. 355–361 (2017)
6. Hu, R., Wen, S., Zeng, Z., Huang, T.: A short-term power load forecasting model based on the generalized regression neural network with decreasing step fruit fly optimization algorithm. Neurocomputing **221**, 24–31 (2017). https://doi.org/10.1016/j.neucom.2016.09.027. https://www.sciencedirect.com/science/article/pii/S092523121631044X

7. Jhana, N.: Hourly energy demand generation and weather (2019). https://www. kaggle.com/datasets/nicholasjhana/energy-consumption-generation-prices-and-weather
8. Kozlov, A.: Daily electricity price and demand data (2020). https://www.kaggle. com/datasets/aramacus/electricity-demand-in-victoria-australia
9. Laouafi, A., Laouafi, F., Boukelia, T.E.: An adaptive hybrid ensemble with pattern similarity analysis and error correction for short-term load forecasting. Appl. Energy **322**, 119525 (2022)
10. Li, D., Gong, Y., Shen, S., Feng, G.: Research and design of meteorological disaster early warning system for power grid based on big data technology. In: 2020 Asia Energy and Electrical Engineering Symposium (AEEES), pp. 658–662 (2020). https://doi.org/10.1109/AEEES48850.2020.9121375
11. Li, J., Ren, Y., Fang, S., Li, K., Sun, M.: Federated learning-based ultra-short term load forecasting in power internet of things. In: Proceedings of the 2020 IEEE International Conference on Energy Internet (ICEI), pp. 63–68 (2020). https://doi. org/10.1109/ICEI49372.2020.00020
12. Lv, L., Wu, Z., Zhang, J., Zhang, L., Tan, Z., Tian, Z.: A VMD and LSTM based hybrid model of load forecasting for power grid security. IEEE Trans. Ind. Inf. **18**(9), 6474–6482 (2022). https://doi.org/10.1109/TII.2021.3130237
13. Nobre, J., Neves, R.F.: Combining principal component analysis, discrete wavelet transform and XGBoost to trade in the financial markets. Expert Syst. Appl. **125**, 181–194 (2019). https://doi.org/10.1016/j.eswa.2019.01.083. https://www. sciencedirect.com/science/article/pii/S0957417419300995
14. Patterson, J., Gibson, A.: Deep Learning: A Practitioner's Approach. O'Reilly Media, Inc. (2017)
15. Peng, L., Lv, S.X., Wang, L., Wang, Z.Y.: Effective electricity load forecasting using enhanced double-reservoir echo state network. Eng. Appl. Artif. Intell. **99**, 104132 (2021). https://doi.org/10.1016/j.engappai.2020.104132. https://www. sciencedirect.com/science/article/pii/S0952197620303699
16. Quan, H., Srinivasan, D., Khosravi, A.: Short-term load and wind power forecasting using neural network-based prediction intervals. IEEE Trans. Neural Netw. Learn. Syst. **25**(2), 303–315 (2014). https://doi.org/10.1109/TNNLS.2013.2276053
17. Silva, G.C., Silva, J.L.R., Lisboa, A.C., Vieira, D.A.G., Saldanha, R.R.: Advanced fuzzy time series applied to short term load forecasting. In: Proceedings of the 2017 IEEE Latin American Conference on Computational Intelligence (LA-CCI), pp. 1–6 (2017). https://doi.org/10.1109/LA-CCI.2017.8285726
18. Vapnik, V.: An overview of statistical learning theory. IEEE Trans. Neural Netw. **10**(5), 988–999 (1999). https://doi.org/10.1109/72.788640
19. Wang, C., Wang, C.: The study of the existing problems and processing measures based on power grid scheduling and safe operation. In: Proceedings of the 2018 Chinese Control and Decision Conference (CCDC), pp. 5194–5198 (2018). https:// doi.org/10.1109/CCDC.2018.8408034
20. Wang, X., Fang, F., Zhang, X., Liu, Y., Wei, L., Shi, Y.: LSTM-based short-term load forecasting for building electricity consumption. In: Proceedings of the 2019 IEEE 28th International Symposium on Industrial Electronics (ISIE), pp. 1418–1423 (2019). https://doi.org/10.1109/ISIE.2019.8781349
21. Xu, J.: Research on power load forecasting based on machine learning. In: Proceedings of the 2020 7th International Forum on Electrical Engineering and Automation (IFEEA), pp. 562–567 (2020). https://doi.org/10.1109/IFEEA51475.2020.00121
22. Ye, N., Liu, Y., Wang, Y.: Short-term power load forecasting based on SVM. In: Proceedings of the World Automation Congress 2012, pp. 47–51 (2012)

# Threat Detection and Analysis

# AST2Vec: A Robust Neural Code Representation for Malicious PowerShell Detection

Han Miao[1,2], Huaifeng Bao[1,2], Zixian Tang[1,2], Wenhao Li[1,2], Wen Wang[1(✉)],
Huashan Chen[1], Feng Liu[1,2], and Yanhui Sun[3]

[1] Institute of Information Engineering, Chinese Academy of Sciences, Beijing, China
wangwen@iie.ac.cn
[2] School of Cyber Security, University of Chinese Academy of Sciences,
Beijing, China
[3] China Assets Cybersecurity Technology Co., Ltd., Beijing, China

**Abstract.** In recent years, PowerShell has become a commonly used carrier to wage cyber attacks. As a script, PowerShell is easy to obfuscate to evade detection. Thus, they are difficult to detect directly using traditional anti-virus software. Existing advanced detection methods generally recover obfuscated scripts before detection. However, most deobfuscation tools can not achieve precise recovery on obfuscated scripts due to emerging obfuscation techniques. To solve the problem, we propose a robust neural code representation method, namely AST2Vec, to detect malicious PowerShell without de-obfuscating scripts. 6 Abstract Syntax Tree (AST) recovery-related statement nodes are defined to identify obfuscated subtrees. Then AST2Vec splits the large AST of entire PowerShell scripts into a set of small subtrees rooted by these 6 types of nodes and performs tree-based neural embeddings on all extracted subtrees by capturing lexical and syntactical knowledge of statement nodes. Based on the sequence of statement vectors, a bidirectional recursive neural network (Bi-RNN) is modeled to leverage the context of statements and finally produce vector representation of scripts. We evaluate the proposed method for malicious PowerShell detection through extensive experiments. Experimental results indicate that our model outperforms the state-of-the-art approaches.

**Keywords:** PowerShell scripts · malware detection · abstract syntax trees · representation learning · Bi-RNN

## 1 Introduction

PowerShell, the scripting language and shell framework installed by default on most Windows computers, is becoming a favored attack tool by malware. It provides access to most major functions of the operating system, which makes it an ideal candidate for many abused purpose such as reconnaissance, gaining

© The Author(s), under exclusive license to Springer Nature Switzerland AG 2023
M. Yung et al. (Eds.): SciSec 2023, LNCS 14299, pp. 207–224, 2023.
https://doi.org/10.1007/978-3-031-45933-7_13

persistence in the attacked system, communicating with a command and control server or downloading a payload. What's more, as a script, it is easy to obfuscate and difficult to detect with traditional security tools. According to a report from Symantec, over 95% of scripts using PowerShell were found to be malicious [10].

Many methods have been proposed to detect malicious PowerShell scripts. These methods can be categorized into two types: dynamic detection and static detection. Dynamic analysis often executes samples in an isolated environment and records their behaviors. However, there are many techniques to circumvent this method, such as sandbox awareness, which can hinder the execution of dynamic analysis. Static analysis often extracts static features such as strings for further detection, including two representative methods: 1) rule-based methods, which identify if the script matches one of the malware rules among the library collected in advance; 2) learning-based methods, which train machine learning or deep learning models using different static features, such as textual, token and AST node features [1,2,6,19] and verify new scripts based on this model. However, these approaches can be evaded by obfuscated scripts because obfuscation can easily change features mentioned above.

To tackle the issues, PowerShell deobfuscation work have been proposed. PSDEM [20], PSDecode [21] and PowerDrive [22] and PowerDecode [23] design a set of regular expressions to match obfuscated script pieces. Li et al. [4] identify obfuscated script pieces using a machine learning based classifier and conduct recovery through a simulation-based way. Invoke-Deobfuscation [3] utilizes recoverable nodes on AST to identify obfuscated pieces and implements variable tracing to mitigate the challenge above. However, all these approaches have some limitations. Regex expression based approaches are efficient, yet they often encounter syntax errors and semantics inconsistency due to lacking context and wrong replacement. Simulation-based approaches like Invoke-Deobfuscation obtain better results than regex expression based ones, but time-consuming and failed to deal with some sophisticated cases.

Thus, detecting malicious PowerShell scripts without deobfuscation can be considered as a promising approach to solve the problem. Intuitively, the idea is feasible for multiple reasons.

- In analogy to encrypted traffic analytics, many state-of-the-art works in this field like ETA [5] are typical content-independent approaches that only consider the observable metadata in structure level of traffic fragments and ignore the encrypted data information. Another similar task is software clone detection. Code obfuscation techniques are also used to prevent software clone detection. Some structure-based methods like ASTNN [7] have been proposed to solve the problem and achieve good results. These two works inspired us that structure may contain a large amount of useful information, which is not inferior to lexical information. We can formulate malicious PowerShell scripts detection as a similar task.
- Most new PowerShell scripts recycle large chunks of source code from existing scripts with some changes and additions. These scripts are similar in structure and functions, and just different in specific parameters. If the training data

were sufficient to contain malicious obfuscation scripts, the trained model would identify the similar malicious samples, no matter whether these samples were obfuscated or not.

In this paper, we propose a robust approach for representing PowerShell scripts that do not have to be deobfuscated, called AST2Vec, which splits the large AST of entire PowerShell scripts into a set of small subtrees and performs tree-based neural embeddings on all extracted subtrees. The contributions of this paper can be summarized as follows:

- We propose an efficient, effective and robust neural PowerShell scripts representation, which can capture the code fragments about recovering obfuscated pieces.
- Based on the new neural representation method, we design the first malicious PowerShell detection system without deobfuscation.
- The result shows that detection based on AST2Vec representation was enough to achieve a 3-fold cross-validation accuracy of 96.7% on obfuscated dataset.

The rest of this paper is organized as follows: Related work and preliminaries are introduced in Sect. 2 and Sect. 3, respectively. Section 4 presents the methodology. Section 5 describes the implementation and results of experiments. Discussion is presented in Sect. 6. Finally, we conclude the paper in Sect. 6.

## 2 Related Work

### 2.1 Obfuscation Techniques

Obfuscation techniques are often used by malware-writers to obfuscate their code and hinder static analysis. Compression, encryption, and encoding are some of the most common obfuscation methods used by threat actors. Multiple methods are often used in tandem to evade a wider variety of cyber security tools at the initial point of intrusion.

**String-Based Obfuscation.** String based obfuscation techniques typically include encoding and encryption [13–15], which are common in scripts and Java code. It rewrites the names of various elements of the code, such as variables, functions, and classes, into meaningless names. This greatly reduces the readability of the code, making it impossible for readers to guess its purpose based on the name.

**Logical Obfuscation.** Logical obfuscation is another type of obfuscation technique targeted at disrupting the control flow and the data flow, which changes the code into a functionally equivalent but less understandable form by rewriting parts of the logic in the code. It is commonly used to obfuscate binary samples. LLVM is a tool for anti-sample reverse analysis [29]. It uses multiple optimizers to achieve a variety of logical obfuscation effects, such as code control flow flattening, false control flow, instruction replacement, etc. Analyzing samples that are confused by logic is very difficult.

## 2.2 Detection of Obfuscated Malware

In recent years, static analysis techniques [20–23], dynamic analysis techniques [30], as well as machine learning [1,4] and deep learning [2,19,24] have been widely used in malicious code detection and have achieved good results. However, detecting obfuscated malicious code remains a major challenge, as obfuscation techniques can make logically and functionally similar code appear completely different. Therefore, additional work is needed to assist traditional detection methods. Nicola Ruaro et al. [28] proposed a technology based on Symbolic Execution to disambiguate advanced malicious Excel files. Overall, the task of detecting malicious code without resolving confusion is very difficult.

## 2.3 Deobfuscation Techniques

**Dynamic Analysis Based Deobfuscation Techniques.** Dynamic analysis often executes samples in an isolated environment and records their behaviors [30]. However, there are many techniques that can hinder the execution of dynamic analysis, such as sandbox detection. Besides, benign scripts and malicious scripts quite frequently exhibit the same behaviors and functionality so without supplemental context or meta-data, it's hard to discern their intent on the surface by dynamic analysis.

**Statistic Analysis Based Deobfuscation Techniques.** Static analysis identifies obfuscated data and the corresponding recovery algorithm, which is usually very difficult. Common deobfuscation techniques based on statistic analysis can be divided into two types: regex expression based and AST based. Regex expression based tools process scripts at the level of strings and ignore the syntax of script pieces so that they cannot identify obfuscation pieces precisely. Li et al. [4]identify obfuscated script pieces using a machine learning based classifier and AST features. However, due to lacking context and wrong replacement, their tool approach often encounters syntax errors and semantics inconsistency. Invoke-Deobfuscation [3] utilizes recoverable nodes on AST to identify obfuscated pieces and implements variable tracing to mitigate the challenge above. However, it does not consider the situation of loop obfuscation and custom encryption.

# 3   Preliminaries

## 3.1   Abstract Syntax Tree

Abstract Syntax Tree (AST) is a hierarchical tree-like structure that captures the structure of a program's syntax, without including details such as formatting, comments, or other non-essential information. The nodes in the tree represent the program's constructs, such as expressions, statements, functions, classes, and other language-specific constructs. As illustrated in Fig. 1(b), nodes of an AST are corresponding to constructs and symbols of the PowerShell script. The high-level abstraction makes ASTs widely used in software engineering and programming language analysis fields.

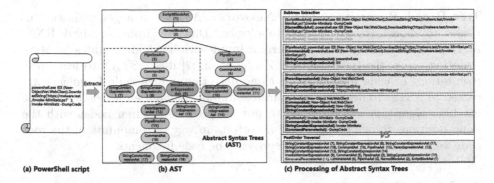

(a) PowerShell script    (b) AST    (c) Processing of Abstract Syntax Trees

Fig. 1. An example of AST extraction.

Some studies [19,24] use ASTs in machine learning or deep learning based methods for malicious PowerShell detection by traversing or encoding ASTs as an entire or extracting statistical characteristics of trees. We shall be using ASTs as well but split the large AST of one code fragment into a set of small trees at the statement level for the next steps. The lower part of Fig. 1(c) illustrates the post order traversal of AST to get a sequence which destroys the original syntactic structure of source code. The upper part represents the newly proposed preprocessing method in this paper. Details will be covered in Sect. 3.

### 3.2 Tree-Based Neural Networks

Recently, many methods have proposed to take ASTs as input of Tree-based Neural Networks (TNNs) [7,8,18]. Given a tree, TNNs learns its vector representation by recursively computing node embeddings in a bottom-up way. Tree-based CNN (TBCNN) [8] and Tree-based Recursive Neural Networks (ASTNN) [7] are representative works in the field of tree-based neural networks.

**Tree-Based Convolutional Neural Network.** TBCNN is a tree-based convolutional neural network which uses CNN over tree structures for supervised learning such as source code classification [8]. The most important module of it is an AST-based convolutional layer, which applies a set of fixed-depth feature detectors by sliding over entire ASTs. This procedure can be formulated by:

$$y = tanh(\sum_{i=1}^{n} W_{conv,i} \cdot x_i + b_{conv})$$

where $x_1, \cdots, x_n$ are the vectors of nodes within each sliding window, $W_{counv,i}$ is the parameter matrices and $b_conv$ is the bias. TBCNN adopts a bottom-up encoding layer to integrate some global information for improving its localness. Although nodes in the original AST may have more than two children, TBCNN treats ASTs as continuous full binary trees because of the fixed size of convolution.

**Tree-Based Recursive Neural Network.** ASTNN is a generalization of RNNs to model tree-structured topologies. Different from standard RNN, ASTNN [7] recursively combines current input with its children states for state updating across the tree structure. Zhang et al. [7] uses ASTNN to learn representations of code fragments for clone detection, where code fragments are parsed to ASTs. To deal with the variable number of children nodes, a dynamic batching algorithm was proposed to put all possible children nodes with the same positions to one group. After a bottom-up way of computation, the root node vectors of ASTs are used to represent the code fragments.

# 4    Methodology

## 4.1    Framework

In this section, we introduce our malicious PowerShell detection approaches based on AST-based neural representation method (AST2Vec). The overall framework of our AST2Vec based detection is shown in Fig. 2. At a high level, the process of detection can be divided into three stages.

1) **Extraction of ASTs.** During the first stage, we parse a PowerShell script into an AST.
2) **Representation learning.** Then, we design a preorder traversal algorithm to split each AST to a sequence of subtrees, as illustrated in Fig. 1(c). All subtrees are encoded by the Subtree Encoder to vectors, denoted as $e_1, \cdots, e_t$. We then use Bidirectional Gated Recurrent Unit [16] (Bi-GRU), to model the context of the subtrees. The hidden states of Bi-GRU are sampled into a single vector by pooling, which is the representation of the scripts. Details of AST2Vec are illustrated in Fig. 3.
3) **Classifier training.** Finally, we employ the embedding as a first layer inputs in a deep neural network trained (using the labeled instances of the training set) to detect malicious PowerShell code.

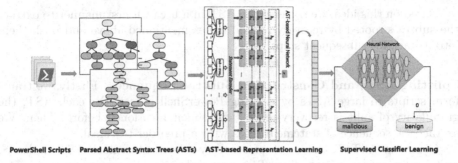

**Fig. 2.** The general framework of AST-based neural network for malicious PowerShell detection.

## 4.2 Splitting ASTs and Constructing Subtree Sequences

**The Granularity of Subtrees.** We adopt Microsoft's official library *System. Management.Automation.Language* to obtain ASTs for PowerShell. There are 71 types of PowerShell's AST nodes in total, such as PipelineAst, CommandAst, CommandExpressionAst, etc. As illustrated in Fig. 1(b), the parser returns an AST with a ScriptBlockAst type of root. A typical script with sizes of several Kilobytes can have thousands of nodes in AST, which means thousands of subtrees and thus makes it time-consuming to consider all subtrees.

To solve the problem, we only consider the below 6 types of subtrees, subtrees roots of PipelineAst, UnaryExpressionAst, BinaryExpressionAst, ConvertExpressionAst, InvokeMemberExpressionAst and SubExpressionAst type. This choice has two advantages:

1) **Significantly reduce the count of subtrees.** Through our observation, the numbers of nodes for certain node types are typically increased during the obfuscation process. For example, string reordering will add several ParenExpressionAst nodes and StringConstantExpressAst nodes to AST, which makes the AST more redundant. Fortunately, nodes like PipelineAst, UnaryExpressionAst, BinaryExpressionAst, ConvertExpressionAst, Invoke-MemberExpressionAst and SubExpressionAst are related to node recovery in obfuscated scripts [3]. We call these 6 types of subtrees, recovery-related statement subtrees. Subtrees rooted by these nodes may include abundant nodes like StringConstantExpressAst, which means extracting them as an entire subtree will significantly reduce the count of meaningless subtrees.

2) **Improve the interpretability of subtrees.** At the AST level, the deobfuscation process can be considered as converting subtrees in an obfuscated script to corresponding ones in the target script. So extracting subtrees rooted by recovery-related statement nodes and representing them as an entire will improve the interpretability of subtrees.

Based on this idea, we traverse the AST in a breadth-first manner to extract the subtrees rooted by recovery-related nodes mentioned above and push them into a stack for subsequent steps.

**Splitting ASTs and Constructing Subtree Sequences.** Firstly, we transform scripts to large ASTs by existing PowerShell parser. For each AST, the granularity of split is recovery-related statement mentioned before. Then, We extract the sequence of statement trees with a preorder traversal.

Given an AST $T$ and a set of recovery-related statement AST nodes $S$, each statement node $s \in S$ in $T$ corresponds to one recovery-related statement of scripts. We treat ScriptBlockAst as a special Statement node, thus $S = S \cup \{ScriptBlockAst\}$. As shown in Fig. 1, there're many nested statements in ASTs. To get subtrees without non-overlapping, we reserve the nested statements such as Pipeline and InvokeMemberExpression statements in the body of the parent tree and copy them as the header of descendants. All the descendants of statement node $s \in S$ is denoted by $D(s)$. For any $d \in D(s)$, if there exists one path from $s$ to $d$ through one node $p = d$, it means that the node $d$ is included by one statement in the body of statement $s$. We call node $d$ one substatement node of $s$. Then a subtree rooted by the statement node $s \in S$ is the tree consisting of node $s$ and all of its descendants, excluding its substatement nodes' descendants in $T$. For example, the subtree rooted by PipelineAst is surrounded by dashed lines in Fig. 1(b), which includes nodes such as "CommandAst", two "StringConstantExpressionAst" and "InvokeMemberExpressionAst" and excludes the nodes of "ParenExpressionAst" and its sibling and descendant nodes in the body. The node "InvokeMemberExpressionAst" is both in the body of PipelineAst statement tree and in the descendants as a header. In this way, one large AST can be split to a sequence of multi-way subtrees.

The splitting of ASTs is straightforward by a traverser which visits each node through the ASTs in a depth-first walk in preorder. Then, a constructor recursively creates a subtree to sequentially add to the subtree sequences. Such a practice guarantees that one new subtree is appended by the order of the scripts. Finally, we get the sequence of subtrees as the raw input of AST2Vec.

### 4.3   Encoding Multi-way Subtrees

In this section, we introduce how to encode recovery-related statements mentioned above on multi-way subtrees. There are two challenges: learning vector representations of statements and designing the batch processing algorithm to encode subtrees with different number of children nodes.

**Statement Vectors.** Given the subtrees, we design a Bi-RNN based statement encoder to learn vector representations of statements.

Firstly, we obtain all the symbols by postorder traversal of ASTs as the corpus for unsupervised representation training used word2vec [25]. The trained

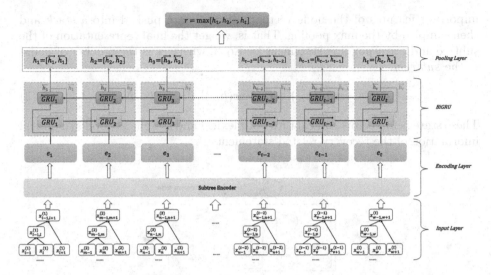

**Fig. 3.** The architecture of AST-based neural network.

embeddings of symbols are served as initial parameters in the subtree encoder, which are illustrated in the bottom in Fig. 3.

Taking the subtree rooted by the node of PipelineAST in Fig. 1(b) surrounded by dashed lines as an example, the encoder traverses the subtree and recursively takes the symbol of current node as new input to compute, together with the hidden states of its children nodes. This is illustrated in the lower two level in Fig. 4, in the subtree, the three children nodes–two StringConstantExpressionAst and a InvokeMemberExpressionAst enrich the meaning of CommandAst and the CommandAst transfer the merged information to the upper PipelineAST node.

Specifically, given a subtree $t$, let $n$ denote a non-leaf node and $C$ denote the count of its children nodes. At the beginning, with the pre-trained embedding parameters $W_\varepsilon \in \mathbb{R}^{|V| \times d}$ where $V$ is the vocabulary size and $d$ is the embedding dimension of symbols, the lexical vector of node $n$ can be obtained by:

$$v_n = W_\varepsilon^\top x_n \tag{1}$$

where $x_n$ is the one-hot representation of symbol $n$ and $v_n$ is the embedding. Next, the vector representation of node $n$ is computed by the following equation:

$$h = f(W_n^\top v_n + \sum_{i \in [1,C]} h_i + b_n) \tag{2}$$

where $W_n \in \mathbb{R}^{d \times k}$ is the weight matrix with encoding dimension $k$, $b_n$ is a bias term, $h_i$ is the hidden state for each child $i$, $h$ is the updated hidden state, and $f$ is the activation function such as $tanh$ or the identity function. We use the identity function in this paper. Similarly, we can recursively compute and optimize the vectors of all nodes in the subtree $t$. In addition, in order to determine the most important features of the node vectors, all nodes are pushed into a stack and then sampled by the max pooling. That is, we get the final representation of the subtree and corresponding statement by Eq. 3, where $N$ is the number of nodes in the $subtree_t$.

$$e_t = [max(h_{i1}), \cdots, max(h_{ik})], i = 1, \cdots, N \tag{3}$$

These statement vectors can capture both lexical and statement-level syntactical information of the recovery-related statements.

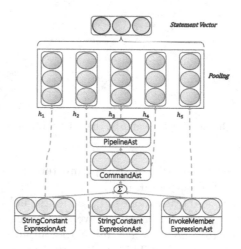

**Fig. 4.** The subtree encoder.

**Batch Processing.** There is a problem when batch processing on multiway subtrees, since the number of children nodes varies for the parent nodes in the same position of one batch. For example, given two parent nodes st1 with 3 children nodes and st2 with 2 children nodes in Fig. 5, directly calculating Eq. 2 for the two parents in one batch is impossible due to different $C$ values. To tackle this problem, we design an algorithm that dynamically processes batch samples.

Intuitively, although parent nodes have different number of children nodes, the algorithm can dynamically detect and put all possible children nodes with the same positions in groups, and then speed up the calculations of Eqs. 2 of each group in a batch way by leveraging matrix operations. An example is illustrated in Fig. 5.

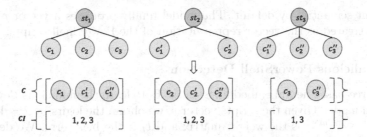

**Fig. 5.** An example of dynamically batching children nodes.

### 4.4 Representing the Sequence of Subtrees

After getting the sequences of subtrees vectors, we exploit GRU [17] to track the context of scripts.

Given a PowerShell script, suppose there are $T$ subtrees extracted from its AST and let $Q \in \mathbb{R}^{T \times k} = [e_1, \cdots, e_t, \cdots, e_T], t \in [1, T]$ denote the vectors of encoded subtrees in the sequence. At time $t$, the transition equations are as follows:

$$
\begin{aligned}
r_t &= f(W_r e_t + U_r h_{t-1} + b_r) \\
z_t &= f(W_z e_t + U_z h_{t-1} + b_z) \\
\tilde{h}_t &= tanh(W_h e_t + r_t \odot (U_h h_{t-1}) + b_h) \\
h_t &= (1 - z_t) \odot h_{t-1} + z_t \odot \tilde{h}_t
\end{aligned}
\tag{4}
$$

where $r_t$ is the reset gate to control the influence of previous state, $z_t$ is the update gate to combine past and new information, $\tilde{h}_t$ is the candidate state and used to make a linear interpolation together with previous state $h_{t-1}$ to determine the current state $h_t$. $W_r, W_z, W_h, U_r, U_z, U_h \in \mathbb{R}^{k \times m}$ are weight matrices and $b_r, b_z, b_h$ are bias terms. $\odot$ denotes multiplication by elements. After iteratively computing hidden states of all time steps, the sequential context of these statements can be obtained.

In order to further enhance the capability of the recurrent layer for capturing the dependency information, we adopt a bidirectional GRU [16], where the hidden states of both directions are concatenated to form the new states as follows:

$$
\begin{aligned}
\overrightarrow{h_t} &= \overrightarrow{GRU}(e_t), t \in [1, T] \\
\overleftarrow{h_t} &= \overleftarrow{GRU}(e_t), t \in [T, 1] \\
h_t &= [\overrightarrow{h_t}, \overleftarrow{h_t}], t \in [1, T]
\end{aligned}
\tag{5}
$$

Similar to the subtree encoder, the most important features of these states are then sampled by the max pooling or average pooling. Considering the importance of different statements are intuitively not equal, for example, InvokeMemberExpressionAst statements may contain more functional information than StringConstantExpressionAst, thus we use max pooling for capturing the most important semantics by default. The model finally produces a vector $r \in R^{2m}$, which is treated as the vector representation of the PowerShell scripts.

### 4.5  Malicious PowerShell Detection

We use tree-based neural embeddings on ASTs to learn the malicious PowerShell detection model. Given the scripts vector $r$, we obtain the logits by $\hat{x} = W_l r + b_l$, where $W_l \in \mathbb{R}^{2m \times 2}$ is the weight matrix and $b_l$ is the bias term. We define the loss function as the widely used cross-entropy loss:

$$J(\Theta, \hat{x}, y) = \sum \left( -log \frac{exp(\hat{x}_y)}{\sum_j exp(\hat{x}_j)} \right) \tag{6}$$

## 5  Experiments and Results

### 5.1  Dataset Description

To evaluate our method, we create a collection of malicious and benign, obfuscated and nonobfuscated PowerShell samples. They come from all possible download sources that can have PowerShell scripts, e.g., GitHub, shared academic data, security blogs, etc. The overall information for datasets is listed in Table 1.

**Malicious Samples.** We get 3,346 obfuscated malicious PowerShell scripts from the author of Invoke-Deobfuscation [3] and download all 619 PowerShell instances from 'MalwareBazaar' website [12]. Besides, 4079 nonobfuscated malicious instances are obtained from unit42's blogs [11]. Finally, we removed invalid and duplicate PowerShell instances utilizing the syntax information and textual features of instances. After preprocessing, we ultimately get 8,005 malicious PowerShell samples.

**Benign Samples.** There are substantial PowerShell samples in PowerShellCorpus [26] created by Daniel Bohannon and most of them are benign. We select 8000 benign ones from the corpus by the help of VT [27]. To balance the proportion of obfuscated and nonobfuscated samples, we obfuscate half of them using popular obfuscation tools, e.g., Invoke-Obfuscation [13], PowerSploit [14], Empire [15], etc.

### 5.2  Experiment Settings

In order to obtain ASTs for PowerShell, we adopt Microsoft's official library *System.Management.Automation.Language* to parse PowerShell scripts to ASTs. We trained embeddings of AST nodes using word2vec [25] with Skip-gram algorithm and set the embedding size to be 128. The hidden dimension of subtree encoder and bidirectional GRU is 100. We set the batch size to 32 and a maximum of 20 epochs. The threshold is set to 0.5 for malicious PowerShell detection. We randomly divide the dataset into three parts, of which the proportions

Table 1. Overall information for datasets.

| Datasets | # Scripts | # Malicious instances | % Obfuscation |
|---|---|---|---|
| Invoke-Deobfuscation | 3,346 | 3,346 | 100 |
| MalwareBazaar | 580 | 580 | 51.72 |
| Unit42 | 4,079 | 4,079 | 0 |
| PowerShellCorpus(sub-fraction) | 8,000 | 0 | 50 |

are 60%, 20%, 20% for training, validation and testing. We use the optimizer AdaMax with learning rate 0.002 for training. All experiments are conducted on a computer with Intel Core i7-10750H CPU and 16 GB of memory for all experiments.

### 5.3   Evaluation on Detection Task

To evaluate the effectiveness of malicious PowerShell detection, we use the test accuracy metric, which computes the percentage of correct classifications for the test set. In this section, we compare the detection accuracy with state-of-the-art approaches in different groups including deobfuscation-based, simple AST features based and character-based approaches as follows. Experimental results are provided in Table 2 The best results are shown in bold.

**Deobfuscation-Based.** Li et al. [4] designed a light-weight deobfuscation approach for PowerShell scripts. Building upon this deobfuscation method, they further designed a semantic-aware PowerShell attack detection system. We reproduced the approach in our obfuscated and mixed datasets.

**Simple AST Features Based.** The detection method of Rusak et al. [19] used only two features: depth and number of nodes per PowerShell AST and was conducted on nonobfuscated scripts. This method didn't fully mining the potential information of abstract syntax trees. We reproduced the approach in our nonobfuscated, obfuscated and mixed datasets.

**Token-Based.** Hendler et al. [2] proposed a detection approach which apply token-based features for detection. The original design support several classifiers, and we only choose the combination of a 3-CNN and traditional 3-gram from which with the best results on their paper to reproduce. [6] was also proposed by Hendler et al. This method first gets deobfuscated scripts from AMSI, then extracts tokens and characters from scripts, uses Word2Vec for token embedding, and finally models convolutional neural networks and a Bi-LSTM to classify PowerShell scripts.

**Table 2.** Comparison with state-of-the-art detection approaches in accuracy.

| Detection approaches | Obfuscated scripts | Nonobfuscated scripts | Mixed Scripts |
|---|---|---|---|
| Li et al. [4] | – | 92.3% | 92.2% |
| Rusak et al. [19] | 0.0% | 85.7% | 9.6%2 |
| Hendler et al. [2] | 12.1% | 95.1% | 34.7% |
| ASMI-based [6] | 30.2% | 96.5% | 40.3% |
| AST2Vec | **96.7%** | **97.1%** | **96.4%** |

## 5.4   Results

Based on the evaluation on the above task, we aim to investigate the following research questions:

**RQ1: How does our approach perform in malicious PowerShell scripts detection?** In the task of malicious PowerShell scripts detection, the samples are strictly balanced among the malicious and benign classes and our evaluation metric is the test accuracy. Experimental results are provided in Table 2. The best results are shown in bold.

As shown, except for deobfuscation-based method, both the AST-based and character-based detection approaches will be bypassed by obfuscation. Once these scripts are deobfuscated, our results show that these two previous approaches can achieve similar accuracy rates with our approach. However, we would like to note that since the features used by character-based detection approach are manually crafted, they can be more easily evaded compared to our structural-aware approach. To show this, we simply mix benign pieces into malicious samples at the granularity of script lines, which changes AST structure and character distributions without affecting the script behavior. In Table 2, we call them "mixed scripts". As shown, these mixed scripts can greatly decrease the accuracy rates of simple AST features based and character-based detection approaches, but cannot affect that of our approach.

**RQ2: What are the effects of different design choices for the proposed model?** We conduct experiments to study how different design choices affect the performance of the proposed model on the detection task. As shown in Table 3, we consider the following design choices:

– **Splitting granularities of ASTs.** There are many ways to split a large AST into different sequences of non-overlapping small trees. Besides our semantic level splitting, another possible way is to extract all nodes of the AST(AST-Nodes) as special "trees". In comparative experiment, we adopt post order traversal to extract all nodes of the AST. After splitting, the follow-up encoding and bidirectional GRU processing are the same as those in our model. We can see that our semantic-based splitting outperform extreme splitting

**Table 3.** Comparison between the proposed model and its design alternatives.

| Description | Malicious PowerShell detection F1 score (%) |
|---|---|
| Bi-LSTM instead of Bi-GRU | 97.2 |
| GRU instead of Bi-GRU | 95.4 |
| One-hot instead of Word2Vec | 96.0 |
| Removing Pooling-I | 96.6 |
| Removing Pooling-II | 94.2 |
| AST-Node(PostOrder traversal) | 92.2 |
| AST2Vec | **97.4** |

approaches of AST-Nodes. Our model achieves a better performance, as analyzed in Sect. 3, this is because it balances a good trade-off between the size of subtree and the richness of syntactical and semantic information.

– **Initial representation of nodes.** The word2vec [25] is used to learn unsupervised vectors of the symbols, and the trained embeddings of symbols are served as initial parameters in the subtree encoder. We extract all ASTs from the total PowerShell scripts and train embeddings on them. The qualitative results are summarized in a dendrogram in Fig. 4. It shows the relationships of embeddings with similar ones. Notably, the TryStatement and CatchClause node types are neighbors, as well as Command and CommandParameter. Afterward, we will use these embeddings as initial parameters in the subtree encoder to enrich the semantic information of en, since one would expect such commands to serve similar functions in scripts. There are 71 types of PowerShell's AST nodes in total. If replacing word2vec embeddings by one-hot representation, the results indicate that overall one-hot representation decrease the F1 score by 1.4%. This shows that the initial representation of AST nodes plays a significant role in detection.

– **GRU.** We use GRU in the recurrent layer of our proposed model by default. If replacing GRU by LSTM, the results indicate that overall LSTM has a slightly poor with GRU. We prefer GRU in our model since it achieves more efficient training.

– **Pooling layer.** In our model, we use the max pooling on subtrees in the statement encoder (Pooling-I) and the max pooling layer on the statement sequences after the recurrent layer (Pooling-II) as described in Sect. 3. We study whether the two pooling components affect the performance or not by removing them and directly using the last layer hidden states. From the Table 3, we can see that the pooling on statement sequences provides a comparatively significant performance boost, whereas pooling on subtrees matters little. This shows that different statements of the same scripts actually have different weights (Fig. 6).

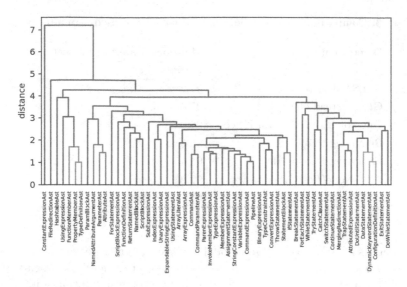

**Fig. 6.** Dendrogram of node types and their relationships in the Learning Node Representations experiment.

# 6    Discussion

## 6.1    Generality of Our Approach

Although our subtree-based representation approach in this paper is developed for PowerShell, its design does not require specific features of PowerShell. As long as the obfuscation is to cover the semantics by hiding the script pieces as strings, our approach can be applied to achieve effective representation and detection. As far as we know, JavaScript and WebShell utilize similar obfuscation techniques.

Moreover, our method requires only a parser and an unmodified interpreter for the target language, both of which typically have official tools available. Therefore, the only extra work required to construct a detection system for a new language is to collect the samples of the obfuscated subtrees and train the attack detector.

## 6.2    Necessity of Obfuscation for Malicious PowerShell Detection

The results in Table 2 shows that our approach performs well on both obfuscated and nonobfuscated datasets. There are several reasons accounting for this. For one thing, most new PowerShell scripts recycle large chunks of source code from existing scripts with some changes and additions, whether the source code is obfuscated. These scripts are similar in structure and functions, and differ only in specific parameters. If these scripts are parsed to ASTs, they will have similar, even same patterns in ASTs. For another, our proposed approach learns vector representations of scripts fragments, which can capture the syntactical knowledge and naturalness of semantic statements of the PowerShell well. Thus, If the training data are sufficient to contain malicious obfuscation scripts, the trained model would identify the similar malicious samples, no matter whether these samples are obfuscated or not.

## 7  Conclusion

In this paper, we design the first effective and feasible detection approach for malicious PowerShell scripts without deobfuscation. To address the key challenge of robustly representing the PowerShell scripts both obfuscated and nonobfuscated, we design a novel AST-based representation method that could capture the recovery-related statement at the level of subtrees. Building upon the representation method, we design the PowerShell attack detection system. Based on a collection of 16,005 PowerShell scripts, our detection model is shown to be both robust and effective.

**Acknowledgments.** This work was supported by the National Key R&D Program of China with No. 2021YFB3101402.

## References

1. Fang, Y., Zhou, X., Huang, C.: Effective method for detecting malicious PowerShell scripts based on hybrid features. Neurocomputing **448**, 30–39 (2021)
2. Hendler, D., Kels, S., Rubin, A.: Detecting malicious PowerShell commands using deep neural networks. In: Proceedings of the 2018 on Asia Conference on Computer and Communications Security (2018)
3. Chai, H., Ying, L., Duan, H., Zha, D.: Invoke-deobfuscation: AST-based and semantics-preserving deobfuscation for PowerShell scripts. In: 2022 52nd Annual IEEE/IFIP International Conference on Dependable Systems and Networks (DSN), pp. 295–306 (2022)
4. Li, Z., Chen, Q.A., Xiong, C., Chen, Y., Zhu, T., Yang, H.: Effective and lightweight deobfuscation and semantic-aware attack detection for PowerShell scripts. In: Proceedings of the 2019 ACM SIGSAC Conference on Computer and Communications Security (2019)
5. Blake, A., David, M.: Identifying encrypted malware traffic with contextual flow data. In: Proceedings of the 2016 ACM Workshop on Artificial Intelligence and Security, AISec 2016, pp. 35–46 (2016)
6. Hendler, D., Kels, S., Rubin, A.: AMSI-based detection of malicious PowerShell code using contextual embeddings. In: Proceedings of the 15th ACM Asia Conference on Computer and Communications Security (2019)
7. Zhang, J., Wang, X., Zhang, H., Sun, H., Wang, K., Liu, X.: A novel neural source code representation based on abstract syntax tree. In: Proceedings of the 41st International Conference on Software Engineering, pp. 783–794. IEEE Press (2019)
8. Mou, L., Li, G., Zhang, L., Wang, T., Jin, Z.: Convolutional neural networks over tree structures for programming language processing (2015)
9. Mikolov, T., Karafiát, M., Burget, L., Cernock, J., Khudanpur, S.: Recurrent neural network based language model. In: Interspeech, Conference of the International Speech Communication Association, Makuhari, Chiba, Japan, September (2015)

10. ISTR Living off the land fileless attack techniques. https://www.symantec.com/content/dam/symantec/docs/security-center/whitepapers/istr-living-off-the-land-and-fileless-attack-techniques-en.pdf. Accessed 11 Apr 2023
11. karttoon, psencmds (2019). https://github.com/pan-unit42/iocs/commits/master/psencmds. Accessed 13 Dec 2019
12. MalwareBazaar. https://bazaar.abuse.ch/
13. Bohannon, D.: Invoke-obfuscation - powershell obfuscator. https://github.com/danielbohannon/Invoke-Obfuscation
14. Powersploit - a powershell post-exploitation framework. https://github.com/PowerShellMafia/PowerSploit
15. Empire - a PowerShell and python post-exploitation agent. https://github.com/EmpireProject/Empire
16. Tang, D., Qin, B., Liu, T.: Document modeling with gated recurrent neural network for sentiment classification. In: Proceedings of the 2015 Conference on Empirical Methods in Natural Language Processing, pp. 1422–1432 (2015)
17. Bahdanau, D., Cho, K., Bengio, Y.: Neural machine translation by jointly learning to align and translate. arXiv preprint arXiv:1409.0473 (2014)
18. Wei, H.-H., Li, M.: Supervised deep features for software functional clone detection by exploiting lexical and syntactical information in source code. In: Proceedings of the 26th International Joint Conference on Artificial Intelligence, pp. 3034–3040. AAAI Press (2017)
19. Rusak, G., Al-Dujaili, A., O'Reilly, U.M.: AST-based deep learning for detecting malicious PowerShell. In: ACM CCS (2018)
20. Liu, C., Xia, B., Yu, M., Liu, Y.: PSDEM: a feasible de-obfuscation method for malicious PowerShell detection. In: IEEE ISCC (2018)
21. Psdecode - PowerShell script for deobfuscating encoded PowerShell scripts. https://github.com/R3MRUM/PSDecode
22. Ugarte, D., Maiorca, D., Cara, F., Giacinto, G.: PowerDrive: accurate de-obfuscation and analysis of PowerShell malware. In: Perdisci, R., Maurice, C., Giacinto, G., Almgren, M. (eds.) DIMVA 2019. LNCS, vol. 11543, pp. 240–259. Springer, Cham (2019). https://doi.org/10.1007/978-3-030-22038-9_12
23. Malandrone, G.M., Virdis, G., Giacinto, G., Maiorca, D.: PowerDecode: a Power-Shell script decoder dedicated to malware analysis. In: ITASEC (2021)
24. Gao, Y., Peng, G., Yang, X.: PowerShell malicious code family classification based on deep learning. J. Wuhan Univ. (Nat. Sci. Ed.) **68**(1), 8–16 (2022). https://doi.org/10.14188/j.1671-8836
25. Mikolov, T., Sutskever, I., Chen, K., Corrado, G.S., Dean, J.: Distributed representations of words and phrases and their compositionality. In: Advances in Neural Information Processing Systems, pp. 3111–3119 (2013)
26. Bohannon, D., Holmes, L.: Revoke-Obfuscation: PowerShell Obfuscation Detection Using Science (2017). https://www.fireeye.com/blog/threatresearch/2017/07/revoke-obfuscation-powershell.html
27. VirusTotal. https://www.virustotal.com/
28. Ruaro, N., Pagani, F., Ortolani, S., Kruegel, C., Vigna, G.: SYMBEXCEL: automated analysis and understanding of malicious excel 4.0 macros. In: 43rd IEEE Symposium on Security and Privacy, SP 2022, San Francisco, CA, USA, 22–26 May 2022, pp. 1066–1081. IEEE (2022)
29. LLVM. https://www.llvm.org/
30. Cozzi, E., Graziano, M., Fratantonio, Y., Balzarotti, D.: Understanding Linux malware. In: 2018 IEEE Symposium on Security and Privacy (S&P), pp. 161–175 (2018)

# Real-Time Aggregation for Massive Alerts Based on Dynamic Attack Granularity Graph

Haiping Wang[1,2] , Binbin Li[1(✉)], Tianning Zang[1,2], Yifei Yang[1], Zisen Qi[1,2],
Siyu Jia[1,2], and Yu Ding[1,2]

[1] Institute of Information Engineering, Chinese Academy of Sciences, Beijing, China
{wanghaiping,libinbin}@iie.ac.cn
[2] School of Cyber Security, University of Chinese Academy of Sciences,
Beijing, China
wanghaiping20@mails.ucas.ac.cn

**Abstract.** To perceive the overall cyber security situation of the target
network, cyberspace situational awareness platforms collect billions of
alerts from IDS, IPS, firewalls, probes, and third-party systems in real-
time. It is urgent to aggregate these multi-source, massive, real-time,
heterogeneous, dynamic, strong timeliness alerts to help analysts find
valuable clues as quickly as possible. In this paper, we propose a novel
real-time alert aggregation approach. It is based on a dynamic attack
granularity graph model combined with a dynamic threshold update
algorithm, which not only can effectively solve the redundancy problem of
massive multi-source data, but also can extract valuable alert aggregation
from the data flood. To evaluate our approach, we conduct experiments
using the public datasets *Suricata* (81600 alerts) and a real 24-hour online
dataset (1204753 alerts). We use evaluation metrics including *Aggregation Rate(AR), Simplicity Metric(SM) and Time Delay(TD)*, and select
three common alert aggregation algorithms to perform a comparative
test in a simulated real-time situation. The experiment shows that our
approach achieves more than 98% aggregation rate, reduces data complexity by more than 82%, and has stronger robustness.

**Keywords:** Cyber Security Situation Awareness · Massive Alerts
Aggregation · Real-time · Granularity Graph · Dynamic Threshold

## 1 Introduction

The existing cyber security situation awareness system needs to collect a large
number of alert logs from network traffic monitoring devices such as IDS, IPS,
firewalls, probes and third-party systems. The billions of alerts are multi-source,
massive, real-time, heterogeneous, dynamic, and time-sensitive, which contain a
lot of noise and redundant information. The valuable information is discrete and
difficult to extract.

Existing solutions can be summarized into the following two aspects: solutions based on expert knowledge or experience, and algorithms based on machine

© The Author(s), under exclusive license to Springer Nature Switzerland AG 2023
M. Yung et al. (Eds.): SciSec 2023, LNCS 14299, pp. 225–243, 2023.
https://doi.org/10.1007/978-3-031-45933-7_14

learning and data mining; The effectiveness and alert types coverage rate based on expert knowledge depend on the comprehensiveness and correctness of the rules, and cannot handle alert logs with dynamically increasing types in real application scenarios. The algorithms based on distance calculation, similarity mining, machine learning and deep learning are inefficient and cannot adapt to massive data traffic.

Backbone network cyber security situation awareness scenario requires aggregating massive multi-source heterogeneous alerts in real-time. The challenges are as follows:

- **Massive and Multi-Source:** More than billions of alert logs collected from multiple devices or systems need to be aggregated each day.
- **Numerous and Dynamic Alert Types:** Thousands or even more alert types are dynamically changing and difficult to maintain experts' knowledge manually. New alert types are generated when the monitoring engines are upgraded or security operations are performed.
- **Attack Granularity Relationship Varies:** The roles and relationships between attackers and victims dynamically change in different attack procedures. The aggregation algorithms based on fixed rules cannot adapt to complex and changeable scenarios.
- **Real-time Aggregation Requirements:** It is necessary to find out the start time and end time of each attack procedure in real-time. The common real-time algorithms based on time window cannot determine the start time and end time of an attack procedure. The complex offline algorithm cannot meet the real-time performance.

By observing a large number of security alerts and analyzing their distribution characteristics, we find out that the time interval between two adjacent alerts in the same attack procedure is much shorter than the time interval between two attack procedures. Besides, there are multiple attacker-victim relationships such as one-to-many, many-to-one, many-to-many and one-to-one. Different attackers or victims usually belong to the same segment C network address during the same attack behavior. Thus in this paper, we propose a real-time unsupervised alert aggregation algorithm based on lightweight attack granularity graph. The main contributions of this paper include four aspects:

- We propose a granularity graph model for caching attack relationships, which supports real-time maintenance of multiple possible attack procedures. The attack procedures are associated with IP addresses to maximize the aggregation of valuable information.
- Based on the idea of normal distribution anomaly detection, we design an aggregation threshold dynamically updating algorithm. It solves the problem of low aggregation rate caused by differences in the attack procedures, while ensuring the timeliness of the operation, making this approach capable of handling massive alert traffic.
- Based on the above two points, a new real-time alert aggregation solution is proposed, which can solve the massive alerts aggregation problem in real-time.

The solution can ensure a high level of aggregation capability, significantly reduce data complexity, coalesce valuable data, and resist the impact of noisy data.

- To evaluate our approach, we implemented our aggregation algorithm based on Flink [4] and use the public datasets *Suricata* and a real-world online dataset. We evaluate the aggregation rate, simplicity and timeliness of the algorithm results by some metrics and conduct comparative experiments with other common algorithms to illustrate the advantages and disadvantages of this approach.

The paper is organized as follows: Sect. 2 details the related work in alert aggregation. Sections 3 and 4 describe the details of attack granularity graph model and aggregate methodology. Section 5 provides the design of experiments and results. Section 6 illustrates our implementation based on Kafka and Flink [4]. Section 7 concludes the paper.

## 2   Related Work

In recent decades, alert aggregation and correlation problems have attracted the attention of many researchers. Existing alert aggregation approaches can be categorized as *similarity-based, statistical-based, knowledge-based, machine learning-based, evolutionary-based* and *time series-based* [1].

### 2.1   Similarity-Based Approach

The similarity-based [2,6] approach is defined as a measure to find the similarity between two alerts or alert clusters. This approach clusters similar alerts in time to reduce the number of alerts and increase its ability to discover the known attacks. The approach can discriminate between false positive, redundant, and invalid alerts, whereas the types of attacks do not need to be defined. The similarity between alerts can be modeled based on attributes with simple distance calculation algorithms, such as HASH, cosine.

### 2.2   Statistical-Based Approach

The statistical-based approach attempts to aggregate the alert data with similar statistical characteristics by analyzing and modeling the rules of alerts. Relevant research work mainly focuses on alert feature statistics [8,14,23] or probability distribution statistics [2,19]. Alert feature statistics based methods vectorization of alert attributes to calculate the mean square error, and the logs with the same or similar mean square error are aggregated. Probability distribution statistics based methods assume that the same distribution are aggregated together, otherwise, they belong to different sets.

## 2.3  Knowledge-Based Approach

Compared with alert similarity calculation based on statistical method, [6] uses expert knowledge to propose the concept of clustering stability. In this method, alert field similarity and judgment rules are defined by the expert system, and the basis for alert aggregation is provided by human-computer interaction. Based on expert experience, reference [24] designed a decision tree, the problem of alert logs aggregation is transformed into a classification problem based on decision tree.

Existing works of this approach are based on the knowledgebase of attack definitions, which have been divided into two components: scenario [5,21] and prerequisites/consequences [22].

## 2.4  Machine Learning-Based Approach

Typically, alert logs detected by the same type of attack have the same or similar characteristics (or attributes). Therefore, by comparing the similar features of alert attributes, it is feasible to implement alert log aggregation based on clustering algorithms.

The clustering algorithm realizes the vectoring of alert log data through feature attribute selection, distance calculation formula, similarity function, weight and other operations. Then select a suitable algorithm for clustering to achieve the goal of alert aggregation. Debar [7] proposes an implicit component to complete feature extraction. Mohamed [15] performs feature extraction by calculating the distance between the MD5 hash value of the destination IP address, alert ID and timestamp.

Alert logs are divided into groups or classes in an unsupervised way by clustering. Algorithms Based on clustering mainly include: distance and density based algorithms [8,23], hierarchical Clustering [11,12] and artificial neural network [10,13].

## 2.5  Evolutionary-Based Approach

Evolutionary algorithms can effectively solve many complex problems that traditional algorithms cannot solve. Typical evolutionary algorithms for alert aggregation includes genetic algorithm GA and artificial immune system.

Evolutionary algorithms [6] based aggregation methods can better solve the problems that the mean square error method is difficult to obtain reasonable clustering centers, they have strong anti-noise ability, and can effectively filter false alerts, but this method requires a suitable training set.

**Table 1.** Comparison From Dynamic Types, Massive Logs, RealTime Support

| Approaches | Realtime | Big Data Support | AlertType | References |
|---|---|---|---|---|
| similarity-based | Realtime or Batch | Easy to support | Dynamic | [2,6] |
| statistical-based | Realtime or Batch | Depends on the scale of indicators' associated data | Dynamic | [2,8,14,19,23] |
| knowledge-based | Realtime or Batch | Depends on the Complexity of knowledge | Fixed | [5,6,21,22,24] |
| machine learning-based | Batch | Difficult to support | Dynamic | [8,10–13,23] |
| evolutionary-based | Batch | Difficult to support | Dynamic | [6] |
| time series-based | Realtime or Batch | Easy to support | Dynamic | [1] |

## 2.6   Time Series-Based Approach

Algorithms based on time series usually aggregate alert data in a certain size sliding time window. The problem of setting the time window is one of the difficulties. How big or small the window is will affect the effectiveness of the alert aggregation. If the time window is set too large, alerts belonging to multiple attack behaviors will be aggregated together, the number of false aggregations will be increased; If the time window is too small, the alert logs caused by the same attack behaviors cannot be aggregated.

Considering the characteristics of time series network attacks, some researchers try to improve the method based on a single fixed sliding window. The main improvement ideas include: (1) using two time windows of different sizes; (2) using both fixed time window and sliding time window; (3) Sum of fixed time window and perturbation coefficient of variation. However, the size of fixed window and sliding window needs to be determined subjectively.

In the application scenarios where alert types change dynamically and massive alert logs are detected in real time, the above algorithms cannot meet the real-time requirement of alert aggregation, nor can identify complex attack relationships such as one-to-many, many-to-one and many-to-many. Table 1 compares the existing algorithms in terms of real-time performance, support for massive data and adaptability to dynamic changes of alert types.

## 3   Attack Graph Model

According to the actual application scenario, in addition to the basic performance of reducing data redundancy, an outstanding alerts aggregation solution requires aggregating valuable and relevant data. Research shows that valuable and relevant alerts are always not centralized. They tend to appear scattered, submerged in the flood of data, and therefore difficult to capture. To better capture the relevant aggregatable alerts, we design the following granularity graph model.

### 3.1   Graph Model

Figure 1 shows the granularity graph model, containing four types of vertices and two types of directed edges. The vertices in the graph are divided

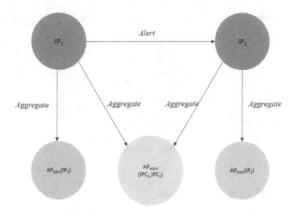

**Fig. 1.** Light Weight Attack Graph Model.

into two categories: one is the IP vertices, where a vertex represents an IP address(attacker IP *aIP* or victim IP *vIP*), and the other is granular state vertices, which store the possible attack procedures associated with the IP vertices. The granular state vertices include three types:

- $AP_{o2m}$: One-to-Many attack procedure. An attacker IP simultaneously attacks multiple victim IPs in a period of time. This type of vertices captures the relevant alerts triggered by the same source IP.
- $AP_{m2o}$: Many-to-One attack procedure. Multiple attacker IPs simultaneously attack a victim IP in a period of time. This type of vertices captures the relevant alerts triggered by the same destination IP.
- $AP_{m2m}$: Many-to-Many attack procedure. Multiple attacker IPs in a C-segment network simultaneously attack multiple victim IPs in another C-segment network in a period of time. This type of vertices captures the relevant alerts triggered by one C-segment network to another C-segment network.

Note that the One-to-One attack procedure can be considered as the special form of any of the above three types, so it is not repeated in this work. These three types of granularity states encompass almost all roles of an IP vertice perspective: as a single source, as a single target, as a member of the attacker group, or as a member of the victim group. Note that more complex attack procedures are possible, but in theoretical terms, any complex attack procedure can be disassembled into a combination of the above three types of procedures. We adopt this approach to preserve the maximum diversity of aggregated forms of an alert, as well as to maximize the value of aggregated alert data. As a cost, this approach generates redundant information inside the graph, but we find a solution, which is described in detail in Sect. 4.4.

The edges in the graph are also divided into two categories: one is the alert edges, where an edge represents an alert, and the other is aggregate edges, where

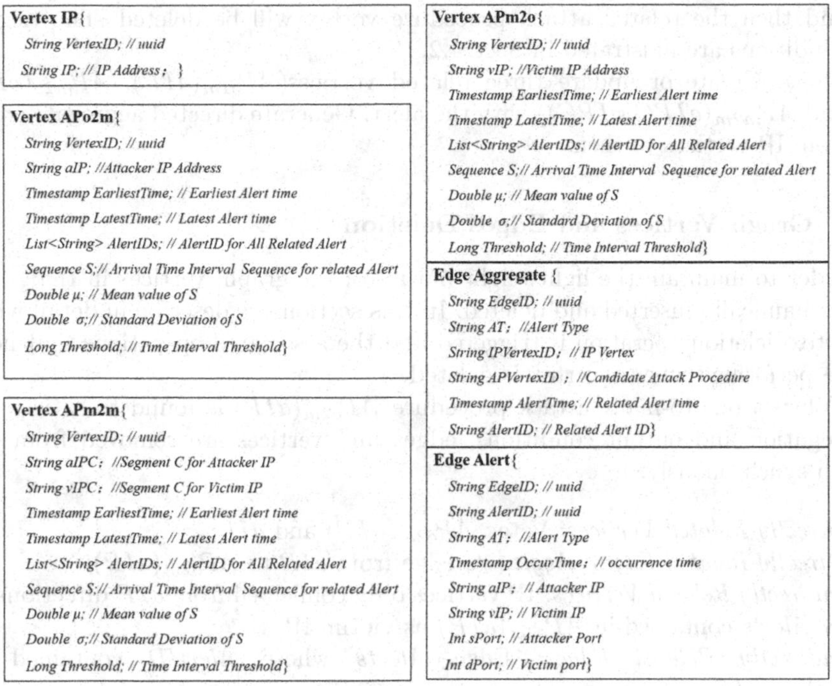

```
Vertex IP{
    String VertexID; // uuid
    String IP; // IP Address ; }
```

```
Vertex APo2m{
    String VertexID; // uuid
    String aIP; //Attacker IP Address
    Timestamp EarliestTime; // Earliest Alert time
    Timestamp LatestTime; // Latest Alert time
    List<String> AlertIDs; // AlertID for All Related Alert
    Sequence S;// Arrival Time Interval  Sequence for related Alert
    Double μ; // Mean value of S
    Double σ;// Standard Deviation of S
    Long Threshold; // Time Interval Threshold}
```

```
Vertex APm2m{
    String VertexID; // uuid
    String aIPC; //Segment C for Attacker IP
    String vIPC; //Segment C for Victim IP
    Timestamp EarliestTime; // Earliest Alert time
    Timestamp LatestTime; // Latest Alert time
    List<String> AlertIDs; // AlertID for All Related Alert
    Sequence S;// Arrival Time Interval  Sequence for related Alert
    Double μ; // Mean value of S
    Double σ;// Standard Deviation of S
    Long Threshold; // Time Interval Threshold}
```

```
Vertex APm2o{
    String VertexID; // uuid
    String vIP; //Victim IP Address
    Timestamp EarliestTime; // Earliest Alert time
    Timestamp LatestTime; // Latest Alert time
    List<String> AlertIDs; // AlertID for All Related Alert
    Sequence S;// Arrival Time Interval  Sequence for related Alert
    Double μ; // Mean value of S
    Double σ;// Standard Deviation of S
    Long Threshold; // Time Interval Threshold}
```

```
Edge Aggregate {
    String EdgeID; // uuid
    String AT; //Alert Type
    String IPVertexID; // IP Vertex
    String APVertexID ; //Candidate Attack Procedure
    Timestamp AlertTime; // Related Alert time
    String AlertID; // Related Alert ID}
```

```
Edge Alert{
    String EdgeID; // uuid
    String AlertID; // uuid
    String AT; //Alert Type
    Timestamp OccurTime; // occurrence time
    String aIP; // Attacker IP
    String vIP; // Victim IP
    Int sPort; // Attacker Port
    Int dPort; // Victim port}
```

**Fig. 2.** Vertex and Edge Structure.

an edge represents an aggregation operation. The structures of the vertices and edges are shown in Fig. 2.

## 3.2 Gragh Initialization and Update

We initialize the granularity graph when the first alert arrives. Firstly, information of the alert(alert ID, alert type, time, attacker IP, source port, victim IP, destination port, etc.) is extracted. Generally, we use *(ID, AT, ts, aIP, sPort, vIP, dPort)* to denote, *aIPC* and *vIPC* are segment C IP addresses of *aIP* and *vIP*. Secondly, we update the graph with the following steps:

- *Step1:* Extract IP information and insert into the graph model. Retrieve existing vertices according to the *aIP* and *vIP*, otherwise generate new vertices.
- *Step2:* Insert the alert into the graph model. Generate a directed alert edge between two IP vertices.
- *Step3:* Query and find related vertices $AP_{m2m}(aIPC, vIPC)$, $AP_{o2m}(aIP)$ and $AP_{m2o}(vIP)$. Update the time interval thresholds and determine whether the aggregation mode trigger conditions are met. Update the graph time according to the latest alert time, and judge whether three types of attack procedure vertices associated with the IP in alert can be triggered by the alert time. If a vertex is triggered, it will be output as an aggregated result

and then the related attack procedure vertex will be deleted. The trigger conditions are illustrated in Sect. 4.2.

- *Step4:* Create or update three related vertices($AP_{o2m}(aIP)$, $AP_{m2o}(vIP)$ and $AP_{m2m}(aIPC, vIPC)$) with the alert. Generate directed aggregate edges from IP vertices to them.

### 3.3   Gragh Vertices and Edges Deletion

In order to maintain the lightweight nature of the graph, vertices in the graph are dynamically inserted and deleted. In this section, we describe in detail when a vertex deletion operation is triggered and the associated operations that need to be performed when a vertex is deleted.

When a one-to-many attack procedure $AP_{o2m}(aIP)$ is found to satisfy the aggregation and output conditions, edges and vertices are removed from the graph synchronously:

- *Directly Related Vertices:* Vetex $AP_{o2m}(aIP)$ and $aIP$;
- *Directly Related Edges: Aggregate* edge from $aIP$ to $AP_{o2m}(aIP)$;
- *Indirectly Related Vertices:* IP vertices only connected from $aIP$ directionally by *Alerts* contained in $AP_{o2m}(aIP)$ as victim IP.
- *Indirectly Related Edges:* Edge *Alerts* whose *AlertID* contained in $AP_{o2m}(aIP)$;

Similarly, when candidate attack procedure $AP_{m2o}(vIP)$ or $AP_{m2m}(aIP, vIP)$ reaches aggregation conditions, corresponding IP vertices and *Alerts* edges are also deleted;

## 4   Aggregate Methodology

### 4.1   Methodology Framework

Figure 3 shows our alert aggregation framework based on granularity graph. The solution presented in this paper consists of the following four components:

- **Lightweight Attack Graph:** The multiple digraph model is used to cache the attack relationship between the attackers and the victims associated with unaggregated security alerts. Each type of alert corresponds to a dynamically updated lightweight graph. See details in Sect. 3.2.
- **Threshold Training:** Based on the idea of normal distribution anomaly detection, we train the time interval threshold for all the candidate attack procedures $AP_{o2m}$, $AP_{m2o}$, $AP_{m2m}$.
- **Aggregate in Realtime:** When a new alert arrives, we check whether the related $AP_{o2m}$, $AP_{m2o}$, $AP_{m2m}$ meets the aggregate conditions in real-time as shown in Eq. 1. If yes, we output the attack procedure as an aggregation result; Otherwise, we update the attack graph. See details in Sect. 4.5 and Algorithm 1.

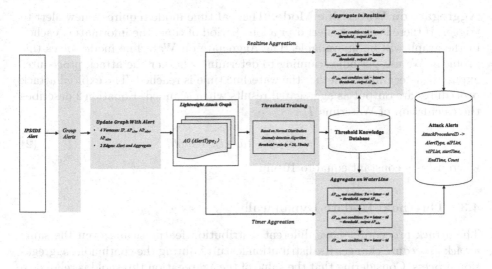

**Fig. 3.** Aggregation Framework Based on Lightweight Attack Graph.

- **Aggregate on WaterLine:** For long-standing untriggered attack procedures, we use the waterline mechanism to periodically output the definitely expired attack procedures to reduce the scale of the graph. The aggregation process will be triggered by conditions as shown in Eq. 2.

## 4.2 Aggregate Mode

The aggregation algorithm proposed in this paper includes two types of trigger mode: *Aggregate in RealTime* and *Rggregate on WaterLine*.

**Aggregate in RealTime Mode.** Alert aggregation operation is triggered once a new security alert log is received. According to $aIP_k$ and $vIP_k$ contained in the new alert log $AL_k$, we obtained three suspected attack procedure candidates: $AP_{o2m}(aIP_k)$, $AP_{m2o}(vIP_k)$, $AP_{m2m}(aIPC_k, vIPC_k)$. Based on the time interval threshold decision algorithm shown in Eq. 1, we successively determine whether $AP_{o2m}(aIP_k)$, $AP_{m2o}(vIP_k)$ or $AP_{m2m}(aIPC_k, vIPC_k)$ meet the aggregation conditions. If yes, output these related alerts and update the corresponding graph.

$$ts_k - latest > threshold \tag{1}$$

where $ts_k$ is the occurrence time of $AL_k$; *latest* is the latest time attribute of $AP_{m2m}(aIPC_k, vIPC_k)$, $AP_{o2m}(aIP_k)$, $AP_{m2o}(vIP_k)$, *threshold* are their arrival time interval threshold (Section 4.3 shows the calculation and update algorithm for *threshold*).

**Aggregate on WaterLine Mode.** The real-time mode requires a new alert to trigger. If there is no new alert over a long period of time, the information cached in the graph will be unavailable. The Aggregate on Waterline mode solves this problem. We use periodic scanning to determine whether the attack procedures expire by determining whether the waterline time is reached. The expired attack procedures are output as aggregated results when scanned. Equation 2 describes the calculation of Waterline $T_w$.

$$T_w = latest - td - threshold \tag{2}$$

where $td$ is a constant equal to 10 min.

### 4.3  Threshold Update Dynamically

The attack procedures have different distribution features, and even the same attack procedure changes its distribution features during the continuous aggregation process. Considering that the value of the aggregation threshold is related to the distribution features of the attack procedure itself, we propose an aggregation threshold updating algorithm to solve the problem of adapting the aggregation threshold to each attack procedure.

The aggregation threshold updating algorithm is based on the normal distribution algorithm for anomaly detection. In this work, the value of the time interval between sequential alerts is recorded. The distribution of the time intervals approximates the normal distribution, then the confidence level of the normal distribution in the range of $[\mu - 2\delta, \mu + 2\delta]$ is 95.45%. Beyond this range, it has a 95.45% probability of belonging to a new attack procedure. The calculation of the threshold is shown in Eq. 3:

$$threshold = \min{(\mu + 2\delta, 10\,\text{min})} \tag{3}$$

where we set 10 min as the upper limit to prevent the threshold from being too large.

Based on the above algorithm, the aggregation thresholds are different for different attack procedures. The threshold value is larger for groups with more discrete time distribution, and smaller for groups with tighter time distribution. This algorithm can adapt to the different situations of different attack procedures and get more reasonable aggregation results. Besides, the dynamic threshold as the key to aggregation avoids the problem of over-aggregation and ensures the timeliness of the aggregation algorithm at the same time.

### 4.4  Selection and Deletion Strategy

In our approach, alert aggregation is related to three types of attack procedure vertices($AP_{o2m}$, $AP_{m2o}$, $AP_{m2m}$). In the continuous aggregation process, especially for alerts that first arrive, we indiscriminately aggregate alerts into the 3 types of vertices. The reason for the operation, as already described, is to maximize the aggregation of associated valuable alert data. However, the cost of it is that there are 3 copies of the same alert data in our granularity graph. If nothing is done about them, not only it is a huge burden on the system, but also the output aggregation results are duplicated, which increases the data volume for nothing. To solve this problem, we propose a selection and deletion strategy as the following:

**Selection Strategy:** The selection strategy mainly addresses the attribution of the current alerts in the three types of attack procedures. In this method, we use a majority-first selection strategy, i.e., alerts are preferentially attributed to the attack procedure with a higher volume of data.

**Deletion Strategy:** The deletion strategy is to solve the data duplication problem in our alert aggregation approach. The deletion strategy is that when an attack procedure is triggered, all alerts under that attack procedure are subject to a selection strategy. If the alert selects this triggered attack procedure, the data of that alert from other associated attack procedures is deleted. If the alert does not select this triggered attack procedure, the alert is deleted in it.

### 4.5  Aggregation Procedure

In this section, we use pseudo-code to illustrate the details of our aggregation algorithm based on the attack granularity graph. $AP_{m2m}$, $AP_{o2m}$ and $AP_{m2o}$ are the minimum granularity units for alert aggregation. Each aggregation result corresponds to a suspected attack procedure or phase, which contains alert type $AT$, attacker IPs $aIPList$, victim IPs $vIPList$, attack start time $Start$ and end time $End$, number of related alert logs $Count$. The aggregation processes of $AP_{m2m}$, $AP_{o2m}$ and $AP_{m2o}$ are similar. We take $AP_{m2m}$ as an example to introduce the specific aggregation process as shown in Algorithm 1.

---

**Data**: Alert $AL_k$, Attack Graph $AG$
**Result**: Attack Procedure: $AT, aIPList, vIPList, Start, End, Count$
1  Extract $aIP_k$ and $vIP_k$ from $AL_k$, calculate $aIPC_k$ and $vIPC_k$ ;
2  $AP_{m2m}\ AP = AG.getAP_{m2m}(aIPC_k, vIPC_k)$;
3  **if** $AP\ != null$ **then**
4  $\quad$ $ts_k = AL_k.ts$; $lastest = AP.latest$; $Threshold = AP.Threshold$;
5  $\quad$ **if** $ts_k - lastest > Threshold\ or\ lastest - td - Threshold \leq 0$ **then**
6  $\quad\quad$ **Output** $AP$;
7  $\quad\quad$ Delete all vertices and edges related with $AP$ in $AG$;
8  $\quad\quad$ $AP_{m2m}\ newAP = $ new $AP_{m2m}(aIPC_k, vIPC_k)$;
9  $\quad\quad$ Initialize $newAP$ with alert $AL_k$; $AG.insert(newAP)$;
10 $\quad$ **else**
11 $\quad\quad$ Insert $AL_k$ into $AP$;
12 $\quad\quad$ Update $S$, $\mu, \delta, Threshold$ for $AP$;
13 $\quad\quad$ Update $AG$ with $AP$;
14 $\quad$ **end**
15 **else**
16 $\quad$ $AP_{m2m}\ newAP = $ new $AP_{m2m}(aIPC_k, vIPC_k)$;
17 $\quad$ Initialize $newAP$ with alert $AL_k$; $AG.insert(newAP)$;
18 **end**

**Algorithm 1:** Aggregation Process for $AP_{m2m}$

# 5  Implementation

In this section, we focus on the implementation of our real-time alert aggregation approach. In terms of application requirements for real-time cyber security situation awareness of backbone network traffic, the implementation method of alert aggregation must meet the following requirements:

**Real-time:** A real-time computing framework should be used to realize the aggregation algorithm to ensure real-time aggregation of alert logs for cyber security situation awareness.

**Large-scale data volume:** The computing framework must meet the processing requirements of massive alert data and adapt to the scenario where the number of alert logs surges.

**Parallel computing:** To ensure efficient computing performance, we need to use a parallel computing technology system.

**Memory management:** An ideal memory management method needs to be found to prevent memory overflow.

To meet the above metrics and to solve some of the problems of the actual application, we choose the architecture of "*Hadoop* [16] + *Kafka* [17] + *Flink* + *MySQL* + *Zookeeper*" to implement our system. We have deployed 8 *Kafka* nodes, 41 *Yarn NodeManager* nodes, 2 *MySQL* nodes and 3 *Zookeeper* nodes in the cluster. The *Flink* program is started using *Flink on Yarn* mode, which enables resource scheduling and management.

To solve the real-time and stability problems of large-scale data, we connect the large-scale data from the data source to some Kafka topic and then consume it as a consumer. This can avoid the problem of system instability caused by data explosion. We set the number of partitions of the Kafka topic to 8 and the number of replicas to 3 to ensure the integrity of the data and the stability of the whole system.

To improve the efficiency of data processing, we apply the parallel computing in *Flink*. We set the parallelism to 5, which means that the core business of our *Flink* program is sliced into 5 arithmetic subtasks at the same time. In parallel computing, the problem of unbalanced data flow from different subtasks may occur. We apply *rebalance()* function in *Flink* so that the upstream data can be sent down in a balanced way to maximize the computational efficiency.

Alert logs received from security vendors have a large amount of redundant information, much of which is not relevant to our alert aggregation business. It is necessary to do some extraction steps to reduce the space occupied by the data, reduce memory consumption and improve the performance of the system. We use *Flink flatMap()* function to extract attributes such as alert ID *ID*, alert type *AT*, occurrence time *ts*, attacker IP *aIP*, source port *sPort*, victim IP *vIP*, destination port *dPort*.

In this work, we maintain an attack granularity graph for each alert type and do alert aggregation based on this graph model in parallel. We use *Flink keyBy()* to group alert logs by alert type.

We overwrite *Flink process()* function to implement the core processing operations of our aggregation algorithm and cache attack granularity graph. We use locks to prevent conflicts between data processing processes. A multi-threaded timer is implemented inside the *process()* function to periodically output expired aggregated data. This allows us to clean up the redundant "branches" of the granularity graph model at regular intervals, and in conjunction with the garbage collection mechanism, to achieve memory management effects to prevent memory overflow. We also set the number of memory for the Flink application to 30GB for Job Manager and 30GB for Task Manager to improve the memory tolerance. Our implementations have proven to be able to handle large-scale data access and processing implementations.

## 6   Experiments and Results

To evaluate the feasibility and effectiveness of our approach, we exploit public datasets *Suricata* [20] and real-world IDS data to address the following questions:

- **Aggregation Capability:** What is the aggregation capability of our approach for massive real-time alerts?
- **Complexity reduction:** Whether the data aggregated by our approach is conducive to manual observation? How much complexity has been reduced?
- **Real-time performance:** What is the timeliness of our approach?
- **Granularity distinction:** How does our approach distinguish the attack procedures' granularity, and what is the specific effect?

### 6.1   Experiment DataSets

We set up two experiments with real-time and large-scale alerts. In the first experiment, public datasets *Suricata* [20] were converted into a real-time data source. We design a 10-hour real-time alerts transmitter to simulate sending *Suricata* [20] alerts from 13:30 on November 3, 2018 to 23:30 on November 3, 2018 in real time, for a total of 81600 alerts; In the second experiment, we used a 24 h period from 17:30 on February 14, 2023 to 17:30 on February 15, 2023 real-world IDS real-time alerts, for a total of 1204753 alerts.

### 6.2   Baselines

We compare our approach with three approaches, FTW [9,18], ASW and DBSCAN [25], which are all common alerts aggregation approaches. FTW aggregates alerts based on *aIP, dIP* and *AT* within fixed time window. We set the fixed time threshold of FTW to 5 min, whose rationality has been verified [9]. ASW processes real-time alerts through sliding window, and dynamically expands the window by detecting whether the proportion of current aggregatable alerts in the window reaches the threshold. The threshold of ASW is set to 0.5. Using DBSCAN to aggregate alerts requires establishing a quantitative model of alert attributes and forming feature vector [18]. DBSCAN aggregates alerts based on the density between feature vectors [25].

## 6.3  Environment

The experiments are conducted in a Hadoop cluster. We divide four Yarn queues with the same resources and deploy four algorithms in four Yarn queues respectively to avoid the impact of resource preemption among algorithms ensuring they are running in the same environment; In terms of data sources, we send the experimental data to Kafka in a real-time manner, and the four algorithms consume the experimental data as four different consumer groups, which ensure the four algorithms receive the same real-time data.

## 6.4  Metrics

We conduct comparative evaluation from three aspects: *Aggregation Rate(AR)*, *Simplicity Metric(SM)* and *Time Delay(TD)*.

*Aggregation Rate(AR)*: $AR$ is used to measure the ratio between aggregated alerts and original alerts, reflecting the aggregation degree of data under the algorithm. As shown in Eq. 4, M indicates the amount of original alerts, N indicates the amount of results.

$$AR = 1 - \frac{N}{M} \tag{4}$$

*Simplicity Metric(SM)*: Paper [3] quantifies the degree of manual observation for the aggregation results of each attack procedure by using the simplicity metric. By analyzing the relationship between attacker IPs and victim IPs for all candidate attack procedures output by aggregation algorithm, we evaluate the consistency between experiment results with the real attacks scenario such as many-to-one, one-to-many, and many-to-many attacks. See Eq. 5 for details.

$$SM = \frac{1}{M} \sum_{k=1}^{k=N} (aIP_k + vIP_k) \tag{5}$$

*Time Delay (TD)*: The timeliness performance of an aggregating process is evaluated by the time interval between the output timestamp $t_k$ and the last alert timestamp $LT_k$, and the TD expression is given in Eq. 6.

$$TD = \frac{1}{N} \times \sum_{k=1}^{k=N} (ts_k - LT_k) \tag{6}$$

## 6.5  Evaluation Results

To answer the research questions, we adopt the three evaluation indicators in Sect. 6.4: $AR$ evaluates the aggregation capability of algorithms; $SM$ evaluates the artificial observability and the complexity reduction degree of results; $TD$ evaluates the timeliness of aggregating process. The experiment results are shown in Table 2.

**Aggregation Capability.** In the first experiment, we compare the AR of four algorithms based on the public datasets *Suricata*. As shown in Table 2, AR of

**Table 2.** Experimental Results Comparison based on Public and Online Datasets

| DataSets | Algorithm | FTW | ASW | DBSCAN | Our Approach |
|----------|-----------|-----|-----|--------|--------------|
| Suricata | Aggregation Rate | 0.9479 | 0.9249 | 0.9968 | 0.9900 |
| Suricata | Simplicity Metric | 0.0520 | 0.0751 | 0.0060 | 0.0678 |
| Suricata | Time Delay/s | 69.935 | 147.509 | 16113.772 | 268.823 |
| Online data | Aggregation Rate | 0.5449 | 0.3975 | 0.8007 | 0.9810 |
| Online data | Simplicity Metric | 0.4551 | 0.6025 | 0.3084 | 0.1791 |
| Online data | Time Delay/s | 42.470 | 44.287 | 59943.921 | 902.315 |

our approach reaches 99.00%, second only to DBSCAN, ranking first during the real-time algorithms. Compared with FTW and ASW, AR of our approach is 4.21% and 6.51% higher respectively. Compared with DBSCAN, the gap of AR is only 0.68%, which shows that the aggregation capability of our approach is considerable. In the second experiment, we compare the four algorithms based on the online IDS alerts. Online IDS data contains large quantity of noise data, which impact on the performance of common aggregation approaches, drowning the truly valuable data in noise and reducing aggregation rates. As shown in Table 2, the aggregation rate of FTW in the second experiment is 54.49%, ASW is 39.75%, and DBSCAN is 80.07%, all of which have a certain degree of degradation in aggregation effect due to the presence of noise compared with the first experiment. However, the AR of our approach is still maintained at an amazingly high level of 98.10%, which means that the noise of the real IDS data has no significant effect on our algorithm, which fully illustrates the robustness and outstanding aggregation capability of our algorithm.

**Table 3.** Examples for One2Many, Many2One,Many2Many Attack Procedure

| Attack Mode | Alerts | Aggregation Results: Attack Procedures |
|-------------|--------|-----------------------------------------|
| One to Many | 11620 | AlertType:2010, start time: 2023-02-14 20:41:55 000; end time: 2023-02-14 20:45:12 000, Attack Events:102, Attacker: *.*.9.99, victims:*.*.84.206-231; *.*.94.173-174; *.*.56.229-232;*.*.56.179-184;*.*.244.226-229; 106.*.*.230-233; *.*.220.224-227;*.*.118.101-105;*.*.149.108... |
| Many to One | 10312 | AlertType: 2048, start time:2023-02-14 20:43:31 000 end time: 2023-02-14 20:47:28 000, Attack Events: 112, Attacker:*.*.37.139; *.*.37.218; *.*.37.145; *.*.37.86; *.*.37.248; *.*.130.167; *.*.122.16;*.*.108.155; victim: *.*.213.58 |
| Many to Many | 17966 | AlertType: 2010, start time:2023-02-14 17:28:56 000 end time: 2023-02-15 05:23:08 000, Attack Events:787, Attacker: *.*.9.99-117; victims:*.*.25.106; *.*.25.180 |

**Complexity Reduction:** In the first experiment, we compare the simplicity metric of the four algorithms. As shown in Table 2, based on public datasets *Suricata*, the data complexity after aggregation by the four algorithms decreases by 94.8%, 92.49%, 99.40% and 93.22%, respectively. This result indicates that in the first experiment, the data aggregated by the four algorithms have significant observability and a substantial decrease in complexity. While in the second experiment, the data complexity after FTW aggregation decreases only 54.49%, ASW decreases by 39.75%, and DBSCAN decreases by 69.26% due to the presence of noisy data. The data complexity after aggregation by our method decreases by 82.09%, which is the best performance. It shows that based on the online IDS real-time alerts, the aggregated data of our approach obviously has better observability and significantly reduced data complexity, with stronger robustness compared with other common aggregation approaches.

**Real-Time Performance:** In two experiments, we observe the *Time Delay* of the four algorithms. Note that DBSCAN, being an offline aggregation algorithm, theoretically needs to wait for the whole window before aggregation, and therefore the *Time Delay* is larger. Among the three real-time algorithms, our approach is slightly deficient in real-time performance. This occurs because we cannot judge from the data itself whether it is noisy or not. Our approach sacrifices some real-time performance by setting a waiting time for all data to ensure that no relevant data is lost. The larger the Time Delay of the aggregated results of our method when there is more noise in the data, which is also verified in this experiment. Through our analysis and evaluation, the Time Delay of our approach is acceptable in practical engineering applications, and it can also be reduced to improve the real-time performance by lowering the waiting time threshold according to the business requirements. **Granularity distinction:** Considering the association characteristics between alerts, our approach divides the alerts aggregation into three granularity modes: one-to-many, many-to-one and many-to-many. Among them, the two granularity modes of one-to-many and many-to-one describe the behavior of an IP address interacting with multiple unrelated IP addresses over a period of time, of which the most typical security event is a DDOS attack; The granularity mode of many-to-many describes the behavior of single/multiple IP addresses in the same network segment interacting with single/multiple IP addresses in another network segment over a period of time, such as a common information leakage event. The granularity statistics of the aggregation results of this method based on 24-hour online IDS real-time alerts are shown in Table 3 below. As shown in Table 3, the aggregated data can reflect the alerts relationship, the attacks number and the time interval characteristics. The distinction between different granularity modes is obvious.

# 7    Conclusion

Multi-source, massive, heterogeneous, dynamic, and strongly time-sensitive alert flooding problem has been a key issue in network security situational awareness. In this paper, we propose a real-time alert aggregation solution based on the dynamic attack granularity graph. It has a powerful aggregation capability to reduce the redundancy of alert data information, and dynamically aggregates alerts according to three types of attack procedures: one-to-many, many-to-one, and many-to-many. Through an aggregation threshold dynamically updating algorithm to adapt each attack procedure, it significantly reduces the complexity of alerts and makes the aggregated results have better observability. To evaluate our approach, we conduct comparative experiments using the public datasets *Suricata*(81600 alerts) and a 24-hour real online dataset(1204753 alerts). The experiments show that our approach performs a high aggregation rate of over 98% and reduces the data complexity by over 82%. When dealing with real data with high noise, our approach performs more outstandingly and maintains a high level of performance with stronger robustness compared to common alert aggregation algorithms. At the end of the experiment, we give statistics and examples, reflecting the intuitiveness of the aggregation results.

# References

1. Albasheer, H., et al.: Cyber-attack prediction based on network intrusion detection systems for alert correlation techniques: a survey. Sensors **22**(4), 1494 (2022). https://doi.org/10.3390/s22041494
2. Valdes, A., Skinner, K.: Probabilistic alert correlation. In: Lee, W., Mé, L., Wespi, A. (eds.) RAID 2001. LNCS, vol. 2212, pp. 54–68. Springer, Heidelberg (2001). https://doi.org/10.1007/3-540-45474-8_4
3. de Alvarenga, S.C., Barbon, S., Miani, R.S., Cukier, M., Zarpelão, B.B.: Process mining and hierarchical clustering to help intrusion alert visualization. Comput. Secur. **73**, 474–491 (2018). https://doi.org/10.1016/j.cose.2017.11.021
4. Carbone, P., Katsifodimos, A., Ewen, S., Markl, V., Haridi, S., Tzoumas, K.: Apache flink: stream and batch processing in a single engine. IEEE Data Eng. Bull. **38**(4), 28–38 (2015). http://sites.computer.org/debull/A15dec/p28.pdf
5. Cheung, S., Lindqvist, U., Fong, M.W.: Modeling multistep cyber attacks for scenario recognition. In: 3rd DARPA Information Survivability Conference and Exposition (DISCEX-III 2003), Washington, DC, USA, 22–24 April 2003, pp. 284–292. IEEE Computer Society (2003). https://doi.org/10.1109/DISCEX.2003.1194892
6. Cuppens, F.: Managing alerts in a multi-intrusion detection environment. In: 17th Annual Computer Security Applications Conference (ACSAC 2001), New Orleans, Louisiana, USA, 11–14 December 2001, pp. 22–31. IEEE Computer Society (2001). https://doi.org/10.1109/ACSAC.2001.991518
7. Debar, H., Wespi, A.: Aggregation and correlation of intrusion-detection alerts. In: Lee, W., Mé, L., Wespi, A. (eds.) RAID 2001. LNCS, vol. 2212, pp. 85–103. Springer, Heidelberg (2001). https://doi.org/10.1007/3-540-45474-8_6
8. Fatma, H., Mohamed, L.: A two-stage technique to improve intrusion detection systems based on data mining algorithms. In: 2013 5th International Conference on Modeling, Simulation and Applied Optimization (ICMSAO), pp. 1–6 (2013). https://doi.org/10.1109/ICMSAO.2013.6552542
9. Husák, M., Cermák, M., Lastovicka, M., Vykopal, J.: Exchanging security events: which and how many alerts can we aggregate? In: 2017 IFIP/IEEE Symposium on Integrated Network and Service Management (IM), Lisbon, Portugal, 8–12 May 2017, pp. 604–607. IEEE (2017). https://doi.org/10.23919/INM.2017.7987340

10. Wang, J.-X., Wang, Z.-Y., Dai, K.: A PCA-LVQ model for intrusion alert analysis. In: Mehrotra, S., Zeng, D.D., Chen, H., Thuraisingham, B., Wang, F.-Y. (eds.) ISI 2006. LNCS, vol. 3975, pp. 715–716. Springer, Heidelberg (2006). https://doi.org/10.1007/11760146_102

11. Julisch, K.: Mining alarm clusters to improve alarm handling efficiency. In: Seventeenth Annual Computer Security Applications Conference, pp. 12–21 (2001)

12. Julisch, K.: Info, claims: clustering intrusion detection alarms to support root cause analysis. ACM Trans. Inf. Syst. Secur. **6**, 443–471 (2003). https://doi.org/10.1145/950191.950192

13. Kumar, M., Siddique, S., Noor, H.: Feature-based alert correlation in security systems using self organizing maps. In: Dasarathy, B.V. (ed.) Data Mining, Intrusion Detection, Information Security and Assurance, and Data Networks Security, Orlando, Florida, USA, 13 April 2009. SPIE Proceedings, vol. 7344, p. 734404. SPIE (2009). https://doi.org/10.1117/12.820000

14. Man, D., Yang, W., Wang, W., Xuan, S.: An alert aggregation algorithm based on iterative self-organization. Procedia Eng. **29**, 3033–3038 (2012). https://doi.org/10.1016/j.proeng.2012.01.435. https://www.sciencedirect.com/science/article/pii/S1877705812004456. 2012 International Workshop on Information and Electronics Engineering

15. Mohamed, A.B., Idris, N.B., Shanmugum, B.: Alert correlation using a novel clustering approach. **2212**(12747443), 720–725 (2012). https://doi.org/10.1109/CSNT.2012.212

16. Nandimath, J., Banerjee, E., Patil, A., Kakade, P., Vaidya, S.: Big data analysis using apache hadoop. In: IEEE 14th International Conference on Information Reuse & Integration, IRI 2013, San Francisco, CA, USA, 14–16 August 2013, pp. 700–703. IEEE Computer Society (2013). https://doi.org/10.1109/IRI.2013.6642536

17. Noac'h, P.L., Costan, A., Bougé, L.: A performance evaluation of Apache Kafka in support of big data streaming applications. In: Nie, J., et al. (eds.) 2017 IEEE International Conference on Big Data (IEEE BigData 2017), Boston, MA, USA, 11–14 December 2017, pp. 4803–4806. IEEE Computer Society (2017). https://doi.org/10.1109/BigData.2017.8258548

18. Raftopoulos, E., Dimitropoulos, X.A.: IDS alert correlation in the wild with edge. IEEE J. Sel. Areas Commun. **32**(10), 1933–1946 (2014). https://doi.org/10.1109/JSAC.2014.2358834

19. Benferhat, S., Boudjelida, A., Tabia, K., Drias, H.: An intrusion detection and alert correlation approach based on revising probabilistic classifiers using expert knowledge. Appl. Intell. **38**(15), 520–540 (2013). https://doi.org/10.1007/s10489-012-0383-7

20. Suricata: Suricata open source IDS (2020). https://suricata-ids.org/

21. Tan, T.K., Darken, C.J.: Learning and prediction of relational time series. Comput. Math. Organ. Theory **21**(2), 210–241 (2015). https://doi.org/10.1007/s10588-015-9182-0

22. Templeton, S.J., Levitt, K.E.: A requires/provides model for computer attacks. In: Zurko, M.E., Greenwald, S.J. (eds.) Proceedings of the 2000 Workshop on New Security Paradigms, Ballycotton, Co., Cork, Ireland, 18–21 September 2000, pp. 31–38. ACM (2000). https://doi.org/10.1145/366173.366187

23. Tjhai, G.C., Furnell, S., Papadaki, M., Clarke, N.L.: A preliminary two-stage alarm correlation and filtering system using SOM neural network and k-means algorithm. Comput. Secur. **29**(6), 712–723 (2010)

24. Zhang, Y., Huang, S., Wang, Y.: IDS alert classification model construction using decision support techniques. In: 2012 International Conference on Computer Science and Electronics Engineering, vol. 1, pp. 301–305 (2012). https://doi.org/10.1109/ICCSEE.2012.242
25. Zhao, N., et al.: Understanding and handling alert storm for online service systems. In: Rothermel, G., Bae, D. (eds.) ICSE-SEIP 2020: 42nd International Conference on Software Engineering, Software Engineering in Practice, Seoul, South Korea, 27 June–19 July 2020, pp. 162–171. ACM (2020). https://doi.org/10.1145/3377813.3381363

# Decompilation Based Deep Binary-Source Function Matching

Xiaowei Wang[1,2,3,4], Zimu Yuan[1,2,3,4](✉), Yang Xiao[1,2,3,4],
Liyan Wang[1,2,3,4], Yican Yao[1,2,3,4], Haiming Chen[5], and Wei Huo[1,2,3,4]

[1] School of Cyber Security, University of Chinese Academy of Sciences,
Beijing, China
{wangxiaowei,yuanzimu,xiaoyang,wangliyan,yaoyican,huowei}@iie.ac.cn
[2] Institute of Information Engineering, Chinese Academy of Sciences, Beijing, China
[3] Key Laboratory of Network Assessment Technology, Chinese Academy of Sciences,
Beijing, China
[4] Beijing Key Laboratory of Network Security and Protection Technology,
Beijing, China
[5] Institute of Software, Chinese Academy of Sciences, Beijing, China
chm@ios.ac.cn

**Abstract.** Binary and source matching is vital for vulnerability detection or program comprehension. Most existing works focus on library matching (coarse-grained) by utilizing some simple features. However, they are so coarse-grained that high false positives occur since developers tend to reuse source code library partly. These shortcomings drive us to perform fine-grained matching (i.e., binary and source function matching). At the same time, due to the enormous differences between the form of binary and source functions, function matching (fine-grained) meets huge challenges. In this work, inspired by the decompilation technique and advanced neural networks, we propose tool, a **D**ecompilation based deep **B**inary-**S**ource function **M**atching framework. Specifically, we take the triplet features from both *binary pseudo-code* and *source code* functions as input, which are extracted from code property graph and can represent both the syntactic and semantic information. In this way, the binary and source functions are represented in the same feature space so to ease the matching model to learn function similarity. For the matching model, we adopt a self-attention based siamese network with contrastive loss. Experiments on two datasets, $R0$ and $R3$, show that our tool achieves consistent improvements than other methods, which demonstrate the effectiveness of our self-attention based matching model, and our triplets features can well capture the two kinds of code functions. Our work improves the accuracy of binary and source code matching, which in turn enables us to better address security issues such as vulnerability detection and program comprehension.

**Keywords:** Function Matching · Binary-Source Function Matching · Deep Learning

© The Author(s), under exclusive license to Springer Nature Switzerland AG 2023
M. Yung et al. (Eds.): SciSec 2023, LNCS 14299, pp. 244–260, 2023.
https://doi.org/10.1007/978-3-031-45933-7_15

# 1   Introduction

Binary-source matching is an important task in computer science, which can be applied in various applications such as vulnerability detection [9], plagiarism detection [29], malware detection [30] and reverse engineering [32].

Existing works mainly focus on binary-source library matching [9,14,16]. These works extract a part of constant features which are not changed in the process of compiling from both source and binary projects. These features include strings, integers, export functions, string arrays, integer arrays and so on. Then, various matching algorithms (such as TF-IDF [33] based) are applied on these features to match binary and source projects. However, there are two main disadvantages of binary-source library matching works. First, these features may not exist in binary projects with removing features intentionally, which lead to false positives and false negatives. Second, they are too coarse-grained and these features are not adequate for fine-grained (function level) matching, while it is important.

**Listing 1.1.** Example source code function

```
1 void even_or_odd()
2 {
3     int input = get_input();
4     if (input % 2 == 0)
5         printf("%d is an even number!\n", input);
6     else
7         printf("%d is an odd number!\n", input);
8 }
```

Therefore, we try to focus on more fine-grained matching, i.e., binary-source function matching. However, binary-source function matching meets huge challenges since the form of binary and source code functions are totally different. On the one hand, binary function is represented by machine code, assembly code or control flow graph (CFG), while source code function is represented by statements, abstract syntax tree (AST), CFG or program dependency graph (PDG). On the other hand, even both functions are represented by CFGs, the nodes in the graph are different. Figure 1(a) and Fig. 1(b) show CFGs of source code function and corresponding binary function in Listing 1.1. Node in Fig. 1(a) is a statement while node in Fig. 1(b) is composed by a sequence of assembly code. It is not easy to match a statement to a sequence of assembly code without debug information. Besides, the number of nodes in two functions' CFG are not same, which means we do not know how to map nodes in source to nodes in binary, since one node in source can map to two or more nodes in binary, and vice versa. For example, the second and the third node in Fig. 1(a) are equivalent on semantic of the first node in Fig. 1(b). At the same time, if the conditional check statement in source function is composed by two or more conditions, there would be more than one node in the CFG of binary function while one node in the CFG of source code function. In details, the number of conditions in one conditional check of source function corresponds to the number of nodes in binary's CFG. Therefore, how to select the features to present the source and binary functions

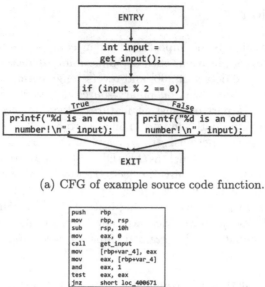

(a) CFG of example source code function.

(b) CFG of example binary function.

**Fig. 1.** An example of CFG for source and binary function.

is important and not easy, which makes it to be a huge challenge to perform the function matching between source and binary.

**Listing 1.2.** Example pseudo-code function

```
1 int even_or_odd()
2 {
3   int result; // eax
4   unsigned int v1; // [rsp+Ch] [rbp-4h]
5
6   v1 = get_input();
7   if ( v1 & 1 )
8     result = printf("%d is an odd number!\n", v1);
9   else
10    result = printf("%d is an even number!\n", v1);
11  return result;
12 }
```

In this paper, to solve above issues, we propose a Decompilation based deep Binary-Source function Matching (DBSM) framework, which is built upon the function-level matching and needs no compilation from source code projects into binaries. Specifically, our approach consists of the following steps. *Step-1:*

Binary functions are lifted to pseudo-code with the aid of IDA-Pro[1] toolkit. For example, Listing 1.2 is the resulted pseudo-code form of the binary function in Fig. 1(b). Obviously, pseudo-code is now more similar to the source code (shown in Listing 1.1) compared with the original assembly code. *Step-2:* Given the source code and pseudo-code function as the input, we apply a robust parser Joern [13] to parse the codes and generate a code property graph which merges AST, CFG and PDG into a joint data structure for each code. *Step-3:* Triplets are extracted from code property graph of source code and pseudo-code function, which can represent the syntax and semantic information. Triplets are represented as $(src, path, tgt)$, where $src$ and $tgt$ are nodes in the code property graph while $path$ is the relationship between $src$ and $tgt$. *Step-4:* Above triplet features of source code and pseudo-code function are finally fed into a siamese network for function similarity matching. Specifically, we design a self-attention mechanism based siamese neural network, and we adopt the advanced contrastive loss for the matching model training.

We conduct experiments on two datasets, $R0$ and $R3$, for the task evaluation. These datasets are collected from fifteen real world source code projects with compiling them manually in two different settings, which can be useful and typical resources to evaluate our DBSM. The results show that DBSM achieves strong performances and significantly outperforms other baseline models. Specially, we obtain a high 89.6 Recall@1 on $R0$ and 82.2 on $R3$.

In a summary, our contributions are as follows:

- We propose a binary and source function matching framework DBSM by leveraging triplets of source code and pseudo-code, which is lifted from binary assembly.
- We extract triplets from AST, CFG and DDG that can represent the syntax and semantic information for each code function.
- We build a self-attention based siamese neural network, and apply the contrastive loss on the triplets for the function similarity matching.
- Experiments on two datasets show that our DBSM method can obtain superior performance. Especially, DBSM achieves a high Recall@1 score of 89.6 and 82.2 on $R0$ and $R3$.

## 2   Related Work

Our work is related to the binary-source matching on both library level and function level. In this section, we briefly introduce these related works.

### 2.1   Library Matching

It is very common for software developers to integrate source code projects into target software, which can accelerate the development. Developers must comply with the license of source code projects, and vulnerabilities in these projects may

---

[1] https://www.hex-rays.com/products/ida/.

affect target software. At the same time, most target software are binaries (close sourced). Therefore, there are many works trying to detect source code projects in binaries.

Three techniques (detection using strings, detection using compression and detection using binary deltas) are proposed in BAT [16] to detect source code projects in target binaries. OSSPolice [14] selects string literal and exported function as features and applies hierarchical matching algorithm to detect source code projects in android app binaries. B2SFinder [9] chooses seven features which are not changed drastically during the compilation and employs a weighted feature matching algorithm to achieve the task. LibID [15] converts library identification problem into a Binary Integer Programming (BIP) problem.

These works can detect source code projects used in target binaries precisely and efficiently. However, these methods are too coarse-grained since software developers may reuse part of source code projects or even some functions only. Moreover, these features are not suitable for function-level matching since there are not these features in many functions. Therefore, we need to extract more fine-grained features which can represent function.

## 2.2 Function Matching

There are three types of function matching: binary-binary, source-source and binary-source matching.

There are tens of works proposed for solving binary-binary function matching problems. INNEREYE-CC [17] computes similarity beyond function pairs inspired by neural machine translation. DEEPBINDIFF [18] proposes an unsupervised program-wide code representation learning technique for binary diffing with extracting semantic information by leveraging NLP techniques and performing TADW algorithm to generate basic block embeddings. $\alpha$-diff [19] employs three semantic features (intra-function features, in/out degree in call graph features and imported function features) to address function similarity problem with DNN model [37]. These works require users to compile all source code projects into binaries first, which is very time-consuming or even impossible. For one hand, only approximately quarter can be compiled automatically. [9] For the other hand, it is impossible to compile old or complex source code projects manually since it relies on compiling environment deeply.

For source-source function matching, Tree-based convolutional neural network [21] is used for classifying programs according to functionality. CCD [22] introduces a joint code representation that applies fusion embedding technique to learn hidden syntactic and semantic features and takes DNN as a classifier to detect code clone. TreeCaps [23] fuses capsule networks with tree-based convolutional neural networks together for code clone detection task. Different from these code clone works on source code only, we focus on function matching between binary and source code. There is not rich syntactic and semantic information in binary function.

As for binary-source function matching, to the best of our knowledge, there is only one work focus on the task. CodeCMR [20] takes function matching

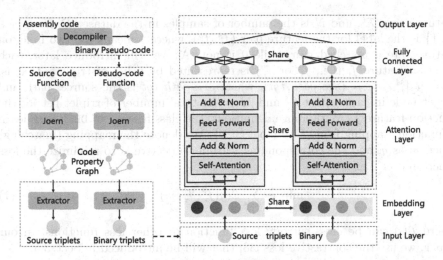

**Fig. 2.** The overview of our DBSM framework.

between binary and source code as an end-to-end cross-modal retrieval task. It adopts DPCNN (resp. GNN) for source code (resp. binary) feature extraction and exploits LSTM to capture two kinds of code literals. In a word, CodeCMR utilizes four different neural networks for the task, bug it is very complex.

## 3 Methodology

Figure 2 shows the overview of DBSM, which contains four modules. The *Decompilation* module (Sect. 3.2) lifts assembly code to pseudo-code. The *Code Property Graph Generation* module (Sect. 3.3) takes source and binary pseudo-code functions as inputs and generates code property graph with the aid of robust parser Joern. The *Feature Extraction* module (Sect. 3.4) takes code property graph of binary and source functions as input, and generates a list of triplets as features for each function. The *Self-attention based Matching Model* module (Sect. 3.5) takes features of binary and source code function as inputs, and trains a model to match binary and source code functions.

### 3.1 Problem Formulation

Existing works focus on binary to source library matching (coarse-grained) or function matching at binary level only. Instead, we perform binary to source function matching, which is more fine-grained. The binary and source function matching task is defined as a matching problem. We first describe the necessary notations used in this paper for our DBSM approach.

We define the input dataset $D = \{(f_i^s, f_i^b; y_i)\}$, $i \in \{1, 2, ..., N\}$, where the $f_i^s \in \mathcal{F}^s$ and $f_i^b \in \mathcal{F}^b$ represents the $i$-th sample of (source feature, binary feature) pair from the source function feature space $\mathcal{F}^s$ and binary function feature space $\mathcal{F}^b$, and $N$ is the number of samples in the dataset $D$. Here $y_i \in \{0, 1\}$ is the label for the feature pair $(f_i^s, f_i^b)$, where 1 means the two function features are a matched pair while 0 is not. As we will introduce later, each function feature $f_i$ (i.e., $f_i^s$ or $f_i^b$) is represented by a list of triplets, that is, $f_i = [t_j]_{j=1}^M$, $t_j = (src, path, tgt)$, where $src, path, tgt$ is the source, path and target node in each triplet $t_j$, and $M$ is the total number of triplet list for the function feature $f_i$. With above definitions, the classification of DBSM is to learn a function mapping from $\mathcal{F}$ to label $\{0, 1\}$. We denote the mapping (matching) function as $g$, and the corresponding learning objective is to minimize the loss function over $g$ as:

$$\min \frac{1}{N} \sum_{i=1}^N \mathcal{L}(g(f_i^s, f_i^b; y_i)), \tag{1}$$

where $\mathcal{L}(\cdot)$ can be cross-entropy loss function or other loss functions. In our paper, we use the contrastive loss [38] that will be introduced in Sect. 3.5.

## 3.2 Decompilation

Binary functions are usually represented as machine code, assembly code or CFG. However, much high-level (semantic) information is lost when compiling source code to binary. In other words, there are enormous differences between binary and source code functions, such as the form. Fortunately, binary analysis tool such as IDA-Pro[2] or Ghidra[3] can decompile assembly code to pseudo-code (decompilation), which means the high-level information can be recovered in some degree. For example, Listing 1.2 is the result of decompilation from Fig. 1(b) by IDA-Pro. Intuitively, it is now much easier to match source function and pseudo-code function since the latter is more similar to the former than other binary function forms.

## 3.3 Code Property Graph Generation

Existing works perform program analysis on source code with various levels, such as token-level, statement-level, function-level, or different representations, such as AST, CFG, PDG. The code property graph (a joint data structure includes AST, CFG and PDG) is proved to be useful in many applications such as vulnerability detection [13,24,25], because the syntactic and semantic relationships can be captured by the code property graph. In details, AST represents the relationship of tokens, which shows the syntactic information, while CFG and PDG represents the relationships among statements, which show the semantic information.

Given the source or pseudo-code function as the input, we apply a robust parser Joern [13] to parse the code and generate the corresponding code property graph. Figure 3 is the code property graph of Listing 1.1. The nodes with color

---

[2] https://www.hex-rays.com/products/ida/.
[3] https://ghidra-sre.org/.

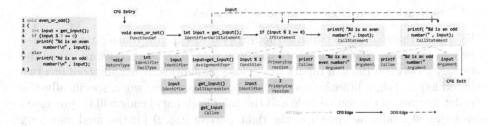

**Fig. 3.** An example of the Code Property Graph for the function. (Color figure online)

light yellow are the statement (CFG and DDG nodes), and nodes with color light green are AST nodes, where the darker is leaf nodes and the brighter is non-leaf nodes. The solid lines are CFG edges, and the dotted lines with color brown (resp. light green) are DDG (resp. AST) edges.

### 3.4 Feature Extraction

From the code property graph, we are now able to extract a list of triplets i.e., $(src, path, tgt)$, where $src$ and $tgt$ are nodes and $path$ is the edge in the graph. For nodes in CFG and DDG, each edge and its corresponding nodes are extracted as triplets. The value of nodes is the *attributes* (highlighted with the red color in the node). For non-leaf nodes in AST, only the *attributes* are taken into consideration, too. While for leaf nodes in AST, both the *attributes* and *values* (highlighted with the black color in the node) are considered. The value of *path* depends on what the edge belongs to, which can be $ast\_x$, $cfg\_x$ and $ddg\_x$, where $x$ represents the count of same triplet $(src, path, tgt)$. For example, if there are two same triplet $(s1, ast, t1)$, then we aggregate them as $(s1, ast2, t1)$. With proposed strategy, syntactic information and semantic information are all covered. In addition, features used in library matching (such as strings and integers) are also extracted.

We take Fig. 3 as an example to clearly illustrate the feature extraction procedure. The triplet (IdentifierDeclStatement, ddg2, CallStatement) is extracted since there are two same triplets (IdentifierDeclStatement, ddg, CallStatement) in DDG, which means that two CallStatements use the same value assigned at IdentifierDeclStatement. It depicts the flow of data from one statement to the others, which represent the semantic information. In AST, the triplet (FunctionDef, ast1, ReturnType) is extracted from the most left two nodes. Since these two nodes are non-leaf nodes, only the value of attributes are taken into consideration. For leaf nodes in AST, the triplet (Argument, ast1, "%d is an odd number!") is extracted, which represents that there is a string literal named "%d is an odd number!" in function. The string feature is used in binary-source library matching, which contributes a lot.

### 3.5 Self-attention Based Siamese Network

We utilize the siamese network structure [40] to capture the relation between the two functions. Siamese network has been proven to be effective to model the similarity between the two inputs from close modalities. It usually contains two shared left and right branches to process the inputs. In our work, specifically, the triplet features of the source code and the pseudo (binary) code will be processed by a shared model (as shown in the right part of Fig. 2) for the final matching learning.

To process the triplet features, we carefully choose outstanding deep learning models. Inspired from the outstanding performances achieved by the self-attention [34] based networks, such as Transformer [26] in natural language processing and computer vision, we adopt the self-attention mechanism to build a siamese network so as to model the triplet features for our matching model. We first feed the list of triplets as input and map them into low-dimensional embedding vectors. Then they will be processed by the self-attention layer to learn the representations. Finally the representations of the two code functions are put into the last layer for matching. The detailed descriptions of each module are in the follows.

*Embedding Layer.* The embedding layer is designed to transform the input elements into fixed low-dimensional vectors. As we introduced before, the triplet feature is consisted of three elements, i.e., $(src, path, tgt)$, and the $src$ and $tgt$ elements are mapped into a same embedding space $E_{st} \in \mathcal{R}^{V_{st} \times d}$, where $V_{st}$ is the vocabulary size of the $src$ and $tgt$ elements, $d$ is the size of the embedding vector. Similarly, the $path$ element is mapped into embedding space $E_p \in \mathcal{R}^{V_p \times d}$, where $V_p$ is the vocabulary size of all $path$ elements. For each triplet feature $t_j$, we denote the mapped embedding vectors as $(e^j_{src}, e^j_{path}, e^j_{tgt})$, then we simply concatenate the three embeddings as the embedding vector for the triple feature, $e_{t_j} = [e^j_{src}; e^j_{path}; e^j_{tgt}] \in \mathcal{R}^{3d}$. Therefore, for one function $f_i$ (see Sect. 3.1), the list of number of $M$ triplet features are embedded by $E_i = [e_{t_1}, e_{t_2}, ..., e_{t_M}] \in \mathcal{R}^{M \times 3d}$.

Following [26], we also add positional embedding to capture the relative order information of triplets in the function feature. Specifically, each position has its own position embedding $E_p = [p_1; p_2; ...; p_N] \in \mathcal{R}^{N \times d}$, and the position embeddings are learnt with other model parameters. The position embedding is summed up with the triplet feature embedding $E_i$ to formulate the final input representation of the model.

*Attention Layer.* The attention layer contains a self-attention sub-layer and a feed-forward sub-layer.

The self-attention [34], or called intra-attention, is an attention mechanism designed to relate different positions of a single sequence in order to compute the representation of the sequence. In our scenario, we regard the list of the triplet features as a sequence. We follow the introduction in [26] to build the scaled dot-product self-attention layer. Specifically, it is computed by:

$$\texttt{attn}(Q,K,V) = softmax(\frac{(QW_Q)(KW_K)^T}{\sqrt{d}})(VW_V), \qquad (2)$$

where $Q, K, V$ represent the Query, Key and Value matrices, which are the list of triplet feature representations outputted by the above introduced embedding layer. The $W_Q, W_K, W_V$ are the three learnable parameter matrices. The scaled item $\sqrt{d}$ is used to prevent the dot-products growing large in magnitude.

After the self-attention, we add a position-wise feed-forward sub-layer to further process non-linear activation:

$$\texttt{FFN}(x) = max(0, xW_1 + b_1)W_2 + b_2, \qquad (3)$$

where $W_1, W_2$ and $b_1, b_2$ are the corresponding weight matrices and bias vectors. The input feature $x$ is the feature output from the attention sub-layer.

Note that for each sub-layer, we also apply a residual connection [35] and a layer normalization [36] operation. Therefore, to make a summary, suppose the input feature to the attention layer is $x$ and the output feature is $\hat{x}$, the whole computation is:

$$\begin{aligned}
\hat{x} &= \texttt{LN}(\texttt{attn}(x) + x), \\
x &= \texttt{LN}(\texttt{FFN}(\hat{x}) + \hat{x}),
\end{aligned} \qquad (4)$$

where $\texttt{LN}$ is the layer normalization operation.

*Output and Loss.* After obtaining the representations for each function features from the attention layer, we now build one last fully-connected layer to transform them into hidden states $h^s$ (source code) and $h^b$ (binary code), and finally we perform the similarity matching.

To train this similarity matching network, we can choose different loss functions, for example, the cross-entropy loss function [39], the contrastive loss function [38]. Here we adopt the contrasitve loss function built on the vector distance. Essentially, contrastive loss is evaluating how good a job the siamese network is distinguishing between the function pairs, which is acknowledged to be a better learning objective. Specifically, the loss function is:

$$\mathcal{L} = \frac{1}{N}\sum_{i=1}^{N}(\frac{1}{2}(y_i * d_i^2 + (1-y_i)*max\{(m-d_i),0\}^2)), \qquad (5)$$

where $y_i \in \{0, 1\}$ is the label of for $i$-th function pair, $d_i$ is the Euclidean distance between the corresponding two features $h_i^s$ and $h_i^b$, and $m$ is the margin value used to control the learning distance.

**Table 1.** The statistics about the open source projects we collected for our dataset $R0$ and $R3$. "LoC" stands for the lines of code in each function, "Avg. Source" means the average number for the source code statistics. The symbols "$\triangle$" and "$\Diamond$" indicate the average number for the binary $R0$ and $R3$, respectively.

| Projects | Domain | # Functions | LoC (Avg. Source) | LoC ($\triangle$) | LoC ($\Diamond$) |
|---|---|---|---|---|---|
| Zlib | Compression | 13 | 49.3 | 61.9 | 50.3 |
| UnRAR | Compression | 120 | 98.2 | 103.0 | 101.4 |
| Expat | XML Parser | 371 | 70.0 | 76.1 | 76.3 |
| Jasper | Image Processing | 760 | 66.1 | 68.7 | 69.3 |
| LibPNG | Image Processing | 867 | 87.0 | 94.6 | 93.9 |
| TCPdump | Packet Analyser | 901 | 94.6 | 96.1 | 108.2 |
| OpenJPEG | Image Processing | 1615 | 79.8 | 82.5 | 85.1 |
| LibTIFF | Image Processing | 1923 | 90.5 | 92.2 | 93.5 |
| FreeType2 | Font Processing | 4289 | 116.9 | 125.0 | 128.4 |
| OpenSSL | Encryption | 4620 | 81.2 | 84.3 | 87.2 |
| VLC | Video Processing | 5139 | 61.5 | 63.4 | 65.0 |
| Sqlite | Database | 6489 | 167.0 | 174.6 | 197.6 |
| CURL | Network | 7047 | 177.7 | 183.3 | 202.0 |
| ImageMagick | Image Processing | 10557 | 255.5 | 242.3 | 275.8 |
| Radare2 | Binary Analyser | 19201 | 90.6 | 90.6 | 90.6 |
| Total | – | 63912 | – | – | – |

## 4 Experiments

### 4.1 Datasets Preparation

To verify the effectiveness of DBSM in matching source code and binary functions, we conduct experiments on two dataset $R0$ and $R3$ collected by ourselves. Concretely, the datasets are compiled from real popular open source projects. We chose open source projects that satisfies the following criteria. First, they are implemented by C/C++ language since DBSM is designed to match functions between C/C++ source code projects and corresponding binaries. Second, they need to cover diverse application domains so that the generality of DBSM can be evaluated. With above two criteria, we chose fifteen open source projects. Table 1 summarizes their statistics. The application domains we cover include compression, xml parser, image processing, packet analyser, font processing, encryption, video processing, database, network, binary analyser, which should be diverse enough to show the generalization ability of our DBSM. After collecting above open source projects, we compile them to binaries manually with different optimization levels, and we remove duplicated functions to save the unique function instance. Then, with the aid of IDA-Pro toolkit, each pseudo-code function for the binary code is extracted and mapped to the corresponding source code function. In total, we finally collect 63,912 pairs of functions. The average LoC in the source code functions range from 49 to 255, while the number of functions

**Table 2.** Experiments results on two datasets $R0$ and $R3$.

| Model | GCC-x64-O0 ($R0$) | | | GCC-x64-O3 ($R3$) | | |
|---|---|---|---|---|---|---|
| | Recall@1 | Recall@3 | Recall@5 | Recall@1 | Recall@3 | Recall@5 |
| Raw code + TextCNN | 56.3 | 71.4 | 75.3 | 44.7 | 55.9 | 60.0 |
| Raw code + DPCNN | 51.1 | 70.6 | 76.5 | 32.5 | 50.1 | 57.8 |
| Raw code + Attention | 63.7 | 79.9 | 84.3 | 42.1 | 57.6 | 64.3 |
| Triplets + LSTM | 83.2 | 94.0 | 95.4 | 77.3 | 90.7 | 93.1 |
| Triplets + TextCNN | 87.5 | 96.8 | 97.7 | 81.2 | 93.9 | 95.5 |
| Triplets + DPCNN | 78.2 | 95.1 | 96.9 | 69.6 | 89.9 | 93.2 |
| **Our model** | **89.6** | **97.4** | **98.0** | **82.2** | **94.2** | **95.9** |

ranges from 13 to 19,201. We take binaries compiled with compiler GCC, architecture x64 (64 bits) and optimization level O0 as dataset $R0$, similarly for the optimization level O3 as dataset $R3$. After processing the data, the vocabulary size of the $V_{st}/V_p$ for $R0$ is about $315k/218k$, while for $R_3$ is $296k/248k$. We randomly split each dataset to be training set, validation set, and test set, and there are about 80% of pairs of functions as training data (51,130 pairs), 10% as validation data (6,391 pairs) and the rest 10% as testing data (6,391 pairs).

## 4.2 Baseline Methods

To show the effectiveness of our proposed method DBSM, we compare with several different baseline methods. As we introduced before, since there are almost no works studying the function level matching problem, we design baseline methods from different aspects, in order to fully test the effectiveness of each component (i.e., triplet features and attention model) in our framework. Specifically, the compared baselines including the following types: (1) The binary transformed pseudo code function and source code function are both fed into the network with their raw code representations, without any preprocessing and feature extractions. (2) The binary transformed pseudo code function and the source code function are both processed by our DBSM method. Besides, we also conduct experiments on various model architectures, including the LSTM [41], TextCNN [42], DPCNN [43] models, which are also used in other similar applications [20], for a better comparison with our attention based models.

## 4.3 Metric

We introduce the Recall value as our evaluation metric. Recall is for the proportion of actual positives that are identified correctly. Specifically, we use Recall@n (i.e., Recall@1, Recall@3, Recall@5) score to measure the Recall score among the top-$n$ predicted pairs.

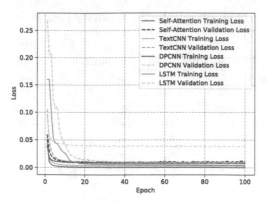

(a) Training/validation loss curves on $R0$.

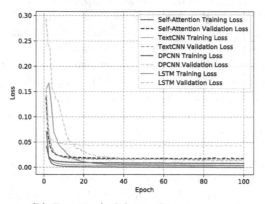

(b) Training/validation loss curve on $R3$.

**Fig. 4.** Training and validation loss curves of different models on $R0$ and $R3$.

### 4.4 Implementation

For the siamese network parameters, the size of the embedding layer $d$ is 128, and the corresponding triplet emebdding size $3d$ is 384. The hidden dimension for FFN layer is 512, and for the last fully-connected layer, it is also 384. For the model training, we use initial learning rate 0.001 and the Adam optimizer [44] with default setting to perform the model optimization. Dropout is used with value 0.1. During training, the batch size is set to be 1024, and we set the max number of triplets for each function to be 200. For training hardware, we train out model for total 100 epochs on single GTX 1080 GPU card.

### 4.5 Main Results

In Table 2, we show the results of our method. Besides, we also show the compared baseline models. From the table, we can see that our method achieves a strong performance on both $R0$ and $R3$ datasets, and over all evaluation metrics,

Recall@1/3/5. Specifically, we obtain 89.6 Recall@1 on $R0$ and 82.2 Recall@1 on $R3$ dataset. Compared with different baseline models, our DBSM clearly outperform all the other methods, no matter with what kind of feature inputs or model networks.

## 4.6  Training Analysis

To give a better understanding of the training process of our method, we plot the loss curves on training and validation sets of our model and other methods. Here we mainly compare with the different model structures on both $R0$ and $R3$ datasets. The results are visualized in Fig. 4(a) and Fig. 4(b). From the figures, we can observe that our self-attention based model has a fast training convergence rate than other models, and the validation loss curve also proves that our model can achieve better performances.

## 4.7  Components Analysis

In this section, we compare DBSM with two types of baseline models (introduced in Sect. 4.2) to illustrate the contribution of triplet feature extraction and attention model.

*Contribution of Triplet Features.*  To evaluate the effect of triplet features, we feed raw code (source and pseudo-code function) into TextCNN, DPCNN and attention models on character level. [20] mentioned that using character code as input is more robust and faster than parsing the code into an AST. We conduct experiments on our dataset and show the results in the first group of lines in Table 2. We can see that DBSM outperform above the raw code based methods on all Recall@n (n=1,3,5) metrics. Specifically, it improved Recall@1 by 33.3, 38.5 and 25.9 on $R0$, while improving 37.5, 49.7 and 40.1 on $R3$. Therefore, these results can clearly demonstrate that triplet features can better capture source and binary function information at both syntactic and semantic level, while the character level features only contains low-level information.

*Contribution of Attention Model.*  To evaluate the superiority of our attention model, we simply replace the attention model with other model structures (LSTM, TextCNN, DPCNN), while keep other components in our framework DBSM unchanged. The experimental results are shown in the second group of lines in Table 2. From these numbers, we can observe that DBSM outperforms LSTM, TextCNN and DPCNN with respect to Recall@1 by 6.4, 2.1 and 11.4 on $R0$, while improving 4.9, 1.0 and 12.6 on $R3$. These improvements clearly demonstrate that our self-attention based model has strong ability to process the triplet features and capture the relationship between these triplet features.

## 5 Conclusion

In this paper, we propose a framework DBSM to solve binary and source function matching problem. Our approach takes advantage of decompilation technique to represent binary functions with high level (semantic) information. Besides, we extract triplets from code property graph, where the features are at syntactical and semantic level. Furthermore, we apply a self-attention based siamese network to process the triple features and perform matching learning task. Our evaluation results on real world datasets have demonstrated that, DBSM can match binary and source function well compared with other approaches.

**Acknowledgments.** The authors would like to thank the anonymous reviewers for their helpful feedback on an earlier version of this paper. This work is partly supported by National Key R&D Program of China under Grant #2022YFB3103900, Strategic Priority Research Program of the CAS under Grant #XDC02030200 and Chinese National Natural Science Foundation (Grants #62032010, #62202462).

## References

1. Langley, P.: Crafting papers on machine learning. In: Langley, P. (ed.) Proceedings of the 17th International Conference on Machine Learning (ICML 2000), pp. 1207–1216. Morgan Kaufmann, Stanford (2000)
2. Mitchell, T.M.: The need for biases in learning generalizations. Computer Science Department, Rutgers University, New Brunswick, MA, Technical Report (1980)
3. Kearns, M.J.: Computational complexity of machine learning. Ph.D. dissertation, Department of Computer Science, Harvard University (1989)
4. Michalski, R.S., Carbonell, J.G., Mitchell, T.M. (eds.): Machine Learning: An Artificial Intelligence Approach, vol. I. Tioga, Palo Alto (1983)
5. Duda, R.O., Hart, P.E., Stork, D.G.: Pattern Classification, 2nd edn. John Wiley and Sons, Hoboken (2000)
6. Author, N.N.: Suppressed for anonymity (2021)
7. Newell, A., Rosenbloom, P.S.: Mechanisms of skill acquisition and the law of practice. In: Anderson, J.R. (ed.) Cognitive Skills and Their Acquisition, vol. 1, pp. 1–51. Lawrence Erlbaum Associates Inc., Hillsdale (1981)
8. Samuel, A.L.: Some studies in machine learning using the game of checkers. IBM J. Res. Dev. **3**(3), 211–229 (1959)
9. Feng, M., et al.: B2sfinder: detecting open-source software reuse in cots software. In: 2019 34th IEEE/ACM International Conference on Automated Software Engineering (ASE), pp. 1038–1049 (2019)
10. Ding, S.H.H., Fung, B., Charland, P.: Asm2vec: boosting static representation robustness for binary clone search against code obfuscation and compiler optimization. In: 2019 IEEE Symposium on Security and Privacy (SP), pp. 472–489 (2019)
11. Feng, Q., Zhou, R., Xu, C., Cheng, Y., Testa, B., Yin, H.: Scalable graph-based bug search for firmware images. In: Proceedings of the 2016 ACM SIGSAC Conference on Computer and Communications Security (2016)
12. Ida pro (2020). https://www.hex-rays.com/products/ida/

13. Yamaguchi, F., Golde, N., Arp, D., Rieck, K.: Modeling and discovering vulner-abilities with code property graphs. In: Proceedings of the IEEE Symposium on Security and Privacy, pp. 590–604 (2014)
14. Duan, R., Bijlani, A., Xu, M., Kim, T., Lee, W.: Identifying open-source license violation and 1-day security risk at large scale. In: Proceedings of the 2017 ACM SIGSAC Conference on Computer and Communications Security (2017)
15. Zhang, J., Beresford, A., Kollmann, S.A.: Libid: reliable identification of obfus-cated third-party android libraries. In: Proceedings of the 28th ACM SIGSOFT International Symposium on Software Testing and Analysis (2019)
16. Hemel, A., Kalleberg, K., Vermaas, R., Dolstra, E.: Finding software license vio-lations through binary code clone detection. In: MSR 2011 (2011)
17. Zuo, F., Li, X., Zhang, Z., Young, P., Luo, L., Zeng, Q.: Neural machine translation inspired binary code similarity comparison beyond function pairs. ArXiv:1808.04706 (2019)
18. Li, X.: Learning program-wide code representations for binary diffing. In: NDSS (2020)
19. Liu, B., et al.: α diff: cross-version binary code similarity detection with dnn. In: 2018 33rd IEEE/ACM International Conference on Automated Software Engineer-ing (ASE), pp. 667–678 (2018)
20. Yu, Z., Zheng, W., Wang, J., Tang, Q., Nie, S., Wu, S.: Codecmr: cross-modal retrieval for function-level binary source code matching. In: NeurIPS (2020)
21. Mou, L., Li, G., Zhang, L., Wang, T., Jin, Z.: Convolutional neural networks over tree structures for programming language processing. In: AAAI (2016)
22. Fang, C., Liu, Z., Shi, Y., Huang, J., Shi, Q.: Functional code clone detection with syntax and semantics fusion learning. In: Proceedings of the 29th ACM SIGSOFT International Symposium on Software Testing and Analysis (2020)
23. Bui, N.D.Q., Yu, Y., Jiang, L.: Treecaps: tree-based capsule networks for source code processing. ArXiv:2009.09777 (2020)
24. Zhou, Y., Liu, S., Siow, J., Du, X., Liu, Y.: Devign: effective vulnerability identi-fication by learning comprehensive program semantics via graph neural networks. In: NeurIPS (2019)
25. Xiao, Y., et al.: {MVP}: detecting vulnerabilities using patch-enhanced vulnera-bility signatures. In: 29th {USENIX} Security Symposium ({USENIX} Security 20), pp. 1165–1182 (2020)
26. Vaswani, A., et al.: Attention is all you need. In: NIPS (2017)
27. Xu, X., Liu, C., Feng, Q., Yin, H., Song, L., Song, D.: Neural network-based graph embedding for cross-platform binary code similarity detection. In: Proceedings of the 2017 ACM SIGSAC Conference on Computer and Communications Security, pp. 363–376 (2017)
28. Feng, Q., Zhou, R., Xu, C., Cheng, Y., Testa, B., Yin, H.: Scalable graph-based bug search for firmware images. In: Proceedings of the 2016 ACM SIGSAC Conference on Computer and Communications Security, pp. 480–491 (2016)
29. Luo, L., Ming, J., Wu, D., Liu, P., Zhu, S.: Semantics-based obfuscation-resilient binary code similarity comparison with applications to software and algorithm plagiarism detection. IEEE Trans. Softw. Eng. 43(12), 1157–1177 (2017)
30. Xu, D., Ming, J., Wu, D.: Cryptographic function detection in obfuscated binaries via bit-precise symbolic loop mapping. In: 2017 IEEE Symposium on Security and Privacy (SP), pp. 921–937. IEEE (2017)
31. Zuo, F., Li, X., Young, P., Luo, L., Zeng, Q., Zhang, Z.: Neural machine translation inspired binary code similarity comparison beyond function pairs. arXiv preprint arXiv:1808.04706 (2018)

32. David, Y., Partush, N., Yahav, E.: Similarity of binaries through re-optimization. In: Proceedings of the 38th ACM SIGPLAN Conference on Programming Language Design and Implementation, pp. 79–94 (2017)
33. Ramos, J., et al.: Using TF-IDF to determine word relevance in document queries. In: Proceedings of the First Instructional Conference on Machine Learning, vol. 242, no. 1, pp. 29–48. Citeseer (2003)
34. Lin, Z., et al.: A structured self-attentive sentence embedding. In: International Conference on Learning Representations (2017)
35. He, K., Zhang, X., Ren, S., Sun, J.: Deep residual learning for image recognition. In: Proceedings of the IEEE Conference on Computer Vision and Pattern Recognition (CVPR) (2016)
36. Ba, J.L., Kiros, J.R., Hinton, G.E.: Layer normalization. arXiv preprint arXiv:1607.06450 (2016)
37. Larochelle, H., Bengio, Y., Louradour, J., Lamblin, P.: Exploring strategies for training deep neural networks. J. Mach. Learn. Res. **10**(1), 1–40 (2009)
38. Hadsell, R., Chopra, S., LeCun, Y.: Dimensionality reduction by learning an invariant mapping. In: 2006 IEEE Computer Society Conference on Computer Vision and Pattern Recognition (CVPR 2006), vol. 2, pp. 1735–1742. IEEE (2006)
39. De Boer, P.-T., Kroese, D.P., Mannor, S., Rubinstein, R.Y.: A tutorial on the cross-entropy method. Ann. Oper. Res. **134**(1), 19–67 (2005)
40. Bromley, J., Guyon, I., LeCun, Y., Säckinger, E., Shah, R.: Signature verification using a "siamese" time delay neural network. In: Proceedings of the 6th International Conference on Neural Information Processing Systems, pp. 737–744 (1993)
41. Hochreiter, S., Schmidhuber, J.: Long short-term memory. Neural Comput. **9**(8), 1735–1780 (1997)
42. Kim, Y.: Convolutional neural networks for sentence classification. In: Proceedings of the 2014 Conference on Empirical Methods in Natural Language Processing (EMNLP), pp. 1746–1751 (2014)
43. Johnson, R., Zhang, T.: Deep pyramid convolutional neural networks for text categorization. In: Proceedings of the 55th Annual Meeting of the Association for Computational Linguistics, vol. 1: Long Papers, pp. 562–570 (2017)
44. Kingma, D.P., Ba, J.: Adam: a method for stochastic optimization. arXiv preprint arXiv:1412.6980 (2014)

# Event-Based Threat Intelligence Ontology Model

Peng Wang[✉], Guangxiang Dai, and Lidong Zhai

19 Shucun Road, Haiden District, Beijing, China
wangpeng3@iie.ac.cn

**Abstract.** Cyber Threat Intelligence (CTI) has become an essential part of contemporary threat detection and response solutions. However, threat intelligence is facing challenges such as lack of unified standards, low efficiency of aggregation, difficulties in widely sharing, and low level of formalization in large-scale applications, which limits its potential in threat detection and response. In response to these challenges, this paper proposes an event-based threat intelligence ontology model based on a thorough analysis of existing threat intelligence standards, aiming to address the urgent need for efficient threat intelligence aggregation and human-machine application. Firstly, the ontology model leverages the semantic characteristics of events to reorganize the elements of threat intelligence, enabling humans to make quicker decisions, simplifying the hierarchical structure for automation processing, while being compatible with existing standards to promote intelligence sharing. Secondly, it combines the skeleton method and Formal Concept Analysis (FCA) method to achieve semi-automated construction, which can improve the efficiency and level of formalization, and aiding in the automated correlation analysis. Finally, we evaluate the proposed ontology and validates its effectiveness with specific instance data, hoping to provide inspiration and reference for other researchers.

**Keywords:** Threat Detection and Response · Threat Intelligence · Event-based Ontology · Intelligence Aggregation · Correlation Analysis · Intelligence Sharing

## 1 Introduction

As the cyberspace confrontation becomes increasingly intense, the direction of security operation is changing from passive defense to active defense, which is characterized by continuous threat detection and response, and timely and accurate warning for assets. To achieve this goal, threat intelligence is essential. At present, domestic and foreign security companies have established their own threat intelligence platforms. However, due to the differences in the level of technology, fields and standards of each, coupled with market competitions, these platforms can not achieve large-scale convergence and sharing. To promote the sharing of threat intelligence, the industry has proposed a series of standards and specifications, involving the unified description and exchange of threat intelligence among different entities. The mainstream related standards include CyboX (Cyber Observable eXpression), CAPEC ( Common Attack Pattern Enumeration and

© The Author(s), under exclusive license to Springer Nature Switzerland AG 2023
M. Yung et al. (Eds.): SciSec 2023, LNCS 14299, pp. 261–282, 2023.
https://doi.org/10.1007/978-3-031-45933-7_16

Classification), OpenIOC (Open Indicator of Compromise), STIX (Structured Threat Information eXression, Structured Threat Information eXression), China's proposed information security technology cyber security threat information format specification (GB/T 36643–2018), TAXII (Trusted Automated eXchange of Indicator Information of Trusted Automated eXchange of Indicator Information), etc. Based on these standards, the academic community has proposed and constructed corresponding threat intelligence sharing models and platforms [1]. These standards have promoted the sharing of threat intelligence to a certain extent, but due to the mechanism factors such as trust barriers and difficulty in allocation of benefits [2] and protection of privacy [3], vendors are not willing to share high-value intelligence, and the actual sharing is not effective. In addition, there are difficulties in producing and aggregating threat intelligence on the basis of these standards. Firstly, although some standard represented by STIX has strong expressive capability, it also has high complexity and can not reach a usable level in terms of automatic extraction accuracy. Secondly, the low level of formalization makes it difficult to conduct automatic correlation analysis, which is not conducive to the deep application of intelligence.

To facilitate the convergence and sharing of threat intelligence on a large scale, some researchers have proposed some ontology-based models and concepts to organize threat intelligence. An ontology is a formal description of important concepts shared in a specific domain, which provides a consistent framework and semantic model for individuals with different backgrounds and purposes by reducing conceptual and terminological ambiguities, thus enabling ubiquitous understanding and communication of information. Therefore, compared with various existing threat intelligence exchange languages, the ontology-based information sharing approach is more responsive to the needs of event information and knowledge organization in the modern cybersecurity domain [4].

Threat intelligence can be divided into human-readable threat intelligence and machine-readable threat intelligence from the perspective of user role. Regardless of the type, it should facilitate the role to understand the intelligence quickly so that it can make a swift decision. Domestic and foreign researchers generally believe that events have natural semantic properties and there are intrinsic connections among events, and building ontology models centered on events can facilitate the analysis of internal factors of events and the reasoning of relationships among events [5]. Ultimately it is conducive to promoting semantic retrieval and knowledge sharing [6].

Based on the analysis above, an event-based threat intelligence ontology model is proposed for threat detection and response scenarios that require efficient threat intelligence aggregation and human-machine co-application. Firstly, the ontology model uses the semantic characteristics of events to reorganize the elements of threat intelligence, which helps human to understand and make decisions quickly, and simplifies the expression hierarchy and improves the degree of structure, which facilitates the automated processing by machines, while it is easy to maintain compatibility with existing standards and promote the sharing of intelligence; secondly, the model combines the skeleton method and Formal Concept Analysis (FCA) method to achieve semi-automated construction, which improves the construction efficiency and formalization level of the model and helps to automate the correlation analysis of intelligence; finally, this paper evaluates the ontology and verifies the effectiveness of the model with specific instance data, hoping to provide reference for other researchers.

The next chapters of this paper are organized as follows:

Section 2 reviews and summarizes the work related to the construction of threat intelligence ontology. Then Sect. 3 describes the detailed building process of the event-based ontology, and Sect. 4 introduces the application method of this ontology. Finally, Sect. 5 summarizes the full work and proposes the research direction in the future.

# 2 Related Works

## 2.1 Ontology Construction Research

From the perspective of the degree of manual involvement, the current ontology construction methods can be categorized into three types: manual construction, semi-automated construction and automatic construction. However, there is no highly effective method for automatic construction, so the first two methods will be mainly introduced.

### (1) Manual Construction

At present, the mature manual construction methods include the skeleton method, TOVE (Toronto Virtual Enterprise) method, cyclic five-step method and six-step method. Among these methods, the skeleton method and six-step method have evolutionary and optimal evaluation steps, which help reuse and enhance the value of existing ontologies [7], and are therefore more commonly used. Manual construction methods rely on experts in this field and have high accuracy, but they are also subjective, costly and have poor portability.

Taking the skeleton method for an example, it generally needs to go through the main steps, which include application goal and scope determination, ontology analysis, ontology representation, and ontology evaluation. The first step is mainly used to determine the application area and scope of the ontology, which requires sorting out the specific knowledge of the domain according to the application scenario. Then, a conceptual model is formed through the ontology analysis step, and a normalized application model is created through the ontology representation step. Finally, after evaluation and correction, an ontology is constructed.

Among the steps, the ontology representation step involves how to express the ontology model, and the commonly used modeling meta-language for ontology models mainly includes concept classes, relations, functions, axioms, and instances [4]. Concept classes represent the set of all objects that conform to the concept, including common information such as properties and behaviors; relations refer to the logical or interactive relations between concept classes, such as inclusion relations, usage relations, etc.; functions can be regarded as special interactive relations between classes, where n-1 elements can uniquely determine the n-th element, and are often used in knowledge inference; axioms refer to eternal truth assertions, which are the basis of inference rules in the conceptual system; instances are the concretization of concept classes, which have all the properties and behaviors specified by the concepts and are influenced by conceptual relations. In the step of ontology evaluation, there is no unified evaluation system. Yue Lixin et al. [8]

selected five indicators of completeness, clarity, consistency, scalability and compatibility to evaluate multiple ontologies constructed by different methods at home and abroad, and Boeker et al. [9] proposed three indicators: usability, structure and functionality to evaluate ontologies.

(2) Semi-automated Construction

Semi-automated construction methods are mainly based on manual construction and automate some part of the steps to reduce the construction cost and subjectivity. There are mainly statistical-based methods and deep natural language processing-based methods. Statistical-based construction methods mainly use clustering, word frequency statistics, word co-occurrence analysis and other techniques for ontology element extraction and inter-element relationship mining, which use simple natural language processing techniques and are not ideal for relationship extraction; while deep natural language processing-based construction methods use semantic analysis techniques such as lexical annotation, syntactic analysis, dependency analysis, and semantic annotation, which can more effectively mine the relationship between elements, but these methods are difficult to apply to multiple domains because of the high requirements for training models [10].

In addition to this, there are many scholars working on the semi-automated construction of ontologies based on the formal concept analysis (FCA) approach. Formal concept analysis theory [9], a tool for data analysis and rule extraction from a formal context, improves automation by automating the construction of concept lattices to compensate for the tedious ontology hierarchical structure construction process. At present, this method has been widely applied in many fields. But it suffers from op-erational limitations when targeting multi-source heterogeneous data [10]. Liu Ting et al. [11] proposed a semi-automated construction method of coal mining face ontology CFOCFCA (Coal Face Ontology Construction based on FCA) based on the character-istics of coal mining face. Sun Li et al. [12] proposed a subject word list and FCA-based maritime ontology construction method, which merges structured resources (subject word list) and unstructured resources (text) to construct ontologies, extending the coverage of ontologies.

With the rapid development of Large Language Model (LLM) in recent years, LLM-based ontology construction methods have received widespread attention. For example, Milena T et al. [15] proposed to extract valuable information from unstructured text by automated means, to assist in the construction of knowledge ontologies. In addition, a series of LLM-based information extraction techniques have been proposed [16, 17, 25], which have to some extent facilitated the automated construction of ontologies. However, the LLM-based methods are currently limited by the lack of annotated data in large model training and have not been widely used.

## 2.2 Ontology Research of Threat Intelligence

Traditional researches of threat intelligence ontology can be divided into generalized ontologies and specialized ontologies. The generalized ontologies model the main concepts in the threat intelligence domain and focus on the representation. For example, Gao Jian et al. [18] constructed a threat intelligence ontology model that can be shared, reused, and extended based on STIX2.0 standard, and used the knowledge graph to visually represent the important elements in intelligence and relationship between them, which helps intelligence analysts to make analytical decisions, but it's limited by the high complexity of representation and the difficulty of practical implementation. Specialized ontologies model sub-domains of threat intelligence and focus on applications. For example, Christian R et al. [19] proposed the ontology model MALOnt2.0 for capturing malware threat intelligence from heterogeneous data sources while constructing the malware threat knowledge graph MalKG, but it only supports graph query functions and is lack of inference and analysis capabilities. Yeboah-Ofori A et al. [20] proposed a cyber attack ontology to improve security based on cyber supply chain security, but the ontology only supports first-order logical queries in terms of application and doesn't contain inference capabilities. Sánchez-Zas C et al. [21] proposed an ontology for real-time risk management and cyber situational awareness that defines and validates a series of inference rules, but it lacks automatic response capability. Syed R et al. [22] proposed a network security vulnerability ontology that integrates vulnerability information from multiple sources and has a wide coverage. In addition, the authors designed an alerting system based on this ontology, which was evaluated to have a good performance, but the system is not fully automated.

Event ontology is a representation method for event knowledge, and there are different representation models in different research fields, which are generally divided into three categories: representation models based on conceptual hierarchy, logical hierarchy and event hexadecimal. In the third model, the event consists of six elements: action, object, time, environment, assertion and language, and the action element is the core, which can describe the event dynamically. The relationship between event elements can be described in detail and is more commonly used at present. In the research related to event-based threat intelligence ontology, there are no mature and systematic research results at home and abroad. Li Wenxiong et al. [23] studied network attack behavior and attack events from the attack case perspective and constructed a network attack case ontology, but the event elements were incomplete and lacked inference analysis capability. Yazid Merah et al. [24] proposed an ontology for risk detection, considering both security events and threat intelligence, and developed a network risk detection framework based on inference rules, but the application scope is limited to query and retrieval.

In conclusion, the existing threat intelligence ontology models, whether traditional or event-based ontology models, mainly focus on the expression, sharing, reuse and expansion of intelligence, and focus on the application of inference rules for retrieval and query, with little on intelligence correlation analysis and automated response.

## 3   Ontology Model Construction

In order to remedy these shortcomings and improve the dynamic semantic expression and reasoning ability of threat intelligence, the improved skeleton method is adopted to construct the threat intelligence domain ontology, in which the formal concept analysis method is used to improve the level of automation of ontology construction. The overall process is shown in Fig. 1.

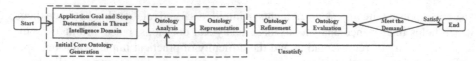

**Fig. 1.** Flow chart of Event-based Threat Intelligence Ontology Construction

The first three steps are defined as the Initial Core Ontology Generation Module, which is completed by domain experts. The Ontology Refinement step acquires the implicit information in the threat intelligence data automatically, and then combines the expert experience to refine the ontology.

### 3.1 Initial Core Ontology Generation

**Application Goal and Scope Determination.** The ontology designed in this paper is oriented toward both security operators for asset-protection-oriented security operations and numerous intelligence providers for facilitating the dissemination of threat intelligence. Specifically, the ontology model is applied to the following three aspects: intelligence aggregation, intelligence correlation analysis and intelligence sharing.

(1) Intelligence aggregation: Before being used for security operations, the threat intelligence ontology first needs to be able to store and manage intelligence, and other forms of intelligence should be easily converted according to the ontology model.
(2) Intelligence correlation analysis: For asset protection scenarios, the ontology should have intelligence correlation analysis capability, such as combining existing intelligence data to analyze threat information related to assets, including attacker information, asset vulnerabilities, countermeasures, etc.
(3) Intelligence sharing: For many intelligence providers, the ontology should have efficient intelligence sharing capability.

**Ontology Analysis.** In order to meet the application requirements mentioned above, this paper uses the event elements as the main line to organize threat intelligence, and each element of the event is elevated to the top level. With the principle that "each piece of intelligence is a (group of) event", each piece of intelligence should contain the following information shown in Table 2 of Appendix.

Based on the information requirements above, a conceptual model diagram of event-based threat intelligence ontology is developed as follows (Fig. 2).

**Ontology Representation.** Combining the elements in the conceptual model, and using object-oriented design philosophy, each of the elements above is abstracted into an event class. In this paper, the event-based threat intelligence ontology (ETIO) is defined formally as follows:

**Definition 1.** ETIO ::= {TECs, ECs, As, Rs, Rules}.

Among them, TECs is the set of top-level event classes, ECs is the set of classes other than top-level event classes for future expansion of the event ontology, As is the

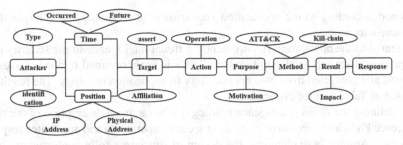

**Fig. 2.** Conceptual Model Diagram of Event-based Threat Intelligence Ontology

set of attributes of each event class, Rs is the set of relationships between event classes and between events, and Rules is the set of inference rules.

Considering the complexity of the event elements, the TECs are further defined formally in this paper as follows:

**Definition 2** TECs:: = {Attacker, Time, Location, Target, Action,
Motivation, TTP, Result, CourseOfAction}

Among them, Target describes the attacked object (also represents the defender); Action describes the type of attack in this event; TTP describes the attacker's attack method, including technique and tactics, tools, and process; Result describes the attack result; and CourseOfAction describes the response measures.

**Definition 3** As:: = {Name, Type, Country, Time, Network location,
Geographocal position, Location regularity, Version, Number, Tactical
objective, Straregic target, Influence degree, Reliability, Description}

**Definition 4** Rs:: = {R_event, R_event_class}
R_event:: = {Result, Follow, Co_occurrence}
R_event_class:: = {Has, Occurre_in, Include, Use, Aim_at,
Cause, Belong_to, Locate_in}

Here, R_event refers to the relationship between events, R_event_class refers to the relationship between event class.

**Definition 5.** Rules ::= {Rules_as, Rules_other}.

Here, Rules_as denotes the inference rule between attributes of event class and Rules_other indicates other forms of inference rules.

Through the formal definitions above, a semantic foundation based on event elements is laid for threat intelligence, in which the set of relationships and inference rules can be defined according to the application scenarios of the ontology in a targeted and complementary manner.

Oriented to the application requirements of threat intelligence in the security operation process, each top-level event class in TECs is further refined into subclasses and attributes, and converted from concept ontology to application ontology. The results are presented in Table 5 of the Appendix.

By defining the threat intelligence ontology, a hierarchical description of the threat intelligence knowledge required for guiding security operations and automated response is achieved. Among these elements, the design of subclasses fully conforms to object-oriented thought and is highly reusable, thereby reducing the number of layers and complexity.

## 3.2 Ontology Refinement

Traditional ontology construction methods rely heavily on human involvement, with a large degree of subjective influence, poor scalability and high construction costs. To tackle these problems, we combine the formal concept analysis method and expert experience to semi-automatically refine the ontology (Fig. 3).

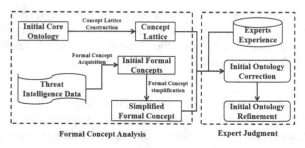

**Fig. 3.** Flow Chart of Ontology Refinement

Among them, the formal concept analysis part relies on the machine to automate the implementation, which can achieve the effect of human-machine collaboration.

**Formal Concept Analysis (FCA).** The theory of FCA is based on the mathematization of concepts and their hierarchies, and mathematical means are used to represent objective knowledge, thus weakening the subjective influence of ontology builders. In the FCA approach, the concept lattice is able to describe the hierarchy among concepts in an essentially clear manner and uncover the implicit information in the data [12], as well as facilitate human understanding. Therefore, in this paper, FCA will be used to assist experts in refining the ontology.

*Formal Concept Acquisition and Simplification.* Since there is a large amount of structured and unstructured data in the threat intelligence domain, which is real-time and may contain implicit information, the FCA approach is considered to automate the mining of implicit information related to event elements in the data. Based on this idea, this paper extracts the objects and attributes contained in the threat intelligence data by using textual information extraction techniques, and then combines thematic models to simplify the formal concepts.

### (1) Formal Concept Acquisition

ChatGPT has become a hot topic due to its powerful text generation capability recently. Wei X et al. [25] proposed ChatIE, which transformed the zero-sample information extraction task into a two-stage framework with multiple rounds of answering questions, and evaluated three tasks of relationship extraction, named entity recognition, and event extraction, and the experimental results showed that on two languages and six datasets, ChatIE achieves rather good results. Inspired by this, this paper considers using ChatGPT to extract objects and attributes from textual data in the threat intelligence domain. Specifically, we select two blogs, one report and one news media report that are highly relevant to cyber security as data sources, and select "cyber attack" as the keyword for information extraction, and finally obtain 117 initial formal concepts. Considering that these formal concepts are not only large in number but also uneven in quality, it is necessary to simplify the formal concepts before providing them to experts for analysis.

### (2) Formal Concept Simplification

The LDA (Latent Dirichlet Allocation) topic model is an unsupervised learning algorithm that can efficiently process document data and classify numerous documents into topics according to probability distributions while displaying topic words. Therefore, in this paper, we consider to simplify formal concepts by using topic words in text. We select the same data and use the method based TF-IDF [26] and set the number of topics and the number of subject words to 4 and 50, respectively. Then we use ChatGPT to further sieve out the subject words that are not related to "cyber attack", and finally obtain a subject word list with a length of 32.

We define the simplification rule as follows: for each formal concept, we count the number of topic words (denoted as N), and set the threshold (denoted as keys), if $N \leq$ keys, we delete the formal concept, otherwise we keep it. After the experimental analysis, when keys are 2, 3 and 4, the number of simplified formal concepts is 46, 16 and 8 respectively. Finally we choose keys $= 4$, and the set of formal concepts is shown on the github repository[1].

*Concept Lattice Construction.* Firstly, the initial core ontology is transformed into the formal context K1 by defining the following rules: i) the bottom concept class (or the top concept class if there is no subclass) of all ontologies is selected as the object in the formal context; ii) the attributes in the ontology are selected as the attributes in the formal context. Next, we use the method proposed by Lindig C et al. [27] to automate the construction of the concept lattice L1 corresponding to the formal context K1, which will be provided to the experts for ontology correction.

### Expert Judgment

*Initial Ontology Correction.* Domain experts can be subjective in the process of constructing initial core ontologies, especially in the selection of attributes. In FCA, there

---

[1] https://github.com/LIGHTdgx/ETIO-Extraction-Results/tree/Results.

are three principles [28]: (i) concepts are described by attributes; (ii) attributes determine the hierarchy of concepts; and (iii) when two concepts have the same attributes, these two concepts are considered to be the same. Based on these principles, ontology correction is performed by domain experts according to the following correction principle: for the same attribute of different objects, choose to keep, delete or further divide it; for objects with the same attribute, choose to merge objects or add new attributes to distinguish them (Fig. 4).

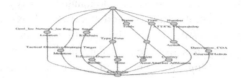

**Fig. 4.** Concept Lattice L1

In the *Concept Lattice Construction* subsection, we constructed the concept lattice L1 corresponding to the initial core ontology, but we also observed that the attacker and the affiliation are grouped into the same formal concept due to the fact that these two concept classes in the Initial Core Ontology have the same attributes. According to the correction principle, we chose to delete attribute "country", add new attribute "experience", and further divide some attributes to make the distinction. Similarly, through a series of corrections, we obtain the corrected formal background K2 and the corresponding concept lattice L2. To make the figure clearer, we number the following attributes "Name of Attacker, Type of Attacker, Name of Asset, Type of Asset, Name of Affiliation, Type of Affiliation, Type of Time, Number of Observation, Name of Tool, Number of Attack Method, Deployment Difficulty, Number of CourseOfAction, Geographical location, Network location, Tactical Objective, Strategic Target, Degree of Influence" in order from "A" to "Q".

The corrected concept lattice L2 shown in Fig. 5 is more consistent with the definition of the initial ontology. Additionally, an implicit message can be inferred: the ATT/CK becomes a sub-concept of the kill chain and Vulnerability. Through the mapping rule proposed by Wei Lian et al. [29] between ontology and concept lattice, we mapped the modified concept lattice as ontology, and according to the division of abstraction levels in the network attack model, we chose to move ATT/CK into the subclass of the Kill chain.

*Initial Ontology Refinement.* In the subsection of *Formal Concept Acquisition and Simplification*, we obtained a simplified set of formal concepts, which were then analyzed by domain experts and constructed a new concept lattice to obtain the implied information, and finally the ontology was refined.

From the perspective of "object", we consider "AI security" as a generalization. From the perspective of "attribute", the data reflects the different stages of AI technology that can be used to automate cyber attack and defense, so we choose "stage", "deployment difficulty", and "automation" to summarize it. Finally, we obtain a new formal concept:

{"object": "AI security", "attribute": "stage, deployment difficulty, automation"}, which is added to the formal context K2 to build the new concept Lattice L3.

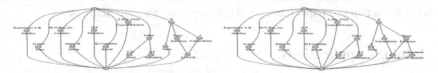

**Fig. 5.** Concept Lattice L2 and L3

We can infer that AI security is a sub-concept of the Kill chain along with the ATT/CK, which means that a cyber-attacker can combine AI techniques with attack techniques in ATT/CK to improve attack efficiency, but the top-level attack steps are still based on the Kill chain.

Finally, the refined concept lattice has been mapped into ontology according to the same mapping rules and so far we have obtained the corrected and refined ontology.

### 3.3  Ontology Evaluation

We use the "usability" and "structure" and "functionality" metrics mentioned in [9] to evaluate the ontology, where usability is concerned with the specific use of the ontology, i.e., whether other people can use the ontology without ambiguity. After the detailed description of the ontology construction process above, it is convenient for others to understand and meet the usability requirements. Structure is concerned with the formal structure of the ontology. Since the ontology defined in this paper is an event-based threat intelligence domain ontology, it is necessary to evaluate whether it encompassed all the information in the threat intelligence dimension and the event dimension, and the following will refer to the ontologies related to threat intelligence and security events for evaluation. Functionality is concerned with ontology applications, and we will discuss it in the next chapter.

*Threat Intelligence-Related Ontology.* Malware ontology [19] (MALOnt2.0), TAL ontology [30] (Threat-Agent-Library) and threat intelligence ontology proposed by Gao et al. [18] were selected for comparison, using the main components in the Information Security Technology Cybersecurity Threat Information Format Specification [31] (GB/T 36643–2018) as evaluation criteria. The results are shown in Table 3 of Appendix. From the table, it's evident that the ontology defined in this paper contains most of the elements in the threat intelligence domain. Compared to the malware ontology which mainly describes the attack mode and attack behavior, the TAL ontology which focuses on describing information related to the threat subject, and the ontology proposed by Gao et al. which aims to describe the specific attack mode and attack indicators.

*Security Event-Related Ontology.* The computer security event ontology [32], the intrusion detection ontology [33], and the network attack case ontology [23] were selected for comparison, and the results are shown in Table 4 of Appendix. From the table, we can see that the ontology defined in this paper has rich elements in the event dimension, which basically covers all elements of cyber security events, and the practitioners of security using this ontology can select different elements for correlation analysis.

Combining the analysis results in Table 3 and Table 4, it can be concluded that the ontology defined in this paper has a wide coverage and a strong richness.

## 4    Ontology Applications

Aiming at threat detection and response for asset protection, the ontology is applied from three perspectives: Intelligence Aggregation, Correlation Analysis, and Intelligence Sharing for multi-source heterogeneous intelligence data, and the flow chart of ontology application is as follows. In the step of intelligence aggregation, for structured data, it can be directly transformed into instance data through field mapping, and for unstructured data, the corresponding field values can be obtained from the text through information extraction technologies (Fig. 6).

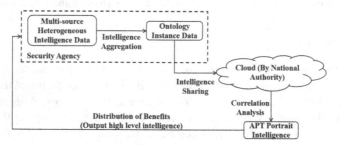

**Fig. 6.** Flow Chart of Ontology Application

### 4.1    Ontology Mapping

Ontology mapping refers to the extraction and transformation of information from data described by other standards. For reasons of corporate interests and privacy protection, most of the structured data in the field of threat intelligence is under control of domestic security vendors. Thus, we have chosen unstructured data as our data source. Next, we will take an APT report of AridViper [34] as an example. The data will be mapped according to the ontology of this paper by ChatGPT and the result will be presented as a conceptual diagram. The attributes of ATT/CK and CourseOfAction are referred to the technical and tactical knowledge base proposed by MITRE [35], and the attributes of attacker experience are referred to the "sophistication" field in the threat body component of the standard [31] (Fig. 7).

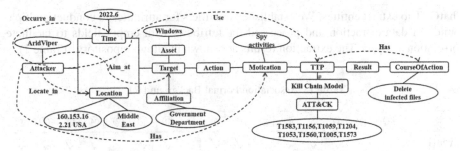

**Fig. 7.** An Instance Data for APT Report

## 4.2 Ontology-Based Intelligence Correlation Analysis

With the rapid development of network security technology, in practical application scenarios, the same attack organization will change its attack characteristics with the improvement of economic strength and technical capability. Therefore, we consider different types of attacks for the same attack organization to conduct correlation analysis, where the different types mainly refer to the attack target and the attack method. Decision implication [36] is a method that could be used for automated association analysis. In this paper, we will use the decision implication method to obtain the decision implication about APT organization by using the attribute values of elements in APT attack event as input, and then construct a portrait of APT organization to enable threat detection and response.

**Instantiation of Inference Rules.** To make the correlation analysis based on different classes of attack events more general, we instantiate the inference rules.

**Definition 6.** Rules_as ::= $\{f(\text{K\_as}) \rightarrow d1, g(d1) \rightarrow d2, h(d2) \rightarrow d3\}$.

**Definition 7.** Rules_as_augment : If $f(A) \rightarrow B, A \in A1, B1 \in B$, then $f(A1) \rightarrow B1$.

**Definition 8.** Rules_as_combine : If $f(A) \rightarrow B, f(A1) \rightarrow f(B1)$, then $f(A \cup A1) \rightarrow B \cup B1$.

Here, function $f$ is the inference rule corresponding to the decision implication method, and its mapping logic is described in [36]. K_as is the formal background set about the attribute set, respectively. Functions $g$ and $h$ are two inference rules based on decision implication with the mapping logic referring to the descriptions in Definition 7 and 8, and d1, d2 and d3 denote different decision implication. In addition, the literature [37] proves the soundness, completeness and non-redundancy of the latter two inference rules. In the following, we will apply these three inference rules to analyze the instance data.

**Instance Analysis.** Five representative event elements are selected: the attack time, attack motive, attack target, attack method and response measures, so that different attack events can be represented based on the attribute values of these elements. We selected eight reports on the "white elephant" APT organization as the data source and used

ChatGPT to extract entities. We established qualifiers for attributes to conduct multiple rounds of data extraction and processed the attributes with empty fields to facilitate correlation analysis. The extraction results are shown in github repository[1] (Table 1).

**Table 1.** Association Formal Background R1

|        | ① | ② | ③ | ④ | ⑤ | ⑥ | ⑦ | ⑧ | ⑨ | ⑩ | ⑪ | ⑫ | ⑬ | ⑭ | ⑮ | ⑯ | ⑰ |
|--------|---|---|---|---|---|---|---|---|---|---|---|---|---|---|---|---|---|
| Event1 |   |   | √ | √ | √ | √ |   | √ | √ | √ | √ | √ |   | √ | √ | √ | √ |
| Event2 | √ |   | √ | √ | √ | √ | √ | √ | √ | √ | √ |   | √ |   | √ | √ |   |
| Event3 |   |   | √ |   |   |   | √ | √ |   |   |   |   |   |   | √ | √ |   |
| Event4 | √ |   | √ | √ | √ | √ | √ | √ | √ | √ |   |   | √ |   | √ | √ |   |
| Event5 |   |   | √ | √ | √ | √ |   |   |   |   |   | √ |   |   | √ | √ |   |
| Event6 | √ | √ | √ | √ | √ | √ | √ | √ | √ |   |   |   | √ | √ | √ |   |   |
| Event7 | √ | √ | √ | √ |   | √ | √ | √ | √ |   |   |   |   |   | √ | √ | √ |
| Event8 |   |   | √ | √ | √ | √ | √ | √ | √ | √ |   |   |   |   | √ | √ |   |

Aiming at threat detection and response, we select the attributes of attack method as the conditional attribute, and the attributes of attack target and response measures as the decision attribute to obtain the association formal background R1. For ease of presentation, we remove the attribute values unique to each event, and number the remaining attribute values "Reconnaissance, Resource Development, Initial Access, Execution, Persistence, Defense Evasion, Discovery, Collection, Command and Control, Exfiltration, CVE-2014–4114, CVE-2015–1641, MetaSploit, LINK, Data, Organization, Improving email precautions" in order from ① to ⑰.

Based on the first inference rule, the decision implication d1 is obtained and we demonstrate one of them:

{Initial Access, Execution, Defense Evasion, Collection, Command and Control} → {Data, Organization, Improving email precautions}

Based on the third inference rule, we get the combined decision implication d2.

{Reconnaissance, Resource Development, Initial Access, Execution, Persistence, Defense Evasion, Discovery, Collection, Command and Control,Exfiltration,CVE-2014–4114,CVE-2015–1641,MetaSploit} → {LINK, Data, Organization, Improving email precautions}

Based on the second inference rule, it is impossible to continue to augment the conditional attributes at this point. Finally, we represent decision implication d2 based on the ontology structure to obtain a portrait of the attack organization consisting of three elements: attack target, attack method and response measures (Fig. 8).

**Evaluation Analysis.** Finally, we evaluate the functionality of the ontology from three aspects: time performance of ontology mapping, completeness of ontology mapping and coverage of the portrait. The completeness represents the proportion of non-missing

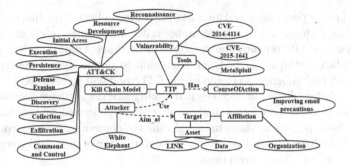

**Fig. 8.** Portrait of "White Elephant" APT Organization

fields in the mapping result (i.e. instance data) while the coverage represents the proportion of instance data transformed into a portrait. The formal definitions are as follows:

We denote $N_i$ (i = 1,2,...,8) as the instance data extracted from the ith APT report, $|N_i|$ as the number of non-empty fields it contains, $N$ as the number of mapping template fields (N = 11), $e$ as the number of leaf nodes in the generated APT organizational portrait graph, $M$ as ontology mapping completeness, and $P$ as the portrait graph coverage.

$$M = \frac{1}{N} \sum_i \frac{|N_i|}{N}$$

$$P = \frac{e}{\sigma \left( \sum_i |N_i| \right)}$$

**Here,** $\sigma$ denotes the de-duplication operation. After calculation, the completeness of ontology mapping is 84.1%, and the coverage of portrait is 26.6%. The mapping process took about 139 s. It can be seen that the mapping method used in this paper has good completeness and time performance. However, due to the small amount of data used in this study (only 8 APT reports), the coverage of the portrait is relatively low.

### 4.3  Ontology-Based Intelligence Sharing

To address the problems of trust barriers, benefit distribution [1] and fake intelligence provision in the current threat intelligence sharing model, a threat intelligence sharing mechanism is proposed based on the ontology model defined in this paper. The mechanism is targeted at security agencies and a CTI cloud served by national authorities, and the main steps are as follows.

(1) Security devices in different agencies only share raw intelligence information and do not involve sensitive intelligence (e.g., vendor product vulnerability information) or private data, which could reduce not only the sensitivity of intelligence but also the problem of sharing trust barriers.

(2) The national authority plays the role as the CTI cloud and is responsible for collecting and pre-processing the raw intelligence provided by each security device into an event-based threat intelligence ontology structure.

(3) The CTI cloud performs correlation analysis of the collected intelligence based on the ontology structure to generate higher-order intelligence (such as the portrait intelligence of APT organization described in **Instance Analysis** Subsect. 4.2). As the correlation analysis methods are continuously supplemented, the types of the higher-order intelligence generated by the CTI cloud will be enriched.

(4) The contribution degree of each original intelligence is calculated according to the contribution in the correlation analysis process of generated higher-order intelligence, while a penalty mechanism is established by combining the negative feedback from the intelligence user. Finally, the total contribution of the security organization is updated by the cloud in real time.

(5) The degree of contribution is used as the main reference for benefit distribution, which can take on various forms, such as awarding, certifications and licenses for enjoying higher-order intelligence subscription services, etc.

## 5  Summary and Outlook

In this paper, an event-based threat intelligence ontology model is proposed to address the urgent need for efficient aggregation of threat intelligence and efficient human-machine applications in threat detection and response scenarios. The semantic characteristics of events are used to reconstruct threat intelligence, which simplifies the expression hierarchy and improves the structure compared with existing standards. Secondly, we propose a semi-automated construction method based on improved skeleton method, which improves the model construction efficiency and formalization level. Then we introduce the application method of the ontology through an example, which can improve the efficiency aggregation and automated analysis level of threat intelligence to a certain extent, and promote threat intelligence sharing on a large scale. There are also some problems, such as the efficiency of human-computer combination in ontology refinement still needs to be further improved. In addition, the quality of the instance data obtained by mapping of unstructured data is affected by the data source and the prompt words used in the extraction process. Moreover, threat intelligence is time-sensitive but the ontology model cannot effectively represent the dynamic evolution of knowledge. The invalid intelligence is still required for manual filtering.

Our future work will focus on the following aspects:

(1) Research more effective automated/semi-automated ontology construction methods to improve the efficiency of ontology construction;

(2) Further improve the set of relations and attributes in the model and develop more effective inference rules to extend the application scope of the ontology;

(3) Research more effective information extraction methods to acquire instance data with higher quality, such as finetuning the GPT model;

(4) Research sharing techniques and mechanisms based on this ontology model, especially for the calculation of contribution degree;

(5) Research more effective evaluation criteria for ontology;

(6) Research threat intelligence correlation analysis method that deals with incomplete instance data referring to the work of Ning Hu et al. [38];

(7) Research dynamic knowledge representation methods based on spatio-temporal information for threat intelligence referring to the work of Jia Y et al. [39].

# Appendix

**Table 2.** Event-based Threat Intelligence Concept Ontology

| Event Elements | Element Description | Content Example |
|---|---|---|
| Human elements | Attacker's identity information (organization and country) | APT 29 |
| Time elements | ① Time of single attack ② Time rule of multiple attacks ③ Predict the occurrence time and probability of attacks | ① 2022/5/12 22:00:23 GMT + 08:00 ② Launch DOS attacks frequently within 30 days ③ The probability of attacking a certain type of asset within 30 days is 70% |
| Location elements | ① Attacker's network address (address pool) ② Attacker's Physical address | ① Network address: 10.10.10.10 ② Physical address: XX country XX Province XX city |
| Object elements | ① Target assets ② Asset owner information | ① Windows 7 PC host ② XX Company |
| Motivational elements | Purpose of attack | Destruction, data theft, remote control |
| Movement elements | Behavior type | Normal access, determined attack, suspicious access |
| Methodological elements | Means of attack | Use the Blue of Eternity vulnerability to launch blackmail attack |
| Result elements | ① Impact degree ② Credibility | ① Serious impact, slight impact, no impact ② Reliability expressed by probability |
| Response elements | Preventive measures and disposal suggestions | Update the patch or upgrade to a higher version of Windows system as soon as possible |

**Table 3.** Event-based Threat Intelligence Concept Ontology

|  | Our Model | MALOnt | TAL | Gao's Model |
|---|---|---|---|---|
| Type of Threat | ✓ | ✓ | ✓ | ✓ |
| Experience of Threat | ✓ |  | ✓ |  |
| Time | ✓ | ✓ |  |  |
| Impacted Assets | ✓ |  |  | ✓ |
| Security Event Status | ✓ |  |  |  |
| Motivation of Threat | ✓ |  | ✓ |  |
| Attack Behavior | ✓ | ✓ |  | ✓ |
| Impact Assessment | ✓ |  |  |  |
| Credibility | ✓ |  |  |  |
| Response Measures | ✓ |  |  | ✓ |
| Observable Data | ✓ | ✓ |  | ✓ |
| Attack Stage | ✓ |  |  | ✓ |
| Attack Method | ✓ | ✓ |  | ✓ |
| Information Source |  | ✓ |  |  |
| Attack Resources |  |  | ✓ |  |
| Vulnerability | ✓ | ✓ |  | ✓ |

**Table 4.** Event-based Threat Intelligence Concept Ontology

|  | Our Model | [26] | [27] | [17] |
|---|---|---|---|---|
| Time | ✓ |  |  |  |
| Location | ✓ |  |  |  |
| Attacker | ✓ | ✓ |  | ✓ |
| Victim | ✓ | ✓ |  | ✓ |
| Method | ✓ | ✓ | ✓ | ✓ |
| Result | ✓ | ✓ | ✓ | ✓ |
| Measure | ✓ |  |  | ✓ |
| Motivation | ✓ | ✓ |  |  |
| Behavior | ✓ | ✓ |  |  |

**Table 5.** Event-based Threat Intelligence Application Ontology

| Event Category | Subcategory | Attribute | Description |
|---|---|---|---|
| Attacker | / | name | Name or code information of the attacker |
| | / | type | Individuals and organizations |
| | / | Country | Country of attacker |
| Time | / | type | Single attack occurrence time, multiple attack occurrence time pattern, attack time prediction |
| | / | time | Single attack occurrence time, predicted time |
| Location | / | network location | The IP address used by the attacker |
| | / | geographical location | Geographic location of the attacker |
| | / | location regularity | Used to describe non-independent events(e.g., multiple attacks of the same kind launched by the same attacker) location patterns, such as multipoint concurrency |
| Target | Asset | name | Specific name of the asset, such as Windows 7 PC |
| | | type | Hardware, link, service, data |
| | | version | Version information corresponding to the target asset |
| | Affiliation | type | Individuals and organizations |
| | | name | The name of Individuals and organizations |
| | | Country | The country of Individuals and organizations |
| Action | / | type | Normal access, determined attack, suspicious access |
| | Observation | number | Numbering of observable behaviors described in STIX |
| Motivation | / | tactical objective | Attack objectives achieved at the tactical level |
| | / | strategic objective | Attack objectives achieved at the strategic level |
| TTP | Kill chain | stage | The seven stages described by the Killchain model |

(*continued*)

**Table 5.** (*continued*)

| Event Category | Subcategory | Attribute | Description |
|---|---|---|---|
| | ATT/CK | number | The number of the attack means described in ATT/CK |
| | Vulnerability | number | The CVE number of the used vulnerability |
| | Tool | name | The name of the used tool |
| Result | / | influence degree | The severity of the attack and the degree of caused impact, 0 - no 1 - minor 2 - moderate 3 - severe |
| | / | credibility | The credibility of this attack intelligence information, a continuous value between 0 and 1, 0 represents the lowest credibility, 1 represents the highest credibility |
| CourseOfAction | / | measure | Response codes taken |
| | | description | A description of the response, such as patching |

# References

1. Karatisoglou, M., Farao, A., Bolgouras, V., Xenakis, C.: BRIDGE: BRIDGing the gap bEtween CTI production and consumption. In: 2022 14th International Conference on Communications (COMM), 16 June 2022, pp. 1–6. IEEE (2022)
2. Lin, Y., Liu, P., Wang, H., et al.: Overview of threat intelligence sharing and exchange in cybersecurity. J. Comput. Res. Dev. **57**(10), 2052 (2020)
3. Sarhan, M., Layeghy, S., Moustafa, N., Portmann, M.: Cyber threat intelligence sharing scheme based on federated learning for network intrusion detection. J. Netw. Syst. Manag. **31**(1), 3 (2023)
4. Chen, J.F., Fan, H.B.: Ontological threat intelligence sharing in cyberspace security. Commun. Technol. **51**(1), 177–183 (2018)
5. Liu, X.F., Fu, J.G., et al.: A comparative study of event-centric ontology models. J. Libr. Inf. Sci. **6**(02), 52–60 (2021)
6. Liu, Q.: Research on Ontology Construction and Application Based on Emergencies-Take the Covid-19 epidemic as an example. Shanxi University, Shanxi (2021)
7. Liu, S., Liu, X., Liu, X.: Overview of event ontology representation model and construction. J. Beijing Inf. Sci. Technol. Univ. **33**(2), 35–40 (2018)
8. Yue, L., Liu, W.: A comparative study of domestic and foreign domain ontology construction methods. Intell. Theory Pract. **39**(8), 119–125 (2016)
9. Astrid, D.R., Martin, B., Ludger, J., et al.: Evaluating the good ontology design guideline (GoodOD) with the ontology quality requirements and evaluation method and metrics (OQuaRE). Plos One **9**(8), e104463 (2014)
10. Ren, F.L., Shen, J.K., et al.: A review for domain ontology construction from text. Chin. J. Comput. **42**(3), 654–676 (2019)
11. Ganter, B., Wille, R.: Formal Concept Analysis. Springer, Berlin (1999)

12. Han, D.J., Gan, T., et al.: Research of ontology construction method based on formal concept analysis. Comput. Eng. **42**(02), 300–306 (2016)
13. Liu, T.: Research on Dynamic Ontology Construction and Reasoning Rules of Minning Face. Taiyuan University of Science and Technology, Taiyuan (2017)
14. Sun, L.: Research on Maritime Ontology Construction Based on Thesaurus and FCA. Dalian Maritime University, Dalian (2010)
15. Trajanoska, M., Stojanov, R., Trajanov, D.: Enhancing Knowledge Graph Construction Using Large Language Models. arXiv preprint arXiv:2305.04676 (2023)
16. Wang, S., Sun, X., Li, X., et al.: Gpt-Ner: named entity recognition via large language models. arXiv preprint arXiv:2304.10428 (2023)
17. Gao, J., Zhao, H., Yu, C., et al.: Exploring the feasibility of chatgpt for event extraction. arXiv preprint arXiv:2303.03836 (2023)
18. Gao, J., Wang, A.: Research on ontology-based network threat intelligence analysis technology. Comput. Eng. Appl. **56**(11), 112–117 (2020)
19. Christian, R., Dutta, S., Park, Y., et al.: An ontology-driven knowledge graph for android malware. In: Proceedings of the 2021 ACM SIGSAC Conference on Computer and Communications Security, pp. 2435–2437 (2021)
20. Yeboah-Ofori, A., Ismail, U.M., Swidurski, T., et al.: Cyberattack ontology: a knowledge representation for cyber supply chain security. In: 2021 International Conference on Computing, Computational Modelling and Applications (ICCMA), pp. 65–70. IEEE (2021)
21. Sánchez-Zas, C., Villagrá, V.A., Vega-Barbas, M., et al.: Ontology-based approach to real-time risk management and cyber-situational awareness. Futur. Gener. Comput. Syst. **141**, 462–472 (2023)
22. Syed, R.: Cybersecurity vulnerability management: a conceptual ontology and cyber intelligence alert system. Inf. Manag. **57**(6), 103334 (2020)
23. Li, W.X., Wu, D.Y., et al.: Research on cyber attack case base model based on onotology. Comput. Sci. **41**(10), 5 (2014)
24. Merah, Y., Kenaza, T.: Ontology-based cyber risk monitoring using cyber threat intelligence. In: Proceedings of the 16th International Conference on Availability, Reliability and Security, pp. 1–8 (2021)
25. Wei, X., Cui, X., Cheng, N., et al.: Zero-shot information extraction via chatting with ChatGPT. arXiv preprint arXiv:2302.10205 (2023)
26. Ge, B., Zheng, W., Yang, G.M., et al.: Microblog topic mining based on a combined TF-IDF and LDA topic model. In: Automatic Control, Mechatronics and Industrial Engineering, pp. 291–296. CRC Press (2019)
27. Lindig, C.: Fast concept analysis. In: Working with Conceptual Structures-Contributions to ICCS 2000, pp. 152–161 (2000)
28. Qian, J.: Research on Approaches of FCA-based Ontology Building and Mapping. National University of Defense Technology, Changsha (2016)
29. Wei, L., Li, D.M., et al.: Research on heterogeneous resource ontology construction based on FCA and Word2vec. Inf. Sci. **35**(3), 69–75 (2017)
30. Mavroeidis, V., Hohimer, R., Casey, T., et al.: Threat actor type inference and characterization within cyber threat intelligence. In:2021 13th International Conference on Cyber Conflict (CyCon), pp. 327–352. IEEE (2021)
31. GB/T 36643–2018. Information security technology—Cyber security threat information format (2018)
32. Howard, J.D., Longstaff, T.A.: A common language for computer security incidents. Sandia National Lab.(SNL-NM), Albuquerque, NM (United States); Sandia National Lab.(SNL-CA), Livermore, CA (United States) (1998)

33. Undercofer, J., Joshi, A., Finin, T., et al.: A target-centric ontology for intrusion detection. In: Workshop on Ontologies in Distributed Systems, held at The 18th International Joint Conference on Artificial Intelligence (2003)
34. The Phantom that Wanders the Middle East - Analysis of Recent Attack Activity by APT Group AridViper. https://www.uu11.com/keji/690217.html. Accessed 26 NOv 2022
35. ATT&CK Matrix for Enterprise. https://attack.mitre.org/. Accessed 25 Oct 2022
36. Zhang, S.X.: Research on Knowledge Representation and Reasoning Based on Decision Implication. Shanxi University, Taiyuan (2021)
37. Yanhui, Z., Deyu, L., Kaishe, Q.: Decision implications: a logical point of view. Int. J. Mach. Learn. Cybern. 5, 509–516 (2014)
38. Ning, H., Tian, Z., Hui, L., Xiaojiang, D., Guizani, M.: A multiple-kernel clustering based intrusion detection scheme for 5G and IoT networks. Int. J. Mach. Learn. Cybern. 12(11), 3129–3144 (2021). https://doi.org/10.1007/s13042-020-01253-w
39. Jia, Y., Gu, Z., Li, A.: MDATA: a new knowledge representation model. Springer, Heidelberg (2021). https://doi.org/10.1007/978-3-030-71590-8

# Web and Privacy Security

# Optimally Blending Honeypots into Production Networks: Hardness and Algorithms

Md Mahabub Uz Zaman[1], Liangde Tao[2], Mark Maldonado[3], Chang Liu[1],
Ahmed Sunny[1], Shouhuai Xu[3], and Lin Chen[1(✉)]

[1] Texas Tech University, Lubbock, TX 79409, USA
lin.chen@ttu.edu
[2] Zhejiang University, Hangzhou 310027, China
[3] University of Colorado Colorado Springs, Colorado Springs, CO 80918, USA

**Abstract.** Honeypot is an important cyber defense technique that can
expose attackers' new attacks (e.g., zero-day exploits). However, the effectiveness of honeypots has not been systematically investigated, beyond
the rule of thumb that their effectiveness depends on how they are
deployed. In this paper, we initiate a systematic study on characterizing
the cybersecurity effectiveness of a new paradigm of deploying honeypots:
blending honeypot computers (or IP addresses) into production computers. This leads to the following Honeypot Deployment (HD) problem: *How
should the defender blend honeypot computers into production computers
to maximize the utility in forcing attackers to expose their new attacks while
minimizing the loss to the defender in terms of the digital assets stored in
the compromised production computers?* We formalize HD as a combinatorial optimization problem, prove its NP-hardness, provide a near-optimal
algorithm (i.e., polynomial-time approximation scheme). We also conduct
simulations to show the impact of attacker capabilities.

**Keywords:** Cybersecurity Dynamics · Honeypot Deployment ·
Approximation Algorithm · Risk Attitude · Combinatorial
Optimization

## 1 Introduction

Cyberspace is complex and extremely challenging to defend because there are
so many vulnerabilities that can be exploited to compromise its components,
including both technological ones (e.g., software or network configuration vulnerabilities) and non-technological ones (e.g., human factors) [29,37]. It would
be ideal if we could prevent all attacks; unfortunately, this is not possible for
reasons that include *undecidability* of computer malware [1]. Not surprisingly,
cyber attacks have caused tremendous damages [26,28].

Honeypot [13,27,35] is a deception technique for luring and exposing cyber
attacks, especially new attacks or zero-day exploits. The basic idea is to set up
fake services that are open in the Internet, meaning that any access to these

© The Author(s), under exclusive license to Springer Nature Switzerland AG 2023
M. Yung et al. (Eds.): SciSec 2023, LNCS 14299, pp. 285–304, 2023.
https://doi.org/10.1007/978-3-031-45933-7_17

fake services can be deemed as malicious and the defender can learn the attacks by monitoring these fake services. The importance and potential of honeypots have attracted a due amount of attention (e.g., [4–6, 20–22, 24, 33, 43]). However, the effectiveness of honeypots has not been systematically characterized. The rule of thumb is that their effectiveness depends on how they are deployed. The traditional way of deploying honeypots is to isolate a set of honeypot computers from any production network. However, it is well known that such honeypots can be easily figured out, and thus evaded, by attackers. One approach to addressing this problem is to "blend" honeypot computers into the production computers of an enterprise network.

**Our Contributions.** This paper makes two contributions. The *conceptual* contribution is to initiate the study on systematically characterizing the effectiveness of blending honeypot computers with production computers, leading to the formalization of a Honeypot Deployment (HD) problem: *How should the defender blend honeypot computers with production computers to maximize the utility of honeypot in forcing attackers to expose their new attacks, while minimizing the loss to the defender in terms of the digital assets stored in the compromised production computers?* One salient feature of the formalization is that it can naturally incorporate attacker's *risk attitude* (i.e., risk-seeking, risk-neutral, or risk-averse). The *technical* contribution is that we show: (i) the decision version of the HD problem is NP-complete; and (ii) we present a near-optimal algorithm to solve it, namely a Polynomial-Time Approximation Scheme (PTAS) by leveraging a given sequence of attacker's preference (i.e., attack priority) resulting from the attacker's reconnaissance process and the attacker's risk attitude. We also conduct simulation studies to draw insights into the aspects which we cannot analytically treat yet, which would shed light on future analytic research.

## 2   Problem Statement

**Intuition.** It is non-trivial to model the Honeypot Deployment (HD) problem, so we start with an intuitive discussion. Consider (for example) an enterprise network with some *production* computers, which provide real business services, and a set of IP addresses. Some IP addresses are assigned to these production computers. The defender deploys some *traditional* defense tools (e.g., anti-malware tools and intrusion-prevention systems) to detect and block recognizable attacks. However, these tools can be evaded by new attacks (e.g., zero-day exploits), which are not recognizable by them, per definition. In order to defend the network against new attacks, the defender can blend some *honeypot* computers into the production ones, meaning that some of the remaining (or unassigned) IP addresses are assigned to honeypot computers, and some IP addresses may not be used at all (in which case we say these IP addresses are assigned to *dummy* computers). Each computer (production, honeypot, and dummy alike) will be assigned one unique IP address. Note that traditional defense tools are still useful because they can detect and block recognizable attacks.

The research is to investigate how to *optimally* assign IP addresses to computers to benefit the defender, where the meaning of optimization is specified as follows. *First*, suppose deploying one honeypot computer incurs a cost to the defender, which is plausible because the honeypot computer does not provide any business-related service. *Second*, when a new attack is waged against a production computer, it incurs a loss to the defender because the digital assets stored in the production computer are compromised and the new attack cannot be blocked by traditional defense tools. *Third*, when a new attack is waged against a honeypot computer, it incurs no loss to the defender but does incur a cost to the attacker because the new attack now becomes recognizable to the defender. In this case, we say a *valid* new attack becomes *invalid* (i.e., no more useful to the attacker). *Fourth*, the usefulness of honeypot computers is based on the premise that the attacker does not know which IP addresses are assigned to honeypot computers; otherwise, the attacker can simply avoid attacking them. In the real world, the attacker often uses a *reconnaissance* process, which can be based on a range of techniques (from social engineering to technical methods), to help determine which IP addresses may be assigned to honeypot computers. The reconnaissance process often correctly detects which IP addresses are assigned to the dummy computers because no attempt is made by the defender to disguise these IP addresses (otherwise, they can be deemed as honeypot computers). The reconnaissance process is not perfect in identifying the honeypot computers, meaning that when the attacker decides to attack a computer, which the attacker deems as a production computer, the computer is actually a honeypot one, causing the attacker to lose the new attack. Because of the uncertainty associated with the outcome of the reconnaissance process, the attacker would decide whether to attack a computer with some probability, which reflects the attacker's reconnaissance capability, the attacker's risk attitude (i.e., risk-seeking, risk-neutral, or risk-averse), and the honeypot computer's capability in disguising itself as a production computer. To accommodate attacker reconnaissance capabilities, we assume the probabilities are given; this is reasonable because deriving such probabilities is orthogonal to the focus of this study.

**Problem Formalization.** Let $\mathbb{N}$ denote the set of positive integers and $\mathbb{R}$ the set of real numbers. For any positive integer $z \in \mathbb{N}$, we define $[z] = \{1, \ldots, z\}$. Suppose the defender is given $n \in \mathbb{N}$ production computers and $n + m$ IP addresses, meaning that $m \in \mathbb{N}$ is the number of IP addresses that can be used to deploy honeypot computers and dummy computers (if applicable). Suppose the defender needs to deploy up-to $m \in \mathbb{N}$ honeypots computers. Each honeypot computer may incur a cost $c \in \mathbb{N}$, which may vary depending on the degree of sophistication embedded into the honeypot computer (i.e., the most sophisticated a honeypot computer, the more difficult for the attacker's reconnaissance to determine whether it is a honeypot or production computer). Suppose the total budget for deploying the honeypot computers is $B \in \mathbb{N}$. This means that the defender will select $h$-out-of-the-$m$ IP addresses for deploying honeypot computers subject to the total cost for deploying the $h$ honeypot computers is at most $B$. Then, $m - h$ dummy computers are respectively deployed at the remaining $m - h$ IP addresses. Recall that the attacker can correctly recognize the dummy computers and will not attack them. We use the term "non-dummy computer" to indicate a computer that is a production or honeypot computer.

For ease of reference, we use "computer $j$" to denote "the computer assigned with IP address $j$, where the computer may be a production, honeypot, or dummy one." Let $v_{j,D}$ be the value of the digital assets stored in computer $j$ (e.g., sensitive data and/or credentials). This leads to a unified representation: an IP address $j \in [n+m]$ is associated with a value $v_{j,D} \in \mathbb{N}$ as assessed by the defender and a cost $c_j \in \mathbb{N}$ to the defender, where

- $v_{j,D} > 0$ if $j$ is a production computer, and $v_{j,D} = 0$ otherwise.
- $c_j > 0$ if $j$ is a honeypot computer, and $c_j = 0$ otherwise.

Suppose the attacker has $r$ valid new attacks which are not recognized by the traditional defense tools employed by the defender, where $r$ represents the attacker's budget. Before using these attacks, the attacker often conducts a reconnaissance process to identify: (i) the value of digital assets stored in computer $j \in [n+m]$, denoted by $v_{j,A} \in \mathbb{N}$, which is the attacker's perception of the ground-truth value $v_{j,D}$ that is not known to the attacker (otherwise, the attacker would already know which computers are honeypot ones); and (ii) the probability or likelihood that a non-dummy computer $j$ is a honeypot computer, denoted by $q_j$, which is the probability that the attacker will *not* attack computer $j$ (i.e., $1 - q_j$ is the probability that the attacker will attack it). The attacker knows which computer is a dummy one and will not attack any dummy computers. To summarize, let $x_j \in \{0,1\}$ be an indicating vector such that $x_j = 1$ if computer $j$ is a production or honeypot computer, and $x_j = 0$ if computer $j$ is a dummy computer, then we can see that the attacker will attack computer $j$ with probability $(1 - q_j)x_j$, which is $1 - q_j$ for a non-dummy computer $j$ and $0$ otherwise. Note that $v_{j,A}$ and $q_j$ together reflect the attacker's reconnaissance capability.

When the attacker attacks computer $j$ which happens to be a production one, the loss to the defender (i.e., the reward to the attacker) is $v_{j,D} > 0$; when the attacker attacks computer $j$ which happens to be a honeypot one, the loss to the defender is $v_{j,D} = 0$ and the attacker's budget decreases by 1 because the new attack now becomes *invalid* (i.e., recognizable to the defender). The attacker cannot wage any successful attack after its $r$ new attacks become invalid.

The research question is to identify the optimal strategy in assigning some of the $m$ IP addresses to honeypot computers under budget constraint $B$ so as to minimize the *expected loss* to the defender. The assignment of IP addresses to the production, honeypot, and dummy computers is called a *defense solution*, which is characterized by a vector $\boldsymbol{x} = (x_1, x_2, \cdots, x_{m+n}) \in \{0,1\}^{n+m}$ where $x_j = 1$ means computer $j$ is a production or honeypot computer, and $x_j = 0$ means it is a dummy computer. Given a fixed *defense solution* $\boldsymbol{x}$, the loss to the defender is defined as the total value of the production computers that are attacked. The loss to the defender is probabilistic because the attacker attacks a non-dummy computer with a probability, meaning that we should consider the *expected loss*. To compute the expected loss, we need to specify the probability distribution of the loss, which depends on the probability distribution of the attacker's decisions on attacking non-dummy computers. To characterize this distribution, we introduce two concepts, *attack sequence* and *attack scenario*; the former is a stepping-stone for introducing the latter, which is used to compute the expected loss. Given a defense solution, we assume the attacker sequentially decides whether to attack computer $j \in [n+m]$ according to probability $q_j$. The order according to which the attacker makes decisions is called *attack sequence*.

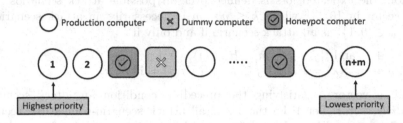

**Fig. 1.** Illustrating the concepts of production, honeypot, and dummy computers and the idea of *attack sequence*.

As illustrated in Fig. 1, we use circles to represent production computers and squares to represent the $m$ IP addresses for which the defender needs to decide whether to deploy a honeypot or dummy computer. An attack sequence represents the attacker's choice of priority in considering which non-dummy computers to attack, where priority depends on the attacker's perception of the value of computer $j$, namely $v_{j,A}$, and the probability $q_j$, and the attacker's risk attitude. We assume such attack sequences are given as input to the present study because attaining these attack sequences is an orthogonal research problem. To define *attack scenario*, we should specify when the attacker stops. The attacker stops when any of the following two conditions hold: (i) after attacking some honeypot computer which causes the attacker's budget to decrease from 1 to 0, meaning that the attacker has no more valid new attack to use; (ii) the attacker finishes attacking all computers it would like to attack (based on its probabilistic decision) even if there are still valid new attacks, meaning that the attacker only considers whether or not to attack a non-dummy computer once, which is plausible. Specifically, given a defense solution $x$ and an attack sequence, the decisions on whether or not to attack computers $j \in [j^*]$ is called an *attack scenario*; that is, an attack scenario $s$ is a binary vector $\pi_s = (\pi_s(1), \pi_s(2), \cdots, \pi_s(j_s)) \in \{0,1\}^{j_s}$, where $\pi_s(j) = 1$ means computer $j$ is indeed attacked and $\pi_s(j) = 0$ otherwise, and $j_s \leq n+m$ is the last computer that is attacked. Note that $\pi_s(j_s) = 1$ by definition. Recall that the attacker attacks computer $j$ with probability $1 - q_j$, meaning $\Pr[\pi_s(j) = 1] = 1 - q_j$ and $\Pr[\pi_s(j) = 0] = q_j$. Define $\Pr[\pi_s]$ as the probability that attack scenario $\pi_s$ occurs, then we have

$$\Pr[\pi_s] = \prod_{j \in [j_s]:\pi_s(j)=1, x_j=1} (1 - q_j) \cdot \prod_{j \in [j_s]:\pi_s(j)=0, x_j=1} q_j.$$

Note that if $x_j = 0$ then computer $j$ is a dummy computer and the attacker will not attack it at all, so it does not contribute to the probability $\Pr[\pi_s]$. Let $\mathcal{P} \subset [n+m]$ be the subset of the IP addresses that are assigned to production computers and $\mathcal{H} := [n+m] \setminus \mathcal{P}$ be the subset of the remaining

IP addresses. If attack scenario $\pi_s$ occurs, then the loss to the defender is the total value of all production computers attacked by the attacker, which is $Loss(\pi_s) = \sum_{j \in \mathcal{H}:\pi_s(j)=1} v_{j,D}$.

Given the loss incurred by a specific attack scenario as shown in the above equation, the expected loss is defined over all possible attack scenarios. Not every vector in $\{0,1\}^k$ where $k \leq n + m$ is necessarily an attack scenario; a vector $\pi \in \{0,1\}^k$ is an attack scenario if and only if

- $k = n + m$ and $|\{j \in [n+m] \cap \mathcal{H} : \pi(j) = 1\}| \leq r$; or
- $k < n + m$, $\pi(k) = 1$ and $|\{j \in [k] \cap \mathcal{H} : \pi(j) = 1\}| = r$.

We call a vector $\pi$ satisfying the preceding condition an *attack scenario-compatible* vector. Let $\mathcal{V}$ be the set of all attack scenario-compatible vectors. Then, the expected loss of the defender with respect to a fixed defense solution is:

$$\mathbb{E}[Loss] = \sum_{\pi \in \mathcal{V}} Loss(\pi) \Pr[\pi]. \tag{1}$$

In summary, we have:

---

**Honeypot Deployment (HD) Problem**

**Input**: There are $n \in \mathbb{N}$ production computers and a budget of $B \in \mathbb{N}$ for the defender. There are $n + m$ IP addresses, which are indexed as $1, \ldots, n + m$. Among these $n + m$ IP addresses, the $n$ production computers are respectively deployed at $n$ pre-determined IP addresses; among the remaining $m$ IP addresses, the defender will select a subset of them to deploy honeypot computers, and the other IP addresses will be assigned to dummy computers, which are known to, and not be attacked by, the attacker. For each computer $j \in [n + m]$, there is an associated value $v_{j,D} \in \mathbb{N}$, where $v_{j,D} = 0$ if $j$ is a honeypot or dummy computer and $v_{j,D} > 0$ otherwise; moreover, there is an associated cost $c_j \in \mathbb{N}$, where $c_j = 0$ if $j$ is a production or dummy computer, and $c_j > 0$ if $j$ is a honeypot computer. The total cost incurred by deploying honeypot computers cannot exceed budget $B$. The attacker has $r \in \mathbb{N}$ valid new attacks. For computer $j \in [n + m]$, the attacker has a perceived value $v_{j,A}$ and a probability $q_j$ that a non-dummy computer $j$ is a honeypot computer. The attacker needs to use one valid new attack to attack a *non-dummy* computer $j$. The attacker attacks a non-dummy computer with probability $1 - q_j$, and does not attack a dummy computer. If the attacker indeed attacks a non-dummy computer $j$, there are two cases: in the case $j$ is a honeypot computer, then the loss to the defender is $v_{j,D} = 0$ and the attacker's budget decreases by 1; in the case $j$ is a production computer, then the loss to the defender is $v_{j,D} > 0$ and the attack's budget remains unchanged. The attacker stops when its budget becomes 0, meaning that all of its $r$ new attacks become invalid, or it has made decisions on whether to attack the $n + m$ computers. **Output**: Decide which of the $m$ IP addresses should be assigned to honeypot computers within budget $B$ so as to minimize the expected loss defined in Eq. (1).

# 3  Hardness and Algorithmic Results

In this section, we study the computational complexity of, and algorithms for solving the HD problem, assuming an attack sequence is given. Without loss of generality, we can index the computers so that the attack sequence is $(1, 2, \cdots, n + m)$. As illustrated in Fig. 1, we use circles to represent production computers and squares to represent the $m$ IP addresses for which the defender needs to decide whether to deploy a honeypot or dummy computer. Recall $\mathcal{P}$ is exactly the set of indices of the circles and $\mathcal{H}$ is exactly that of the squares.

Note that the $v_{j,A}$'s, $q_j$'s together with the attacker's risk attitude decide the priority of the non-dummy computers to the attacker, and thus the attack sequence. Once an attack sequence is fixed, the objective value (i.e., expected loss to the defender) only depends on $v_{j,D}$'s. Since our hardness and algorithmic results are based on a fixed attack sequence, our discussion throughout this section does not involve $v_{j,A}$'s. Hence, we let $v_j = v_{j,D}$ for simplifying notations.

## 3.1  Hardness Result

Now we study the decision version of the HD problem: decide whether or not there exists an assignment of the $m$ IP addresses to honeypot computers with budget $B$ such that the expected loss is no larger than the given threshold $T$.

**Theorem 1.** *The decision version of the HD problem is NP-complete.*

The proof of Theorem 1 is deferred to Appendix A. Here we discuss the basic idea behind the proof. Membership in NP is straightforward. Towards the NP-hardness proof, we reduce from the Subset Product problem. The instance and the solution of the Subset Product problem are given as follows:

---

**Subset Product Problem**
**Input:** $k \in \mathbb{N}$, $S = \{1, \ldots, m\}$, $w = (w_1, \ldots, w_m) \in \mathbb{N}^m$.
**Output:** Is there $S' \subseteq S$ such that $\prod_{i \in S'} w_i = k$?

---

A key fact for the Subset Product problem (which is different from the Subset Sum problem) is that Subset Product is NP-hard even if each $a_i$ is bounded by $m^{O(1)}$. This means we leverage the following Lemma 1 given by Yao [50].

**Lemma 1** ([50]). *Assuming $P \neq NP$, the Subset Product problem can not be solved in $(m w_{max} \log k)^{O(1)}$ time where $w_{max} = \max_i w_i$.*

## 3.2  Algorithmic Results

As a warm-up, we present an exact algorithm, Algorithm 1, to brute forces the optimal solution in exponential time. We show this algorithm can be modified to obtain Algorithm 2 to find a near-optimal solution in polynomial time.

**An Exact Algorithm via Dynamic Programming.** Algorithm 1 essentially branches on whether or not to deploy a honeypot computer for every $t \in \mathcal{H}$,

---

**Algorithm 1.** Dynamic Programming for the HD problem

---

**Input:** $I$ : the attack sequence of IP address

    $q_t$ : the probability that attacker does not attack computer $t$

    $v_t$ : computer's value assigned with IP address $t$

    $c_t$ : cost of deploying honeypot computer at IP address $t$

    $B$ : budget of the defender

**Output:** The assignment of IP address to honeypot computers which minimizes the expected loss of the defender and satisfies the total deployment cost is no greater than $B$.

1: $\widetilde{\mathcal{F}}_0 = \{(0, 1, 0, \ldots, 0)\}$

2: **for** $t = 1$ to $n + m$ **do**

3:     $\widetilde{\mathcal{F}}_t = \emptyset$

4:     **for all** $(t-1, p_0, p_1, \ldots, p_r, e, b) \in \widetilde{\mathcal{F}}_{t-1}$ **do**

5:         **if** IP address $t$ is assigned to a production computer **then**

6:             $\widetilde{\mathcal{F}}_t \leftarrow \widetilde{\mathcal{F}}_t \cup (t, p_0, p_1, \ldots, p_r, e + q_t v_t \sum_{i=0}^{r-1} p_i, b)$

7:         **else**

8:             $\widetilde{\mathcal{F}}_t \leftarrow \widetilde{\mathcal{F}}_t \cup (t, p_0, p_1, \ldots, p_r, e, b)$

9:             **for** $k = 1$ to $r$ **do**

10:                 $p_k' = (1 - q_t) \cdot p_{k-1} + q_t \cdot p_k$

11:             **end for**

12:             $p_0' = q_t p_0$

13:             $\mathcal{F}_h \leftarrow \widetilde{\mathcal{F}}_t \cup (t, p_0', \ldots, p_r', e, b + c_t)$

14:         **end if**

15:     **end for**

16:     eliminate all the dominated states in $\widetilde{\mathcal{F}}_t$

17: **end for**

18: return $\min\{e : (n + m, p_0, p_1, \ldots, p_r, e, b) \in \widetilde{\mathcal{F}}_{n+m}$ and $b \leq B\}$

---

thus the total number of distinct dominated states (e.g., $\sum_t |\widetilde{\mathcal{F}}_t|$) is bounded by $2^{O(m)}$. Hence, its running time is $2^{O(m)}$. Algorithm 1 serves two purposes: (i) it provides a method to recursively compute Eq. (1), while noting that the definition of Eq. (1) involves an exponential number of attack scenarios that cannot be used directly; (ii) it can be combined with rounding techniques to give a polynomial-time approximation scheme, which is the main algorithmic result (Algorithm 2).

We consider the following sub-problem: Let $t \in [n + m]$. Is it possible to deploy honeypot computers at some IP addresses within $[t]$ such that (i) the total cost equals $b$ which is a given constraint, (ii) the expected loss to the defender because of attacks against the production computers in $\mathcal{P} \cap [t]$ equals $e$, and (iii) the probability that the attacker attacks exactly $k$ honeypot computers within $[t]$ is $p_k$ for every $k = 0, 1, \cdots, r$? We denote the sub-problem by a $(r + 4)$-tuple $(t, p_0, p_1, \ldots, p_r, e, b)$.

We define set $\mathcal{F}_t$ as: If the answer to the sub-problem $(t, p_0, p_1, \ldots, p_r, e, b)$ is "yes," then we call $(t, p_0, p_1, \ldots, p_r, e, b)$ as a stage-$t$ state and store it in $\mathcal{F}_t$. Note that $\mathcal{F}_t$ can be computed recursively as follows. Suppose we have computed $\mathcal{F}_{t-1}$. Each stage-$(t-1)$ state gives rise to stage-$t$ states as follows:

- If $t \in \mathcal{P}$, i.e., IP address $t$ is associated with a production computer, then the stage-$(t-1)$ state $(t-1, p_0, p_1, \ldots, p_r, e, b)$ gives rise for one stage-$t$ state $(t, p_0, p_1, \ldots, p_r, e', b)$ where $e' = e + (1-q_t)v_t \sum_{k=1}^{r-1} p_k$. We explain this equation as follows. By the definition of a state, $(t-1, p_0, p_1, \ldots, p_r, e, b) \in \mathcal{F}_{t-1}$ implies that $p_r$ is the probability that the attacker has attacked $r$ honeypot computers within $[t-1]$, and thus the attacker cannot attack anymore. If this event happens, we have $e' = e$; otherwise (with probability $\sum_{k=1}^{r-1} p_k = 1 - p_r$) the attacker is able to attack the production computer at IP address $t$ which has a value $v_t$, and the attacker attacks it with probability $1 - q_t$. This leads to $e' = ep_r + (e + v_t)(1 - q_t) \sum_{k=1}^{r-1} p_k = e + (1 - q_t)v_t \sum_{k=1}^{r-1} p_k$.
- If $t \in \mathcal{H}$, then we have two options, i.e., we either assign a honeypot computer or a dummy computer to IP address $t$. If we assign a dummy computer, then the stage-$(t-1)$ state $(t-1, p_0, p_1, \ldots, p_r, e, b)$ gives rise to stage-$t$ state $(t, p_0, p_1, \ldots, p_r, e, b)$; if we assign a honeypot computer, then $(t-1, p_0, p_1, \ldots, p_r, e, b)$ gives rise to stage-$t$ state $(t, p_0', p_1', \ldots, p_r', e, b + c_t)$ where $p_0' = p_0 q_t$ and $p_k' = p_k q_t + p_{k-1}(1 - q_t)$ for $1 \leq k \leq r$.

**Definition 1.** *We say stage-$t$ state $(t, p_0, p_1, \ldots, p_r, e, b)$ dominates $(t, p_0, p_1, \ldots, p_r, e, b')$ if it holds that $b < b'$.*

Denote by $\widetilde{\mathcal{F}}_t \subseteq \mathcal{F}_t$ the set of all stage-$t$ states which are not dominated by any of the other stage-$t$ states. Let $x^*$ denote the optimal solution of the HD problem. Let $\mathcal{H}(x^*)$ be the set of IP addresses that are assigned to all honeypot computers. Consider $\mathcal{H}(x^*) \cap [t]$, which is the set of IP addresses in $[t]$ that are assigned to a honeypot computer in the optimal solution $x^*$. Denote by $b_t(x^*)$ the total cost of deploying honeypot computers in $\mathcal{H}(x^*) \cap [t]$. Denote by $p_{t,i}(x^*)$ the probability that the attacker has attacked exact $i$ honeypot computers within $\mathcal{H}(x^*) \cap [t]$. Denote by $e_t(x^*)$ the expected loss to the defender from computers in $[t]$. The following lemma demonstrates that the optimal solution $x^*$ can be determined from $\widetilde{\mathcal{F}}_{n+m}$; its proof is deferred to Appendix B.

**Lemma 2.** *For each optimal solution $x^*$ of the HD problem and each $t \in [1, n + m]$, there exists some stage-$t$ state $(t, p_{t,0}(x^*), \ldots, p_{t,r}(x^*), e_t(x^*), b) \in \widetilde{\mathcal{F}}_t$ such that $b \leq b_t(x^*)$.*

**A Polynomial-Time Approximation Scheme (PTAS).** Now we design a PTAS (i.e., Algorithm 2) for the HD problem by modifying Algorithm 1. The key idea is to reduce the total number of states that need to be stored during the dynamic programming. We start with a high-level description of Algorithm 2. Let $\xi = \epsilon/2(n+m)$. Define $\Gamma_\xi = \{[0], (0, 1], (1, 1+\xi], \ldots, ((1+\xi)^{\gamma-1}, (1+\xi)^\gamma]\}$ where $(1+\xi)^{\gamma-1} < nv_{\max} \leq (1+\xi)^\gamma$. Define $\Lambda_\xi = \{[0], (0, (1+\xi)^{-\gamma}], ((1+\xi)^{-\gamma}, (1+\xi)^{-\gamma+1}], \ldots, ((1+\xi)^{-1}, 1]\}$. The high dimensional area $[0, 1]^{r+1} \times [0, (1+\xi)^\gamma]$ is then divided into a collection of boxes where each box $\mathcal{I} \in \Lambda_\xi^{r+1} \times \Gamma_\xi$. In each box, only one representative state will be constructed and stored in $\widehat{\mathcal{F}}_t$. $\widehat{\mathcal{F}}_t$ is computed recursively in two steps: (i). Given $\widehat{\mathcal{F}}_{t-1}$, each of its state gives rise to stage-$t$ states following the same formula as Algorithm 1 (see line 5 to line

13 of Algorithm 2). Here $\mathcal{S}_t$ is introduced as a temporary set that contains all the stage-$t$ states computed from $\widehat{\mathcal{F}}_{t-1}$. (ii). Within each box $\mathcal{I}$, if $\mathcal{S}_t$ contains multiple states, then only the state with the minimal value in coordinate $b$ will be kept. All other states are removed. By doing so we obtain $\widehat{\mathcal{F}}_t$ from $\mathcal{S}_t$. Details of Algorithm 2 are presented below. The rest of this subsection is devoted to proving the following theorem.

**Theorem 2.** *Algorithm 2 gives an* $(1+\epsilon)$-*approximation solution for the HD problem and runs in* $(\frac{n+m}{\epsilon})^{O(r)} \log(v_{\max})$ *time where* $v_{\max} = \max_i v_i$.

To prove Theorem 2, we need the following lemma that estimates the error accumulated in the recursive calculation of Algorithm 2 and illustrates the relationship between $\widetilde{\mathcal{F}}_t$ and $\widehat{\mathcal{F}}_t$.

**Lemma 3.** *For each* $(t, p_0, p_1, \ldots, p_r, e, b) \in \widetilde{\mathcal{F}}_t$, *there exists* $(t, \hat{p}_0, \hat{p}_1, \ldots, \hat{p}_r, \hat{e}, \hat{b}) \in \widehat{\mathcal{F}}_t$ *such that* $\hat{b} \leq b$, $(1-\xi)^t \hat{e} \leq e \leq \hat{e}$, *and for* $i = 0, \ldots, r$ *it holds that* $(1-\xi)^t \hat{p}_i \leq p_i \leq \hat{p}_i$.

---

**Algorithm 2.** Improved Dynamic Programming for the HD problem

---

**Input:** $I, q_t, v_t, c_t, B$

**Output:** The assignment of IP address to honeypot computers which minimizes the expected loss of the defender and satisfies the total deployment cost is no greater than $B$.

1: $\widehat{\mathcal{F}}_0 = \{(0, \ldots, 0)\}$
2: **for** $t = 1$ to $n + m$ **do**
3:      $\mathcal{S}_t = \emptyset$
4:      **for all** $(t-1, p_1, \ldots, p_r, e, b) \in \widehat{\mathcal{F}}_{t-1}$ **do**
5:          **if** IP address $t$ is assigned to a production computer **then**
6:              $\mathcal{S}_t \leftarrow \mathcal{S}_t \cup (t, p_0, p_1, \ldots, p_r, e + q_t v_t \sum_{i=0}^{r-1} p_i, b)$
7:          **else**
8:              $\mathcal{S}_t \leftarrow \mathcal{S}_t \cup (t, p_0, p_1, \ldots, p_r, e, b)$
9:              **for** $k = 1$ to $r$ **do**
10:                  $p'_k = (1 - q_t) p_{k-1} + q_t \cdot p_k$
11:              **end for**
12:              $\mathcal{S}_t \leftarrow \mathcal{S}_t \cup (t, q_t p_0, p'_1, \ldots, p'_r, e, b + c_t)$
13:          **end if**
14:      **end for**
15:      $\widehat{\mathcal{F}}_t = \emptyset$
16:      **for all** Box $\mathcal{I} \in (\Lambda_\xi)^{r+1} \times \Gamma_\xi$ **do**
17:          **for** $i = 0$ to $r$ **do**
18:              $\hat{p}_i = \max\{p_i : (h, p_0, p_1, \ldots, p_i, \ldots, p_r, e, b) \in \mathcal{S}_t \cap \mathcal{I}\}$
19:          **end for**
20:          $\hat{e} = \max\{e : (h, p_0, p_1, \ldots, p_r, e, b) \in \mathcal{S}_t \cap \mathcal{I}\}$
21:          $\hat{b} = \min\{b : (h, p_0, p_1, \ldots, p_r, e, b) \in \mathcal{S}_t \cap \mathcal{I}\}$
22:          $\widehat{\mathcal{F}}_t \leftarrow \widehat{\mathcal{F}}_t \cup (t, \hat{p}_0, \hat{p}_1, \ldots, \hat{p}_r, \hat{e}, \hat{b})$
23:      **end for**
24: **end for**
25: **return** $\min\{\hat{e} : (n, \hat{p}_0, \hat{p}_1, \ldots, \hat{p}_r, \hat{e}, \hat{b}) \in \widehat{\mathcal{F}}_{n+m} \text{ and } \hat{b} \leq B\}$

---

*Proof.* We prove this by induction. Clearly, Lemma 3 holds for $t = 1$. Suppose it holds for $t = \ell - 1$, i.e., for each $(\ell - 1, p_0, p_1, \ldots, p_r, e, b) \in \widetilde{\mathcal{F}}_{\ell-1}$, there exists $(\ell - 1, \hat{p}_0, \hat{p}_1, \ldots, \hat{p}_r, \hat{e}, \hat{b}) \in \widehat{\mathcal{F}}_{\ell-1}$ such that $\hat{b} \leq b$, $(1 - \xi)^{k-1}\hat{e} \leq e \leq \hat{e}$, and for $i = 0, \ldots, r$ it holds that $(1 - \xi)^{k-1}\hat{p}_i \leq p_i \leq \hat{p}_i$. We prove Lemma 3 for $t = \ell$.

Note that the recursive computation is the same as Algorithm 1, the only difference is that we replace the accurate value $p_k$'s with the approximate value $\hat{p}_k$'s. The rounding error will accumulate through the calculation, but will not increase too much in each step through the following two observations: (i) For any $\alpha \in [0, 1]$ and any $j \in [1, r]$ it holds that

$$(1 - \xi)^{\ell-1}(\alpha\hat{p}_{j-1} + (1 - \alpha)\hat{p}_j) \leq \alpha p_{j-1} + (1 - \alpha)p_j \leq \alpha\hat{p}_{j-1} + (1 - \alpha)\hat{p}_j.$$

(ii) For any $\beta \in \mathbb{R}_{\geq 0}$ it holds that

$$(1 - \xi)^{\ell-1}[\hat{e} + \beta\sum_{i=1}^{r-1}\hat{p}_i] \leq e + \beta\sum_{i=1}^{r-1}p_i \leq \hat{e} + \beta\sum_{i=1}^{r-1}\hat{p}_i.$$

Hence, we know that for each $(\ell, p_0, p_1, \ldots, p_r, e, b) \in \widetilde{\mathcal{F}}_\ell$, there exists $(\ell, p'_0, p'_1, \ldots, p'_r, e', b') \in \mathcal{S}_\ell$ such that $b' \leq b$, $(1 - \xi)^{\ell-1}e' \leq e \leq e'$, and for $i = 0, \ldots, r$ it holds that $(1 - \xi)^{\ell-1}p'_i \leq p_i \leq p'_i$.

We know that within each box $\mathcal{I}$, only one representative state $(t, \hat{p}_0, \ldots, \hat{p}_r, \hat{e})$ will be constructed and stored in $\widehat{\mathcal{F}}_t$. By the definition of $\Lambda_\xi$ and $\Gamma_\xi$, we know that for each $(k, p'_0, p'_1, \ldots, p'_r, e', b') \in \mathcal{S}_t \cap \Gamma$ there exists $(t, \hat{p}_0, \hat{p}_1, \ldots, \hat{p}_r, \hat{e}, \hat{b}) \in \widehat{\mathcal{F}}_t$ such that $\hat{b} \leq b'$, $(1 - \xi)\hat{e} \leq e' \leq \hat{e}$, and for $i = 0, \ldots, r$ it holds that $(1 - \xi)\hat{p}_i \leq p'_i \leq \hat{p}_i$. Thus, Lemma 3 is proved. □

Now we are ready to prove Theorem 2.

*Proof of Theorem 2.* We first estimate the overall error incurred in Algorithm 2. Let $\xi = \epsilon/2(n + m)$. Since $1 - (n + m)\xi \leq (1 - \xi)^{n+m}$, it is easy to verify that $(1 - \xi)^{-n-m} \leq 1 + \epsilon$. According to Lemma 3, Algorithm 2 gives an $(1 + \epsilon)$-approximation solution for the HD problem, i.e., the expected loss of the defender is no larger than $(1 + \epsilon)$ times the minimum expected loss of the defender.

Now we estimate the overall running time. The total number of distinct dominated states (e.g., $\sum_t |\widehat{\mathcal{F}}_t|$) is bounded by $O((n+m)|\Lambda_\xi|^{r+1}|\Gamma_\xi|)$. We know that $|\Lambda_\xi| \leq O(\frac{n+m}{\epsilon})$. In the meantime, $|\Gamma_\xi| = \log_{1+\xi}(nv_{\max}) \leq O(\frac{n+m}{\epsilon}\log(v_{\max}))$ where $v_{\max} = \max_i v_i$. Overall, Algorithm 2 runs in $(n + m)^{r+3}\log(v_{\max})/\epsilon^{r+2}$ time. Hence, Theorem 2 is proved. □

## 4   Experiment

**Simulation Parameters.** In our simulation, we set the number of production computers as $n = 255$, $m \in \{15, 20, 25, 30\}$, the number of attacker's new attacks as $r \in \{5, 10, 15\}$, and the defender's budget as $B = \{1000, 2000, 3000, 4000\}$. The defender's perceived value for each production computer $j$, $v_{j,D}$, is generated uniformly at random within $[50, 2000]$ (while noting that $v_{j,D} = 0$ if $j$ is no production computer); the attacker's perceived value for each non-dummy computer $j$, $v_{j,A}$, is also generated uniformly at random within $[50, 2000]$. The cost for deploying honeypot computer $j$ is generated uniformly at random within $[50, 200]$. The probability $q_j \in [0, 1]$ that the attacker believes $j$ is a honeypot computer will be set in specific experiments. To conduct a fair comparison between the expected losses incurred in different parameter settings, we normalize them via *expected relative loss*, which is the ratio between the expected loss and the summation of all $v_{j,D}$, leading to a normalized range $[0, 1]$.

### 4.1　Expected Losses Under Different Attack Sequences

Now we study how different attack sequences may affect the optimal objective value of the HD problem. Recall that attack sequence is an input to our algorithms. Given an attack sequence, we can apply Algorithm 2 to compute a near-optimal defense solution to the defender, which leads to essentially the smallest expected loss the defender can hope for. Consequently, the smallest expected loss, i.e., the optimal objective value, reflects how destructive the attacker is when it chooses a certain attack sequence.

While the attack sequence can be arbitrary, we are interested in the attack sequences that are likely to be adopted by a rational attacker. Note that the attacker observes $q_j$ and $v_{j,A}$ for each computer $j \in [n + m]$. Intuitively, the attacker needs to weigh between the potential gain $v_{j,A}$ and the risk that the attacker cannot get this gain, namely probability $q_j$. Therefore, the attacker's risk attitude determines the attack sequence. Leveraging ideas from economics, we study three types of risk attitudes of the attacker (see, e.g. [16,19]): (i) *risk-seeking*, meaning that the attacker wants to maximize its revenue fast; (ii) *risk-averse*, meaning that the attacker wants to minimize its chance of losing a valid new attack; (iii) *risk-neutral*, meaning that the attacker acts in between risk-seeking and risk-averse. A formal definition of risk-attitude depends on the notion of *utility function*, denoted by $u_j$. We focus on a broad class of utility functions, known as exponential utility [8,34], which is defined as:

$$u_j = \begin{cases} \frac{1 - e^{-\alpha v_{j,A}}}{\alpha}, & \text{if } \alpha \neq 0 \\ v_{j,A}, & \text{if } \alpha = 0 \end{cases}$$

where $\alpha$ is the *coefficient of absolute risk aversion*, which, roughly speaking, measures how much the attacker is willing to sacrifice the expected value $v_{j,A}$ in order to achieve perfect certainty about the value it can receive. If $\alpha > 0$, then the attacker is risk-aversion; if $\alpha = 0$, the attacker is risk-neutral; if $\alpha < 0$, the attacker is risk-seeking. With the utility function, the attacker (with risk attitude specified by $\alpha$) will rank (or prioritize) the non-dummy computers based on the non-increasing order of the expected utility value $(1 - q_j) \cdot u_j$ and use these tanks to formulate an attack sequence.

In our experiment, we choose $\alpha \in \{-0.05, -0.005, 0, 0.005, 0.05\}$, where $\alpha = -0.05$ means the attacker is strongly risk-seeking and $\alpha = 0.05$ means the attacker is strongly risk-averse. For each $\alpha$, we generate 4,560 instances. Figure 2

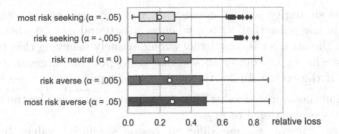

**Fig. 2.** Expected relative loss with respect to different risk-attitude (i.e., $\alpha = -0.05$ for most risk-seeking, $\alpha = -0.005$ for risk-seeking, $\alpha = 0$ for risk-neutral, $\alpha = 0.005$ for risk-averse, $\alpha = 0.05$ for most risk-averse).

uses the standard box-plot to summarize the expected relative loss with respect to different values of $\alpha$. Recall that for box-plot, the left and right boundary of the rectangle respectively corresponds to the 25th and 75th percentile; the line in the middle marks the 50th percentile or median; the small empty circle within each box is the mean value; the black dots are outliers.

**Insight 1.** *Under exponential utility, the expected relative loss with respect to a risk-seeking attacker has a smaller variance than that of a risk-averse attacker, meaning that defending against a risk-seeking attacker is more predictable.*

Insight 1 is counter-intuitive at first glance because risk-averse attackers are, by definition, more deterministic or prefer less variance. However, it can be understood as follows: risk-averse attackers are very sensitive to the $q_j$'s. Among the randomly generated instances, we observe that the attack sequences of a risk-averse attacker can vary substantially for two instances with similar $q_j$'s; by contrast, the attack sequence of a risk-seeking attacker does not. Since different attack sequences can cause significant changes to the expected relative loss, the expected relative loss of a risk-seeking attacker has a smaller variance in general.

## 4.2   Expected Loss w.r.t. Attacker's Reconnaissance Capability

An attacker's is reflected by the $q_j$'s and $v_{j,A}$'s. For a perfectly capable attacker, it holds that $v_{j,A} = v_{j,D}$ (i.e., the attacker can correctly obtain the value of the digital assets in computer $j$), $q_j = 0$ for each production computer $j$, and $q_j = 1$ for each honeypot computers $j$. For a specific attacker, we measure its reconnaissance capability by comparing it with the perfect attacker, namely b comparing two sequences: the sequence of the expected values perceived by an arbitrary attacker, $(a_j)_{j=1}^{n+m}$ where $a_j = (1 - q_j) \cdot v_{j,A}$; the sequence of the expected value perceived by the perfectly capable attacker, $(v_{j,D})_{j=1}^{n+m}$. We measure the similarity between these two sequences by treating them as $(n + m)$-dimensional vectors and using the *cosine similarity* metric widely used in data science [40]. The cosine similarity between vector $A = (a_j)_{j=1}^{n+m}$ and $D = (v_{j,D})_{j=1}^{n+m}$ is defined as:

$$S_C(A, D) = \frac{A \cdot D}{\|A\|\|D\|}.$$

If the cosine similarity is 0, it means that the two vectors are orthogonal to each other in the sense that $a_j > 0$ when $v_{j,D} = 0$ and $a_j = 0$ when $v_{j,D} > 0$. In this case, the attacker is completely wrong, namely believing that production computers are honeypot computers and that honeypot computers are production computers. If the cosine similarity equals to 1, then $\frac{a_j}{\sum_{j=1}^{n+m} a_j} = \frac{v_{j,D}}{\sum_{j=1}^{n+m} v_{j,D}}$ for all $j$, meaning the expected value of each computer perceived by the attacker is almost always proportional to $v_{j,D}$.

In our experiment, we use different cosine similarity values by generating the $q_j$'s in a "semi-random" fashion (because drawing $q_j$'s uniformly at random from $[0,1]$ always yields a large cosine similarity). More specifically, we generate the $q_j$'s according to normal distribution $\mathcal{N}(x, 0.1)$ where $x \in \{0.1, 0.25, 0.5, 0.75, 0.9\}$, and for each $x$ we use $\mathcal{N}(x, 0.1)$ to generate 20% of the $q_j$'s.

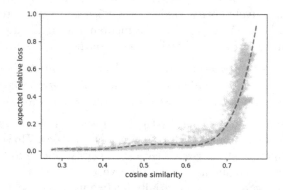

**Fig. 3.** Attacker's reconnaissance capability vs. the expected loss to the defender.

Figure 3 plots the experimental result, showing that the expected relative loss increases marginally when the cosine similarity is below a certain threshold, but increases sharply when it is above a certain threshold.

**Insight 2.** *Blending honeypots into production computers is extremely effective when the attacker's reconnaissance capability is below a threshold.*

## 5   Limitations

This study has a number of limitations. First, we assume that the attacker attacks computers in an independent fashion. In practice, the attacker may re-evaluate its perception of both $v_{j,A}$ and $q_j$ after attacking a computer. This is possible because the attacker will receive feedback from attacking a production computer that is different from attacking a honeypot computer. This poses an outstanding open problem for future research: How should we extend the model to incorporate this kind of feedback? Second, we assume that the new attacks available to the attackers are equally capable of attacking any production computer and will be applicable to any honeypot computer. The former is a simplifying assumption because new attacks may have different capabilities and incur different costs (e.g., a zero-day exploit against an operating system would be more powerful and expensive than a zero-day exploit against an application program). The latter is also a simplifying assumption because, for example, an exploit against Microsoft Windows is not applicable to Linux. Future research needs to investigate how to extend the model to accommodate such differences. Third, we formalize the HD problem in a "one-shot" fashion, meaning that the honeypot computers, once deployed, are never re-deployed (i.e., their IP addresses never change after deployment). The effectiveness of honeypots would be improved by dynamically adjusting the locations of the honeypot computers.

# 6  Related Work

**Prior Studies Related to Honeypots.** From a conceptual point of view, the present study follows the Cybersecurity Dynamics framework [47–49], which aims to rigorously and quantitatively model attack-defense interactions in cyberspace. From a technical point of view, honeypot is a cyber deception technique. We refer to [3,18,38,41,44,53] for cyber deception in a broader context. We divide prior studies on honeypots into three families based on their purposes.

The first purpose is to study how to leverage honeypots to detect new attacks (e.g., [4–6,22,24,33]). These studies typically assume that honeypot computers and production computers belong to two different networks; this isolation renders honeypot's utility questionable because it is easy for attackers to determine the presence of such honeypot networks or honeynets. The present study falls under this thrust of research but advocates blending honeypot computers into production computers. Moreover, our study is through an innovative lens, which is to maximize the utility of honeypot in forcing attackers to expose their new attacks, while minimizing the loss to the defender in terms of its digital assets stored in the compromised production computers. To the best of our knowledge, this is the first study on modeling and analyzing the utility of honeypots.

The second purpose is to study how to prepare or use honeypots [2,7,10, 17,20,21,23,25,32,43,45]. For example, some studies are geared toward making honeypot computers and production computers look the same to disrupt attackers' reconnaissance process [2,23,25,32]; some studies are geared toward deploying honeypots to defend networks with known vulnerabilities (e.g., [7]); some studies focus on making honeypot self-adaptive to attacks [20,21,43]. Our study is different from these studies for at least three reasons. (i) Putting into the terminology of our study, these studies can be understood as treating the $v_{j,A}$'s and the $q_j$'s as their goal of study. Whereas, we treat the $v_{j,A}$'s and $q_j$'s as a stepping-stone for characterizing the utility of honeypots in forcing attackers to expose their new attacks, which are not known to the defender. This means that these studies, which lead to honeypot computers with various degrees of sophistication, can be incorporated into our model to formulate a more comprehensive framework. (ii) These studies are dominated by game-theoretic models. By contrast, we use a combinatorial optimization approach. This difference in approach can be justified by the difference in the goals because we focus on characterizing the utility of honeypots in forcing attackers to expose their new attacks, while minimizing the loss to the defender in terms of digital assets. (iii) Some of these studies assume that the vulnerabilities are known (but unpatched). By contrast, we can accommodate both known (but unpatched) and unknown vulnerabilities (e.g., zero-day vulnerabilities unknown to the defender).

The third purpose is to study how to leverage honeypot-collected data to forecast cyber threats [12,15,30,42,46,51,52]. These studies lead to innovative statistical or deep learning models which can accurately forecast the number or the type of incoming attacks. However, these studies leverage traditional honeypot deployments mentioned above, namely that honeypot computers belong to a different network than the production network. By contrast, we investigate how to optimally blend honeypot computers into production networks, which would enable of more realistic forecasting results [39].

**Prior Studies Related to Our Hardness and Algorithmic Results.** We study the defender's optimization problem given a stochastic attacker. This problem is closely related to the bi-level optimization problem [9,11,14,31,36]. However, these results are all for a deterministic follower (attacker), while the HD problem studied in this paper involves a stochastic attacker who makes decisions in a probabilistic way. We are not aware of approximation algorithms for bi-level optimization problems where the follower (attacker) is stochastic.

# 7    Conclusion

Honeypot is an important cyber defense technique, especially in forcing attackers to expose their new attacks (e.g., zero-day exploits). However, the effectiveness of honeypots has not been systematically investigated. This motivated us to formalize the Honeypot Deployment (HD) problem as one manifestation of understanding the effectiveness of blending honeypot computers into production computers in an enterprise network. We show that the HD problem is NP-hard, provide a polynomial time approximation scheme to solve it, and present experimental results to draw further insights. The limitations mentioned above represent interesting open problems for future research.

**Acknowledgement.** This work was supported in part by NSF Grants #2122631, #2115134 and #2004096, and Colorado State Bill 18-086.

# A    Proof of Theorem 1

*Proof of Theorem* 1. Given an arbitrary instance of Subset Product, we construct an instance of the HD problem as follows: there is only one production computer (i.e., $n = 1$) with value 1. There are $m + 1$ IP addresses in total, where the production computer is the last computer (i.e., computer $m + 1$). The defender can deploy honeypot computers at IP addresses from 1 to $m$. In particular, deploying honeypot computer at IP address $i$ costs $c_i = \lfloor \log(w_i)M \rfloor / M$. The defender's budget is $B = \lceil \log(k)M \rceil / M$ where $M = k(m + 1)$. The probability is $q_i = 1/w_i$ for non-dummy computer $i \in [m]$, and $q_{m+1} = 0$. Set parameter $r = 1$ and the threshold of the expected loss to the defender as $T = 1/k$.

Note that although $\log k$ and $\log w_i$'s are not rational numbers, for the purpose of determining the value of (e.g.) $\lfloor \log(w_i)M \rfloor$, it suffices to compute $\log w_i$ up to a precision of $O(1/M)$, which can be done in $O(\log M)$ time. It is easy to verify that the input length of the HD problem is $O(m \log k + m \log w_{max})$ where $w_{max} = \max_i w_i$.

Consider the expected loss of the defender. Since the value of the production computer is 1, $\mathbb{E}(Loss)$ equals the probability that computer $m + 1$ is attacked. Since $r = 1$, the attacker can attack computer $m + 1$ only if it does not attack any computer from 1 to $m$, which equals to

$$\prod_{i \in [m], \text{a honeypot computer is deployed at } i} \frac{1}{w_i}.$$

Suppose the answer to the instance of the Subset Product instance is "yes", then we know that there exists $S'$ such that $\prod_{i \in S'} w_i = k$. Deploying honeypot computers at IP addresses $i$ where $i \in S'$, we know the total cost to the defender equals

$$\sum_{i \in S'} \frac{\lfloor M \log w_i \rfloor}{M} \leq \sum_{i \in S'} \log w_i \leq \log k \leq \frac{\lceil M \log k \rceil}{M} = B,$$

which is within budget $B$. Meanwhile, the expected loss is

$$\prod_{i \in S'} \frac{1}{w_i} = 1/k.$$

Hence, the minimum expected loss for the HD instance is no larger than $1/k$, i.e., the answer for the HD instance is "yes".

Suppose the answer for the HD instance is "yes" (i.e., the minimum expected loss to the defender is no larger than $1/k$). Let $S' \subseteq [m]$ be the IP addresses where the defender deploys honeypot computers, we know that

$$\sum_{i \in S'} \log(w_i) - m/M \leq \sum_{i \in S'} c_i \leq B \leq \log(k) + 1/M, \tag{2}$$

and

$$\prod_{i \in S'} 1/w_i \leq 1/k. \tag{3}$$

By Combining Eq. (2) and Eq. (3), we know that

$$k \leq \prod_{i \in S'} w_i \leq k \cdot 2^{(m+1)/M} < k + 1.$$

Consequently, $\prod_{i \in S'} w_i = k$. Hence, the answer to the Subset Product instance is "yes". $\qquad \square$

# B    Proof of Lemma 2

*Proof.* We prove this by induction. It is easy to verify Lemma 2 for $t = 1$. Suppose it holds for $t = \ell - 1$, that is, there exists some $(\ell - 1, p_{\ell-1,0}(x^*), \ldots, p_{\ell-1,r}(x^*), e_{\ell-1}x^*), b) \in \tilde{\mathcal{F}}_{\ell-1}$ such that $b \leq b_{\ell-1}(x^*)$, we prove Lemma 2 for $t = \ell$. We distinguish two cases based on computer $\ell$ and the optimal solution $x^*$:

- If $\ell \in \mathcal{P}$, then for the optimal solution $x^*$ it holds that $p_{\ell,k}(x^*) = p_{\ell-1,k}(x^*)$ for $0 \leq k \leq r$, $e_\ell(x^*) = e_{\ell-1}(x^*) + (1 - q_\ell)v_\ell \sum_{k=1}^{r-1} p_{\ell-1,k}$, and $b_\ell(x^*) = b_{\ell-1}(x^*)$. Given that $(\ell - 1, p_{\ell-1,0}(x^*), \ldots, p_{\ell-1,r}(x^*), e_{\ell-1}(x^*), b) \in \tilde{\mathcal{F}}_{\ell-1} \subseteq \mathcal{F}_{\ell-1}$, according to the recursive computation of $\mathcal{F}_\ell$ from $\mathcal{F}_{\ell-1}$, $(\ell, p_{\ell,0}(x^*), \ldots, p_{\ell,r}(x^*), e_\ell(x^*), b_\ell(x^*)) \in \mathcal{F}_\ell$. Hence, there exists some $(\ell, p_{\ell,0}(x^*), \ldots, p_{\ell,r}(x^*), e_\ell(x^*), b) \in \tilde{\mathcal{F}}_\ell$ such that $b \leq b_\ell(x^*)$ by the definition of $\tilde{\mathcal{F}}_\ell$.
- If $\ell \in \mathcal{H}$, we further distinguish two sub-cases: (i) If computer $\ell$ is a dummy computer in $x^*$, then $p_{\ell,k}(x^*) = p_{\ell-1,k}(x^*)$ for $0 \leq k \leq r$, $e_\ell(x^*) = e_{\ell-1}(x^*)$, and $b_\ell(x^*) = b_{\ell-1}(x^*)$. (ii) If computer $\ell$ is a honeypot computer in $x^*$, then we have $p_{\ell,0}(x^*) = q_\ell p_{\ell-1,0}(x^*)$, $p_{\ell,k}(x^*) = p_{\ell-1,k}(x^*)q_\ell + (1 - q_\ell)p_{\ell-1,k-1}(x^*)$ for $1 \leq k \leq r$, and $e_\ell(x^*) = e_{\ell-1}(x^*)$, $b_\ell(x^*) = b_{\ell-1}(x^*) + c_t$. In both sub-cases, the recursive computation of $\mathcal{F}_\ell$ from $\mathcal{F}_{\ell-1}$ implies that $(\ell, p_{\ell,0}(x^*), \ldots, p_{\ell,r}(x^*), e_\ell(x^*), b_\ell(x^*)) \in \mathcal{F}_\ell$. Hence there exists $(\ell, p_{\ell,0}(x^*), \ldots, p_{\ell,r}(x^*), e_\ell(x^*), b) \in \tilde{\mathcal{F}}_\ell$ such that $b \leq b_\ell(x^*)$ by the definition of $\tilde{\mathcal{F}}_\ell$.

Hence, Lemma 2 is true for all $t \in [n + m]$.    $\square$

# References

1. Adleman, L.M.: An abstract theory of computer viruses. In: Goldwasser, S. (ed.) CRYPTO 1988. LNCS, vol. 403, pp. 354–374. Springer, New York (1990). https://doi.org/10.1007/0-387-34799-2_28
2. Aggarwal, P., Du, Y., Singh, K., Gonzalez, C.: Decoys in cybersecurity: an exploratory study to test the effectiveness of 2-sided deception. arXiv preprint arXiv:2108.11037 (2021)
3. Al-Shaer, E., Wei, J., Kevin, W., Wang, C.: Autonomous Cyber Deception. Springer, Heidelberg (2019). https://doi.org/10.1007/978-3-030-02110-8
4. Almotairi, S., Clark, A., Mohay, G., Zimmermann, J.: A technique for detecting new attacks in low-interaction honeypot traffic. In: Proceedings of International Conference on Internet Monitoring and Protection (2009)
5. Almotairi, S.I., Clark, A.J., Mohay, G.M., Zimmermann, J.: Characterization of attackers' activities in honeypot traffic using principal component analysis. In: Proceedings of IFIP International Conference on Network and Parallel Computing (2008)
6. Anagnostakis, K.G., Sidiroglou, S., Akritidis, P., Xinidis, K., Markatos, E.P., Keromytis, A.D.: Detecting targeted attacks using shadow honeypots. In: USENIX Security Symposium (2005)
7. Anwar, A.H., Kamhoua, C.A., Leslie, N., Kiekintveld, C.D.: Honeypot allocation games over attack graphs for cyber deception. In: Game Theory and Machine Learning for Cyber Security (2021)

8. Camerer, C.F., Loewenstein, G., Rabin, M.: Advances in Behavioral Economics. Princeton University Press, Princeton (2004)
9. Caprara, A., Carvalho, M., Lodi, A., Woeginger, G.J.: A complexity and approximability study of the bilevel knapsack problem. In: International Conference on Integer Programming and Combinatorial Optimization, IPCO (2013)
10. Carroll, T.E., Grosu, D.: A game theoretic investigation of deception in network security. Secur. Commun. Netw. **4**(10), 1162–1172 (2011)
11. Chen, L., Zhang, G.: Approximation algorithms for a bi-level knapsack problem. Theor. Comput. Sci. **497**, 1–12 (2013)
12. Chen, Y., Huang, Z., Xu, S., Lai, Y.: Spatiotemporal patterns and predictability of cyberattacks. PLoS One **10**(5) (2015)
13. Cohen, F.: The use of deception techniques: honeypots and decoys. Handb. Inf. Secur. **3**(1), 645–655 (2006)
14. Dempe, S., Richter, K.: Bilevel programming with knapsack constraints. Central Eur. J. Oper. Res. (2000)
15. Fang, X., Xu, M., Xu, S., Zhao,: A deep learning framework for predicting cyber attacks rates. EURASIP J. Inf. Secur. (2019)
16. Galinkin, E., Carter, J., Mancoridis, S.: Evaluating attacker risk behavior in an internet of things ecosystem. In: GameSec (2021)
17. Garg, N., Grosu, D.: Deception in honeynets: a game-theoretic analysis. In: IEEE SMC Information Assurance and Security Workshop (2007)
18. Han, X., Kheir, N., Balzarotti, D.: Deception techniques in computer security: a research perspective. ACM Comput. Surv. **51**(4), 1–36 (2018)
19. Hillson, D., Murray-Webster, R.: Understanding and managing risk attitude (2007)
20. Huang, L., Zhu, Q.: Adaptive honeypot engagement through reinforcement learning of semi-markov decision processes. In: GameSec (2019)
21. Huang, L., Zhu, Q.: Farsighted risk mitigation of lateral movement using dynamic cognitive honeypots. In: GameSec (2020)
22. Kreibich, C., Crowcroft, J.: Honeycomb: creating intrusion detection signatures using honeypots. ACM SIGCOMM Comput. Commun. Rev. **34**(1), 51–56 (2004)
23. Kulkarni, A.N., Fu, J., Luo, H., Kamhoua, C.A., Leslie, N.O.: Decoy allocation games on graphs with temporal logic objectives. In: GameSec (2020)
24. Li, Z., Goyal, A., Chen, Y., Paxson, V.: Towards situational awareness of large-scale botnet probing events. IEEE Trans. Inf. Forensics Secur. **6**(1), 175–188 (2010)
25. Miah, M.S., Gutierrez, M., Veliz, O., Thakoor, O., Kiekintveld, C.: Concealing cyber-decoys using two-sided feature deception games. In: Hawaii International Conference on System Sciences, HICSS (2020)
26. Morgan, S.: Cybercrime to cost the world $10.5 trillion annually by 2025 (2020). https://cybersecurityventures.com/cybercrime-damages-6-trillion-by-2021/
27. Nawrocki, M., Wählisch, M., Schmidt, T.C., Keil, C., Schönfelder, J.: A survey on honeypot software and data analysis. arXiv preprint arXiv:1608.06249 (2016)
28. NYSDFS: Solarwinds cyber espionage attack and institutions' response (2021). https://www.dfs.ny.gov/system/files/documents/2021/04/solarwinds_report_2021.pdf
29. Pendleton, M., Garcia-Lebron, R., Cho, J.H., Xu, S.: A survey on systems security metrics. ACM Comput. Surv. **49**(4), 1–35 (2016)
30. Peng, C., Xu, M., Xu, S., Hu, T.: Modeling and predicting extreme cyber attack rates via marked point processes. J. Appl. Stat. **44**(14), 2534–2563 (2017)
31. Pferschy, U., Nicosia, G., Pacifici, A.: A stackelberg knapsack game with weight control. Theor. Comput. Sci. **799**, 149–159 (2019)

32. Píbil, R., Lisỳ, V., Kiekintveld, C., Bošanskỳ, B., Pěchouček, M.: Game theoretic model of strategic honeypot selection in computer networks. In: GameSec (2012)
33. Portokalidis, G., Bos, H.: Sweetbait: zero-hour worm detection and containment using low-and high-interaction honeypots. Comput. Netw. **51**(5), 1256–1274 (2007)
34. Pratt, J.W.: Risk aversion in the small and in the large. In: Uncertainty in Economics (1978)
35. Provos, N., et al.: A virtual honeypot framework. In: USENIX Security (2004)
36. Qiu, X., Kern, W.: Improved approximation algorithms for a bilevel knapsack problem. Theor. Comput. Sci. **595**, 120–129 (2015)
37. Rodriguez, R.M., Xu, S.: Cyber social engineering kill chain. In: SciSec (2022)
38. Rowe, N.C., Rrushi, J., et al.: Introduction to cyberdeception (2016)
39. Sun, Z., Xu, M., Schweitzer, K., Bateman, R., Kott, A., Xu, S.: Cyber attacks against enterprise networks: characterization, modeling and forecasting. In: Proceedings of SciSec 2023 (2023)
40. Thearling, K.: An introduction to data mining. Direct Mark. Maga. (1999)
41. Thomas, S.: Cyber deception: building the scientific foundation (2016)
42. Trieu-Do, V., Garcia-Lebron, R., Xu, M., Xu, S., Feng, Y.: Characterizing and leveraging granger causality in cybersecurity: framework and case study. ICST Trans. Secur. Saf. **7**(25), 1–18 (2021)
43. Wagener, G., State, R., Engel, T., Dulaunoy, A.: Adaptive and self-configurable honeypots. In: IFIP IEEE International Symposium on Integrated Network Management (IM) (2011)
44. Wang, C., Lu, Z.: Cyber deception: overview and the road ahead. IEEE Secur. Priv. **16**(2), 80–85 (2018)
45. Wang, S., Pei, Q., Wang, J., Tang, G., Zhang, Y., Liu, X.: An intelligent deployment policy for deception resources based on reinforcement learning. IEEE Access **8**, 35792–35804 (2020)
46. Xu, M., Hua, L., Xu, S.: A vine copula model for predicting the effectiveness of cyber defense early-warning. Technometrics **59**(4), 508–520 (2017)
47. Xu, S.: Cybersecurity dynamics: a foundation for the science of cybersecurity. In: Lu, Z., Wang, C. (eds.) Proactive and Dynamic Network Defense, vol. 74, Springer, Heidelberg (2019). https://doi.org/10.1007/978-3-030-10597-6_1
48. Xu, S.: The cybersecurity dynamics way of thinking and landscape (invited paper). In: ACM Workshop on Moving Target Defense (2020)
49. Xu, S.: Sarr: a cybersecurity metrics and quantification framework (keynote). In: International Conference Science of Cyber Security (SciSec 2021), pp. 3–17 (2021)
50. Yao, A.: New algorithms for bin packing. J. ACM **27**(2) (1980)
51. Zhan, Z., Xu, M., Xu, S.: Characterizing honeypot-captured cyber attacks: statistical framework and case study. IEEE Trans. Inf. Forensics Secur. **8**(11), 1775–1789 (2013)
52. Zhan, Z., Xu, M., Xu, S.: Predicting cyber attack rates with extreme values. IEEE Trans. Inf. Forensics Secur. **10**(8), 1666–1677 (2015)
53. Zhu, M., Anwar, A.H., Wan, Z., Cho, J.H., Kamhoua, C.A., Singh, M.P.: A survey of defensive deception: approaches using game theory and machine learning. IEEE Commun. Surv. Tutor. **23**(4), 2460–2493 (2021)

# WebMea: A Google Chrome Extension for Web Security and Privacy Measurement Studies

Mengxia Ren, Joshua Josey, and Chuan Yue(✉) (iD)

Colorado School of Mines, Golden, CO 80401, USA
{mengxiaren,jjosey,chuanyue}@mines.edu

**Abstract.** Web measurement is an integral part of web security and privacy analysis. A large-scale web measurement usually relies on a crawler. Browser-based crawlers are often preferred and they are typically implemented through one of two approaches: the DevTools approach or the browser extension approach. It is often not clear to researchers which approach should be taken to implement a crawler since the analysis of these two approaches is limited. In this paper, we analyze and compare the web measurement capability of the Google Chrome DevTools approach and the Google Chrome extension approach to help researchers make decisions. We find that both approaches have the capability to satisfy primary web measurement requirements, although their APIs and the complexity of using their APIs are different. We point out such differences to help researchers more easily choose or take the appropriate approach. Taking the former approach, Puppeteer is a crawler that can be readily used by researchers. However, it is difficult for researchers to find a good baseline Google Chrome extension to start with their extension-based crawler construction. Therefore, we further build Web-Mea as a baseline Google Chrome extension that can measure multiple types of web data. Researchers can use or customize WebMea according to their web measurement requirements. WebMea is applicable to different web measurement studies, and it can help make web measurement studies more reproducible and replicable.

**Keywords:** Web Measurement · Tool · Security · Privacy · Browser Extension

## 1 Introduction

Many large-scale web security and privacy measurement studies rely on crawlers for data collection. Some researchers use or customize existing crawlers, while others implement their own crawlers. Properly building, choosing, and configuring crawlers can help improve the reproducibility and replicability of web measurement studies; meanwhile, browser-based crawlers are often preferred due to their better integration with browsers and correspondingly their advanced measurement capabilities [2,3,13].

© The Author(s), under exclusive license to Springer Nature Switzerland AG 2023
M. Yung et al. (Eds.): SciSec 2023, LNCS 14299, pp. 305–318, 2023.
https://doi.org/10.1007/978-3-031-45933-7_18

Browser-based crawlers are typically implemented through one of two approaches: the DevTools approach or the browser extension approach. Dev-Tools is a set of developer tools provided by a given browser for web testing and debugging, and it is often wrapped as a set of DevTools protocol APIs [7,9]. In the DevTools approach, researchers use DevTools protocol APIs to implement a crawler. A browser extension is a software program for extending the browser's functionalities [6,10]. In the browser extension approach, researchers implement a crawler (which is a browser extension) by using extension APIs and providing other required files such as the manifest and resource files. Because of their different features, these two approaches are applicable to different scenarios. However, it is often not clear to researchers which is the more appropriate approach to constructing a crawler since the analysis of these two approaches is limited.

In this paper, we study these two approaches in the context of the Google Chrome browser because of its popularity. We first outline the primary requirements of the crawlers for web security and privacy measurement studies. We then analyze and compare the web measurement capability of these two Google Chrome browser-based approaches. A good crawler implementation approach should have the capability to satisfy all primary web measurement requirements. We find that the Google Chrome DevTools approach and the Google Chrome extension approach rely on Chrome DevTools Protocol (CDP) APIs [9] and Chrome extension APIs [10] respectively to implement a crawler, and the complexities of using their APIs are different. However, both approaches have the capability to satisfy all primary web measurement requirements. To help researchers take the appropriate approach, we further point out the advantages, limitations, and applicable scenarios of both approaches.

Meanwhile, taking the Google Chrome DevTools approach, researchers can easily find a baseline Puppeteer crawler which is one popular browser automation tool controlling Google Chrome through CDP APIs [11]. However, there is no good baseline Google Chrome extension for researchers to start with their extension-based crawler construction because of the complexity of using the Google Chrome extension APIs. Therefore, we further build WebMea, a Google Chrome extension that can measure both web content and HTTP(S) traffic, as a baseline Google Chrome extension for researchers to construct their extension-based crawlers. Researchers can use or customize WebMea according to their web measurement requirements. Our contributions in this paper are as follows:

- We compare the web measurement capability of two Google Chrome browser-based crawler implementation approaches to help researchers more easily choose or take the appropriate approach.
- We build WebMea as a baseline Google Chrome extension for researchers to construct their extension-based crawlers. WebMea can be used for different web measurement studies because it can measure multiple types of web data, which can help make web security and privacy measurement studies more reproducible and replicable.

The rest of this paper is organized as follows. Section 2 reviews the related work. Section 3 presents the primary web measurement requirements, the overview of the Google Chrome DevTools approach and the Google Chrome extension approach, and the comparison of the web measurement capability of these two approaches. Section 4 describes the design, implementation, and customization of WebMea. Section 5 concludes this paper.

## 2   Related Work

Browser-based crawlers are popularly used in web security and privacy measurement studies because of their better integration with browsers and correspondingly their advanced measurement capabilities [2,3,13]. Some browser-based crawlers are built through the DevTools approach. Taking this approach, Puppeteer is a crawler that can be readily used by researchers [1,15]. Din et al. implemented PERCIVAL, a browser-embedded ad blocker for the Chromium-based browsers, in which Puppeteer was used for collecting ad and non-ad images of webpages [1]. Smith et al. constructed SugarCoat, a system for automatically generating privacy-preserving JavaScript (JS) replacements, in which Puppeteer was used for extracting JS scripts [15]. Some browser-based crawlers are implemented through the browser extension approach. Englehardt et al. constructed OpenWPM, a web privacy measurement framework for collecting cookies, response metadata, and behavior of scripts, in which data collection is performed by a Firefox extension [4]. Roesner ct al. implemented TrackingObserver, a Google Chrome extension that collects trackers' in-browser behaviors to perform third-party tracking analysis [16]. However, both browser extensions are more applicable for web privacy measurement studies, and they also have limited capabilities on extracting web content and simulating user interactions.

The reproducibility and the replicability of web security and privacy measurement studies are often impacted by experimental setups including crawlers and crawler configurations. Ahmad et al. conducted comparative empirical crawler evaluations to understand how the choice of crawlers impacts the web measurement studies [2]. They found that the characteristics of network traffic and application data measured by different crawlers are not the same, which impacts web measurement results and conclusions. Demir et al. researched the reproducibility and replicability of current web measurement studies, and performed a large-scale web measurement study to show that different experimental settings (e.g., browsers, regions, and user interactions) impact the overall results of web measurement studies [3]. They also found that experimental setups were insufficiently documented in many analyzed web measurement studies. Jueckstock et al. explored how configurations of crawlers affect the web security and privacy measurement studies [13]. They observed that under different vantage points and browser configurations of crawlers, measurement bias exists in the HTTP traffic, known ad/tracking domains, and JS context. Therefore, researchers should ensure that their crawlers match the real-world user experience and can be reproduced or customized.

Overall, existing studies did not analyze or compare the two major approaches for browser-based crawler construction; therefore, we provide such an analysis to help researchers (especially beginning researchers) more easily choose or take the appropriate approach. Meanwhile, there is no good baseline Google Chrome extension for researchers to start with their extension-based crawler construction; therefore, we build WebMea which can measure multiple types of data for different studies and can simulate user interactions for matching the real-world user experience. Researchers can use or customize WebMea according to their web measurement requirements. We also expect that WebMea can help make web security and privacy measurement studies more reproducible and replicable.

## 3    Comparing the Two Crawler Implementation Approaches

A good crawler implementation approach should have the capability to satisfy all primary web measurement requirements. In this section, we first outline these primary requirements. We then review, analyze, and compare the Google Chrome DevTools approach and the Google Chrome extension approach. To further illustrate the differences between these two approaches, we also provide case studies for each approach in this section.

### 3.1    Primary Requirements for Crawlers

We outline the primary requirements of crawlers for web security and privacy measurement studies as follows.

**Automated Crawling.** Crawlers should be able to automatically visit webpages in a particular sequence according to a supplied URL list, and can further identify and visit a specific number of subpages (with the same domain or subdomain) of the visited webpages. Automated crawling is an essential requirement for large-scale web security and privacy analysis.

**HTTP(S) Traffic Interception and Modification.** Based on the need of the studies, crawlers should be able to control HTTP(S) requests and responses. Specifically, crawlers should be able to intercept requests, add requests, redirect requests, remove requests, and modify request body content and request headers. Crawlers should also be able to intercept responses and control them such as modifying response body content and response headers. For example, content security policy analysis and web tracking analysis usually need crawlers to measure the HTTP(S) traffic of crawled webpages.

**Webpage Content Extraction and Modification.** Crawlers should have the capabilities of extracting web content, adding web elements to a webpage, removing web elements from a webpage, and modifying attributes of web elements. For example, web attack analyses (e.g., phishing detection) often require crawlers to extract or modify web content.

**Simulation of User Interactions.** Crawlers should be able to simulate user interactions such as clicking on links or scrolling down the webpages. The simulation of user interactions is an important part of setting up an experimental environment that can match the real-world user experience.

## 3.2   Google Chrome DevTools and Extension Approaches

The DevTools approach and the browser extension approach are based on different designs. DevTools is a set of developer tools provided by a given browser for creating, testing, or debugging websites or web applications [5,8]. DevTools can either be built in a browser or used through the DevTools protocol APIs. Google Chrome provides both the built-in Chrome DevTools and the Chrome DevTools Protocol (CDP) APIs [9] for developers. However, the built-in Chrome DevTools is not often applied to large-scale web measurement studies since it is built for in-browser webpage debugging. In the Google Chrome DevTools approach, researchers use the CDP APIs to construct their crawlers. Google Chrome divides the CDP APIs into 46 domains (e.g., DOM domain, Page domain, and Network domain). Each domain includes a number of APIs and events it supports and generates. Puppeteer is a good baseline crawler that takes this approach for controlling Google Chrome in web security and privacy measurement studies [1,15].

Extensions are software programs for extending browsers' functionalities. Google Chrome extensions are event-based programs that rely on three major web techniques: HTML, CSS, and JS [10]. Google Chrome provides 76 extension APIs to help extensions control the browser and interact with crawled webpages. In the Google Chrome extension approach, researchers construct an extension as a crawler by using the Google Chrome extension APIs and providing other required files. A Google Chrome extension can be composed of the manifest file, background scripts, content scripts, an options page, and UI elements. The manifest file is used for configuring extension settings and permissions. The background scripts can monitor events triggered by the browser, and Google Chrome extensions interact with the browser through the background scripts. The content scripts can run on crawled webpages, and Google Chrome extensions interact with webpages through the content scripts. The options page allows users to adjust extension options. The UI elements help users interact with Google Chrome extensions.

The capability of the Google Chrome DevTools approach and the Google Chrome extension approach depends on the capability of the CDP APIs and the extension APIs, respectively. We draw Fig. 1 to summarize the major categories of capabilities of these two types of APIs. These two types of APIs have some common capabilities but also have some different capabilities. They both have the capabilities of configuring browser settings, interacting with webpages, controlling browser actions, controlling web content, controlling HTTP(S) traffic, auditing webpages, and emulating different page environments. Other major capabilities of the CDP APIs are controlling media, controlling background services, and measuring web performance. Other major capabilities of the Google Chrome extension APIs are implementing text-to-speech (TTS) engines, extending DevTools panels, managing Chrome OS, and managing extensions and web applications installed or running in the browser. Although the CDP APIs and the Google Chrome extension APIs focus on web debugging and browser function extension respectively, both of them have the capabilities of interacting with webpages and browsers. Therefore, both approaches can be used to construct crawlers. Note that Google Chrome extensions can also access 15 CDP API domains (e.g., those for webpage interaction, web content control, and HTTP(S) traffic control) through the *chrome.debugger* extension API. They cannot access other CDP API domains for security reasons.

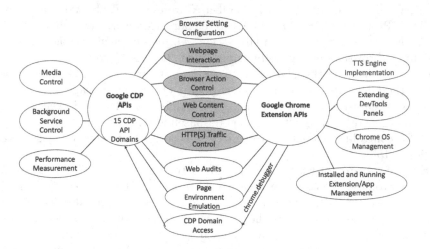

**Fig. 1.** Major Categories of Capabilities of the Chrome DevTools Protocol (CDP) APIs and the Google Chrome extension APIs. Google Chrome extensions can also access 15 CDP API domains through the *chrome.debugger* extension API.

A good crawler implementation approach should have the capability to satisfy all primary web measurement requirements outlined in Sect. 3.1. In other words, it should be capable of controlling browser actions to achieve automated crawling, controlling web content to achieve webpage content extraction and modification, controlling HTTP(S) traffic to achieve HTTP(S) traffic interception and modification, and interacting with webpages to achieve simulations of user interactions. As Fig. 1 shows, the CDP APIs and the Google Chrome extension APIs both have these four categories of web measurement capabilities to satisfy the primary web measurement requirements.

Although the CDP APIs and the Google Chrome extension APIs commonly have the major web measurement capabilities, their granularities of controlling webpages and browsers are different. Therefore, these two approaches have different advantages and are sometimes applicable to different scenarios. With the CDP APIs, the Google Chrome DevTools approach can simulate page environments at a more fine-grained level than the Google Chrome extension approach. For example, the CDP LayerTree domain provides APIs for analyzing and operating on layers, and the CDP DeviceOrientation domain provides APIs for clearing and overriding device orientations. However, these two CDP domains are not accessible to Google Chrome extensions. Researchers can give priority to constructing a crawler through the Google Chrome DevTools approach when page environments need to be very close to the real one. On the other hand, the Google Chrome extension approach can control HTTP(S) requests at a more fine-grained level through the webRequest extension API. This API can monitor a set of events that follow the entire life cycle of an HTTP(S) request. Researchers can give priority to implementing a crawler through the Google Chrome extension approach when their crawlers are required to work closely on requests and responses following the life cycle of requests.

To help researchers meet the primary web measurement requirements in their crawlers, we further summarize the corresponding major APIs of the two approaches in Table 1. The Google Chrome DevTools approach mainly relies on Page domain CDP API and Network domain CDP API to achieve the primary web measurement capabilities, while the Google Chrome extension approach mainly relies on webRequest extension API, tabs extension API, runtime extension API, and content scripts. To further illustrate the differences between these two approaches, we provide a case study for each of them in Sects. 3.3 and 3.4, respectively. Sometimes crawler implementation by taking the Google Chrome extension approach is more complex than taking the Google Chrome DevTools approach since the former needs multiple required files to work together to achieve a web measurement function. Therefore, it will be helpful to have a good baseline Google Chrome extension for researchers to start with their extension-based crawler construction (Sect. 4).

**Table 1.** The major CDP APIs and Google Chrome extension APIs of the two approaches for satisfying the primary web measurement requirements.

| Requirement | Operation | DevTools Approach | Extension Approach |
|---|---|---|---|
| Automated Crawling | Accessing webpages | Page domain CDP API | tabs extension API |
| | Webpage link extraction | Page domain CDP API | Content scripts and runtime extension API |
| HTTP(S) Traffic Interception and Modification | Request interception and modification | Network domain CDP API | webRequest extension API |
| | Response interception and response header modification | Network domain CDP API | webRequest extension API |
| | Response body modification | Network domain CDP API | Network domain CDP API and debugger extension API |
| Webpage Content Extraction and Modification | Webpage content extraction | Page domain CDP API | Content scripts and runtime extension API |
| | Webpage content modification | Page domain CDP API | Content scripts and runtime extension API |
| Simulation of User Interactions | Simulating user experience in the real-word | Page domain, Runtime domain, and Emulation domain CDP APIs | Content scripts and runtime extension API |

## 3.3   DevTools Case Study: Puppeteer

Puppeteer is a Node.js library that provides a high-level API to control headless or real Chrome and Chromium browsers over the CDP APIs [11]. The high-level API contains multiple classes and corresponding class methods to help users easily implement the functions of a crawler. Users can create an instance of a browser and then manipulate it with Puppeteer's class methods which can create CDP sessions for using the CDP APIs. Puppeteer is composed of JS files, declaration files, and source map files. The JS files and declaration files are used for class declarations. The methods of a class are in the corresponding JS file, and the method parameters are in the corresponding declaration file. Source map files are provided for each class to make JS debugging and testing efficient. Because Puppeteer is CDP-based, its capability depends on the capability of the CDP APIs. Therefore, Puppeteer has web testing and debugging capabilities that include navigating webpages, collecting screenshots, automating form submissions, and extracting or modifying data from the DOM.

Puppeteer's high-level interaction with the browser as well as low-level control over data collection make it a good baseline crawler, upon which many other Google Chrome DevTools-based crawlers are built. The crawlers used in the studies of Din et al. [1] and Smith et al. [15] were built upon Puppeteer as we reviewed in Sect. 2. They only used the basic features of Puppeteer without simulating user interactions; meanwhile, they focused on web content extraction and had a limited capability of HTTP(S) traffic measurement. In addition, there is a Google Chrome headless-chrome-crawler [12] built on Puppeteer for crawling dynamic websites, and researchers can customize this crawler through the Puppeteer high-level API and the CDP APIs.

## 3.4   Chrome Extension Case Study: TrackingObserver

The Tracking-Observer Google Chrome extension [16] we reviewed in Sect. 2 has the following primary web measurement functions: (1) crawling a list of websites in an arbitrary link depth, (2) tracking and resetting cookies, and (3) tracking history. All these functions are implemented in a background script and a content script. For crawling a website in an arbitrary link depth, the background script sends a request to the content script to extract a specific number of links from crawled webpages. The content script listens to requests sent from the background script and sends the extracted links back to the background script through the *chrome.extension* API. TrackingObserver tracks cookies by filtering cookie items from HTTP(S) request headers through the webRequest extension API. It can also cancel and remove original HTTP(S) requests through the webRequest extension API. TrackingObserver resets cookies through the *set()* method of the cookies extension API. For tacking history, the *chrome.history.search()* extension method is used for obtaining all history of the last visit time. TrackingObserver takes the advantages of the webRequest extension API to process HTTP(S) requests.

TrackingObserver is more applicable for web privacy measurement studies, and it also has limited capabilities on extracting web content and simulating user interactions. Therefore, using it as a baseline Google Chrome extension for other web security and privacy measurement studies will still need lots of development effort from researchers.

# 4 Proposed Chrome Extension WebMea

In this section, we present the design, implementation, and customization of our proposed Google Chrome extension WebMea.

## 4.1 Design of WebMea

WebMea can collect HTTP(S) requests and corresponding responses, HTML documents, and http-equiv attributes of meta elements from crawled webpages. It can be used in different web security and privacy measurement studies. Figure 2 shows the overview of WebMea. It mainly comprises a background script, four content scripts (i.e., subpage_extract.js, dom_extract.js, meta_extract.js, and page_scroll.js files), and a popup.js file. The background script implements functions that need to interact with the browser. Content scripts implement functions that need to interact with crawled webpages. The popup.js file provides UI for users to interact with WebMea. Researchers can start crawling by clicking the "Start Crawling" button on the popup page. Researchers can save the collected web data either in an in-browser database maintained by Google Chrome or in a remote database supported by Node.js. By default, WebMea saves all collected data in the in-browser database. WebMea provides multiple functions listed as follows to satisfy the primary web measurement requirements.

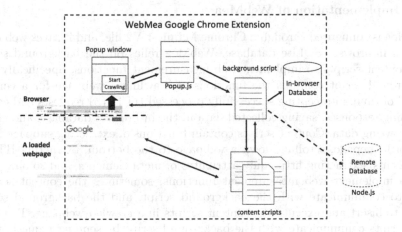

**Fig. 2.** The High-level Overview of WebMea

*Automated Crawling.* Given a crawling website list, WebMea will crawl the homepage of each website and a specific number of randomly selected subpages (with the same domain or subdomain) of each homepage. WebMea will stay for a certain period of time on each crawled webpage for the page rendering and event handling.

*HTTP(S) Traffic Interception and Modification.* When the browser receives HTTP(S) response headers of webpages, WebMea will intercept them. Researchers can extract content security policies [17] deployed in a webpage from its HTTP(S) response headers. WebMea will also intercept all successfully processed HTTP(S) requests and corresponding responses in the process of loading and rendering a webpage. The types of intercepted HTTP(S) requests include "main_frame", "sub_frame", "stylesheet", "script", "image", "font", "object", "XMLHttpRequest", "ping", "csp_report", "media", "websocket", and "other". The intercepted HTTP(S) requests and corresponding responses can be used for various security and privacy analyses such as malicious script analysis and insecure communication analysis.

*Webpage Content Extraction and Modification.* WebMea will extract the HTML document of a webpage when the webpage is completely loaded. Researchers can further process the extracted web content for various security and privacy analyses such as phishing attack analysis and malicious web element analysis. Some content security policies are also deployed through the meta http-equiv attributes thus WebMea also extracts such attributes of webpages.

*Simulation of User Interactions.* WebMea will scroll down each crawled webpage to the bottom for simulating user interactions on viewing more content of a webpage. WebMea can simulate other user interactions such as link clicking and form filling. It can also clear the browsing history, cached data, and cookies after crawling a webpage to simulate stateless browsing scenarios.

### 4.2　Implementation of WebMea

WeaMea is configured through a Chrome Manifest V2 file, and it saves web data into an in-browser SQLite database. WebMea relies on the background script and content scripts to implement web measurement functions. Specifically, the background script contains the functions of crawling a webpage for a certain period of time, intercepting successfully processed HTTP(S) requests and corresponding responses, saving collected data in the in-browser database, and clearing browsing data. Content scripts contain functions of extracting subpage links from a homepage, scrolling down a webpage to the bottom, extracting HTML documents, extracting http-equiv attributes of meta elements, and so on.

To implement web measurement functions, sometimes the content scripts need to communicate with the background script, and the background script needs to insert and execute the content scripts in a crawled webpage. The content scripts communicate with the background script by sending requests to it through the *sendMessage()* method of the runtime extension API. The background script receives requests by monitoring the *onMessage()* event of the runtime extension API. The background script inserts and executes content scripts in a webpage by using the *executeScript()* method of the tabs extension API. We detail the implementation of these web measurement functions as follows.

**Automated Crawling.** By default, WebMea automatically crawls the homepage and a specific number of subpages for each website. The automated crawling is implemented in the background script through the *onUpdated.addListener()*, *update()*, and *query()* methods of the tabs extension API. The *update()* method updates the URL of the current tab. The *onUpdated.addListener()* method monitors the webpage updates in the current tab. The *query()* method can get the information of the webpage in the current tab. When the *onUpdated.addListener()* method observes the webpage update in a tab, the corresponding callback function will be executed to further process the current webpage. The callback function of the *onUpdated.addListener()* method checks whether the current webpage is a homepage. If the webpage is a homepage, the callback function will execute the content script subpage_extract.js which will extract and send back the subpage links of the homepage. Then the callback function will receive and add subpage links to the crawling list. WebMea uses the JS setInterval() method to control the time staying on a webpage.

**HTTP(S) Traffic Interception and Modification.** By default, WebMea can intercept HTTP(S) response headers before responses are processed, and it can intercept successfully processed HTTP(S) requests and corresponding responses after webpages are completely loaded. These functions are implemented in the background script. The interception of HTTP(S) response headers is implemented by monitoring the *onHeadersReceived()* event of the webRequest extension API. The interception of processed HTTP(S) requests and corresponding responses is implemented through monitoring the *onCompleted()* event of the webRequest extension API. For *onHeadersReceived()* and *onCompleted* events, returned data (i.e., HTTP(S) response headers and HTTP(S) requests) are saved in the corresponding "details" parameter of each event. The "filter" parameter is provided in both events for filtering intercepted HTTP(S) response headers and requests.

**Webpage Content Extraction and Modification.** By default, WebMea can extract HTML documents and http-equiv attributes of meta elements from crawled webpages. The web content is extracted by inserting and executing content scripts in the crawled webpage and further processed by the background script. The HTML documents are extracted through the JS *XMLSerializer()* method in the content script dom_extract.js. The meta http-equiv attributes are extracted through the *getAttribute()* method of the HTML Element class in the content script meta_extract.js. The dom_extract.js and meta_extract.js scripts wrap the extracted HTML documents and meta http-equiv attributes in requests and send the requests to the background script. The background script receives the requests and will further process the received web content.

*Simulation of User Interactions.* By default, WebMea scrolls down a webpage to the bottom after the webpage is completely loaded. The *scrollingElement* property of the HTML document object implements the scrolling down function in the content script page_scroll.js. Then the background script inserts and executes the page_scroll.js script in the crawled pages. After crawling a webpage, WebMea can clear the browsing history, cached data, and cookies through the remove() method of the browsingData extension API.

### 4.3    Customization of WebMea

Researchers can customize WebMea based on their web measurement requirements. Some customization examples are as follows.

*Customizing Automated Crawling.* Researchers can specify their desired number of crawled subpages in the subpage_extract.js script. They can change the crawling sequence of webpages by modifying the callback function of the *onUpdated.addListener()* method of the tabs extension API in the background script.

*Customizing HTTP(S) Traffic Interception and Modification.* Researchers can cancel HTTP(S) requests by monitoring the *onBeforeRequest* event, the *onAuthRequired* event, or the *onHeadersReceived* event of the webRequest API. They can redirect an HTTP(S) request by monitoring the *onBeforeRequest()* event and the *onHeadersReceived()* event of the webRequest API. They can intercept, add, modify, and delete HTTP(S) request headers by monitoring the *onBeforeSendHeaders()* event of the webRequest API. They can modify HTTP(S) response headers by monitoring the *onHeadersReceived()* event of the webRequest API. They can intercept failed HTTP(S) requests by monitoring the *onErrorOccurred()* event of the webRequest API. They can modify HTTP(S) response body content through the Network domain CDP API which is accessible by using the debugger extension API.

*Customizing Webpage Content Extraction and Modification.* There are two ways to modify the content of a webpage. Researchers can modify webpage content through the Page domain CDP API which is accessible by using the debugger extension API. They can also modify webpage content by using content scripts, in which the methods of the HTML document object such as *createElement()* can be used to modify web content; meanwhile, they need to let the background script insert and execute the content scripts in the webpage.

*Customizing Simulation of User Interactions.* WebMea implements scrolling down webpages by using content scripts. Researchers can implement other user interactions in the same way. For example, to simulate clicking on a webpage element such as a link or a button, researchers can implement a function to trigger a click on the identified webpage element in a content script, and then let the background script insert and execute the content script in the webpage.

*Customizing Dataset Connection.* Researchers can save web data to remote databases. One way is to save web data to a remote database supported by a

Node.js web server. At the server side, researchers need to create a database (e.g., a MongoDB database in our WebMea implementation), use the *connect()* method of Node.js to connect the web server to the database, and process the web data contained in the HTTP(S) requests that are sent from WebMea at the browser side. Based on their preferences, researchers can use other types of database systems instead of MongoDB, and can use other types of web servers instead of relying on Node.js.

## 5   Conclusion

In this paper, we studied two browser-based crawler implementation approaches in the context of the Google Chrome browser: the DevTools approach and the extension approach. We found that both approaches have the capability to satisfy primary web measurement requirements, although their APIs and the complexity of using their APIs are different. The DevTools approach is more appropriate in scenarios where page environments need to be very close to the real one. The extension approach is more appropriate in scenarios where crawlers are required to work closely on HTTP(S) requests and responses following the life cycle of requests. We expect the analysis and comparison of these two approaches can help researchers more easily choose or take the appropriate approach to construct their browser-based crawlers. Meanwhile, we built and presented WebMea as a baseline Google Chrome extension for researchers to easily start with their extension-based crawler construction. WebMea can be used or customized in different web security and privacy measurement studies, and can help make those studies more reproducible and replicable. We have used WebMea in [14] and another project for HTTP(S) traffic, webpage content, and JavaScript measurement. We provide the GitHub link to WebMea at [18].

## References

1. Abi Din, Z., Tigas, P., King, S.T., Livshits, B.: PERCIVAL: making in-browser perceptual ad blocking practical with deep learning. In: Proceedings of the USENIX Annual Technical Conference (ATC) (2020)
2. Ahmad, S.S., Dar, M.D., Zaffar, M.F., Vallina-Rodriguez, N., Nithyanand, R.: Apophanies or epiphanies? How crawlers impact our understanding of the web. In: Proceedings of the ACM Web Conference (WWW) (2020)
3. Demir, N., Große-Kampmann, M., Urban, T., Wressnegger, C., Holz, T., Pohlmann, N.: Reproducibility and replicability of web measurement studies. In: Proceedings of the ACM Web Conference (WWW) (2022)
4. Englehardt, S., Narayanan, A.: Online tracking: a 1-million-site measurement and analysis. In: Proceedings of the ACM CCS (2016)
5. Firefox DevTools (2023). https://firefox-source-docs.mozilla.org/devtools-user/
6. Firefox Extension (2023). https://developer.mozilla.org/en-US/docs/Mozilla/Add-ons/WebExtensions
7. Firefox Remote Protocol (2023). https://wiki.mozilla.org/WebDriver/RemoteProtocol

8. Google Chrome DevTools (2023). https://developer.chrome.com/docs/devtools/
9. Google Chrome DevTools Protocol (2023). https://chromedevtools.github.io/devtools-protocol/
10. Google Chrome Extension (2023). https://developer.chrome.com/docs/extensions/
11. Google Chrome Puppeteer (2023). https://developer.chrome.com/docs/puppeteer/
12. Headless Chrome Crawler (2023). https://github.com/yujiosaka/headless-chrome-crawler
13. Jueckstock, J., et al.: Towards realistic and reproducible web crawl measurements. In: Proceedings of the ACM Web Conference (WWW) (2021)
14. Ren, M., Yue, C.: Coverage and secure use analysis of content security policies via clustering. In: Proceedings of IEEE European Symposium on Security and Privacy (EuroS&P) (2023)
15. Smith, M., Snyder, P., Livshits, B., Stefan, D.: SugarCoat: programmatically generating privacy-preserving, web-compatible resource replacements for content blocking. In: Proceedings of the ACM CCS (2021)
16. TrackingObserver: a browser-based web tracking detection platform (2023). https://trackingobserver.cs.washington.edu/
17. W3C: Content Security Policy Level 3 (2023). https://www.w3.org/TR/CSP3/
18. WebMea (2023). https://github.com/mengxiaren2024/WebMea.git

# Quantifying Psychological Sophistication of Malicious Emails

Rosana Montañez Rodriguez[1], Theodore Longtchi[2], Kora Gwartney[2],
Ekzhin Ear[2], David P. Azari[3], Christopher P. Kelley[3], and Shouhuai Xu[2(✉)]

[1] Department of Computer Science, University of Texas, San Antonio, USA
[2] Department of Computer Science, University of Colorado, Colorado Springs, USA
sxu@uccs.edu
[3] Department of Behavioral Sciences and Leadership, US Air Force Academy,
Colorado Springs, USA

**Abstract.** Malicious emails (including phishing, spam, and scam) are
significant attacks. Despite numerous defenses to counter them, they
remain effective because our understanding of their psychological properties is superficial. This motivates us to investigate the psychological
sophistication, or *sophistication* for short, of malicious emails. For this
purpose, we propose an innovative framework of two pillars: *Psychological Techniques* (PTechs) and *Psychological Tactics* (PTacs). We propose
metrics and grading rules for human experts to assess the sophistication
of malicious emails through PTechs and PTacs. To demonstrate the usefulness of the framework, we conduct a case study based on 200 malicious
emails assessed by four independent graders.

**Keywords:** Malicious emails · psychological sophistication ·
psychological techniques · psychological tactics · cybersecurity metrics

## 1 Introduction

Malicious emails remain effective despite numerous defensive efforts because
existing solutions do not adequately consider psychological factors [13]. This
inspires us to introduce and investigate the notion of *psychological sophistication* of malicious emails to pave the way towards designing effective defenses.
More specifically, we ask and investigate two questions: (i) How can we quantify the psychological sophistication of malicious emails? (ii) How sophistication
varies among different categories (or types) of malicious emails in the real world?

**Our Contributions.** First, we propose an innovative and systematic framework
for quantifying the psychological sophistication, or sophistication for short, of
malicious emails. The framework deconstructs and compares the content of malicious emails through two lenses. At a low-level, we propose identifying the number of psychologically relevant textual and imagery elements in an email message,

---

R. M. Rodriguez and T. Longtchi—Equal contribution.

© The Author(s), under exclusive license to Springer Nature Switzerland AG 2023
M. Yung et al. (Eds.): SciSec 2023, LNCS 14299, pp. 319–331, 2023.
https://doi.org/10.1007/978-3-031-45933-7_19

dubbed *Psychological Techniques* (PTechs), to provide a detailed accounting of the elements employed by an attack. At a high-level, we propose assessing an attacker's overall deliberate thoughtfulness (i.e., effort) in framing malicious content to influence an email recipient, dubbed *Psychological Tactics* (PTacs), to offer insights into an attacker's effort to exploit human fallibility. Second, we demonstrate the usefulness of the framework by applying it to quantify sophistication of 200 malicious emails. This leads to useful insights, including: (i) Phishing emails are psychologically more sophisticated than spam and scam emails because phishing emails contained both higher PTech scores and higher PTac scores. (ii) Emails having low PTac scores also have low PTech scores.

**Ethical Issue.** In consultation with University of Colorado Colorado Springs Internal Review Board (IRB), this study does not need IRB approval because no subjects are part of the study and the emails are provided by a third party.

**Related Work.** Although several studies have discussed the use of psychological content in phishing emails (e.g., [2,6,8]), few studies provide a systematic approach to quantifying it. Heijden and Allodi [23] measure the presence of persuasion elements in phishing emails to identify those that are more likely to succeed. By contrast, we also consider how the email overall message presentation affects success. Nelms et al. [17] identify and categorize psychological tactics to encourage users to download malicious applications. By contrast, we consider a broader set of psychological constructs used in phishing emails. Ferreira and Lenzini [5] systematically quantify psychological content in phishing messages based on low-level psychological elements. By contrast, we incorporate these principles and several other psychological elements, while leveraging phishing emails from the Anti Phishing Working Group (APWG).

**Paper Outline.** Section 2 describes the core concepts. Section 3 presents the framework. Section 4 reports a case study. Section 5 discuss limitations of the present study; Sect. 6 concludes the paper.

## 2    Concepts

**PTech.** A PTech is a concrete (i.e., quantifiable) textual or imagery element that encourages individuals to comply with a malicious email. The following 13 PTechs have been identified in the literature include [13,16]. (1) *Urgency*: The use of textual elements (e.g., "acting now") to trigger a recipient's immediate action [4,24]. (2) *Visual Deception*: The use of visual elements (e.g., logos) or "similar" characters in URL (e.g., replacing'vv' with'w') to project trust [15]. (3) *Incentive and Motivator*: The use of textual elements, such as "free stuff" (incentive) or "help others", to incentivize or motivate a recipient to take action [3,16]. (4) *Persuasion*: The use of textual elements related to Cialdini's principles (e.g., "C-Suite titles," "last chance," or "expert opinion") to encourage a recipient to encourage a behavior [5,17]. (5) *Quid-Pro-Quo*: The use of textual elements (e.g., "Pay an upfront fee") to ask a recipient for a favor in exchange for a bigger reward [22]. (6) *Foot-in-the-Door*: The use of textual elements (e.g., "from our last email/ conversation..." ) to obtain compliance from a recipient via gradually increasing demands [7]. (7) *Trusted Relationship*: The exploitation of an established third-party relationship of trust with the recipient by using textual elements like "John told me about you" to convince a recipient to act [2]. (8) *Impersonation*: The use of a false persona to gain the trust of a recipient by using elements like "I'm billionaire Warren Buffet" [2,5]. (9) *Contextualization*: Referencing current event by using textual elements like "the Pandemic" or "War in Ukraine" [8,15]. (10) *Pretexting*: Providing a motive to establish contact with a recipient by using textual elements like "I am recruiter for XYZ company" [1,8]. (11) *Personalization*: Addressing a recipient using detailed personal information in textual elements such as "Dear John" or "Your credit card ending in..." [11,15]. (12) *Attention Grabbing*: The use of graphical/auditory elements to draw attention to textual elements such as highlighted text, brightly colored buttons, or extra large fonts [6,17]. (13) *Affection trust*: Developing an effective relationship to extort a recipient by using textual elements like "My child is sick and I have no money to pay the treatment" [14].

**PTac.** This is a new concept introduced in this paper. A PTac aims to measure the attacker's effort at crafting and framing an email effectively to prompt a recipient's action. There are 7 PTacs. (1) *Familiarity*: It reflects the attempt of an attacker to engender a positive (and therefore trusting) association with a recipient. Emails of high familiarity may impersonate specific people (e.g., co-workers, bosses, family members, close friends) [1,14]. (2) *Immediacy*: It is an amplifier which uses time as a mechanism to short-cut recipient skepticism or scrutiny for any desired action, for example, by suggesting that promptness, swiftness, or a quick reaction is required [15,17]. (3) *Reward*: It is a clear exchange of something (physical or social) valuable for a recipient. Rewards are often presented as a tangible good (e.g., money) in exchange for action but can also be an offer to improve social standing (e.g., power, authority, prestige) [8,13]. (4) *Threat of Loss*: It is an appeal to a recipient's desire to maintain their current status, prevent a loss (e.g., opportunity) or injury (e.g., damage, pain), or avoid the risk of having something stolen. It has been hypothesized to be more impactful than potential of gain (e.g., reward) [5,8,22]. (5) *Threat to Identity*: It is a recipient's desire to maintain a positive, socially valuable reputation [14,22]. (6) *Claim to Legitimate Authority*: It is intended to leverage respect for legitimate power. The attacker may assume a position of technical expertise or a valuable institutional role, or hold a traditionally respected office [5,22]. (7) *Fit & Form*: It mirrors the expected composition style of an authentic message. An attacker often exploits commonly expected written or visual display format to resonate with the email's apparent sender and purpose [8,20].

## 3 Framework

The framework consists of six components: (i) selecting PTechs and PTacs for assessment; (ii) defining metrics to quantify sophistication of malicious emails;

(iii) designing grading rules and calibration process to guide the grading of malicious email sophistication; (iv) preparing a dataset of malicious emails for expert graders to assess; (v) grading emails in the dataset by expert graders; and (vi) analyzing outcome of the grading process.

## 3.1  Selecting PTechs and PTacs

We propose selecting the PTechs that (i) are known to be used in malicious emails based on research evidence and (ii) require a one-time interaction to be effective. This selection criterion is flexible enough to accommodate future understanding and knowledge (e.g., when new PTechs are discovered in the future). Similarly, we propose selecting PTacs that (i) are known to be used in malicious emails based on research evidence, (ii) are independent of one another, and (iii) reflect the holistic effort of an attacker. Suppose, according to the respective selection criteria, $\ell$ PTechs are selected, denoted by $\{PTech_1, \ldots, PTech_\ell\}$, and $m$ PTacs are selected, denoted by $\{PTac_1, \ldots, PTac_m\}$.

## 3.2  Defining Sophistication Metrics

**Metrics for Measuring PTechs.** Consider a malicious email and a set of $\ell$ PTechs denoted by $\{PTech_1, \ldots, PTech_\ell\}$ as described above. For $PTech_i$ where $1 \leq i \leq \ell$, we propose counting the number of elements with respect to $PTech_i$, leading to an integer score $s'_i$. Then, the sophistication of the malicious email through the lens of the $\ell$ PTechs can be defined as, $s' = \frac{1}{\ell} \sum_{i=1}^{\ell} s'_i$. Since ground truth $s'_i$ is difficult to obtain, we propose to approximate it by using a number of $n$ graders (or evaluators) to count the elements concerning $PTech_i$ while assuring that the graders can count the elements as consistently as possible. For a malicious email, let $s_{i,j}$ denote the count of elements in the email by grader $j$ with respect to $PTech_i$, where $1 \leq j \leq n$ and $1 \leq i \leq \ell$. Then, the sophistication of the email concerning $PTech_i$ can be defined as,

$$S_i = \frac{1}{n} \sum_{j=1}^{n} s_{i,j}. \tag{1}$$

Given $S_i$ for $1 \leq i \leq \ell$, we propose defining the sophistication of the email with respect to the $\ell$ PTechs, denoted by $S_{PTech}$, as:

$$S_{PTech} = \frac{1}{\ell} \sum_{i=1}^{\ell} S_i. \tag{2}$$

**Metrics for Measuring PTacs.** Consider a malicious email and a set of $m$ PTacs denoted by $\{PTac_1, \ldots, PTac_m\}$. Since the ground-truth sophistication reflected by $PTac_i$ is hard to obtain, we propose assessing $PTac_i$ using a rating scale ranging from 1 to $\beta$ (e.g., $\beta = 5$ in our case study), also by a panel of $n$ independent graders, where $p_{i,j}$ denotes the assessment of grader $j$ with respect to $PTac_i$ in a message, where $1 \le j \le n$ and $1 \le i \le m$. The final assessed value of $PTac_i$ can be defined as

$$P_i = \frac{1}{n} \sum_{j=1}^{n} p_{i,j}. \tag{3}$$

The overall PTac-based sophistication of an email can be defined as:

$$S_{PTac} = \frac{1}{m} \sum_{i=1}^{m} P_i/\beta, \tag{4}$$

where $P_i/\beta$ reflects the degree of $PTac_i$ in a malicious email. Now we are ready to define the sophistication of malicious emails.

**Definition 1 (sophistication of malicious email).** *The sophistication of a malicious email is measured as a two-dimensional vector $(S_{PTech}, S_{PTac})$, where $S_{PTech}$ is defined in Eq. (2) and $S_{PTac}$ is defined in Eq. (4).*

Note that Definition 1 operates in the ideal world where every score given by ever grader will be incorporated. In the real world, some grade by some grader may be outlier, which may need to be excluded according to some well-established inclusion criteria. This means that Definition 1, or Eq. (1) and Eq. (3), may need to be amended to accommodate such realistic situations. Specifically, when coping with Eq. (1), which computes the average score $S_i$ of $PTech_i$ by the $n$ graders, we may encounter, for example, grader $j'$ ($1 \le j' \le n$) gives an outlier score $s_{i,j'}$. In this case, $s_{i,j'}$ may be excluded when computing the average score. As a result, Eq. (1) becomes for $1 \le i \le \ell$:

$$S_i = \frac{1}{n-1} \left( \sum_{j=1}^{n} s_{i,j} - s_{i,j'} \right). \tag{5}$$

When there are multiple outliers with respect to $PTech_i$, they all can be removed in the same fashion. Similarly, suppose grader $j^*$ ($1 \le j^* \le n$) gives an outlier score $p_{i,j^*}$ with respect to $PTac_i$. Then, $p_{i,j^*}$ may be excluded when computing the average score, meaning that Eq. (3) now becomes for $1 \le i \le m$:

$$P_i = \frac{1}{n-1} \left( \sum_{j=1}^{n} p_{i,j} - p_{i,j^*} \right). \tag{6}$$

Similarly, when there are multiple outliers with respect to $PTac_i$, they all can be removed in the same fashion. With the amended Eqs. (5) and (6), Eqs. (2) and (4), and thus Definition 1, remain valid.

### 3.3 Designing Grading Rules

To measure PTechs and PTacs, we propose that each grader manually counts the number of psychological elements of each PTech and each PTac exhibited in an email. Given that the interpretation of "element" relies on one's domain expertise, we design *grading rules* to reduce subjectivity during the grading process. To guide the development of grading rules, we propose the following: (i) Initial rules are designed by multiple experts. (ii) These initial rules may be reconciled into a unified set of rules to resolve any discrepancies in the initial rules. (iii) The resulting rules are tested by a group of graders using sample data so that discord or ambiguity in the grading rules that arise during the testing can be documented and addressed. (iv) The grading rules may be further revised to mitigate potential inconsistencies or discrepancies in the grading process. The preceding (iii)-(iv) may be repeated until satisfactory consistency is achieved among graders guided by grading results.

### 3.4   Preparing the Data

Several issues must be addressed when preparing data, including collection and preprocessing. First, to ensure dataset quality (i.e., emails are suitable for quantifying sophistication), we must determine if the emails are malicious. Second, given a set of malicious emails, we must ensure that each email content is rendered similarly on different machines and platforms from a visual point of view. Third, we must ensure that the data preparation process does not cause damage to the research environment. Fourth, malicious emails may contain broken links or missing images needed to complete an email for sophistication assessment. In these cases, we must reconstruct an email by adding the missing links or images.

### 3.5   Calibration and Grading

Calibration mitigates human (including expert) subjectivity in grading. With grading rules on hand, the graders consistently learn how to apply the rules and practice grading using sample emails. For this purpose, we propose the calibration process highlighted in Fig. 1.

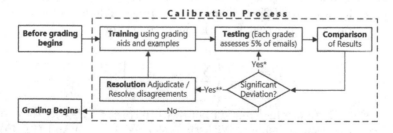

**Fig. 1.** The calibration process includes presenting initial grading rules to graders, training, and testing. Results are compared according to a defined threshold, such as a Krippendorff's Alpha or Kalpha or simply $\alpha$. A value lower than the threshold may require training before testing. The calibration process is iterative, and the initial grading rules may be refined when reconciling the discrepancy among graders. Grading begins at the end of the calibration process.

The process has 4 steps. (i) Training: Graders learn how to grade emails using the initial grading rules and grading aid, where the latter demonstrates the application of the former. Graders can ask questions (e.g., what would count as an "element" for a specific PTech), resolve disagreements, and collectively build a shared understanding of the assessment methods through consensus. (ii) Testing: Each grader evaluates an email sample (e.g., 5% of a data corpus) from the study dataset. This sample is excluded from the study. (iii) Comparison: Results obtained from the Testing step are compared for agreement. We propose to use a reliable method to measure the agreement between the graders. For example, *Krippendorff's* (i.e., Kalpha or $\alpha$) [10] is a reliability coefficient that measures the agreement among raters. Kalpha supports categorical, ordinal, and interval data, and it is robust in the light of any potential missing data. A Kalpha ($\alpha$) with $0.667 \leq \alpha < 0.8$ is considered an acceptable value, meaning that the data is statistically reliable to draw conclusions. Kalpha $\alpha = 1$ means a perfect agreement among graders. When $\alpha < 0.667$, resolution is necessary and conducted in the next step. (iv) Resolution: when $\alpha < 0.667$, we propose using a consensus-building technique to reach an agreement, such as the Delphi standard consensus technique [9]. Steps (iii)-(iv) may be repeated until consistent grading is achieved. Once calibration is completed, the graders can start to grade the rest emails in the dataset (i.e., those not used in the calibration process). After the grading, outliers grades are excluded with respect to each PTech and each PTac, as per Eqs. (5) and (6) based on a predefined inclusion criteria.

### 3.6 Analysis

The analysis may be geared towards answering research questions, which are often proposed based on researchers' insights from some unique perspective. Examples of research questions include: How sophistication varies among different categories of malicious emails in the real world? Further research questions can include: Which PTechs and PTacs are most commonly employed in real-world malicious emails?

## 4   Case Study

### 4.1   Selecting PTechs and PTacs

The PTech selection criteria described in the framework prompt us to select the following 8 PTechs based on the selection criteria described in Sect. 3.1: (i) *urgency*, (ii) *incentives and motivators*, (iii) *attention grabbing*, (iv) *personalization*, (v) contextualization, (vi) *persuasion*, (vii) *impersonation*, and (viii) *visual deception*. Moreover, the PTac selection criteria described in the framework prompt us to select all the 7 PTacs presented in Sect. 2.

## 4.2   Instantiating Sophistication Metrics

As descried in the framework, the effect of each PTech can be quantified by counting the occurrence of psychological elements associated with a PTech in the email. However, the PTac metric scale $\beta$ described in the framework needs to be instantiated to a specific scale value. For this purpose, we propose using the Likert scale $[0, 5]$ (i.e., $\beta = 5$ in the terminology of the framework) which is scaling method commonly used in psychological studies [12]: '0' for *no measurable* application of a PTac (i.e., the attacker does not employ any PTac); '1' for *minimal* application of a PTac (i.e., the attacker does consider the PTac but neither applied it clearly nor consistently); '2' for *light* application of a PTac (i.e., the attacker considers the PTac, but with inconsistency, confusion, or lapses/errors in their approach); '3' for a *moderate* application of a PTac (i.e., the attacker clearly applies the PTac but may still have inconsistencies in their approach); '4' for a *significant* application of a PTac (i.e., the attacker clearly and consistently applies the PTac with minimal errors or lapses); '5' for an *extraordinary* application of a PTac (i.e., the attacker expertly and diligently crafts their message to apply this PTac in a cohesive and thoughtful way). Note that the Likert scale or $\beta = 5$ is just a specific choice, but there could be other choices of interest.

## 4.3   Designing Grading Rules

To ensure consistency of grading, we develop an in-depth grading aid. The aid includes detailed definitions and real-world emails explaining the grading rationale for each PTech and PTac. We also develop a quick reference of psychological element (i.e., key terms) associated with a particular PTech (Table 1) and a set of examples of emails that are graded with respect to the PTechs and PTacs. Figure 2 shows a specific example.

**Table 1.** A sample of PTech grading rules, which provide graders with an extensive list of examples for each PTech.

| PTech | Examples elements from Emails |
|---|---|
| *Urgency* | "call me now" / "Last chance to save your social life" |
| *Visual Deception* | PayPal logo, IRS logo / Replacing 'fbi.gov' with 'fbi.gov.net' |
| *Incentives & Motivators* | "Your refund notice" / "looking for a part-time assistant,(...) 3 h a week, (...)$400 per week" |
| *Persuasion* | **Commitment** - "We are grateful for you past generosity" |
| *Impersonation* | "Yours sincerely, Warren Buffet" / Phone/Fax number |
| *Contextualization* | "Your UW.edu account..." / "Emergency Covid-19 tax relief" |
| *Personalization* | "Hi Wendy" / "Important message intended for John Doe" |
| *Att. Grabbing* | "CLICK HERE" / "Safety Measures.pdf" |

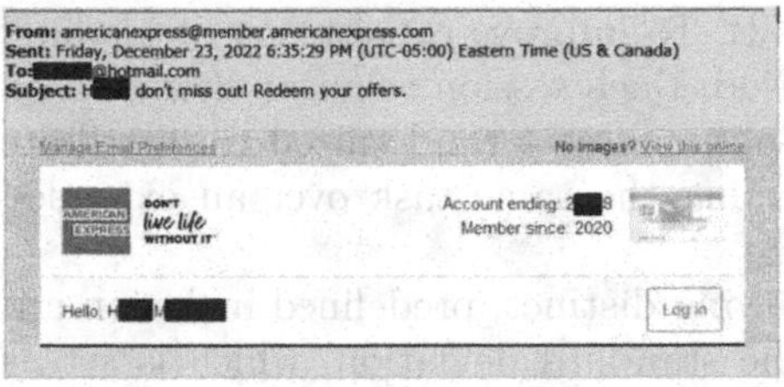

(a) Email screenshot

| PT | Count | | Framing Construct | Value |
|---|---|---|---|---|
| Urgency | 0 | | Familiarity | 4 |
| Visual Deception | 0 | | Immediacy | 1 |
| Incentives & Motivators | 1 | | Reward | 2 |
| Persuasion | 1 | | Threat of Loss | 0 |
| Impersonation | 1 | | Threat to Identity | 0 |
| Contextualization | 0 | | Claim to legitimate | 4 |
| Personalization | 3 | | Authority | |
| Attention Grabbing | 5 | | Fit and Form | 4 |

(b) Grading outcome of email in Fig.2a

**Fig. 2.** Example of grading outcome, where a grader evaluates a screenshot of email in Fig. 2a(redacted for publication, but not for grading). Figure 2b is the grade of the email in Fig. 2a with respect to each PTech and PTac.

## 4.4 Preparing Data

We prepare a dataset of 200 randomly selected emails using the APWG Reported Phishing module API. To increase the chance of collecting true-positive malicious emails, we select emails submitted by US-CERT (Computer Emergency Response Team), a reputable source. The collected email sample (200 emails) consists of 55.0% phishing (110 emails), 31.5% scam (63 emails), and 13.5% spam (27 emails). Emails are selected using random dates between September 1, 2021, and August 31, 2022. The select emails are reconstructed by appending the raw email header and body into a .eml file. Emails are restored with missing elements (e.g., broken images) and are sanitized by removing embedded warnings and email duplicates. An email client (i.e., reader software) is used to display emails. Email screenshots are used to ensure a consistent display of an email content during the grading. Emails are inventoried and categorized as follows: (i) phishing emails, which are the ones that require a one-time interaction for victimization and include a link or an attachment;(ii) scam emails, which are the ones that require multiple interactions for victimization via phone call or email exchange, or request personal information, while noting that scam emails do not include links or attachments; and (iii) spam emails, which are the ones that are non-malicious and do not obscure information, but usually intended to sell a product or service. To further differentiate between email categories, we examine email links using the ScamPredictor algorithm developed by ScamWatcher [21]. The algorithm is a machine learning classifier based on website characteristics known to be indicators of malicious sites [19].

## 4.5 Calibration and Grading

Each email is graded by four cybersecurity PhD students.They all conduct a calibration exercise as depicted in Fig. 1.

Once the calibration exercise is completed, grading is conducted as follows. Emails are presented as pop-up windows in a survey developed on the Qualtrics platform. The evaluation is split into two self-paced sessions to minimize grader's fatigue. Each session consists of 100 emails. To improve consistency of grading, each session is completed within 24 h period. Each session requires about 5 h on average. Within each session, the order of emails are randomized to distribute grading variations introduced by performing the same task over an extended period of time.

Grades are examined for outliers following distinct, predefined inclusion criteria for PTechs and PTacs based on the standard deviation, which is set as 1.5. Outliers are addressed using the procedure outlined in Sect. 3.5. Overall, we exclude 170 (5.67%) out of 1,600 PTech ratings and 216 (7.2%) out of 1,400 for PTacs ratings, with 153 of the 200 emails having at least one outlier grade removed. The Kalpha for this study is 0.822 for PTech and 0.768 for PTacs after outliers removal.

## 4.6 Analysis

To understand how PTechs and PTacs may differ among the three categories of malicious emails, we use the Z-Score method [18] to normalize the PTech scores to a scale comparable to the PTac scores. Figure 3 shows that phishing emails have higher normalized PTech and PTac scores than the other two categories of malicious emails, respectively. This leads to:

**Insight 1.** *Phishing emails are psychologically more sophisticated than spam and scam emails from the points of view of both PTech and PTac.*

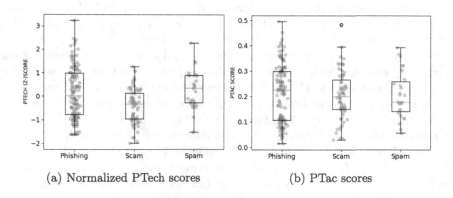

(a) Normalized PTech scores                    (b) PTac scores

**Fig. 3.** Boxplots of the normalized PTech scores and the original PTac scores.

To further explore the relationship between PTech-incurred sophistication and PTac-incurred sophistication, we look at the size of the intersection set of the two sets of emails corresponding to PTech and PTac, respectively. At Q1, the size of the intersection set is: 17.27% for phishing, 22.22% for scam, and 51.85% for spam. At Q3, the size of the intersection set is: 15.45% for phishing, 3.17% for scam, and 7.40% for spam. This leads to:

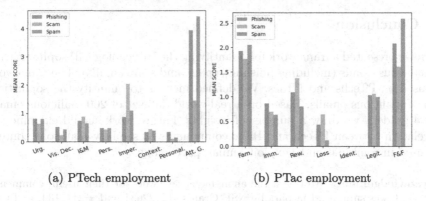

(a) PTech employment                    (b) PTac employment

**Fig. 4.** Comparing the mean score (the $y$-axis) of each PTech and PTac

**Insight 2.** *Less sophisticated emails from the PTac perspective are also less sophisticated from the PTech perspective.*

Figure 4 plots the employment of PTechs and PTacs. From this, we draw:

**Insight 3.** Attention Grabbing *is the most widely employed PTech, and* Fit & Form *and* Familiarity *are the two most widely employed PTac.*

## 5   Limitations

First, the framework has three limitations: (i) The framework reflects our understanding of the factors that can reflect the psychological sophistication of malicious emails, namely PTechs and PTacs. There may be other psychological factors that need to be considered, which can be accommodated by extending our framework. (ii) The selection criteria we propose may not be perfect, meaning that the select PTechs and PTacs may not be complete or systematic enough. Fortunately, the framework can be easily extended to accommodate other PTechs and/or PTacs of interest. (iii) The grading rules may need to be refined, to more consistently ensure high levels of concurrence in human assessment of email content. Second, the dataset has five limitations. (i) The dataset may not be representative because we only collected and used emails from APWG even though it is arguably the most reputable source in the world. (ii) The dataset is small as we only analyzed 200 emails. (iii) The inclusion criteria need improvement to provide a mathematically robust approach to address and resolve outlier assessments. (iv) We admit the potential issue of 'informed' graders as the graders are also the ones that design and revise the grading rules. This may affect the validity of the experimental results. (v) While the framework can accommodate any reasonable definitions of PTech and PTac, it would be ideal to assure that the PTechs and PTacs are independent.

# 6  Conclusion

We have presented a framework for quantifying the (psychological) sophistication of malicious emails (including phishing, spam, and scam emails). The framework is based on PTechs and PTacs. We defined metrics to quantify the sophistication of malicious emails. Based on a real-world dataset of 200 malicious emails and 4 graders, we draw a number of insights. Future work should examine the correlation between PTech and PTac components to see how malicious content are used in combination to exploit human psychology.

**Acknowledgement.** We thank the anonymous reviewers for their useful comments. This work was supported in part by NSF Grants #2122631 and #2115134, and Colorado State Bill 18-086. Approved for Public Release; Distribution Unlimited. Public Release Case Number 23-1373. The first author is also affiliated with The MITRE Corporation, which is provided for identification purposes only and is not intended to convey or imply MITRE's concurrence with, or support for, the positions, opinions, or viewpoints expressed by the authors.

# References

1. Al-Hamar, M., Dawson, R., Guan, L.: A culture of trust threatens security and privacy in Qatar. In: 2010 10th IEEE International Conference on Computer and Information Technology, pp. 991–995. IEEE (2010)
2. Allodi, L., Chotza, T., Panina, E., Zannone, N.: The need for new antiphishing measures against spear-phishing attacks. IEEE Secur. Priv. **18**(2), 23–34 (2019)
3. Beckmann, J., Heckhausen, H.: Motivation as a function of expectancy and incentive. In: Heckhausen, J., Heckhausen, H. (eds.) Motivation and Action, pp. 163–220. Springer, Cham (2018). https://doi.org/10.1007/978-3-319-65094-4_5
4. Chowdhury, N.H., Adam, M.T., Skinner, G.: The impact of time pressure on cybersecurity behaviour: a systematic literature review. Behav. Inf. Technol. **38**(12), 1290–1308 (2019)
5. Ferreira, A., Lenzini, G.: An analysis of social engineering principles in effective phishing. In: Workshop on Socio-Technical Aspects in Security and Trust (2015)
6. Flores, W.R., Holm, H., Nohlberg, M., Ekstedt, M.: Investigating personal determinants of phishing and the effect of national culture. Inf. Comput. Secur. **23**, 178–199 (2015)
7. Freedman, J.L., Fraser, S.C.: Compliance without pressure: the foot-in-the-door technique. J. Pers. Soc. Psychol. **4**(2), 195 (1966)
8. Goel, S., Williams, K., Dincelli, E.: Got phished? Internet security and human vulnerability. J. Assoc. Inf. Syst. **18**(1), 2 (2017)
9. Grime, M.M., Wright, G.: Delphi method. Wiley StatsRef Stat. Ref. Online **1**, 16 (2016)
10. Gwet, K.L.: On the krippendorff's alpha coefficient. Manuscript submitted for publication (2011). Accessed 2 Oct 2011
11. Jagatic, T.N., Johnson, N.A., Jakobsson, M., Menczer, F.: Social phishing. Commun. ACM **50**(10), 94–100 (2007)
12. Jebb, A.T., Ng, V., Tay, L.: A review of key Likert scale development advances: 1995–2019. Front. Psychol. **12**, 637547 (2021)

13. Longtchi, T., Rodriguez, R.M., Al-Shawaf, L., Atyabi, A., Xu, S.: SoK: why have defenses against social engineering attacks achieved limited success? arXiv preprint arXiv:2203.08302 (2022)

14. Montañez, R., Atyabi, A., Xu, S.: Social engineering attacks and defenses in the physical world vs. cyberspace: a contrast study. In: Cybersecurity and Cognitive Science, pp. 3–41. Elsevier (2022)

15. Montañez, R., Golob, E., Xu, S.: Human cognition through the lens of social engineering cyberattacks. Front. Psychol. **11**, 1755 (2020)

16. Montañez Rodriguez, R., Xu, S.: Cyber social engineering kill chain. In: Su, C., Sakurai, K., Liu, F. (eds.) SciSec 2022. LNCS, vol. 13580, pp. 487–504. Springer, Cham (2022). https://doi.org/10.1007/978-3-031-17551-0_32

17. Nelms, T., Perdisci, R., Antonakakis, M., Ahamad, M.: Towards measuring and mitigating social engineering software download attacks. In: 25th USENIX Security Symposium, pp. 773–789. USENIX Association, Austin, TX (2016)

18. Nield, T.: Essential Math for Data Science. O'Reilly Media Inc, Sebastopol (2022)

19. Pritom, M., Schweitzer, K., Bateman, R., Xu, M., Xu, S.: Data-driven characterization and Detection of COVID-19 Themed Malicious Websites. In: IEEE ISI (2020)

20. Rajivan, P., Gonzalez, C.: Creative persuasion: a study on adversarial behaviors and strategies in phishing attacks. Front. Psychol. **9**, 135 (2018)

21. SAS, H.: Scamdoc.com. https://www.scamdoc.com/. Accessed 04 Nov 2023

22. Stajano, F., Wilson, P.: Understanding scam victims: seven principles for systems security. Commun. ACM **54**(3), 70–75 (2011)

23. Van Der Heijden, A., Allodi, L.: Cognitive triaging of phishing attacks. In: 28th USENIX Security Symposium 2019, pp. 1309–1326 (2019)

24. Vishwanath, A., Herath, T., Chen, R., Wang, J., Rao, H.R.: Why do people get phished? Decis. Support Syst. **51**(3), 576–586 (2011)

# SVFL: Secure Vertical Federated Learning on Linear Models

Kaifeng Luo[1], Zhenfu Cao[1,2(✉)], Jiachen Shen[1(✉)], and Xiaolei Dong[1,2(✉)]

[1] Shanghai Key Laboratory of Trustworthy Computing East China Normal University, Shanghai 200062, China
{zfcao,jcshen,dongxiaolei}@sei.ecnu.edu.cn
[2] Research Center for Basic Theories of Intelligent Computing, Research Institute of Basic Theories, Zhejiang Lab, Hangzhou 311121, China

**Abstract.** Federated learning (FL) is a popular technique that enables multiple parties to train a machine learning model collaboratively without disclosing the raw data to each other. A vertically partitioned federated learning configuration is applicable in a variety of real-world scenarios. In this configuration, a comprehensive feature collection is established only when all parties' datasets are merged and only one party has access to the labels. Existing vertical federated learning strategies for linear models are not very practical, since they involve either a trusted third-party authority (TPA) or heavy communication overheads. To address this issue, this paper proposes SVFL, a secure vertical federated learning framework on linear models, which is based on the Verifiable Inner-Product Computation (VIP) protocol. SVFL enables the secure and private training of linear models, as well as the validation of a malicious server's computation. In addition, it decreases the number of communication rounds to 3 and is resistant to collusion attacks. Experiments are done on a variety of real-world datasets from the UCI ML repository, and the results demonstrate that SVFL achieves comparable accuracy to conventional linear models.

**Keywords:** Linear models · Vertical federated learning · Privacy-preserving

## 1 Introduction

### 1.1 Background

Machine learning (ML) has become a critical component in numerous technological advancements and applications in recent years. As ML models evolve, becoming more sophisticated and capable, they necessitate massive amounts of data to achieve optimal levels of accuracy and efficiency. This escalating demand for data has given rise to the phenomenon of isolated data islands, where valuable information is sequestered within organizations, industries, or regions due to privacy concerns, regulations, or competitive interests. Furthermore, in many

© The Author(s), under exclusive license to Springer Nature Switzerland AG 2023
M. Yung et al. (Eds.): SciSec 2023, LNCS 14299, pp. 332–344, 2023.
https://doi.org/10.1007/978-3-031-45933-7_20

instances, the training data employed for ML contain highly sensitive information. The proliferation of recent data breach incidents has heightened privacy concerns associated with the widespread collection and use of personal data. Simultaneously, contemporary regulations such as the European General Data Protection Regulation (GDPR) [10], California Consumer Privacy Act, and the Cybersecurity Law of China, among others, impose additional constraints on the accessibility and application of privacy-sensitive data. This context accentuates the need to develop privacy-preserving ML approaches, allowing organizations to exploit the potential of data while adhering to ethical and legal requirements.

Federated learning (FL) has rapidly emerged as a compelling ML paradigm [16], enabling collaborative model training among multiple parties under the supervision of an aggregator without necessitating the sharing of raw training data. Vertical federated learning (VFL) pertains to collaborative scenarios where individual parties lack access to the full set of features and labels and, therefore, cannot train a model locally using their own datasets. Numerous laudable solutions have been proposed to tackle privacy leakage in the federated training process and have been applied to train ML models in VFL.

Table 1 presents a summary of representative privacy-preserving approaches employed by VFL for linear models, in which methods exhibit several drawbacks. To address these limitations, we propose SVFL, an efficient and practical privacy-preserving framework for VFL of linear models. SVFL only requires communication of 2 rounds of party-to-server (p2s) and 1 round of server-to-server (s2s), eliminating the need for a TPA or party-to-party communication among parties . SVFL enables data owners to cooperatively train a global model and maintain a private partial model of their own. SVFL tolerates up to one malicious server and ensures adherence to agreements. Furthermore, SVFL allows nearly all types of collusive adversaries.

**Table 1.** Comprehensive Comparison of VFL Solution on Linear Models

| | p2p | p2s | s2s | Approach | Gradient | Server |
|---|---|---|---|---|---|---|
| Gascón et al. [11] | ✓ | 2 | 0 | Garbled circuits | Encrypted | - |
| Hardy et al. [12] | ✓ | 2 | 0 | Partial HE | Plaintext | semi-honest |
| Chen et al. [5] | ✗ | 4 | 0 | DP | Approximate | semi-honest |
| Xu et al. [19] | ✗ | 2 | 4 | FE | Plaintext | semi-honest+TPA |
| SVFL (our work) | ✗ | 2 | 1 | Random Mask | Encryped | semi-honest+malicious |

## 1.2 Related Work

FL is promising approach for training ML models on distributed data sets across multiple devices while mitigating data leakage. The majority of the existing work primarily focuses on horizontal federated learning, with a particular emphasis on privacy preservation, security [13], and system robustness [15].

Hardy et al. [12] introduced a two-party VFL scheme for training a privacy-preserving logistic regression model. By applying Taylor approximation to the loss and gradient functions, they were able to utilize the Paillier cryptosystem [7] for privacy-preserving computations. However, the secure aggregation process based on partially additive homomorphic encryption is less efficient in terms of communication and computation costs compared to the approach proposed in this paper. Other linear regression implementations, such as [3,17], support multi-party scenarios but are restricted to datasets with a maximum of 20 features. Meanwhile, [11] addressed the issues of speed and accuracy by employing a novel Conjugate Gradient Descent (CGD) algorithm based on garbled circuits, which resulted in substantial party-to-party communications. These protocols necessitate a considerable number of communication rounds among parties, hindering their deployment in systems with poor connectivity or where communication is restricted to specific entities. In contrast, [5] utilized differential privacy to train a logistic regression model, sacrificing utility due to reliance on approximation-based secure computation or noise perturbation.

The most similar approach to SVFL is [19], which employs functional encryption [1] in VFL. However, the key distinctions between the two methods are as follows: (i) SVFL circumvents the requirement for a TPA and decreases the rounds of s2s communication; (ii) SVFL permits most collusion attacks, whereas Xu et al. [19] assumes non-collusive *Aggregator* and *Parties*; and (iii) SVFL trains private partial models, whereas [19] reveals the global model to the server.

## 2    Preliminaries

### 2.1    Vertical Federated Learning

Vertical Federated Learning represents a powerful approach for constructing ML models to address numerous real-world problems, particularly in situations where a single entity does not possess comprehensive access to all the training features or labels. In VFL scenarios, the dataset $\mathcal{D}$ is distributed amongst multiple parties. Privacy-preserving VFL is, therefore, contingent upon the guarantee that the dataset of each party remains confidential.

***Notation:*** Let $\mathcal{P} = \{p_i\}_{i \in [m]}$ be the set of $m$ parties in VFL setting. Let $\mathcal{D}^{n \times d}$ be the dataset of $n$ records with $d$ features across the party set $\mathcal{P}$, where $X \in \mathbb{R}^d$ represents the feature set and $Y \in \mathbb{R}$ denotes the labels. Each party $p_i$ owns a private sub-dataset $\mathcal{D}_{p_i}$ with $d_i$ features, where $\mathcal{D}_{p_1} = Y \| X_{p_1}$ and $\mathcal{D}_{p_i} = X_{p_i}$ when $p_i$ is a passive party. We assume that except for the identifier features, there are no overlapping training features between any two of the private datasets, and these datasets can be spliced together into a "global" dataset $\mathcal{D}$, which also means $d = \sum_{p_i \in P} d_i$. We consider that only one party owns the labels $Y$, which is called the *active party*, while the others are called *passive parties*. We simply assign $p_1$ as the *active party*.

### 2.2    Key Agreement

A classical Key Agreement scheme is the Diffie-Hellman key agreement [8]. Let $\mathcal{P} = \{p_1, p_2, \ldots, p_n\}$ represent the set of $n$ clients, with each client being assigned a unique index $i \in [1, n]$. Consider $\mathbb{G}$ as a cyclic group with prime order $p$ and generator $g$. Let $hash : \{0, 1\}^* \to \mathbb{G}_p$ denote a cryptographic hash function that maps strings of arbitrary lengths to integers in $\mathbb{Z}_p$. Each client $p_i$ generates a private key $SK_i \in \mathbb{Z}_p$ and a public key $PK_i = g^{SK_i} \in \mathbb{G}$, subsequently uploading $PK_i$ to the cloud server. Each client $p_i$ then retrieves other clients' public keys and computes $AK_{i,j} = hash((PK_j)^{SK_i})$ for $p_i, p_j \in \mathcal{P}$ and $i \neq j$.

## 3 System Overview

This section delineates an overview of the SVFL framework, comprising two primary components: initialization and model training. The SVFL framework features three distinct entities: (i) a server *Host* responsible for the generation and maintenance of public parameters, (ii) a server *Aggregator* tasked with the verifiable gradient computation of partial models, and (iii) a collection of parties $\mathcal{P}$, regarded as data owners.

### 3.1 Threat Model

SVFL is designed to furnish an integrated and verifiable privacy-preserving VFL framework. In contrast to previous works, SVFL permits the utilization of two non-colluding entities to implement the Cloud Service Provider (CSP), thereby enhancing the framework's feasibility. In our threat model, we consider *Host* and data owners $P$ as honest-but-curious and *Aggregator* as malicious. Our security approaches ensure the *Aggregator* completes computation correctly and verifiably without gaining knowledge.

We assume a limited number of parties may collude to infer others' private information. In SVFL, the number of such parties is bounded by $m - 2$ out of $m$ parties. We also assume parties may collude with either *Aggregator* or *Host*, while *Host* does not collude with *Aggregator* or the active party.

### 3.2 The Proposed Framework

Algorithm 1 outlines the steps followed by SVFL. The procedure can be divided into two phases: Initialization and Model Training. In the Initialization phase, the *Host* initializes cryptosystems, distributes public keys and parameters, and engages in key agreement with *Parties*. Different from passive *Parties*, the active party publishes extra encrypted messages for global model loss computation. In the Model Training phase, each party sends an encrypted representation of its local model to the *Aggregator* at the start of each epoch. The *Aggregator* completes all computation and sends encrypted partial gradients to corresponding *Parties* and encrypted loss to the active *Party*.

## 4 SVFL

Since SVFL can be applied to nearly all linear ML models, we demonstrate how SVFL operates on Linear Regression, a classic ML model, as an example. A linear

---

**Algorithm 1:** SVFL Framework

**Input**: maxEpochs, fearture number, security parameter $\kappa$
**Initialization**

1    $Host.Initialize$;
2    **foreach** $p_i \in P$ **do** $p_i.Initialize$

   **foreach** $epoch \in [maxEpochs]$ **do**
     **Model Training**
3       **foreach** $p_i \in P$ **do** $p_i.ModelCalculation$
4       $Host\&Aggregator.GradientCalculation$;
5       **foreach** $p_i \in P$ **do** $p_i.ModelUpdate$

---

regression model computes gradients of its object using a prediction function that can be written as $f(x, \omega) = \omega^T x$, where $x$ and $\omega$ represent the feature vector and the model weights vector, respectively. We consider the problem of solving a linear regression problem when the transcribed data ($X \in R^{n \times d}, Y \in R^n$) is distributed vertically (across columns) among multiple parties.

## 4.1 Verifiable Inner Product Computation

Before delving into SVFL, we introduce a scheme to achieve verifiable inner product computation (VIP) based on secure aggregation [13], secure inner product computation [18] and linearly homomorphic hash function [21]. We consider each *Party* $p_i$ as the owner of a private vector $x_i$ and a public vector $y$.

Our goals are to securely compute the inner product values $u$, use $u$ for verifiable subsequent calculations like $z = \langle u, c \rangle$ and proof $\sigma$, and ensure the result $z$ is only available to corresponding *Parties*.

Initially, *Host* selects a generator $g$ on $\mathbb{Z}_p^*$ with order $N$ and sends it to *Parties*. Each party $p_i$ synchronously computes a vector of witnesses $\tau p_i = H(t_i)$. Subsequently, *Host* computes a witness $\tau_j$ for each $u_j \in u$, and the proof for $\langle u, c \rangle$ as $\sigma$. After *Aggregator* calculates $z = \langle u, c \rangle$, it sends the result $z$ back to Host for verification. Finally, *Host* compares the proof $\sigma$ with $H(z)$ to decide whether $z$ is valid. We restrict access to $z$ to corresponding *Parties* by employing a homomorphic symmetric mask scheme. We integrate these approaches into a verifiable inner product computation scheme for secure vertical federated linear regression. Our instantiation, named $VIP$, consists of the following algorithms:

**VIP.Setup($1^\kappa$):**

Select a tuple of three large prime numbers, $(\beta, q, N)$ satisfying $\beta^9 < q^3 < N$. Select a random number $\alpha$ satisfying $0 < \alpha < \frac{\lfloor \beta/2 \rfloor}{m^2 n \|M\|_\infty}$ and a parameter $rand$ as the range of random numbers satisfying $m^2 n(rand^2 + 2\alpha Mrand) < \lfloor \frac{\alpha^2}{2} \rfloor$, where $\|M\|_\infty$ is the absolute maximum of messages in message space. Randomly select 2 numbers, $k_H$ on $\mathbb{Z}_q^*$ and $k_H'$ on $\mathbb{Z}_N^*$. Let $H(x) = g^x \mod p$ be a collision-resistant homomorphic hash function on $\mathbb{Z}_p$. The hash parameter g is a random element of $\mathbb{Z}_p^*$ with order $N$. Note that public parameters are $pp = (\alpha, \beta, q, N,, rand, p, g)$. Return $(k_H, k_H', pp)$.

**VIP.SYMMask**$(M, k_1, k_2, q_1, q_2, pp)$:

To encrypt an input message vector $M$ of length $l$ with $(k_1, k_2)$ as the secret key, select a random vector $r = \{r_i\}_{i \in [l]}$, where $r_i \leq rand$. Compute $C = k_2(k_1(\alpha M + r \bmod \beta) \bmod q_1) \bmod q_2$ and return $C$.

**VIP.SYMUnMask**$(C, k_1, k_2, q_1, q_2, pp)$:

To decrypt an vector $C$ encrypted by $(k_1, k_2)$, compute $\epsilon = (k_1^{-1}(k_2^{-1}C \bmod q_2) \bmod q_1) \bmod \beta$. Taking into account the sign of the positive and negative, the output message $M$ is given by $\lfloor \frac{\epsilon}{\alpha^2} \rfloor$ if $0 \leq \epsilon < \frac{\beta}{2}$, otherwise $\lfloor \frac{\epsilon - \beta}{\alpha^2} \rfloor$.

**VIP.SecAgg**$(x, y_i, \{AK_{i,j}\}_{j \neq i}, postfix, q_1, pp)$:

A one-time masked vector of $xy_i$ is supposed to be generated by this function. $R = \sum_{j=1, j \neq i}^{n} (-1)^{j > i} R_j \bmod q_1$, where $(-1)^{k > n} = -1$ if $k > n$ and 1 otherwise, and $\{R_j\}_{j \neq i}$ is the uniformly output random vector from a pseudorandom generator (PRG) [2,20] with the one-time seed series $\{AK_{i,j}\}_{j \neq i}$. $R_j \leftarrow PRG(AK_{i,j} \| postfix) \in \mathbb{Z}_{q_1}^l$, where $q_1$ is the order of the group, $l$ is the length of $x$, and postfix differs in each call of this function. Finally get a mask of model $x \cdot y_i$ base on secret sharing by calculating $t = xy_i + R \bmod q_1$ and return $t$.

**VIP.PrfGen**$(\tau, y, pp)$:

For a vector of witnesses $\tau$ and a vector $y$ both of the same length, compute $\sigma = \prod_{i=1}^{m} (\tau_i)^{y_i} \bmod p$ as the partial proof of result.

**VIP.Verify**$(\sigma, z, k_q, k_N, pp)$:

For a proof $\sigma$ and a value $z$, check whether $\sigma = H(z)$ holds. Output 1 if the check is satisfied or 0 otherwise. It's clear to find that $\sigma = \prod_{i=1}^{m} (\tau_i)^{y_i} = \prod_{i=1}^{m} H(x_i)^{y_i} = g^{\sum_{i=1}^{m} x_i y_i} = H(\langle x, y \rangle) (\bmod p)$, proving the correctness of the validator. To recover the result of $\langle x, y' \rangle$ from $(z, k_q, k_N)$, compute $z' = k_q^{-1}(k_N^{-1} \cdot z \bmod N) \bmod q$. To simplify the expression, we use **VIP.Verify**$(\sigma, z, k_q, k_N, pp)$ as component-wise executions of **VIP.Verify**$(\sigma_i, z_i, r_q, r_N, pp)$ for each $\sigma_i \in \sigma$ and $z_i \in z$. In this case, this function returns $z'$ if all executions output 1, or raises an error otherwise.

## 4.2   Initialization

Algorithm 2 consists of *Host.Initialize* and *P.Initialize*. The *Host* initializes cryptographic schemes for the verifiable Inner Product Computation (VIP), generating public parameters for *Parties'* further initialization.

We assume the regularization parameter $\lambda$ and learning rate $\eta$ are public. In the vertically partitioned case, we assume the intersection between datasets owned by different parties is known as a priori. After that,arties execute the Diffie-Hellman Key Agreement protocol, and each $p_i \in \mathcal{P}$ obtains an agreed key set $A_{p_i} = \{AK_{i,j}\}_{j \neq i}$. Each party $p_i \in \mathcal{P}$ symmetrically masks its dataset $\mathcal{D}_{p_i}$ with secret keys $kp_i$ and $k'p_i$, and publishes the ciphertext $C_{p_i}^{\mathcal{D}}$. The active party accesses the global model loss function value for simplification, although any party could obtain it.

---

**Algorithm 2:** Initialization

**Host** *Initialize*

1    $k_H, k'_H, pp \leftarrow$ **VIP.Setup**$(1^\kappa)$;

2    $C_H \leftarrow$ **VIP.SYMMask**$(1^n, k_H, k'_H, q, N, pp)$;

3    send $(C_H, pp)$ to **Aggregator** and each $p_i \in \mathcal{P}$;

 

**Party** $p_i$ *Initialize*

4    Execute Diffie-Hellman Key Agreement protocol to get $A_{p_i} = \{AK_{i,j}\}_{j \neq i}$;

5    Randomly select two numbers $k_{p_i}$ on $\mathbb{Z}^*_\beta$ and $k'_{p_i}$ on $\mathbb{Z}^*_q$;

6    send $C^{\mathcal{D}}_{p_i} \leftarrow$ **VIP.SYMMask**$(\mathcal{D}_{p_i}, k_{p_i}, k'_{p_i}, \beta, q, pp)$ to **Host**,**Aggregator**

7    ; if $p_i$ *is acitve party* then

8      $C_1 \leftarrow$ **VIP.SYMMask**$(1^n, k_{p_i}, k'_{p_i}, \beta, N, pp)$;

9      $C_2 \leftarrow$ **VIP.SYMMask**$(\alpha^n, k^2_{p_i}, k'^2_{p_i}, \beta, N, pp)$;

10      send $(C_1, C_2)$ to each $p_i \in \mathcal{P}$;

---

### 4.3 Model Training

Each training epoch following Algorithm 3 involves two communication rounds between parties and servers without inter-party communication.

In each epoch, the *ModelCalculation* method is called. Each party $p_i$ calculates its partial model output, $u_{p_i}$, using current partial model weights, $\omega_{p_i}$, and datasets, $\mathcal{D}_{p_i}$. The active party appends a vector with labels $y$. Parties compute three obscured partial model information pieces with blinding factors. For future verification, a vector $\tau^u_{p_i}$ is sent to *Host*.

At the beginning of *GradientCalculation*, *Aggregator* obtains $C^u_H$ by aggregating $\{t^u_{H,p_i}\}_{p_i \in \mathcal{P}}$. *Aggregator* calculates masked partial model gradients $C^J_{H,p_i}$ and sends them to *Host* for verification.

*Host* verifies received results $z$ using proof $\sigma$. This process follows line 13–15 in Algorithm 3, with differences in operations on exponential powers. Using proof $\sigma^J_H$, *Host* checks if $\sigma^J_{H,p_i} = H(C^J_{H,p_i})$ and calculates $C^J_{p_i}$ if successful. *Host* then distributes $\{C^J_{p_i}\}_{p \in \mathcal{P}}$ to corresponding parties. Each party $p_i$ unmask $C^J_{p_i}$ using its secret key tuple, obtaining $J(\omega_{p_i})$ to update their partial models.

Using received $\{(t^u_{1,p_i}, t^\omega_{2,p_i})\}_{p_i \in \mathcal{P}}$, *Host* calculates masked global model loss, $C^{\mathcal{L}}$, and sends it to the active party for unmasking.

## 5 Security and Privacy Analysis

In this section, we present proofs that address the security of party input during secure aggregation, the security of the homomorphic hash function, the security of Hidden Linear Functions, and the solution of non-homogeneous systems.

The proof of the security of parties' input to *Host* and *Aggregator* can be found in secure aggregation schemes [6]. Under the DDH assumption, given

---

**Algorithm 3:** Model Training

**Party** $p_i$ *ModelCalculation*

1    **if** $p_i$ *is acitve party* **then** $u_{p_i} \leftarrow \omega_{p_i} \cdot \mathcal{D}_{p_i} - y$ **else** $u_{p_i} \leftarrow \omega_{p_i} \cdot \mathcal{D}_{p_i}$

3    $t^u_{H,p_i} \leftarrow$ **VIP.SecAgg**$(u_{p_i}, C_H[i], A_{p_i}, epoch \| 0, N, pp)$;

4    $\tau^u_{H,p_i} \leftarrow H(t^u_{H,p_i})$;

5    $t^u_{1,p_i} \leftarrow$ **VIP.SecAgg**$(u_{p_i}, C_1[i], A_{p_i}, epoch \| 1, N, pp)$;

6    $t^\omega_{2,p_i} \leftarrow$ **VIP.SecAgg**$(n\lambda \| \omega_{p_i} \|^2, C_2[i], A_{p_i}, epoch \| 2, N, pp)$;

7    **send** $(\tau^u_{H,p_i}, t^u_{1,p_i}, t^\omega_{2,p_i})$ to **Host** and **send** $t^u_{H,p_i}$ to **Aggregator**;

**Aggregator** *GradientCalculation*

8    $C^u_H \leftarrow \sum_{p_i \in \mathcal{P}} t^u_{H,p_i} \bmod N$;

9    **foreach** $p_i \in P$ **do** $C^u_{H,p_i} \leftarrow C^u_H \cdot C^{\mathcal{D}}_{p_i} \bmod N$ **send** $\{C^J_{H,p_i}\}_{p_i \in \mathcal{P}}$ to **Host**;

**Host** *GradientCalculation*

11    **get** $z$ from **Aggregator**;

12    **foreach** $k \in [s]$ **do** $\tau^u_H[k] \leftarrow \prod_{p_i \in \mathcal{P}} \tau^u_{H,p_i}[k] \bmod p$ **foreach** $p_i \in \mathcal{P}$ and $f \in [d]$ **do** $\sigma^J_{H,p_i}[f] \leftarrow$ **VIP.PrfGen**$(\tau^u_H, C^{\mathcal{D}}_{p_i}[f], pp)$

14    $\{C^J_{p_i}\}_{p_i \in P} \leftarrow$ **VIP.Verify**$(\sigma^J_H, z, k_H, k'_H, pp)$;

15    **foreach** $p_i \in \mathcal{P}$ **do** **send** $C^J_{p_i}$ to $p_i$ **send** $C^{\mathcal{L}}_{act} \leftarrow \| \sum_{p_i \in \mathcal{P}} t^u_{1,p_i} \|^2 + \sum_{p_i \in \mathcal{P}} t^\omega_{2,p_i} \bmod N$ to *active party*;

**Party** $p_i$ *ModelUpdate*

17    **get** $C^J_{p_i}$ from **Host**;

18    $J(\omega_{p_i}) \leftarrow$ **VIP.SYMUnMask**$(C^J_{p_i}, k_{p_i}, k'_{p_i}, \beta, q, pp)$;

19    $\nabla \mathcal{L}_{\mathcal{D}}(\omega) \leftarrow \frac{1}{n} J(\omega_{p_i}) + \lambda \omega_{p_i}$;

20    $\omega_{p_i} \leftarrow \omega_{p_i} - \eta \nabla \mathcal{L}_{\mathcal{D}}(\omega)$;

21    **if** $p_i$ *is active party* **then**

22      **get** $C^{\mathcal{L}}_{act}$ from **Host**;

23      $\mathcal{L} \leftarrow$ **VIP.SYMUnMask**$(C^{\mathcal{L}}_{act}, k^2_{p_i}, k'^2_{p_i}, \beta, N, pp)$;

24      $\mathcal{L}_{\mathcal{D}}(\omega) \leftarrow \frac{1}{2n} \mathcal{L}$;

---

masked input $(t^u_{1,p_i}, t^\omega_{2,p_i})$, no adversary has a non-negligible advantage in breaking the masked input $t^u_{1,p_i}$ and $t^\omega_{2,p_i}$ to directly obtain $u_{p_i}$ and $\omega_{p_i}$, respectively.

The formal proof of the security of parties' witnesses can be found in the homomorphic hashing scheme [14]. Under the discrete logarithm assumption, given masked witness pieces $\tau^u_{H,pi}$ and a complete witness $\tau^u_H$, no adversary has a non-negligible advantage in breaking the homomorphic hash value to directly obtain $t^u_H$.

**Definition 1 (Hidden Linear Function Problem).** *The hidden linear function problem is characterized by the function* $f : f(x_1, x_2) = \pi(sx_1 + x_2 \bmod N)$, *in which* $\pi$ *is a permutation on* $X$. *The function maps the group* $G : \mathbb{Z} \times \mathbb{Z}, +$ *to the finite set* $X : \mathbb{Z}_N$, *and is guaranteed to be constant on cosets of the hidden* subgroup $H = (l, -ls) \mid l, s \in X \subset G$. $\mathbb{Z}_N$ denotes the set $\{0, 1, ..., N-1\}$, and $\mathbb{Z}$ represents the integers. The problem is to find $H$ (or a generating set for it), given a black box for $f$.

Boneh and Lipton [4] observed the connection to the hidden subgroup problem and proposed a quantum algorithm for solving the hidden linear function problem. As of yet, no classical algorithm has been discovered that can efficiently solve the Hidden Linear Function Problem within polynomial time.

**Lemma 1 (Security of Symmetric Double-Layered Mask System).** *The symmetric double-layered mask system exhibits ciphertext indistinguishability. For all symmetrically masked messages, denoted as "C"s, they can only be recovered by the role in the lower-right label; for example, only $p_i$ is capable of unmasking $C_{p_i}^J$ to obtain $J_{p_i}$.*

Symmetric double-layered mask system represents an instance of the hidden linear function problem. **VIP.SYMMask** yields $C = k_2(k_1(\alpha m + r \bmod \beta) \bmod q_1) \bmod q_2$ as the function $f$, with $x_1 = \alpha m$, $x_2 = k_1 r \bmod q_1$, $s = k_1$, $N = q_1$, and $\pi$ as the multiplication by $k_2$ modulo $q_2$. This introduces several subtle differences compared to Definition 1, such as the unselectable input $x_2$ of the black box for $f$, and the permutation $\pi$ producing values in a range significantly larger than $\mathbb{Z}_{q_1}$. It is important to note that these variations in Lemma 1 do not simplify the task of solving Definition 1. On the contrary, it becomes more challenging to devise an effective algorithm, as the input $x_2$ is an unknown random number. The value range $X$ of permutation $\pi$ does not impact the hardness, since for each distinct value of $k_2$, $f$ only outputs $q_1$ specific elements on $\mathbb{Z}_{q_2}$.

As described above, with the security properties of parties' input, parties' witnesses and symmetirc double-layered mask system, any adversary cannot obtain any useful information since there are massive random numbers contained in query requests and responses. Furthermore, both *Host* and *Aggregator* learn nothing about the plaintexts data even colluding with some of the parties. Therefore, we have the following theorem.

**Theorem 1 (Privacy Guarantee of SVFL).** *The SVFL maintains the security property such that two non-collusive CSPs gain no knowledge about user-related private data, including the gradient and the global model details. Furthermore, SVFL thwarts any inference attack executed by an adversary aiming to deduce party $p_i$'s input $u_{p_i}$, its local model $\omega_{p_i}$, or its local features $\mathcal{D}_{p_i}$, as well as any intermediate computational details without direct access to them.*

## 6    Experiments and Efficiency Analysis

In this section, we evaluate the performance of SVFL, ensuring a minimum of 256-bit security. We selected $\beta$ with $\|\beta\| = k = 256$, which provides sufficient message space for managing high-dimension data in our experiments. Our experiments were conducted on a desktop equipped with a 2.20 GHz Intel Core

**Table 2.** Datasets used in our experiment for linear regression models.

| ID | Name | $n$ | $d$ |
|----|------|-----|-----|
| 1 | Auto MPG | 392 | 8 |
| 2 | Boston Housing Dataset | 506 | 14 |
| 3 | Energy Efficiency | 768 | 9 |
| 4 | Wine Quality | 1599 | 12 |
| 5 | Communities and Crime | 1994 | 122 |
| 6 | Parkinsons Telemonitoring | 5875 | 22 |
| 7 | Blog Feedback | 52397 | 280 |
| 8 | CT slices | 53500 | 384 |

**Table 3.** Size of $\mathcal{P}$ and feature size of each private data. Errors that the models achieved. Time in milliseconds (ms) per epoch in the training phase of SVFL

| ID | $m$ | $d_i$ | $\text{RMSE}_{LinR}$ | $\text{RMSE}_{SVFL}$ | Party ($p_1$) | Aggregator | Host | Total |
|----|-----|-------|--------------|---------------|----------------|------------|------|-------|
| 1 | 8 | 1 | 3.17 | 3.17(-0.0%) | 1.24 (1.94) | 4.77 | 1.05 | 7.76 |
| 2 | 7 | 2 | 4.64 | 4.67(+0.6%) | 1.40 (2.09) | 10.34 | 1.35 | 13.78 |
| 3 | 3 | 3 | 2.97 | 3.01(+1.4%) | 1.74 (2.50) | 10.33 | 1.93 | 14.76 |
| 4 | 6 | 2 | 0.64 | 0.64(+0.1%) | 2.49 (3.28) | 27.71 | 3.34 | 24.33 |
| 5 | 5 | 20 | 0.13 | 0.13(−1.4%) | 3.63 (4.29) | 326.42 | 5.41 | 336.11 |
| 6 | 7 | 3 | 3.18 | 3.21(+0.6%) | 10.75 (11.44) | 197.39 | 12.33 | 221.17 |
| 7 | 14 | 20 | 30.91 | 31.29(+1.2%) | 65.97 (66.85) | 20582.21 | 143.30 | 20792.37 |
| 8 | 19 | 20 | 8.24 | 8.31(+0.8%) | 79.15 (80.13) | 30963.57 | 171.05 | 31214.76 |

**Table 4.** Bits of data exchange per epoch to train linear regression models in SVFL.

| Entity | Send | Get |
|--------|------|-----|
| Passive $p_i$ | $n(l_p + 2l_N) + l_N$ | $d_i l_q$ |
| $p_1$ | $n(l_p + 2l_N) + l_N$ | $d_i l_q + l_N$ |
| Aggregator | $dl_N$ | $mnl_N$ |
| Host | $dl_q + l_N$ | $(m + d)l_N + mn(l_p + l_N)$ |

i7 8-Core processor and 16 GB RAM. We assessed the performance of SVFL on eight distinct real-world datasets of varying sizes and dimensions from the UCI ML repository [9], as summarized in Table 2.

For each dataset, we randomize the dataset containing $n$ observations and $d$ features, designating approximately 70% for training and 30% for testing in order to determine model accuracy. The training dataset is subsequently distributed among a group of $m$ users, with each party $p_i$ possessing an equal number of data features $d_i$, as outlined in Table 3. As a comparison, we provide a standard linear regression (under a non-VFL setting and without privacy-preserving mechanisms) using the `sklearn` tool, denoted as LinR, for reference. To gauge the predictive error of each model, we employ the root mean squared error (RMSE) metric for each dataset and repeat the experiments five times.

We compare the accuracy of the model obtained by SVFL with the model obtained by LinR for each dataset. SVFL's achieved accuracy is remarkably close to the non-secure counterpart (which is thought to be state-of-the-art accuracy). In Table 3, we also record the average time cost for each epoch, and the experiments show that *Aggregator* takes almost all the computation overload while parties and *Host* take a little. Table 4 displays the number of bits a party must transfer, along with the communication cost for the CSPs, in each epoch to complete the training phase. We denote the length of our parameters like $l_q = \|q\|$.

## 7    Conclusion

In this paper, we introduced SVFL, a novel framework of secure VFL for the privacy-preserving and secure training of linear models. SVFL is built on a double-layered masking system and exploits the properties of homomorphic hash functions. It enables a server to verify the correctness of the final model without disclosing any data or model details. Requiring neither party-to-party communication nor TPA-setting, SVFL achieves a comparable accuracy for linear models while providing privacy and security guarantees. As a promising approach for training privacy-preserving linear models over vertically partitioned datasets, it has the potential to be extended to other types of ML models.

**Acknowledgement.** This work was supported in part by the National Key Research and Development Program of China (Grant No. 2020YFA0712300), in part by the National Natural Science Foundation of China (Grant No. 62132005, 62172162), in part by Shanghai Trusted Industry Internet Software Collaborative Innovation Center. Zhenfu Cao, Jiachen Shen, and Xiaolei Dong are the corresponding authors.

## References

1. Abdalla, M., Bourse, F., De Caro, A., Pointcheval, D.: Simple functional encryption schemes for inner products. In: Katz, J. (ed.) PKC 2015. LNCS, vol. 9020, pp. 733–751. Springer, Heidelberg (2015). https://doi.org/10.1007/978-3-662-46447-2_33
2. Blum, M., Micali, S.: How to generate cryptographically strong sequences of pseudo random bits. In: 23rd Annual Symposium on Foundations of Computer Science (SFCS 1982), pp. 112–117 (1982)
3. Bogdanov, D., Kamm, L., Laur, S., Sokk, V.: Rmind: a tool for cryptographically secure statistical analysis. IEEE Trans. Dependable Secure Comput. **15**(3), 481–495 (2018). https://doi.org/10.1109/TDSC.2016.2587623

4. Boneh, D., Lipton, R.J.: Quantum cryptanalysis of hidden linear functions. In: Coppersmith, D. (ed.) CRYPTO 1995. LNCS, vol. 963, pp. 424–437. Springer, Heidelberg (1995). https://doi.org/10.1007/3-540-44750-4_34

5. Chen, T., Jin, X., Sun, Y., Yin, W.: VAFL: a method of vertical asynchronous federated learning. CoRR abs/2007.06081 (2020). http://arxiv.org/2007.06081

6. Corrigan-Gibbs, H., Wolinsky, D.I., Ford, B.: Proactively accountable anonymous messaging in verdict. In: King, S.T. (ed.) Proceedings of the 22th USENIX Security Symposium, Washington, DC, USA, 14–16 August 2013, pp. 147–162. USENIX Association (2013)

7. Damgård, I., Jurik, M.: A generalisation, a simplification and some applications of paillier's probabilistic public-key system. In: Kim, K. (ed.) PKC 2001. LNCS, vol. 1992, pp. 119–136. Springer, Heidelberg (2001). https://doi.org/10.1007/3-540-44586-2_9

8. Diffie, W., Hellman, M.E.: New directions in cryptography. In: Democratizing Cryptography (1976)

9. Dua, D., Graff, C.: UCI machine learning repository (2017). http://archive.ics.uci.edu/ml

10. European Parliament, Council of the European Union: Regulation (EU) 2016/679 of the European Parliament and of the Council (2016). http://data.europa.eu/eli/reg/2016/679/oj

11. Gascón, A., et al.: Secure linear regression on vertically partitioned datasets. IACR Cryptol. ePrint Arch. **2016**, 892 (2016)

12. Hardy, S., et al.: Private federated learning on vertically partitioned data via entity resolution and additively homomorphic encryption. CoRR abs/1711.10677 (2017). http://arxiv.org/abs/1711.10677

13. Keith, B., et al.: Practical secure aggregation for privacy-preserving machine learning. In: proceedings of the 2017 ACM SIGSAC Conference on Computer and Communications Security, pp. 1175–1191 (2017)

14. Krohn, M., Freedman, M., Mazieres, D.: On-the-fly verification of rateless erasure codes for efficient content distribution. In: IEEE Symposium on Security and Privacy, 2004. Proceedings 2004, pp. 226–240 (2004). https://doi.org/10.1109/SECPRI.2004.1301326

15. Ma, X., Sun, X., Wu, Y., Liu, Z., Chen, X., Dong, C.: Differentially private byzantine-robust federated learning. IEEE Trans. Parallel Distrib. Syst. **33**(12), 3690–3701 (2022)

16. McMahan, H.B., Moore, E., Ramage, D., Arcas, B.A.: Federated learning of deep networks using model averaging. CoRR abs/1602.05629 (2016). http://arxiv.org/abs/1602.05629

17. Nikolaenko, V., Weinsberg, U., Ioannidis, S., Joye, M., Boneh, D., Taft, N.: Privacy-preserving ridge regression on hundreds of millions of records. In: 2013 IEEE Symposium on Security and Privacy, pp. 334–348 (2013)

18. Wang, F., Zhu, H., Lu, R., Zheng, Y., Li, H.: Achieve efficient and privacy-preserving disease risk assessment over multi-outsourced vertical datasets. IEEE Trans. Dependable Secure Comput. **19**(3), 1492–1504 (2022). https://doi.org/10.1109/TDSC.2020.3026631

19. Xu, R., Baracaldo, N., Zhou, Y., Anwar, A., Joshi, J., Ludwig, H.: FedV: privacy-preserving federated learning over vertically partitioned data. In: Proceedings of the 14th ACM Workshop on Artificial Intelligence and Security, pp. 181–192. AISec 2021, Association for Computing Machinery, New York, NY, USA (2021)

20. Yao, A.C.: Theory and application of trapdoor functions. In: 23rd Annual Symposium on Foundations of Computer Science (SFCS 1982), pp. 80–91 (1982). https://doi.org/10.1109/SFCS.1982.45
21. Yao, H., Wang, C., Hai, B., Zhu, S.: Homomorphic hash and blockchain based authentication key exchange protocol for strangers. In: 2018 Sixth International Conference on Advanced Cloud and Big Data (CBD), pp. 243–248 (2018). https://doi.org/10.1109/CBD.2018.00051

# Cryptography and Authentication II

# Multiprime Strategies for Serial Evaluation of eSIDH-Like Isogenies

Jason T. LeGrow[1](✉), Brian Koziel[2], and Reza Azarderakhsh[2]

[1] Department of Mathematics, Virginia Polytechnic Institute and State University, Blacksburg, VA, USA
jlegrow@vt.edu
[2] Department of Electrical Engineering and Computer Science, Florida Atlantic University, Boca Raton, FL, USA
{bkoziel2017,razarderakhsh}@fau.edu

**Abstract.** We present new results and speedups for the large-degree isogeny computations within the extended supersingular isogeny Diffie-Hellman (eSIDH) key agreement framework. As proposed by Cervantes-Vázquez, Ochoa-Jiménez, and Rodríguez-Henríquez, eSIDH is an extension to SIDH and fourth round NIST post-quantum cryptographic standardization candidate SIKE. By utilizing multiprime large-degree isogenies, eSIDH and eSIKE are faster than the standard SIDH/SIKE and amenable to parallelization techniques that can noticeably increase their speed with multiple cores. Here, we investigate the use of multiprime isogeny strategies to speed up eSIDH and eSIKE in *serial* implementations. These strategies have been investigated for other isogeny schemes such as CSIDH. We apply them to the eSIDH/eSIKE scenario to speed up the multiprime strategy by about 10%. When applied to eSIDH, we achieve a 7–8% speedup for Bob's shared key agreement operation. When applied to eSIKE, we achieve a 3–4% speedup for key decapsulation. Historically, SIDH and SIKE have been considerably slower than its competitors in the NIST PQC standardization process. These results continue to highlight the various speedups achievable with the eSIKE framework to alleviate these speed concerns. Though eSIDH and eSIKE are susceptible to the recent devastating attacks on SIKE, our analysis applies to smooth degree isogeny computations in general, and isogeny-based signature schemes which use isogenies of smooth (not necessarily powersmooth) degree.

**Keywords:** Isogeny-based cryptography · large-degree isogeny · post-quantum cryptography

Jason T. LeGrow is funded in part by New Zealand's Ministry of Business, Innovation, and Employment fund UOAX1933 and the Commonwealth of Virginia's Commonwealth Cyber Initiative (CCI), an investment in the advancement of cyber R&D, innovation, and workforce development. For more information about CCI, visit www.cyberinitiative.org. This work was partially performed while the first author was in the Department of Mathematics at the University of Auckland.

ⓒ The Author(s), under exclusive license to Springer Nature Switzerland AG 2023
M. Yung et al. (Eds.): SciSec 2023, LNCS 14299, pp. 347–366, 2023.
https://doi.org/10.1007/978-3-031-45933-7_21

# 1  Introduction

The impending implementation of large-scale quantum computers necessitates the development of post-quantum cryptosystems to ensure the continued safety of our communications systems. One family of post-quantum primitives are supersingular isogeny-based protocols, whose security is based on the presumed hardness of finding isogenies between supersingular elliptic curves (and many variants of this problem). One such protocol was Supersingular Isogeny Diffie-Hellman (SIDH) [11] key exchange, which underlies the NIST post-quantum cryptography (PQC) competition fourth round alternate candidate supersingular isogeny key encapsulation (SIKE) [2] mechanism which was recently broken [5,16,18]. Among post-quantum cryptosystems, SIDH and SIKE were appealing because of their small key sizes, which reduce both the communication and storage components when transferring or storing public keys, respectively. Unfortunately, SIDH and SIKE suffered from slow speeds, but these concerns continue to be alleviated as more applied research accelerated them.

In both SIDH and SIKE, the most time-consuming computational task that each party must perform is to determine the codomain of an isogeny which is of cryptographically-large but smooth degree. This is a large-degree isogeny computation. For the SIKE parameter sets which were considered for standardization [2, Section 1.6], Alice computes isogenies of degree $2^{e_A}$ for $e_A$ approximately between 200 and 400, while Bob computes isogenies of degree $3^{e_B}$ for $e_B$ approximately between 130 and 230. More exotic protocols built on SIDH have parties construct isogenies of the form $\ell^e$ where $\ell$ is a small prime and $e$ is a small exponent which depends on the desired security level; for instance, 3-party key establishment of the kind given in [1,13] uses isogenies of degree $5^{e_C}$ for a third party, Charlie.

Any isogeny $\psi$ of degree $N = \ell_1^{e_1} \ell_2^{e_2} \cdots \ell_n^{e_n}$ defined on an elliptic curve $E$ defined over a field $\mathbb{F}$ of characteristic coprime to $\ell_1, \ell_2, \ldots, \ell_n$ factors as

$$\psi = \psi_{1,1} \circ \cdots \circ \psi_{1,e_1} \circ \psi_{2,1} \circ \cdots \circ \psi_{2,e_2} \circ \cdots \circ \psi_{n,1} \circ \cdots \circ \psi_{n,e_n} \tag{1}$$

where for each $1 \leq j \leq n$ and $1 \leq k \leq e_j$ we have $\deg \psi_{j,k} = \ell_j$. The best known algorithms for computing the codomain of an isogeny of prime degree $\ell$ requires time fully exponential in $\ell$ (cf. Vélu's formulas [19] and the recent improved Vélu's formulas [4]); however, for isogenies of degrees of the form used in SIKE or CSIDH, the factorization given in Eq. (1) naturally yields a polynomial-time algorithm for computing the required isogenies, for constant values of $\ell_1, \ell_2, \ldots, \ell_n$ and polynomially-sized $e_1, e_2, \ldots, e_n$.

The isogenies of small prime degree which are chained together to construct the isogenies of cryptographically-large degree in SIKE and CSIDH are typically computed using Vélu's formulas; to compute an isogeny of degree $\ell_i$, a point $P_i \in E(\mathbb{F}_{p^2})$ of order $\ell_i$ is required, and exactly which point $P_i$ is required at each step depends on the protocol being used and the user's secret key. In the case of SIKE, Alice's secret key is an integer $0 \leq m_A \leq \ell_A^{e_A} - 1$, and (part of) her public ephemeral key is $\psi_A(E_0)$, where $E_0$ is a global public curve and $\psi_A$ is the unique (up to composition with an isomorphism) isogeny whose kernel is $\ker \psi_A = \langle P_A + m_A Q_A \rangle$, for some global public points $P_A, Q_A$ which generate

$E(\mathbb{F}_{p^2})[2^{e_A}]$. This isogeny factors as $\psi_A = \psi_A^{e_A} \circ \psi_A^{e_A-1} \circ \cdots \circ \psi_A^1$ where for each $1 \le j \le e_A$, we have $\ker \psi_A^j = \langle [2^{e_A-j}]\psi_A^{j-1} \circ \psi_A^{j-2} \circ \cdots \circ \psi_A^1 (P_A + m_A Q_A)\rangle$. This suggests a straightforward method to compute $\psi_A$ from $R_A^{(0)} = P_A + m_A Q_A$: first construct $[2^{e_A-1}]R_A^{(0)}$, which generates $\ker \psi_A^1$. Using Vélu's formulas, construct $R_A^{(1)} = \psi_A^1(R_A^{(0)})$. Now construct $[2^{e_A-2}]R_A^{(1)}$, which generates $\ker \psi_A^2$. Continue in this fashion, alternately multiplying by $[2^{e_A-j}]$ and applying $\psi_A^j$, until you have performed all $e_A$ isogenies. This simple technique requires the application of $e_A - 1 = O(e_A)$ 2-isogeny evaluations, and $\sum_{j=1}^{e_A}(e_A - j) = \binom{e_A}{2} = O(e_A^2)$ point doublings, which is clearly polynomial-time in $e_A$.

*Strategies* have been proposed (first in [11] for SIDH and SIKE, and in [15] for CSIDH [6]) as faster alternatives to this naïve method. Formally, a strategy is a Steiner arborescence in a directed grid graph with particular root and terminals (depending on the degree of the isogeny to be computed) and with edge weights which encode the cost of two basic operations: prime-degree isogeny evaluation and scalar multiplication. A strategy corresponds to an algorithm for constructing an isogeny codomain by associating to each vertex a point on an elliptic curve and to each edge a basic operation. A strategy of minimal total weight thus corresponds to a fastest algorithm (from the class of algorithms which are in correspondence with strategies) for constructing the codomain of an isogeny. The straightforward algorithm of the previous paragraph corresponds to the "multiplication-based" strategy; generally, other strategies are more efficient.

*Extended SIDH* (eSIDH) [8] is a protocol derived from SIDH, in which Bob's prime-power-degree isogenies are replaced by isogenies whose degrees are a product of powers of small primes (3 and 5, for all proposed parameter sets in [8]). The most obvious benefit of this change is that Bob's computations are more amenable to parallelization; in particular, the kernel of an $3^{e_B}5^{e_C}$ isogeny can be generated by a point $R_B = P_B + m_B Q_B$ of order $3^{e_B}$ and a point of $R_C = P_C + m_C Q_C$ of order $5^{e_C}$, and these points can be constructed in parallel. To yield the same security level as Alice's $2^{e_A}$-isogeny, it suffices to take $e_B, e_C$ such that $e_B \log_2 3 + e_C \log_2 5 \approx e_A$; in particular, taking $e_B \log_2 3 \approx e_C \log_2 5 \approx \frac{e_A}{2}$, constructing $R_B$ and $R_C$ in parallel takes approximately half as long as computing $R_A$, which saves a significant amount of time over a parallel implementation of SIDH. Aside from parallelization, this method allows Bob to compute his isogeny using two small strategies rather than one large strategy; this is more efficient even in the serial setting than the corresponding computation in SIDH. We refer to the technique of using multiple small strategies as the "split prime" setting.

Unfortunately, this split prime optimization only yields a net benefit in the first round of eSIDH. In contrast, in the second round, expensive scalar multiplications are required to generate the necessary "auxiliary points" used in constructing the corresponding kernel generators. However, as the authors note in [8, Section 3.3], Bob can instead construct a single point $R'_{BC}$ which generates the kernel of his full $3^{e_B}5^{e_C}$-isogeny, and construct that isogeny using a single strategy of size $e_B + e_C$ which is formed from the smaller strategies from round 1. The authors of [8] refer to this as the "CRT-based" (Chinese Remainder Theorem) approach; more broadly, we refer to a single strategy which is used to compute an isogeny of non-prime-power degree as a "multiprime strategy."

*Contributions.* In this work we consider multiprime strategies of a more general form than those of [8], as well as permutations of the multisets of small primes associated to the strategy, as in [15]. Using these more general strategies along with different techniques for secret key selection and kernel generator construction we accelerate Bob's large-degree isogeny by about 10% over the eSIDH implementations of [8] in the serial setting. In eSIDH, this equates to a 7–8% performance boost for Bob's shared secret generation. In eSIKE, this equates to a 3-4% performance boost for key decapsulation. We note that these performance boosts are achieved simply through pre-computation; the rest of the eSIDH/eSIKE algorithm remains the same. Next, we searched for other eSIDH/eSIKE-friendly primes, proposing new primes that our estimates suggest would provide a further 3% improvement in the isogeny computation. Lastly, we implement our multiprime stategies for some eSIDH primes in the eSIDH library to confirm these performance gains. We stress that these techniques apply to eSIDH variants of SIDH-based signatures based on zero-knowledge proofs of isogeny knowledge [12], such as those of [14], which remain secure despite the recent attacks on SIDH-based key establishment.

The rest of this paper is organized as follows: in Sect. 2 we give the necessary mathematical background for SIDH and eSIDH, which we then describe in Sect. 3. In Sect. 4 we discuss strategies abstractly, and then their applications to isogeny-based protocols. Finally, we present our implementation results in Sect. 5 and conclude in Sect. 6.

## 2  Mathematical Background: Isogenies

In this section we give a brief introduction to isogenies and how they are represented in SIDH and eSIDH. The contents of this section are adapted from [11, Section 2]. Note that SIKE and eSIKE are variants of SIDH and eSIDH, respectively, where the order of isogenies (among other things) is modified to achieve IND-CCA2 security. SIKE and eSIKE functions include key generation, key encapsulation, and key decapsulation.

Let $E_1$ and $E_2$ be elliptic curves defined over a finite field $\mathbb{F}_q$. An *isogeny* from $E_1$ to $E_2$ defined over $\mathbb{F}_q$ is a surjective rational map $\psi \colon E_1 \to E_2$ which is also a group homomorphism of $E_1(\overline{\mathbb{F}}_q)$ to $E_2(\overline{\mathbb{F}}_q)$. When an isogeny from $E_1$ to $E_2$ defined over $\mathbb{F}_q$ exists we say that $E_1$ and $E_2$ are $\mathbb{F}_q$-*isogenous*. Since each isogeny $\psi \colon E_1 \to E_2$ has a dual isogeny $\hat{\psi} \colon E_2 \to E_1$, the property of being isogenous is an equivalence relation on the set of elliptic curves defined over $\mathbb{F}_q$.

An *endomorphism* of an elliptic curve $E$ defined over $\mathbb{F}_q$ is an isogeny $\psi \colon E \to E$ defined over $\mathbb{F}_{q^e}$ for some $e \in \mathbb{N}$. The set of endomorphisms $E$, together with the zero map, forms a ring, called the *endomorphism ring* of $E$, and denoted $\mathrm{End}(E)$. The endomorphism ring is isomorphic to an order either in a quadratic number field—in which case we say that $E$ is *ordinary*—or in a quaternion algebra, in which case we say that $E$ is *supersingular*.

# 3   eSIDH Protocol Description

Here we present eSIDH, to give context for the use of strategies in Sect. 4.

Concisely, eSIDH is constructed from SIDH [11] (described in Appendix A) by having Bob compute isogenies of non-prime-power degree. The protocol flow is much the same; the necessary changes are:

1. Fix $\ell_A = 2$, and choose distinct odd primes $\ell_1, \ell_2, \ldots, \ell_n$.
2. The prime takes the form $p = 2^{e_A} \ell_1^{e_1} \cdots \ell_n^{e_n} f - 1$ with $2^{e_A} \approx \ell_1^{e_1} \cdots \ell_n^{e_n}$
3. We require torsion bases $\{P_i, Q_i\}$ for $E[\ell_i^{e_i}]$ for $i = 1, 2, \ldots, n$, along with a basis $\{P_A, Q_A\}$ for $E[2^{e_A}]$. As well, set $P_B = P_1 + P_2 + \cdots + P_n$ and $Q_B = Q_1 + Q_2 + \cdots + Q_n$.
4. In Bob's key generation round, he chooses secrets $\beta_i \in \{0, 1, \ldots, \ell_i^{e_i} - 1\}$ and constructs the points $R_i = P_i + \beta_i Q_i$. He will construct $n$ isogenies, whose kernels are given by

$$\ker \psi_1 = \langle R_1 \rangle$$
$$\ker \psi_i = \langle \psi_{i-1} \circ \cdots \circ \psi_1 (R_i) \rangle \text{ for } i \geq 2.$$

He also sets $\psi_B = \psi_n \circ \cdots \circ \psi_1$. His secret key is $\mathsf{sk}_B = (\beta_1, \ldots, \beta_n)$ and his public key is $\mathsf{pk}_B = (\psi_B(E), \psi_B(P_A), \psi_B(Q_A))$.
5. In Bob's key establishment round, he constructs the points

$$R_i' = \prod_{j \neq i} \ell_j^{e_j} (S_A + \beta_i T_A) \text{ for } i = 1, 2, \ldots, n$$

(where $S_A = \psi_A(P_B)$ and $T_A = \psi_A(Q_B)$) and uses them to construct isogenies $\psi_1', \ldots, \psi_n', \psi_B'$ as in his key generation round. His key is $K_B = j(\psi_B(E_A))$.

*The Structure of Bob's Computations in eSIDH.* As the authors note in [8, Section 3.3], Bob's isogeny construction does not need to be decomposed into prime-power-order isogenies; instead, he can construct a generator $R_B$ of $\ker \psi_B$ by "combining" his secret values $\beta_1, \ldots, \beta_n$ using the Chinese remainder theorem. This is valuable in the second round of eSIDH, since constructing the points $R_i'$ requires many costly point multiplications. This would allow Bob to use a single strategy (which we describe in the next section) to construct his key $K_B$.

Note that this idea works "in reverse" as well; rather than choosing $\beta_1, \ldots, \beta_n$, Bob can choose $\beta \in \{0, 1, \ldots, \ell_1^{e_1} \cdots \ell_n^{e_n} - 1\}$ and then set each $\beta_i = \beta$ mod $\ell_i^{e_i}$.

# 4   Strategies and Their Applications to (e)SIDH

Strategies were first introduced in [11, Section 4.2.2] (as "full, well-formed" strategies) as a method to define algorithms for computing prime-power-degree isogenies. They were later slightly reformulated in [15, Section 2.1] to be more compatible with certain optimization techniques relevant to CSIDH—we present (a slightly modified version of) that formulation here.

**Definition 1 (The Triangle Graphs $T_n$).** *For $n \in \mathbb{N}$, we denote by $T_n$ the directed graph whose vertices and edges are*

$$\mathcal{V}(T_n) = \{\vec{x} \in \mathbb{Z}^2 \ : \ x_1 + x_2 \leq n - 1 \text{ and } x_1, x_2 \geq 0\}$$
$$\mathcal{E}(T_n) = \{(\vec{x}, \vec{y}) \in \mathcal{V}(T_n)^2 \ : \ \vec{y} - \vec{x} \in \{(1, 0), (0, 1)\}\}$$

*We call $T_n$ the* triangle graph *of side $n$.*

**Definition 2 (Steiner Arborescence).** *Let $G$ be a graph, and let $r \in \mathcal{V}(G)$ and $L \subseteq \mathcal{V}(G) \setminus \{r\}$. A* Steiner arboresence *for $(G, r, L)$ is a subgraph $S$ of $G$ such that:*

1. *For each $t \in L$, $S$ contains a directed path from $r$ to $t$, and;*
2. *$S$ contains no undirected cycles.*

*We call $r$ the* root *of the arborescence, and $L$ the* terminals.

**Definition 3 (Strategy).** *A* strategy $S$ *of size $n$ is a Steiner arborescence for $(T_n, r, L_n)$, where $r = (0, 0)$ and $L_n = \{\vec{x} \in \mathcal{V}(T_n) \ : \ x_1 + x_2 = n - 1\}$.*

**Definition 4 (The Join Operator).** *Given two strategies $S_1, S_2$ of sizes $n_1$ and $n_2$, respectively, we define their* join, *denoted $S_1 \# S_2$ to be the strategy of size $n_1 + n_2$ whose edges are*

$$\mathcal{E}(S_1 \# S_2) = \left\{ \big(\vec{x} + (0, n_2), \vec{y} + (0, n_2)\big) \ : \ (\vec{x}, \vec{y}) \in \mathcal{E}(S_1) \right\}$$
$$\sqcup \left\{ \big(\vec{x} + (n_1, 0), \vec{y} + (n_1, 0)\big) \ : \ (\vec{x}, \vec{y}) \in \mathcal{E}(S_2) \right\}$$
$$\sqcup \left\{ \big((x, 0), (x + 1, 0)\big) \ : \ x = 0, 1, \ldots, n_1 - 1 \right\}$$
$$\sqcup \left\{ \big((0, y), (0, y + 1)\big) \ : \ y = 0, 1, \ldots, n_2 - 1 \right\}$$

More intuitively, $S_1 \# S_2$ is the subgraph of $T_{n_1+n_2}$ which contains: A path from $(0, 0)$ to $(n_1, 0)$; A path from $(0, 0)$ to $(0, n_2)$; A copy of $S_1$, shifted $n_2$ units up, and; A copy of $S_2$, shifted $n_1$ units right. We note that the join operator is both non-commutative and non-associative.

Of particular interest are so-called *canonical strategies*:

**Definition 5 (Canonical Strategy).** *A* strategy $S$ *of size $n$ is* canonical *if: $n = 1$, or $S = S_1 \# S_2$, where $S_1$ and $S_2$ are canonical strategies.*

When $S$ is a canonical strategy we let $S^L = S_1$ and $S^R = S_2$ denote its *left* and *right substrategies*, respectively.

Figure 1 depicts a canonical strategy in $T_9$ and highlights its left and right substrategies.

For optimization purposes we will need to assign weights to the edges of a strategy, which will be inherited from an assignment of weights to a triangle graphs. In this work weights are assigned to triangle graphs by *measures*.

**Definition 6 (Measure).** *A* measure *of size $n$ is a triple $M = (\{\ell_i\}_{i=1}^n, f_H, f_V)$ where $\{\ell_i\}_{i=1}^n$ is a sequence of positive numbers, and $f_H, f_V \colon \mathbb{R}_+ \to \mathbb{R}_+$ are a pair of* weight functions.

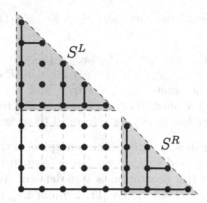

**Fig. 1.** A canonical strategy $S$ of size 9 (black edges) and its left and right substrategies (blue and red shaded regions, respectively) embedded in $T_9$ (vertices, black edges, and dashed grey edges). (Color figure online)

A measure $M = (\{\ell_i\}_{i=1}^n, f_H, f_V)$ assigns weights to $T_n$ as follows:

- Each edge of the form $e = ((x-1,y),(x,y))$ has weight $w_e^{(M)} = f_H(\ell_x)$.
- Each edge of the form $e = ((x,y-1),(x,y))$ has weight $w_e^{(M)} = f_V(\ell_{n-y+1})$.

A strategy $S$ is assigned by $M$ the weights it inherits from $T_n$. The *cost* of a strategy $S$ with respect to measure $M$ is denoted $(S)_M$, defined as

$$(S)_M = \sum_{e \in \mathcal{E}(S)} w_e^{(M)}.$$

### 4.1    Applying Strategies and Measures to Isogeny-Based Protocols

In isogeny-based protocols, strategies are used to define algorithms for computing the codomain of and evaluating smooth-degree isogenies. In this section we discuss the connection between strategies and isogeny algorithms in the contexts of SIDH and CSIDH. We also discuss the relevance of measures in this context.

*Strategies and Measures in SIDH.* Without loss of generality let us consider Alice's computations only; Bob's computations are analogous. For simplicity of notation, we will omit the subscript $A$ in Alice's computations. In SIDH, Alice must construct the codomain of an isogeny $\psi$ of degree $\ell^e$ (in SIKE standardization candidate parameter sets we always have $\ell = 2$, but this is not strictly necessary) whose kernel is $\ker \psi = \langle P + mQ \rangle$ where $P, Q$ are public and generate $E[\ell^e]$, and $m$ is chosen by Alice. As discussed in Sect. 1, this is done by constructing the generators of the kernels of $e$ $\ell$-isogenies, whose kernels are given by

$$\ker \psi_j = \Big\langle \overbrace{[\ell^{e-j}]\psi_{j-1} \circ \psi_{j-2} \circ \cdots \circ \psi_1(P + mQ)}^{R_j} \Big\rangle.$$

For $0 \leq j \leq e$, let $E_j$ denote the curve $\psi_j \circ \psi_{j-1} \circ \cdots \circ \psi_1(E_0)$. We decorate the graph $T_e$ by assigning:

- To each horizontal edge $e = ((x-1, y), (x, y))$ the map $P \mapsto [\ell]P$;
- To each vertical edge $e = ((x, y-1), (x, y))$ the map $P \mapsto \psi_y(P)$;
- To the vertex $(0, 0)$ the point $R_{0,0} = P + mQ$, and;
- To each vertex $(x, y)$ a point $R_{x,y}$ on $E_y$, obtained by applying the maps corresponding to the edges in any path from $(0, 0)$ to $(x, y)$ to the point $R_{0,0}$.

We note that since isogenies are group homomorphisms the maps $\psi_i$ commute with mutlipltication-by-$\ell$, and so any two paths from $(0, 0)$ to $(x, y)$ will yield the same point $R_{x,y}$, so this decoration is well-defined. Any strategy $S$ of size $e$ inherits these decorations from $T_e$. By [11, Lemma 4.2], this decoration corresponds to an algorithm to compute the isogeny $\psi$ with $\ker \psi = \langle P + mQ \rangle$, by first constructing the point $R_{0,0}$, and then applying the maps corresponding to the edges of $S$, in depth-first, bottom-first order, noting that the points $R_{e-y,y-1}$ generate the kernels of the $\psi_y$. This decoration is depicted in Fig. 2.

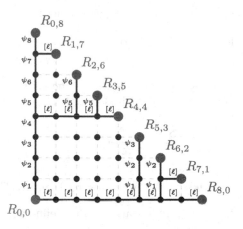

**Fig. 2.** A strategy decorated in the SIDH style. The root point $R_{0,0}$ is labelled and highlighted in red, and the terminals $R_{e-y,y-1}$ for $y = 1, 2, \ldots, e$ are labelled and highlighted in blue. (Color figure online)

If we assign weights to a strategy $S$ by the measure $M = (\{\ell_j\}_{j=1}^{e}, f_H, f_V)$, where $\ell_j = \ell$ for all $j$, and $f_H(\ell)$ and $f_V(\ell)$ are the cost of evaluating $P \mapsto [\ell]P$ and $P \mapsto \psi_j(P)$ for any $j$ (this cost does not depend on $j$, only on the degree of the isogeny, which is $\ell$), then $(S)_M$ is precisely the cost of computing Alice's isogeny $\psi$. Thus the authors of [11] use strategies which have minimal weight with respect to this measure to construct their isogeny construction algorithms.

*Adapting Measures to eSIDH.* Taking $e = e_1 + e_2 + \cdots + e_n$ and the sequence $\{\ell_i\}_{i=1}^{e}$ to be (some permutation of) $\underbrace{\ell_1, \ell_1, \ldots \ell_1}_{e_1}, \underbrace{\ell_2, \ell_2, \ldots \ell_2}_{e_2}, \ldots, \underbrace{\ell_n, \ell_n, \ldots \ell_n}_{e_n}$ as

**Table 1.** Cost of functions in SIDH in quadratic extension field arithmetic

| Operation | M | S | a | Normalized Cost ( S = 0.66 M , a = 0.05 M ) |
|---|---|---|---|---|
| Ladder Step | 7 | 4 | 8 | 10.04 |
| xTPL | 7 | 5 | 10 | 10.8 |
| xQPL | 11 | 6 | 14 | 15.66 |
| eval3Iso | 4 | 2 | 4 | 5.52 |
| eval5Iso | 8 | 2 | 8 | 9.72 |
| get3Iso | 2 | 3 | 13 | 4.63 |
| get5Iso | 6 | 6 | 0 | 9.96 |

described in Sect. 3, then $(S)_M$ is the cost of constructing Bob's isogeny in eSIDH for a prime $p$ of the form $p = 2^{e_A} \ell_1^{e_1} \cdots \ell_n^{e_n} f - 1$.

### 4.2 Optimized Strategies for Multiprime Large-Degree Isogenies

We propose new multiprime strategies for large-degree isogenies in the eSIDH landscape. In particular, this only applies to *Bob's* large-degree isogeny, which is of the form $\ell_1^{e_1} \ell_2^{e_2} \cdots \ell_n^{e_n}$. In eSIDH [8], Bob's large-degree isogeny is of the form $3^{e_B} 5^{e_C}$, where $3^{e_B} \approx 5^{e_C}$ and $2^{e_A} \approx 3^{e_B} 5^{e_C}$. In the first case, $3^{e_B} \approx 5^{e_C}$ so that the eSIDH kernel generations $R_B = P_B + n_B Q_B$ and $R_C = P_C + n_C Q_C$ can be efficiently parallelized into 2 cores to reduce kernel generation latency. The second case, $2^{e_A} \approx 3^{e_B} 5^{e_C}$ so that Alice and Bob perform approximately the same magnitude of isogeny.

eSIDH proposes efficient primes with a similar security as the SIKE parameter levels [2]. These correspond to NIST security levels, ranging from 1 to 5, where SIKE has parameter sets at levels 1, 2, 3, and 5. NIST security level 1 is conjectured to be as hard to break as a brute-force attack on AES128, NIST security level 2 is conjectured to be as hard to break as finding a hash collision in SHA256, NIST security level 3 is conjectured to be as hard to break as a brute-force attack on AES192, and NIST security level 5 is conjectured to be as hard to break as a brute-force attack on AES256. For an initial experiment, we compared the cost of multiprime large-degree isogenies for eSIDH primes $p_{443} = 2^{222} 3^{73} 5^{45} - 1$ and $p_{765} = 2^{391} 3^{119} 5^{81} - 1$.

*Building a Cost Model for the Isogeny Operations.* Similar to computing optimal strategies for SIDH/SIKE, one must first identify the weights of the edges in the strategy graphs. For this purpose, we used the fastest formulas for 3 and 5-isogenies as available in the literature. Luckily, these formulas were available in the literature as they were essential for speeding up CSIDH isogenies of any odd prime. In general, the optimized 3-isogeny formulas come from [9] and 5-isogeny formulas (and higher degree) come from [7]. We summarize the cost of these formulas in terms of finite field arithmetic multiplication, squaring, and addition in Table 1. In the realm of SIDH and SIKE, these finite field operations are in quadratic extension field $\mathbb{F}_{p^2}$. To give a normalized cost of operations, we assumed that the cost of $\mathbb{F}_{p^2}$ squaring is approximately $2/3$ the cost of $\mathbb{F}_{p^2}$ multiplication, and $\mathbb{F}_{p^2}$ addition is approximately $1/20$ the cost of $\mathbb{F}_{p^2}$ multiplication. This generates a single value with which we can compare the cost of a strategy. Other cost models can be made, depending on the selected device. We chose $S = 0.66M$ because there are 2 $\mathbb{F}_p$ multiplication operations in $\mathbb{F}_{p^2}$ squaring and 3 $\mathbb{F}_p$ multiplication operations in $\mathbb{F}_{p^2}$ multiplication.

Table 1 lists the costs of large subroutines in SIDH and SIKE. The large-degree isogeny operation uses small point multiplication operations (xTPL for $Q = 3P$, xQPL for $Q = 5P$), small isogeny evaluations where a point is pushed from one elliptic curve to an isogenous curve (eval3Iso for pushing a point through a 3-isogeny mapping, eval5Iso for pushing a point through a 5-isogeny mapping), and small isogeny computations where you compute an isogenous mapping of a small degree (get3Iso for computing an isogeny mapping of degree 3 from a point of order 3, get5Iso for computing an isogeny mapping from a point of order 5). In order to create a multiprime strategy, the cost of point multiplication and isogeny evaluation creates the weighting of the large-degree isogeny graph. In particular, a horizontal edge is a point multiplication by $\ell$ and a vertical edge is an $\ell$−degree isogeny evaluation. The isogeny computations are computed at the leaf nodes, which is done regardless of strategy.

*Finding Multiprime Strategies.* As in CSIDH, when constructing strategies in the multiprime setting there are two orthogonal algorithmic concepts to optimize: the strategy itself, and the permutation of the list of primes, which determines the order in which the small prime degree isogenies are computed. As shown in [15] for CSIDH, there are efficient techniques to construct optimal strategies for fixed permutations (using dynamic programming, as in [11]) and to construct an optimal permutation of the primes for fixed strategies (using linear programming); however, it is not presently known how to construct globally optimal (permutation, strategy) pairs. For the purposes of this work, we construct (permutation, strategy) pairs using a straightforward randomized alternating algorithm: we choose a random starting permutation, and then alternately optimize our strategy (fixing the permutation) and our permutation (fixing the strategy) until the (permutation, strategy) pair stabilizes. The resulting pair is not globally optimal, in general, so we run 10 trials with different random starting permutations, and then choose the best resulting (permutation, strategy) pair for implementation.

## 4.3   Evaluating the Costs of Multiprime Strategies

Based on the cost models that we proposed in the previous section and the obtained multiprime strategies, we can evaluate the cost of these new strategies and compare them to the state-of-the-art. We note that the implementation is slightly different in the split-prime and multiprime strategy. First, the split-prime uses a different method to generate the secret kernels. In split-prime eSIDH, Bob initially computes two kernel generators $R_B = P_B + n_B Q_B$ and $R_C =$

**Table 2.** SIDH large operation costs for Bob's round functions

| SIDH Round | Operation | Split Prime | Multi Prime | Single Prime |
|---|---|---|---|---|
| | | $p_{443} = 2^{222}3^{73}5^{45} - 1$ | | $p_{434} = 2^{216}3^{137} - 1$ |
| Round 1 | Kernel | 2,219 | 2,209 | 2,179 |
| | Isogeny | 10,479 | 11,027 | 10,757 |
| | Total | 12,698 | 13,236 | 12,935 |
| Round 2 | Kernel | 3,712 | 2,209 | 2,179 |
| | Isogeny | 7,958 | 8,506 | 8,488 |
| | Total | 11,670 | 10,715 | 10,667 |
| | | $p_{765} = 2^{391}3^{119}5^{81} - 1$ | | $p_{751} = 2^{372}3^{239} - 1$ |
| Round 1 | Kernel | 3,795 | 3,775 | 3,795 |
| | Isogeny | 19,297 | 20,392 | 20,345 |
| | Total | 23,092 | 24,168 | 24,141 |
| Round 2 | Kernel | 6,349 | 3,775 | 3,795 |
| | Isogeny | 14,965 | 16,060 | 16,388 |
| | Total | 21,313 | 19,835 | 20,183 |

$P_C + n_C Q_C$, which have order $3^{e_B}$ and $5^{e_C}$, respectively. Thus, the split-prime strategy will first compute a large-degree isogeny of degree $3^{e_B}$ over kernel $R_B$ and then a second large-degree isogeny of degree $5^{e_C}$ over kernel $R_C$ (this order could be flipped if it is faster). The caveat here is that whichever large-degree isogeny is computed first, you have to apply the small degree isogeny evaluations to the other kernel. The multiprime strategy is different in that you compute only a single kernel, $R_{BC} = P_{BC} + n_{BC}Q_{BC}$. This single kernel is then used to compute a large-degree isogeny of order $3^{e_B}5^{e_C}$ using a strategy.

We summarize the total cost of Bob's large-degree isogeny operations over the split prime strategy, multiprime strategy, as well as baseline SIDH/SIKE single prime strategy in Table 2. These are provided for the smallest and largest SIKE parameter sets. The two largest operations are kernel generation, where a double-point multiplication generates the secret kernel, and then the large-degree isogeny where you compute the isogeny over that secret kernel. The split prime strategy generates 2 kernels. This is based on the calculated number of functions used in each of the strategies, where each function cost is taken from Table 1. This estimate does not include the cost of some other functions, such as the finite field inversion or setup, which is expected to be a similar cost amongst SIDH/SIKE primes.

We further break down the cost of operations in Table 3. Here, we list the major SIDH/SIKE operations as well as the functions that are used many times within. For each function, we also specify whether its cost is included in the first or second round of Bob's large-degree isogeny operation. In the first round, Bob can use the public parameters to generate the kernel, whereas the second round uses Alice's public key to generate the kernel. As is specified in the split prime

**Table 3.** SIDH round breakdown of costs in Bob's large operations

| Op | Function | SIDH Round | | Split Prime | Multi Prime | Single Prime |
|---|---|---|---|---|---|---|
| | | R1 | R2 | | | |
| | | | | $p_{443}$ | | $p_{434}$ |
| Kernel | Ladder Step | ✓ | ✓ | 221 | 220 | 217 |
| | xTPL | | ✓ | 73 | 0 | 0 |
| | xQPL | | ✓ | 45 | 0 | 0 |
| Isogeny | xTPL | ✓ | ✓ | 175 | 176 | 466 |
| | xQPL | ✓ | ✓ | 94 | 154 | 0 |
| | eval3Iso | ✓ | ✓ | 382 | 399 | 511 |
| | eval5Iso | ✓ | ✓ | 175 | 124 | 0 |
| | get3Iso | ✓ | ✓ | 73 | 73 | 137 |
| | get5Iso | ✓ | ✓ | 45 | 45 | 0 |
| | R1 eval3Iso | ✓ | | 219 | 219 | 411 |
| | R1 eval5Iso | ✓ | | 135 | 135 | 0 |
| | | | | $p_{765}$ | | $p_{751}$ |
| Kernel | Ladder Step | ✓ | ✓ | 378 | 376 | 378 |
| | xTPL | | ✓ | 119 | 0 | 0 |
| | xQPL | | ✓ | 81 | 0 | 0 |
| Isogeny | xTPL | ✓ | ✓ | 315 | 316 | 913 |
| | xQPL | ✓ | ✓ | 220 | 306 | 0 |
| | eval3Iso | ✓ | ✓ | 684 | 721 | 982 |
| | eval5Iso | ✓ | ✓ | 307 | 259 | 0 |
| | get3Iso | ✓ | ✓ | 119 | 119 | 239 |
| | get5Iso | ✓ | ✓ | 81 | 81 | 0 |
| | R1 eval3Iso | ✓ | | 357 | 357 | 717 |
| | R1 eval5Iso | ✓ | | 243 | 243 | 0 |

strategy in eSIDH [8], Bob's first round can efficiently compute kernels of order $3^{e_B}$ and $5^{e_C}$ because the public parameters include generator points of these orders. For public key size efficiency, Alice applies her large-degree isogeny only over public torsion points of order $3^{e_B}5^{e_C}$. Thus, Bob's second round with the multiprime strategy requires a significant number of point triplings and quintuplings to convert his computed secret kernel of order $3^{e_B}5^{e_C}$ into two secret kernels of order $3^{e_B}$ and $5^{e_C}$. For serial implementations, the magnitude of the Montgomery ladder steps is approximately the same between the three types of strategies. The split strategy suffers significantly in the second round to generate the kernels of correct order.

For the large-degree isogeny operation, we note that the split prime strategy outperforms the multi prime and single prime strategy by about 5% in both rounds 1 and 2. In the view of the strategy, this is to be expected as the split prime strategy begins with a head-start as many of the point multiplications are already computed to get the two kernels. The multiprime strategy is approximately on-par with the single prime strategy, which shows that the inclusion of 5-isogeny operations were efficiently interleaved with the 3-isogeny operations. We note that the R1 eval3Iso and R1 eval5Iso function count is the number of isogeny evaluations applied to the other party's torsion basis, which is only done for the first round.

When we consider the entire round function for SIDH, the split prime strategy is superior for round 1 (key generation), but the multiprime and single prime strategies are superior for round 2 (shared secret generation). The first round's split prime strategy is faster primarily as a result of its more efficient isogeny. However, the second round split prime strategy suffers as it is expensive to generate the kernels. The split prime strategy is about 5% faster for the first round, but the multiprime strategy is about 10% faster for the second round.

### 4.4  Expanding eSIDH to More Primes

In [8], the authors searched for eSIDH-friendly primes by the following criteria:

1. eSIDH primes of the form $p = 4^{e_A} 3^{e_B} 5^{e_C} f - 1$, where $e_A$ is the number of 4-isogenies that Alice performs, $e_B$ is the number of 3-isogenies that Bob performs, $e_C$ is the number of isogenies that Bob performs, and $f$ is a number that makes the number prime.
2. $4^{e_A} \approx 3^{e_B} 5^{e_C}$,
3. $3^{e_B} \approx 5^{e_C}$,
4. $p \equiv \pm \bmod 2^{\gamma w}$, i.e. the prime has many words that are "0xFFFF..." such as for $w = 64$ for efficient modular multiplication.

The first criterion is necessary to guarantee that the $E[4^{e_A}], E[3^{e_B}], E[5^{e_C}] \subseteq E(\mathbb{F}_{p^2})$ (ensuring that the isogenies can be computed using arithmetic in $\mathbb{F}_{p^2}$, rather than a higher degree extension), while the second is required for the protocol to be secure. The third criterion is a heuristic for optimality in the parallel setting, while the fourth yields only minor computational improvements for software processors; in this paper we experiment with how eSIDH performs with multiprime strategies in the serial setting, so we simplify the search for eSIDH serial-efficient primes by removing the third and fourth criteria above. In addition, instead of limiting to only two primes for Bob, we also experiment with the base prime 7.

Another benefit of the eSIDH framework is that it opens up a vast number of primes that can be used and optimized. For instance, the chance of an odd number of the form $p = 4^{e_A} 3^{e_B} f - 1$ is approximately $2/\ln p$. For a 434-bit number as is used as a prime to specify SIKE's NIST level 1 parameter set, there is approximately a 1 in 150 chance that the number is actually prime. With the prime search criterion that $4^{e_A} \approx 3^{e_B}$, this further limits the pool of good parameters. If one isogeny graph is significantly larger than the other, then one party's computations will be asymptotically more expensive.

**Table 4.** Costs of NIST Security Level 1 multiprime strategies considered in this paper in terms of field multiplications, under three standard cost models. In the SIDH/eSIDH landscape, this cost is the "second round" cost.

| Isogeny degree | Cost Model | | |
|---|---|---|---|
| | S = 0.8 M a = 0 M | S = 0.8 M a = 0.05 M | S = 0.66 M a = 0.05 M |
| $3^{73}5^{45}$ | 8858.4 | 9230.9 | 8752.3 |
| $3^{97}5^{27}$ | 8535.2 | 8915.1 | 8430.8 |
| $3^{132}5^{3}$ | 8388.2 | 8794.1 | 8289.5 |
| $3^{91}5^{25}7^{5}$ | 8614.2 | 8990.6 | 8506.3 |
| $3^{93}5^{25}7^{4}$ | 8610.2 | 8988.4 | 8502.7 |
| $3^{109}5^{15}7^{3}$ | 8500.4 | 8888.4 | 8396.5 |

By changing the form of the number to $p = 4^{e_A}3^{e_1}5^{e_2}\ldots\ell_n^{e_n}f - 1$, and using the stipulation $4^{e_A} \approx 3^{e_1}5^{e_2}\ldots\ell_n^{e_n}$, we greatly increase the pool of numbers that can potentially be used as parameter sets. We used this relaxed methodology to find many different eSIDH primes. We highlight some NIST Security Level 1 parameter sets we found in Table 4. Among the three different cost models, we see that the fastest large-degree isogeny $3^{132}5^3$ uses the most 3-isogenies. This is to be expected as the larger isogeny formulas are not nearly as optimized as the 3-isogeny formulas.

For our recommended prime parameter sets, we chose 3- and 5-isogeny primes with a preference for more 3-isogenies. We were also able to ensure that the magnitude of the large-degree isogeny is approximately the same as Alice's large-degree isogeny. In the SIKE parameter set, this is especially imbalanced for $p_{751} = 2^{372}3^{237}$, where $3^{237}$ is over 100 times larger than $2^{372}$. Our proposed eSIDH primes are shown in Table 5. Across the board, we achieve about a 3% performance improvement by going with a multiprime strategy-friendly prime.

## 5   Software Implementation

As a further investigation to the effectiveness of our multiprime strategy, we implemented our multiprime strategies on top of the eSIDH version 2.0 library[1] This library was based on the SIKE team's implementation, SIDH Library v3.2. This eSIDH library features support for SIKE parameter sets $p_{434} = 2^{216}3^{137} - 1$ and $p_{751} = 2^{372}3^{239} - 1$, and eSIDH/eSIKE parameter sets $p_{443} = 2^{222}3^{73}5^{45} - 1$ and $p_{765} = 2^{391}3^{119}5^{81} - 1$. We applied no modifications to the lower-level finite field arithmetic. We ran our modified code on an Intel i7-8650u processor running at 1.9 GHz. All tests were run on a single core with turbo boost disabled.

---

[1] Commit b8f4486 at https://github.com/dcervantesv/eSIDH.

**Table 5.** eSIDH/eSIKE timing results on Intel i7-8650u processor.

| Scheme | Operation | Timings (Mcycles) | | Improvement |
|---|---|---|---|---|
| | | Split Prime | Multiprime | |
| $p_{443} = 2^{222}3^{73}5^{45} - 1$ | | | | |
| eSIDH | Bob R1 | 7.44 | 7.75 | −4.03% |
| | **Bob R2** | **7.00** | **6.47** | **8.24%** |
| eSIKE | Keygen | 7.43 | 7.74 | −4.01% |
| | **Decap** | **13.71** | **13.17** | **4.10%** |
| $p_{765} = 2^{391}3^{119}5^{81} - 1$ | | | | |
| eSIDH | Bob R1 | 27.14 | 28.37 | −4.34% |
| | **Bob R2** | **25.56** | **23.90** | **6.96%** |
| eSIKE | Keygen | 27.14 | 28.39 | −4.42% |
| | **Decap** | **50.47** | **48.83** | **3.36%** |

Our eSIDH and eSIKE timing results are summarized in Table 5. These results summarize the affected SIDH and SIKE operations (only Bob's large-degree isogeny is affected by the changes). As we can see, the multiprime strategy results in a 4% slowdown for Bob's first round, but an 8% speedup for Bob's second round, when considering $p_{443}$. For $p_{765}$, Bob's first round is again 4% slower and his second round is 7% faster. For the SIKE operations, key generation is almost identical to Bob's round 1 in SIDH, resulting in about a 4% slowdown. SIKE key decapsulation then uses Bob's round 2 in SIDH, where we see a 4% speedup for $p_{443}$ and a 3% speedup for $p_{765}$.

*Implementing Multiprime Strategies.* When implementing the multiprime strategy, we used a very similar algorithm as that of SIDH/SIKE. A strategy describes the order in which we traverse from the root of the large-degree isogeny strategy graph to its leaves. The primary difference is that we define a prime list that includes the order in which isogenies are performed. In the single prime strategy this is unnecessary, but is needed in the multiprime isogenies as you are mixing the order of prime isogenies. We defined this list as the order they are to be used in the corresponding strategy. In particular, primes are listed in the order that they are multiplied into the starting point by a strategy. For instance, if we had a small strategy using only 3 s and 5 s, and we had the sublist [3,5], this would mean that the first isogeny would be of degree 3, since we would multiply out 5, then 3, then 3, and then 5, leaving a point of degree 3. The next isogeny would be of degree 5, then degree 3, degree 3, and finally degree 5. In short, the order of the isogenies is the reverse of this list.

We then decomposed the strategy into a single list specifying how many point multiplications by small-degree primes are required to perform to generate a pivot point. Since the prime list is ordered, a strategy may dictate a mix of primes to multiply by to create a pivot point. In Fig. 1, a pivot point is created

**Algorithm 1:** Computing and evaluating a multiprime $\ell_1^{e_1}\ell_2^{e_2}\cdots\ell_n^{e_n}$-isogeny with a strategy

---

    function multiprime_iso

       **Static Parameters:** Small prime numbers $\ell_1, \ell_2, \ldots \ell_n$ and Integers
                 $e_1, e_2, \ldots e_n, e_S = (e_1 + e_2 + \ldots + e_n)$ from public
                 parameters, a list describing an order of point
                 multiplications by small prime
                 $M = (m_1, \ldots, m_{e_S}) \in (\mathbb{N}^+)^{e_S}$, a *strategy*
                 $(s_1, \ldots, s_{e_S-1}) \in (\mathbb{N}^+)^{e_S-1}$

       **Input:** Curve $E_0$ and point $S$ on $E_0$ with exact order $\ell_1^{e_1}\ell_2^{e_2}\cdots\ell_n^{e_n}$

       **Output:** Curve $E = E_0/\langle S \rangle$

1   Initialize empty deque $D$

2   push($D, (e_S, S)$)

3   $E \leftarrow E_0, i \leftarrow 1, h \leftarrow e_S, k = e_S$

4   **while** $D$ *not empty* **do**

5      $(h, R) \leftarrow$ pop($D$)

6      **if** $h = 1$

7          $(E', \phi) \leftarrow$ compute_$\ell$_iso($E, R, \ell = m_k$)

8          Initialize empty deque $D'$

9          **while** $D$ *not empty* **do**

10             $(h, R) \leftarrow$ pull($D$)

11             $R \leftarrow$ evaluate_$\ell$_iso($E', \phi, R, \ell = m_k$)

12             push($D', (h - 1, R)$)

13          $D \leftarrow D', E \leftarrow E', k \leftarrow k - 1$

14      **elif** $0 < s_i < h$

15          push($D, (h, R)$)

16          **for** $j \leftarrow e_S - h$ to $e_S - h + s_i$ **do**

17             $R \leftarrow$ mult_by_$\ell$($R, E, \ell = m_j$)

18          push($D, (h - s_i, R)$), $i \leftarrow i + 1$

19      **else**

20          **Error:** Invalid strategy

21 **return** $E = E_0/\langle S \rangle$

---

and stored whenever there is a vertical edge. These points are then pushed through isogeny evaluations after each isogeny computation. Upon reaching a leaf node, an isogeny would be computed as is specified in the prime list.

Algorithm 1 shows our algorithm for performing a multiprime large-degree isogeny. This is generalized from the SIKE submission's [2] algorithm for computing an $\ell^e$ isogeny. In terms of parameters, the primary difference here is that we have multiple small primes to use. Thus, we have one summation term, $e_S$ to denote the total number of isogenies to perform. Then, the order of the isogenies matter, so this is represented by a list describing the order of the strategy. In total, there are $e_S$ isogenies that need to be performed, so the point multiplication ordered list should have $e_S$ entries, which can further be broken down into $e_B$ entries of $\ell_B$, $e_C$ entries of $\ell_C$, and so on.

In terms of the algorithm flow, we only include one more variable, $k$, which represents the next isogeny to perform. This acts as a reverse iterator of the list $M$, so after computing an $\ell$-isogeny and evaluating each stored pivot point, we decrement the counter in line 13. Otherwise, we now have to identify which $\ell$ we are using at each isogeny operation or point multiplication. For isogeny computation and evaluation, this is variable $k$. For the point multiplication, we are still using the same strategy flow. By this, we mean that say $s_i = 5$. This means that we will perform 5 point multiplications starting at some index in the point multiplication list and iterating through the next 5 entries. We have represented this index by updating the for loop to start at $e_S - h$ as is shown in line 16. $h$ represents the current number of point multiplications that have been applied to the current point, so we subtract this amount from the total number of isogenies to get the current index into the multiplication list.

*Co-Implementing Multiprime and Split-Prime Strategies.* As our results show in Table 5, the split prime approach is ideal for SIDH round 1 operations and the multiprime approach is ideal for SIDH round 2 operations. It is simple to use different large-degree isogeny algorithms for both rounds. The only caveat is that you must calculate a different representation of the private key for each round. For instance, if you start with two private keys to generate the two kernels as part of round 1, then you need to use the Chinese Remainder Theorem to combine these keys. If you start with a single large private key, then you must generate the two smaller keys by performing modulus operations, such as $n_B = n_{BC} \mod \ell_B^{e_B}$ and $n_C = n_{BC} \mod \ell_C^{e_C}$. These can be done efficiently using Montgomery or Barrett reduction [3,17] (with precomputed values), but special care should be taken here to prevent side-channel attacks.

# 6   Conclusions

We applied the concept of multiprime large-degree isogeny strategies to the extended SIDH framework. We see that multiprime strategies can be used to accelerate Bob's large-degree isogeny by about 10% for the balanced eSIDH primes in the serial setting. We applied multiprime strategies to generalized prime forms for Bob, finding new primes that could further accelerate Bob's large-degree isogeny by a further 3%. The beauty of eSIDH is that the generalized form of prime allows for a variety of optimization targets, including parallelization and a prime of a target size. This work continues to push the envelope for performance gains when (e)SIDH parameters are chosen well.

# A   SIDH Protocol Description

Supersingular Isogeny Diffie-Hellman (SIDH) was introduced by De Feo, Jao, and Plût in 2011 [11]. Superficially the protocol resembles the classical

Diffie-Hellman protocol [10], with the base group replaced by a set of elliptic curves, and the group operation replaced with isogeny codomain construction.

**Setup:** We require the following global parameters:

1. A prime $p = \ell_A^{e_A} \ell_B^{e_B} f \pm 1$ where $\ell_A$ and $\ell_B$ are prime, and $\ell_A^{e_A} \approx \ell_B^{e_B}$;
2. A supersingular elliptic curve $E/\mathbb{F}_{p^2}$; and,
3. Four points $P_A, P_B, Q_A, Q_B \in E(\mathbb{F}_{p^2})$ such that $E[\ell_A^{e_A}] = \langle P_A, Q_A \rangle$ and $E[\ell_B^{e_B}] = \langle P_B, Q_B \rangle$.

One party (Alice) will use the $\ell_A^{e_A}$-torsion subgroup, and the other (Bob) will use the $\ell_B^{e_B}$-torsion subgroup.

**Key Generation:** Alice:

1. Selects $\alpha \in \mathbb{Z}/\ell_A^{e_A}\mathbb{Z}$ uniformly at random;
2. Constructs the isogeny $\psi_A \colon E \to E_A = E/\langle P_A + \alpha Q_A \rangle$; and,
3. Constructs the auxiliary points $S_A = \psi_A(P_B)$ and $T_A = \psi_A(Q_B)$.

Alice's private/public keypair is

$$\mathsf{sk}_A = \alpha \text{ and } \mathsf{pk}_A = (E_A, S_A, T_A).$$

Bob proceeds analogously.

**Communication:** The parties exchange their public keys.

**Key Establishment:** Alice computes

$$K_A = j\left(E_B/\langle S_B + \alpha T_B \rangle\right)$$

Bob proceeds analogously to find his key $K_B$. We have $K_A = K_B$.

The protocol is depicted in Fig. 3.

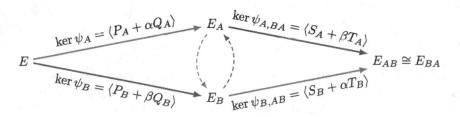

**Fig. 3.** The computations involved in SIDH. Alice follows the solid blue arrows by finding the codomain curve of the indicated isogeny, and follows the dashed blue arrow by reading the message she receives from Bob. Bob analogously follows the red arrows. (Color figure online)

The underlying hard problem of SIDH is the following:

*Problem 1 (Supersingular Decisional Diffie-Hellman Problem).* Let $\phi_A \colon E \to E_A$ be an isogeny with kernel $\langle P_A + \alpha Q_A \rangle$ where $\alpha$ is chosen uniformly at random from $\mathbb{Z}/\ell_A^{e_A}\mathbb{Z}$. Similarly, let $\phi_B \colon E \to E_B$ be an isogeny with kernel $\langle P_B + \beta Q_B \rangle$ where $\beta$ is chosen uniformly at random from $\mathbb{Z}/\ell_B^{e_B}\mathbb{Z}$. Given a tuple

$$(E, E_A, E_B, \phi_A(P_B), \phi_A(Q_B), \phi_B(P_A), \phi_B(Q_A), E_C)$$

where either $E_C = E_{AB} = E/\langle P_A + \alpha Q_A, P_B + \beta Q_B \rangle$ or $E_C$ is sampled uniformly at random from the set of all curves of the form

$$E/\langle P_A + y_A Q_A, P_B + y_B Q_B \rangle$$

where $y_A$ and $y_B$ are chosen with the same conditions as $\alpha$ and $\beta$, respectively, each with probability $\frac{1}{2}$, the *supersingular decisional Diffie-Hellman problem (SSDDH)* is to determine which is the case.

# References

1. Azarderakhsh, R., et al.: Practical Supersingular Isogeny Group Key Agreement. Cryptology ePrint Archive, Report 2019/330 (2019)
2. Azarderakhsh, R., et al.: Supersingular Isogeny Key Encapsulation. Technical report (2020). https://www.sike.org
3. Barrett, P.: Implementing the Rivest Shamir and Adleman public key encryption algorithm on a standard digital signal processor. In: Odlyzko, A.M. (ed.) CRYPTO 1986. LNCS, vol. 263, pp. 311–323. Springer, Heidelberg (1987). https://doi.org/10.1007/3-540-47721-7_24
4. Bernstein, D., et al.: Faster computation of isogenies of large prime degree. Open Book Series **4**, 39–55 (2020)
5. Castryck, W., Decru, T.: An efficient key recovery attack on SIDH. Cryptology ePrint Archive, Paper 2022/975 (2022)
6. Castryck, W., Lange, T., Martindale, C., Panny, L., Renes, J.: CSIDH: an efficient post-quantum commutative group action. In: Peyrin, T., Galbraith, S. (eds.) ASIACRYPT 2018. LNCS, vol. 11274, pp. 395–427. Springer, Cham (2018). https://doi.org/10.1007/978-3-030-03332-3_15
7. Cervantes-Vázquez, D., Rodríguez-Henríquez, F.: A note on the cost of computing odd degree isogenies. Cryptology ePrint Archive, Report 2019/1373 (2019)
8. Cervantes-Vázquez, D., et al.: eSIDH: the revenge of the SIDH. Cryptology ePrint Archive, Report 2020/021 (2020)
9. Costello, C., Hisil, H.: A simple and compact algorithm for SIDH with arbitrary degree isogenies. In: Takagi, T., Peyrin, T. (eds.) ASIACRYPT 2017. LNCS, vol. 10625, pp. 303–329. Springer, Cham (2017). https://doi.org/10.1007/978-3-319-70697-9_11
10. Diffie, W., Hellman, M.E.: New directions in cryptography. IEEE Trans. Inf. Theory **22**(6), 644–654 (1976)
11. De Feo, L., et al.: SIDH Proof of Knowledge. Cryptology ePrint Archive, Paper 2021/1023 (2021)
12. De Feo, L., et al.: Towards quantum-resistant cryptosystems from supersingular elliptic curve isogenies. J. Math. Cryptol. **8**(3), 209–247 (2014)

13. Furukawa, S., et al.: Multi-party key exchange protocols from supersingular isogenies. In: 2018 International Symposium on Information Theory and Its Applications (ISITA), pp. 208–212 (2018)
14. Ghantous, W., et al.: Efficiency of SIDH-based signatures (yes, SIDH). Cryptology ePrint Archive, Paper 2023/433 (2023)
15. Hutchinson, A., LeGrow, J., Koziel, B., Azarderakhsh, R.: Further optimizations of CSIDH: a systematic approach to efficient strategies, permutations, and bound vectors. In: Conti, M., Zhou, J., Casalicchio, E., Spognardi, A. (eds.) ACNS 2020. LNCS, vol. 12146, pp. 481–501. Springer, Cham (2020). https://doi.org/10.1007/978-3-030-57808-4_24
16. Maino, L., Martindale, C.: An attack on SIDH with arbitrary starting curve. Cryptology ePrint Archive, Paper 2022/1026 (2022)
17. Montgomery, P.L.: Modular multiplication without trial division. Math. Comput. **44**(170), 519–521 (1985)
18. Robert, D.: Breaking SIDH in polynomial time. Cryptology ePrint Archive, Paper 2022/1038 (2022)
19. Vélu, J.: Isogénies entre courbes elliptiques. C. R. Acad. Sci. Paris Sér. A-B **273**, A238–A241 (1971)

# Adaptively Secure Constrained Verifiable Random Function

Yao Zan[1,2], Hongda Li[1,2(✉)], and Haixia Xu[1,2]

[1] State Key Laboratory of Information Security, Institute of Information Engineering, Chinese Academy of Sciences, Beijing, China
{zanyao,lihongda,xuhaixia}@iie.ac.cn
[2] School of Cyber Security, University of Chinese Academy of Sciences, Beijing, China

**Abstract.** Constrained Verifiable Random Function (CVRF) is a powerful variant of Pseudorandom Function (PRF). Simply put, CVRF asks the outputs of PRF to be verifiable and the secret key of PRF to be delegatable, thus simultaneously resolving the PRF's trust and "all or nothing" problems. Among the existing constructions of CVRF, the optimal implementation of security, to our knowledge, should be the semi-adaptive security of [SCN 2019] where an adversary can make some queries before issuing its attack target but get critical public information only after the attack. Here we give a generic construction of CVRF that achieves a stronger security, called adaptive security: the adversary has access to this public information at the beginning of the security experiment.

Concretely, we first define a slightly weaker security of CVRF, called single-key security, and prove its existence. Then, using it and Indistinguishability Obfuscation and Partition Scheme, we construct an adaptively secure CVRF. Notably, our proof technique may provide a direction for achieving adaptive security in scenarios related to Indistinguishability Obfuscation, where puncturable techniques have been commonly used before. Beyond this, we analyze the possible implications of our proposed construction in the micro-payment scenario.

**Keywords:** Constrained Verifiable Random Function · Indistinguishability Obfuscation · Partition Technique · Single-key Security

## 1 Introduction

Pseudorandom Function (PRF) is a fundamental primitive in modern cryptography and was introduced by Goldreich, Goldwasser, and Micali [22] in 1984. It is essentially a family of keyed functions, denoted by $F : \mathcal{K} \times \mathcal{X} \to \mathcal{Y}$ for the key space $\mathcal{K}$, domain space $\mathcal{X}$, and range space $\mathcal{Y}$. If an evaluator runs a PRF family, then it can sample a secret key $k$ from the key space $\mathcal{K}$ effectively. For any input $x \in \mathcal{X}$, the PRF evaluator can effectively output a value $y = F(k, x)$. The security of PRF is called pseudorandomness: for any probability polynomial

© The Author(s), under exclusive license to Springer Nature Switzerland AG 2023
M. Yung et al. (Eds.): SciSec 2023, LNCS 14299, pp. 367–385, 2023.
https://doi.org/10.1007/978-3-031-45933-7_22

time (PPT) adversary, the probability of distinguishing $y_* = F(k, x_*)$ from a random value equals at most $\frac{1}{2}$ plus a negligible advantage, even the adversary gets the oracle access to $F(k, \cdot)$ and chooses $x_*$ on which it hasn't computed.

PRF is such a useful tool that can be applied in message authentication, symmetric-key encryption, key derivation, and learning theory (summarized by [9]). Various variants of PRF have been derived to meet different needs. Verifiable Random Function (VRF), introduced by Micali, Rabin, and Vadhan [29] in 1999, resolved the issue of trust in a PRF evaluator, namely, preventing the PRF evaluator from printing a malicious value while claiming that the value is the true output of PRF. In a VRF, the evaluator should (1) disclose some critical public information about the PRF $F(k, \cdot)$, called the public key, (2) for any input $x \in \mathcal{X}$, output not only a value $y = F(k, x)$ but also a non-interactive proof. The receiver, owning the proof and the public key, can verify whether the value $y$ agrees with the output of PRF $F(k, \cdot)$ on input $x$. Without proof, the receiver cannot distinguish $y$ from a random value by a non-negligible advantage.

In 2013, the Constrained Pseudorandom Function (CPRF) was introduced by [10, 11, 24] and resolved the PRF's "all or nothing" problem, partly filling the gap between the evaluation powers of PRF evaluators and the receivers [1]. For any efficiently represented set (i.e. circuit set)[2] $S \subseteq \mathcal{X}$, CPRF allows the PRF evaluator to generate a constrained key, which can be delegated to the receivers or a third party to compute $y = F(k, x)$ only if $x \in S$. In 2014, Fuchsbauer introduced a Constrained Verifiable Random Function (CVRF) [19], which resolves the issue of trust in a CPRF evaluator. The CVRF evaluator not only outputs the public key but for CPRF's any output $y$, a proof must be accompanied with, regardless of whether $y$ is computed by the secret key or the constrained key. At present, the research on CVRF focuses on how to construct it for more diverse constraint sets $S$, or with stronger security. Up till now, the constraints of CVRF are very diverse, such as a puncturable point [3] [27], a bit-fixing set [13, 19], a circuit set [13–15, 19, 25], or a puncturable circuit set that contains a polynomial number of punch points [27], or the Turing Machines [16, 17], etc.

It is well known that in a certain security experiment between an adversary and a challenger, the security depends on the size of the resources available to the adversary. In the case of CVRF, the more resources that the adversary gets before it chooses an attack target $x_*$ (on which it wishes to distinguish the value $y_* = F(k, x_*)$ from a random value without owning the corresponding proof), then the more secure the scheme that can resist such an adversary will be. If security implementations with exponential reduction losses are not taken into

---

[1] In a PRF, the evaluator owns the secret key $k$ of PRF and can compute $y = F(k, x)$ on any $x$ in the domain, the receiver without $k$ can compute nothing except for accessing the oracle $F(k, \cdot)$ in a black-box way.

[2] An efficiently represented set (or a circuit set) $S$ [7]: there is a polynomial *poly* such that $S$ can be represented by a circuit $C_S$ of size $poly(\lambda)$ such that $C_S(s) = 1$ if $s \in S$ and else $C_S(s) = 0$.

[3] Puncturable point: the constrained key can compute the values and proofs on all inputs but except for a certain point, this point is called a puncturable (or punch) point.

account, then the existing constructions satisfy only weak security (see Table 1). For example, in the selective-challenge security experiment, the adversary must output the attack target in the beginning [13,14,17,19,25]. Or the semi-adaptive security: an adversary can make some evaluation queries (using $x \in \mathcal{X}$ to request for a value $y$ and a proof) or constrained key queries (using $S \subseteq \mathcal{X}$ to request for a constrained key) before it issues an attack target but only gets the public key after the attack [27]. Few schemes satisfy adaptive security: an adversary obtains the public key in the beginning and then makes some evaluation or constrained key queries, and finally determines an attack target depending on all the information it has obtained. We consider how to implement this security.

***Contribution.*** We give a generic, more sophisticated, and adaptively secure CVRF construction. Our contribution is divided into two parts. The first part proves the existence of single-key secure CVRF. The second part builds an adaptively secure CVRF using a single-key secure CVRF.

Firstly, like the single-key security of CPRF [2,3,32], we define the single-key security of CVRF. The adversary first uses a puncturable constraint $S' \subseteq \mathcal{X}$ to request for the public key and a constrained key (can be used to compute all values and proofs except for the punch points $x \in S'$), then it chooses a punch point as its attack target. We show that if a CVRF scheme satisfies selective-challenge security, then it also satisfies single-key security for circuit sets that contain a polynomial number of punch points. For unrestricted circuit set $S'$, we propose a construction, which is a natural extension of [25].

Secondly, we construct an adaptive secure CVRF using single-key secure CVRF. Considering the reduction, if there is adversary **A** who can attack the adaptive security with a non-negligible advantage, then we can construct adversary **B** to attack the single-key security. In the proof, **B** needs to emulate the challenger of the adaptive security experiment, to answer all the queries made by **A**. However, **B** can only use a puncturable constraint $S'$ to query its challenger to get the public key and constrained key, which cannot evaluate the values on $x \in S'$. What if **A** makes some queries on $x \in S'$? To overcome it, we rely on the Partition Technique [7,23,28]. In a Partition Scheme, there is the domain space $\mathcal{X}$ and range space $\mathcal{X}'$, the mapping algorithm PAR.Enc can map any $x \in \mathcal{X}$ into $x' \in \mathcal{X}'$. The generation algorithm PAR.Gen can generate a partition set $S' \subset \mathcal{X}'$, such that **A**'s all queries and an attack target in $\mathcal{X}$, after the mapping operation, with a noticeable probability, the mapped queries will fall outside $S'$ while the mapped attack target will fall inside $S'$.

Define the CVRF with adaptive security (i.e. CVRF$^{\mathsf{Adp}}$) on the function family $F : \mathcal{K} \times \mathcal{X} \rightarrow \mathcal{Y}$, the CVRF with single-key security (i.e. CVRF$^{\mathsf{Sig\text{-}key}}$) on the function family $F' : \mathcal{K} \times \mathcal{X}' \rightarrow \mathcal{Y}$. For any input $x \in \mathcal{X}$, CVRF$^{\mathsf{Adp}}$ outputs what the CVRF$^{\mathsf{Sig\text{-}key}}$ outputs on input $x' = $ PAR.Enc$(x)$. For any $S \subseteq \mathcal{X}$, CVRF$^{\mathsf{Adp}}$ first defines a circuit $C$: if $x \in S$, $C$ outputs what the CVRF$^{\mathsf{Sig\text{-}key}}$ outputs on input $x' = $ PAR.Enc$(x)$; if $x \notin S$, $C$ outputs $\bot$; Then using an IO obfuscator generates an obfuscated version of $C$, which is set as the constrained key. This construction inherits the properties of the CVRF$^{\mathsf{Sig\text{-}key}}$.

***Single-Key Proof Technique.*** The single-key proof technology is mainly from [7] where the authors built an adaptively secure VRF from IO, partition scheme, and Verifiable Function Commitment [7]. We find that this technology can also be used in CVRF-related constructions [16,17,25,27], where the puncturable technology and IO have been used. For the puncturable technique, a reduction algorithm $\mathbf{B}$ can ask for a punctured key, with which $\mathbf{B}$ can successfully simulate the challenger's actions except for the computing power at a punch point. The punch point is referred to as the attack target $x_*$ in the case of CVRF. However, to achieve CVRF's adaptive security, $x_*$ is always given at the end while $\mathbf{B}$ wants the punctured key at the beginning. This will motivate $\mathbf{B}$ to guess $x_*$ with an inverse-exponential (in the input length) probability, and hence directly leads to an exponential reduction loss. The single-key technique is more general than the puncturable technique, in which a punch point is replaced by a puncturable set. To finish the reduction, the reduction algorithm $\mathbf{B}$ can run the partition scheme's PAR.Gen algorithm beforehand to generate a partition set $S'$ then uses $S'$ to obtain a constrained key, which can be used to compute all value-proofs except for the inputs $x \in S'$. The partition scheme ensures that, with a noticeable probability (larger than an inverse-polynomial probability), the attack target chosen by the adversary who attacks CVRF adaptive security will fall within $S'$ but its queries fall outside $S'$. This technique can reduce the above exponential level reduction loss to a polynomial level.

**Table 1.** Existing Constructions of CVRF

| No. | Cite | Constraints | Assumption | Security |
|-----|------|-------------|------------|----------|
| 1 | [13,19] | Bit-fixing, Circuit | M-DDH | Sel-cha, Adp(exp) |
| 2 | [25] | Circuit | IO, P-PRF, Com | Sel-cha |
| 3 | [14,15] | Circuit | FE, P-PRF | Sel-cha |
| 4 | [27] | Puncturable Point | IO, Subgroup hiding | Semi-adp(poly) |
| 5 | [27] | Puncturable Circuit-Poly | IO, P-VRF | Semi-adp(poly) |
| 6 | [16,17] | Turing Machine | IO, P-PRF, $\cdots$ | Sel-cha |
| 7 | **Ours** | Puncturable Circuit-Poly | CVRF$^{\text{Sel-cha}}$ | Sig-key(poly) |
| 8 | **Ours** | Puncturable Circuit | IO, CPRF, Com | Sig-key |
| 9 | **Ours** | Circuit-Poly | PAR, IO, CVRF$^{\text{Sig-key}}$ | Adp(poly) |

***Related Works.*** In Table 1, we show the existing constructions of CVRF, the symbols are described by constraints[4], assumptions[5]: CVRF with selective-challenge

---

[4] Bit-fixing: the set that all strings match a vector $v \in \{0,1,?\}^*$ at all coordinates except for '?'. Circuit (-Poly): Any efficient representation set (-that contains a polynomial number of points). Puncturable Point (Circuit or Circuit-Poly): Any set contains all points in the domain except for a single point (circuit set or circuit set that contains a polynomial number of punch points).

[5] M-DDH: Multi-linear Decisional Diffie-Hellman assumption. IO: Indistinguishability Obfuscation. P-PRF(VRF): Puncturable PRF (VRF). Com: Commitment Scheme. FE: Functional Encryption. CVRF$^{\text{Sel-cha(Sig-key)}}$.

(single-key) security. PAR: Partition Scheme., and security[6]. Without considering the adaptive security under an exponential level reduction loss, the optimal security implementation to our knowledge should be the semi-adaptive security [27]. Our adaptively secure CVRF uses a similar constructing process to No.5 of Table 1, that is, using IO and a weaker level of CVRF to generate a higher level of CVRF (P-VRF is a special case of CVRF). The difference lies in we increase a map operation PAR.Enc projecting the input from the domain space of the weaker CVRF into a new domain space. It is known that this partition algorithm PAR.Enc can be any binary error-correcting code or even an identity function [7], so its effect on the CVRF efficiency is negligible and thus our construction is cost-effective.

*Discussion of Applications.* In essence, CVRF resolves the trust issue in CPRF, thus it might be used in the applications related to CPRF to improve the robustness of the system, such as identity-based key exchange [1,10], policy-based key distribution [10], broadcast encryption [1,10], functional encryption (attribute-based encryption) [3,12,18], etc. Moreover, CVRF allows the VRF's secret key to be delegated, thus it might be used in VRF-related application scenarios to reduce key management, such as zero-knowledge-related systems [26], E-Cash [6], and proof of stake consensus mechanisms of blockchain systems [4,21]. There are some known applications of CVRF, for example, [11] showed that a CVRF scheme for a constraint class $\mathcal{S}$ implies a functional signature scheme for the same family; [27] proposed a possible application, micro-payments [30]. In Sect. 6, we take micro-payment as an example to show what the possible implications of our proposed CVRF are.

## 2   Preliminaries

The following are the notations and four tools that will be used in this paper.

**Notations.** Let $\lambda$ be the security parameter. We say $negl(\lambda)$ is a negligible function if for $\forall$ polynomial $p$, $\exists N$ such that $negl(\lambda) < \frac{1}{p(\lambda)}$ for all $\lambda > N$. For a set $S$, $\overline{S}$ is referred to as the complementary set of $S$, $|S|$ represents the number of element $s$ satisfying $s \in S$, $s \leftarrow S$ is referred to as sampling $s$ from $S$ randomly. For an integer $n$, $[n]$ is referred to as the set $\{1, \cdots, n\}$, $[0, n]$ is referred to as the set $\{0, 1, \cdots, n\}$. For binary bit $m$ and $r$, $m\|r$ is referred to the concatenation $m$ and $r$.

**Definition 1 (Constrained Pseudorandom Function [10,11,24]).** *A keyed function $F : \mathcal{K} \times \mathcal{X} \to \mathcal{Y}$ is called a Constrained Pseudorandom Function (CPRF) about a collection of sets with efficient representation $\mathcal{S} = \{\mathcal{S}_\lambda \subseteq 2^{\mathcal{X}}\}_{\lambda \in \mathbb{N}}$ if there are polynomial-time algorithms $\Pi_{CPRF}=$(CPRF.Gen, CPRF.Cons, CPRF.CEval):*

- *CPRF.Gen$(1^\lambda) \to k$: The setup algorithm takes as input the security parameter $1^\lambda$ and outputs a secret key $k \in \mathcal{K}$. On input $x$, there is $y = F(k, x)$.*

---

[6] Sel-cha: Selective-challenge security. Adp: Adaptive security. (exp or poly): with an exponent or polynomial level reduction loss. Semi-adp: Semi-adaptive security. Sig-key: Single-key security.

- CPRF.Cons$(k, S) \rightarrow k_S$: *The key generation algorithm takes as input a master key $k$ and a set $S \in \mathcal{S}_\lambda$ and outputs a constrained key $k_S$.*
- CPRF.CEval$(k_S, x) \rightarrow y$: *The evaluation algorithm takes as input a constrained key $k_S$ and an input $x \in \mathcal{X}$ and outputs a value $y \in \mathcal{Y} \cup \{\perp\}$.*

*which satisfies the following two properties. 1.* **Correctness:** *For all $\lambda \in \mathbb{N}, S \in \mathcal{S}_\lambda$ and $x \in \mathcal{X}$, $k \leftarrow$ CPRF.Gen$(1^\lambda)$, $k_S \leftarrow$ CPRF.Cons$(k, S)$, there is*

- *If $x \in S$, there is* CPRF.CEval$(k_S, x) = \perp$.
- *If $x \notin S$, there is* CPRF.CEval$(k_S, x) = F(k, x)$.

*2.* **(Single-key) Pseudorandomness** [7]: *For any PPT adversary $A$, consider the security experiment $\mathsf{Exp}_A^{CPRF}(1^\lambda)$ as follows:*

1. *$A$ submits a set $S \in \mathcal{S}_\lambda$ to the challenger.*
2. *The challenger samples $k \leftarrow$ CPRF.Gen$(1^\lambda)$, computes a constrained key $k_S \leftarrow$ CPRF.Cons$(k, S)$ and sends $k_S$ to $A$.*
3. *The challenger receives an attack target $x_*$ from $A$ ($A$ is valid if its attack target $x_* \in S$). Then the challenger sets $y_*^0 = F(k, x_*)$, $y_*^1 \leftarrow \mathcal{Y}$, samples $b \leftarrow \{0, 1\}$, and returns $y_*^b$ to $A$.*
4. *$A$ outputs a guess $b'$. The experiment outputs 1 if $b' = b$, else 0.*

*Define $A$'s advantage of guessing $b' = b$ as $\mathsf{Adv}_A^{CPRF} = |Pr[\mathsf{Exp}_A^{CPRF}(1^\lambda) = 1] - \frac{1}{2}|$, we say $F$ is a single-key secure CPRF if $\mathsf{Adv}_A^{CPRF} \leq negl(\lambda)$.*

**Definition 2 (Non-interactive Commitments [8]).** *A non-interactive commitment scheme consists of a polynomial-time commitment algorithm* Com *and in which $p_1, p_2, p_3$ are polynomials in $\lambda$:*

- Com$(m; r) \rightarrow c$: *it takes a message $m \in \{0, 1\}^{p_1}$ and a randomness $r \in \{0, 1\}^{p_2}$ as inputs, outputs a commitment $c \in \{0, 1\}^{p_3}$.*

*which satisfies the following two requirements:*

1. **Perfectly Binding:** *For every security parameter $\lambda \in \mathbb{N}$ and string $c \in \{0, 1\}^{p_3}$, there exists at most one $m \in \{0, 1\}^{p_1}$ such that $c$ is a commitment to $m$: $\forall \lambda \in \mathbb{N}, r_0, r_1 \in \{0, 1\}^{p_2}$, if Com$(m_0; r_0) =$ Com$(m_1; r_1)$, then $m_0 = m_1$.*
2. **Computationally Hiding:** *For any sequence $\mathcal{I} = \{\lambda \in \mathbb{N}, m_0, m_1 \in \{0, 1\}^{p_1}\}$ and any $(\lambda, m_0, m_1) \in \mathcal{I}$, there is*

$$\{c_0 : r \leftarrow \{0, 1\}^{p_2}, c_0 \leftarrow \mathsf{Com}(m_0; r)\} \approx_c \{c_1 : r \leftarrow \{0, 1\}^{p_2}, c_1 \leftarrow \mathsf{Com}(m_1; r)\}.$$

**Definition 3 (Indistinguishability Obfuscation [5,20]).** *A PPT algorithm $\mathcal{O}$ is called an indistinguishability obfuscator for a circuit class $\mathcal{C} = \{\mathcal{C}_\lambda\}_{\lambda \in \mathbb{N}}$ if the following conditions are satisfied:*

1. **Completeness.** *For all security parameters $\lambda \in \mathbb{N}$, for all $C \in \mathcal{C}_\lambda$, for all inputs $x$, we have that $Pr[\widehat{C}(x) = C(x) : \widehat{C} \leftarrow \mathcal{O}(\lambda, C)] = 1$.*

2. **Indistinguishability.** *For any (not necessarily uniform) PPT adversaries* **Samp** *and* **D**, *there exists a negligible function* $negl(\lambda)$ *such that the following holds: For all security parameters* $\lambda \in \mathbb{N}$, *if* $Pr[|C_0| = |C_1| \wedge \forall x, C_0(x) = C_1(x) : (C_0, C_1, \sigma) \leftarrow$ **Samp**$(1^\lambda)] > 1 - negl(\lambda)$, *then we have:*

$$\left| Pr[\boldsymbol{D}(\sigma, \mathcal{O}(\lambda, C_0)) = 1] - Pr[\boldsymbol{D}(\sigma, \mathcal{O}(\lambda, C_1)) = 1] \right| \leq negl(\lambda)$$

**Definition 4 (Partition Scheme [7]).** *Let* $n, n'$ *be polynomially bounded functions,* $\tau < 1$ *be an inverse-polynomial function,* $\mathcal{S}' = \{S' \subseteq 2^{\{0,1\}^{n'(\lambda)}}\}_{\lambda \in \mathbb{N}}$ *be a collection of sets with efficient representation. A partition scheme* PAR *parameterized by* $(n, n', \tau, \mathcal{S}')$ *consists of the following polynomial-time algorithms* $\Pi_{PAR}$=(PAR.Enc, PAR.Gen):

- PAR.Enc$(x) \rightarrow x'$: *A deterministic encoder that maps any* $x \in \{0,1\}^{n(\lambda)}$ *to* $x' \in \{0,1\}^{n'(\lambda)}$.
- PAR.Gen$(1^\lambda, Q, \delta) \rightarrow S'$: *A probabilistic sample algorithm that given security parameter* $1^\lambda$, *integer* $Q$, *and balance parameter* $\delta$ *outputs a set* $S' \in \mathcal{S}'$, *interpreted as a partition* $(S', \overline{S'})$ *of* $\{0,1\}^{n'(\lambda)}$.[7]

*Fix* $\lambda, Q \in \mathbb{N}$, $\delta < 1$. *Let* $\mathcal{E}$ *be a distribution on pairs* $(X, x_*)$ *such that* $X := (x_1, \cdots, x_Q) \in \{0,1\}^{n(\lambda) \times Q}$ *and* $x_* \in \{0,1\}^{n(\lambda)} \backslash X$. *We define the probability that* $(X, x_*)$ *are split by the sampled partition:*

$$C_\mathcal{E}(\lambda, Q, \delta) := \Pr \left[ x'_* \in S', x' \in \overline{S'} \;\middle|\; \begin{array}{l} (X, x_*) \leftarrow \mathcal{X}, x'_* = PAR.Enc(x_*), \\ x' = PAR.Enc(x_i) \text{ for any } x_i \in X, \\ S' \leftarrow PAR.Gen(1^\lambda, Q, \delta). \end{array} \right].$$

*For every* $\lambda, Q \in \mathbb{N}$, $\delta < 1$, *and any two distributions* $\mathcal{E}, \mathcal{E}'$ *as above, we require:*

1. **Probable Partitioning:** $C_\mathcal{E}(\lambda, Q, \delta) \geq \tau(\lambda, Q, \delta^{-1}) = (\frac{\delta}{Q \cdot \lambda})^{O(1)}$.
2. **Balance:** $1 - \delta \leq \frac{P_\mathcal{E}(\lambda, Q, \delta)}{P_{\mathcal{E}'}(\lambda, Q, \delta)} \leq 1 + \delta$.

The partition sets can support the sub-string matching sets over a binary alphabet [7,23,28] or polynomial alphabet [7], or universal hashing [7].

# 3 Definition of CVRF: Algorithms and Properties (Security)

**Definition 5 (Constrained Verifiable Random Functions [13]).** *A function* $F : \mathcal{K} \times \mathcal{X} \rightarrow \mathcal{Y}$ *is said to be a Constrained Verifiable Random Function (CVRF) about a set system* $\mathcal{S} = \{\mathcal{S}_\lambda \subseteq 2^\mathcal{X}\}_{\lambda \in \mathbb{N}}$ *if there exists a constrained key space* $\mathcal{K}'$, *a proof space* $\mathcal{P}$, *and five PPT algorithms* $\Pi_{CVRF}$=(Setup, Eval, Constrain, CEval, Verify):

---

[7] As [7] showed "the set $S'$ has efficient representation in terms of $\lambda$ and does not grow with $Q, \delta^{-1}$, which are arbitrary polynomials in $\lambda$ that depend on the adversary.".

– $Setup(1^\lambda) \rightarrow (pk, sk)$: *The setup algorithm takes the security parameter $1^\lambda$ as input and outputs a pair of public-secret keys $(pk, sk)$.*
– $Eval(sk, x) \rightarrow (y, \pi)$: *The evaluation algorithm takes the secret key $sk$, an input $x \in \mathcal{X}$ as inputs, outputs the value-proof pair $(y, \pi) \in \mathcal{Y} \times \mathcal{P}$.*
  *For convenience, this algorithm is written as a value-function $F : \mathcal{K} \times \mathcal{X} \rightarrow \mathcal{Y}$ and a proof-function $P : \mathcal{K} \times \mathcal{X} \rightarrow \mathcal{P}$ separately: given $sk$ and $x$, there is $y = F(sk, x), \pi = P(sk, x)$.*
– $Cons(sk, S) \rightarrow sk_S$: *The constrained algorithm takes the secret key $sk$ and a set $S \in \mathcal{S}_\lambda$ as inputs and outputs a constrained key $sk_S \in \mathcal{K}'$.*
– $CEval(sk_S, x) \rightarrow (y, \pi)$: *The constrained evaluation algorithm takes a constrained key $sk_S$ and an input $x$ as inputs and outputs a value-proof pair $(y, \pi) \in \mathcal{Y} \times \mathcal{P} \cup \{(\bot, \bot)\}$.*
– $Verify(pk, x, y, \pi) \rightarrow 1/0$: *The verify algorithm takes the public key $pk$, an input $x$, a function value $y$, and a proof $\pi$ as inputs, and outputs a bit.*

*which satisfies the following four properties:* 1. **Provability:** *For all $\lambda \in \mathbb{N}$, $(pk, sk) \leftarrow Setup(1^\lambda)$, any $S \in \mathcal{S}_\lambda$, $sk_S \leftarrow Cons(sk, S)$, $x \in \mathcal{X}$, and $(y, \pi) \leftarrow CEval(sk_S, x)$, it holds that:*

– *If $x \in S$, then $y = F(sk, x)$ and $Verify(pk, x, y, \pi) = 1$.*
– *If $x \notin S$, then $(y, \pi) = (\bot, \bot)$.*

2. **Uniqueness:** *For every $x$ there is at most one value $y$ for which there exists proof that $F(sk, x) = y$. For all $\lambda \in \mathbb{N}$, all $(pk, sk) \leftarrow Setup(1^\lambda)$ and $x \in \mathcal{X}$, there does not exist a tuple $(y_0, y_1, \pi_0, \pi_1)$ such that $y_0 \neq y_1$, and $Verify(pk, x, y_0, \pi_0) = 1$ and $\mathrm{Verify}(pk, x, y_1, \pi_1) = 1$.*

3. **Constraint-Hiding.** *This notion ensures that the proof computed with a constrained key should be distributed like the proof computed with the actual secret key. Formally, for all $\lambda \in \mathbb{N}$, all $(pk, sk) \leftarrow Setup(1^\lambda)$, all $S \in \mathcal{S}_\lambda$, all $sk_S \leftarrow Cons(sk, S)$, the following holds: the second output, $\pi$, of $CEval(sk_S, x)$, is identical to the output of $P(sk, x)$.*

4. **(Adaptive) Pseudorandomness/Adaptive Security Experiment.** *For any PPT adversary $\boldsymbol{A}$, consider the security experiment $\boldsymbol{Expt}_{\boldsymbol{A}}^{CVRF}(1^\lambda)$ as:*

1. *The challenger first generates $(pk, sk) \leftarrow Setup(1^\lambda)$, and returns $pk$ to $\boldsymbol{A}$.*
2. *The challenger provides the following oracles to $\boldsymbol{A}$:*
   – $\mathcal{O}_{Eval}$: *On input $x \in \mathcal{X}$, it returns $(y, \pi) \leftarrow Eval(sk, x)$.*
   – $\mathcal{O}_{Cons}$: *On input a set $S \in \mathcal{S}_\lambda$, it returns $sk_S \leftarrow Cons(sk, S)$.*
3. *The challenger receives an attack target $x_*$ from $\boldsymbol{A}$ ($\boldsymbol{A}$ is valid if its attack target $x_*$ has not been queried in step 2). The challenger computes $y_*^0 = F(sk, x_*)$ and $y_*^1 \leftarrow \mathcal{Y}$, samples $b \leftarrow \{0, 1\}$ and sends $y_*^b$ to $\boldsymbol{A}$.*
4. *$\boldsymbol{A}$ outputs a guess $b'$ of $b$. The experiment outputs 1 if $b' = b$ else outputs 0.*

*Define $\boldsymbol{A}$'s advantage of guessing $b' = b$ as $Adv_{\boldsymbol{A}}^{CVRF}(1^\lambda) = |Pr[\boldsymbol{Expt}_{\boldsymbol{A}}^{CVRF}(1^\lambda) = 1] - \frac{1}{2}|$, we say $F$ is an adaptively secure CVRF if $Adv_{\boldsymbol{A}}^{CVRF}(1^\lambda) \leq negl(\lambda)$.*

# 4 The Existence of Single-Key Secure CVRF

In this section, we first define the single-key security of CVRF, then prove its existence by providing an implied relationship and construction separately. Because most of the construction of CVRF has realized selective-challenge security, we naturally consider whether these achievements can be used for single-key security. We found that there is an implicit relationship between the two types of security, which we explain in Sect. 4.1. However, this relationship only implies single-key security for the circuit set that contains a polynomial number of punch points. To build an adaptively secure CVRF, the single-key secure CVRF should support any circuit set. We present the construction in Sect. 4.2, by replacing the Puncturable PRF of [25] with a more general primitive, called CPRF.

**Definition 6 (Single-key Security of CVRF [27]).** *F is a single-key secure CVRF if it satisfies the definition in Definition 5 except that, the provability and security are replaced by the new provability and single-key security:*

*Provability: For all $\lambda \in \mathbb{N}$, $(pk, sk) \leftarrow Setup(1^\lambda)$, any $S \in \mathcal{S}_\lambda$, $sk_S \leftarrow Cons(sk, S)$, $x \in \mathcal{X}$, and $(y, \pi) \leftarrow CEval(sk_S, x)$, there is*

- *If $x \in S$, then $(y, \pi) = (\perp, \perp)$.*
- *If $x \notin S$, then $y = F(sk, x)$, $Verify(pk, x, y, \pi) = 1$.*

*Single-key Pseudorandomness/Security: For any PPT adversary $\boldsymbol{A}$, consider the experiment $\boldsymbol{Expt}_A^{Sig\text{-}key\text{-}CVRF}(1^\lambda)$ as follows:*

1. *The challenger first receives a set $S \in \mathcal{S}_\lambda$ from $\boldsymbol{A}$.*
2. *The challenger then runs $(pk, sk) \leftarrow Setup(1^\lambda)$, $sk_S \leftarrow Cons(sk, S)$, then sends $(sk_S, pk)$ to $\boldsymbol{A}$.*
3. *The challenger receives an attack target $x_*$ from $\boldsymbol{A}$ ($\boldsymbol{A}$ is valid if its attack target $x_* \in S$), then it computes $y_*^0 = F(sk, x_*)$, $y_*^1 \leftarrow \mathcal{Y}$, samples $b \leftarrow \{0, 1\}$ and returns $y_*^b$.*
4. *$\boldsymbol{A}$ outputs a guess $b'$ of $b$. The experiment outputs 1 if $b' = b$ else outputs 0.*

*Define $\boldsymbol{A}$'s advantage $Adv_A^{Sig\text{-}key\text{-}CVRF}(\lambda) = |\Pr[\boldsymbol{Expt}_A^{Sig\text{-}key\text{-}CVRF}(1^\lambda) = 1] - \frac{1}{2}|$, we say F is a single-key secure CVRF if $Adv_A^{Sig\text{-}key\text{-}CVRF}(\lambda) \leq negl(\lambda)$. The algorithms are described as $\Pi_{Sig\text{-}key\text{-}CVRF} = (Setup_{key}, Eval_{key}, Cons_{key}, CEval_{key}, Verify_{key})$.*

## 4.1 For Circuit Set Containing $poly(\lambda)$ Punch Points

**Theorem 1.** *If a CVRF scheme that supports any circuit set satisfies the selective-challenge security, then it also satisfies the single-key security in which the adversary only queries a circuit set that contains a polynomial number (i.e. $p(\lambda)$) of punch points.*

**Definition 7 (Selective-challenge Security of CVRF [19]).** *An F is a selectively-challenge secure CVRF if it satisfies the definition in Definition 5 except the adaptive security is replaced by **selective-challenge security**: for any PPT adversary $\boldsymbol{A}$, consider the experiment $\boldsymbol{Expt}_A^{Sel\text{-}cha\text{-}CVRF}(1^\lambda)$ as follows:*

1. A first sends an attack target $x_*$ to the challenger. The challenger generates $(pk, sk) \leftarrow Setup(1^\lambda)$, and returns $pk$ to $\mathbf{A}$.
2. The challenger provides the following oracles to $\mathbf{A}$:
   - $\mathcal{O}_{Eval}$: On input $x$, it returns $(y, \pi) \leftarrow Eval(sk, x)$.
   - $\mathcal{O}_{Cons}$: On input $S \in \mathcal{S}_\lambda$, it returns $sk_S \leftarrow Cons(sk, S)$.
3. The challenger computes $y_*^0 = F(sk, x_*)$, $y_*^1 \leftarrow \mathcal{Y}$, samples $b \leftarrow \{0,1\}$ and sends $y_*^b$ to $\mathbf{A}$. $\mathbf{A}$ is valid if for its per query $x \neq x_*$, any set $S$ s.t. $x_* \notin S$.
4. $\mathbf{A}$ outputs a guess $b'$ of $b$. The experiment outputs 1 if $b' = b$; else outputs 0.

Define $\mathbf{A}$'s advantage as $Adv_A^{Sel\text{-}cha\text{-}CVRF}(1^\lambda) = |\Pr[\boldsymbol{Expt}_A^{Sel\text{-}cha\text{-}CVRF}(1^\lambda) = 1] - \frac{1}{2}|$, we say $F$ is selective-challenge secure CVRF if $Adv_A^{Sel\text{-}cha\text{-}CVRF}(1^\lambda) \leq negl(\lambda)$.

*Proof.* If there is a PPT adversary $\mathbf{A}$, who can with a non-negligible advantage attack CVRF single-key security, then we can construct a PPT adversary $\mathbf{B}$ to attack the selective-challenge security. Since $\mathbf{A}$ can only query a circuit set that contains $poly(\lambda)$ punch points and its attack target $x^*$ is one of the punch points, $\mathbf{B}$ after receiving this set can with a probability $\frac{1}{poly(\lambda)}$ guess $x^*$ correctly and thus successfully taking advantage of $\mathbf{A}$'s ability of guessing $b' = b$, thus there is $Adv_\mathbf{B} = \frac{1}{poly(\lambda)} \cdot Adv_\mathbf{A}$. Because of the selective-challenge security, there is $Adv_\mathbf{B} \leq negl(\lambda)$, then $Adv_\mathbf{A} \leq poly(\lambda) \cdot negl(\lambda)$ is also negligible.

### 4.2   For Any Circuit Set

Let $\lambda$ be the security parameter, $p_1, p_2, p_3$ are polynomials in $\lambda$, CVRF on a function family $F' : \mathcal{K} \times \mathcal{X} \to \{0,1\}^{p_1}$ and a proof space $\mathcal{P}' = \{0,1\}^{p_2}$, a collection of circuit sets $\mathcal{S} = \{\mathcal{S}_\lambda \subseteq 2^\mathcal{X}\}_{\lambda \in \mathbb{N}}$, and the following ingredients:

* CPRF: On a function family $F : \mathcal{K} \times \mathcal{X} \to \{0,1\}^{p_1+p_2}$, a collection of sets $\mathcal{S}$ as above, and the algorithms $\Pi_{CPRF} = (CPRF.Gen, CPRF.CEval, CPRF.Cons)$.
* Commitment Scheme: Com: $\{0,1\}^{p_1} \times \{0,1\}^{p_2} \to \{0,1\}^{p_3}$ takes as input a value $m \in \{0,1\}^{p_1}$, $r \in \{0,1\}^{p_2}$ and outputs $c \leftarrow Com(m; r)$.
* Indistinguishability Obfuscator: $\mathcal{O}$

The algorithms of CVRF $\Pi_{key} = (Setup_{key}, Eval_{key}, Cons_{key}, CEval_{key}, Verify_{key})$:

- $\mathbf{Setup_{key}}(1^\lambda)$: On input $1^\lambda$, the setup algorithm runs $CPRF.Gen(1^\lambda) \to k$ and generates a program $PubK_k$ as in Fig. 1, outputs $(sk = k, pk = \mathcal{O}([PubK_k]))$. Here $pk$ is an obfuscated circuit generated by the obfuscator $\mathcal{O}$.

---

**Constants:** $k$;     **Inputs:** $x$
01 Compute $t = F(k, x)$, parse $t = m\|r$ such that $m \in \{0,1\}^{p_1}$, $r \in \{0,1\}^{p_2}$.
02 Output $c = Com(m; r)$.

---

**Fig. 1.** The Program Description of $PubK_k$

- **Eval$_{key}$**$(sk, x)$: On input the secret key $sk = k$ and an input $x \in \mathcal{X}$, the evaluation algorithm first computes $t = F(k, x)$ and parses $t = m\|r$ such that $m \in \{0, 1\}^{p_1}$, $r \in \{0, 1\}^{p_2}$, then outputs $y = m$, $\pi = r$.

  For convenience, this algorithm is also written as a value-function $F'$ and a proof-function $P'$ separately: given $sk = k$ and $x$, after making the same calculation as above, there is $y = F'(sk, x) = m$, $\pi = P'(sk, x) = r$.

- **Cons$_{key}$**$(sk, S)$: On input the secret key $sk = k$ and a set $S \in \mathcal{S}_\lambda$, the constrained algorithm computes CPRF.Cons$(k, S) \rightarrow k_S$ and defines a program ConK$_{k_S}$ as in Fig. 2, then outputs $sk_S = \mathcal{O}([\mathsf{ConK}_{k_S}])$. Here $sk_S$ is an obfuscated circuit generated by the obfuscator $\mathcal{O}$.

- **CEval$_{key}$**$(sk_S, x)$: On input $sk_S$ and an input $x \in \mathcal{X}$, the constrained evaluation algorithm outputs $(y, \pi) = sk_S(x)$.

- **Verify$_{key}$**$(pk, x, y, \pi)$: On input of the public key $pk$, an input $x$, a function value $y$, and a proof $\pi$, the verify algorithm checks whether $pk(x) = \mathsf{Com}(y; \pi)$, if yes, outputs 1; else 0.

---

| **Constants:** $k_S$;    **Inputs:** $x$ |
|---|
| 01 Compute $t = \mathsf{CPRF.CEval}(k_S, x)$ and check if $t = \perp$, output $(y, \pi) = (\perp, \perp)$; |
| If not, parse $t = m\|r$ for $m \in \{0, 1\}^{p_1}$, $r \in \{0, 1\}^{p_2}$, output $(y, \pi) = (m, r)$. |

**Fig. 2.** The Program Description of ConK$_{k_S}$

**Theorem 2.** *If $\mathcal{O}$ is an indistinguishability obfuscator, $F$ is a single-key secure CPRF, and Com is perfect binding, computationally hiding commitment scheme, then the above construction is a single-key secure CVRF as in Definition 6.*

*Proof.* The provability property can be verified directly from the correctness of CPRF and the completeness of IO. The uniqueness property follows from the perfectly binding property of the commitment scheme. If there exists a public key $pk$, an input $x \in \mathcal{X}$ and values $(y_0, \pi_0), (y_1, \pi_1) \in \{0, 1\}^{p_1} \times \{0, 1\}^{p_2}$, such that $y_0 \neq y_1 \wedge \mathsf{Verify}_{key}(pk, x, y_0, \pi_0) = 1 \wedge \mathsf{Verify}_{key}(pk, x, y_1, \pi_1) = 1$, then there must be $y_0 \neq y_1 \wedge \mathsf{Com}(y_0; \pi_0) = \mathsf{Com}(y_1; \pi_1)$, which contradicts the assumption of Com's perfectly binding property. The constraint-hiding property is also obvious, since in our construction, the proof generated by a secret key is the same as the proof generated by a constrained key.

We define the following experiments to prove single-key security. In each experiment, **A** is referred to as a valid adversary whose attack target satisfies $x_* \in S$ where $S$ is the puncturable single-key query.

**Expt$_0$:** This experiment is identical to the experiment of Definition 6:

1. The challenger receives a set $S \in \mathcal{S}_\lambda$ from **A**.
2. The challenger runs CPRF.Gen$(1^\lambda) \rightarrow k$, CPRF.Cons$(k, S) \rightarrow k_S$, then
   (a) it generates a program PubK$_k$ as in Fig. 1 and sets $pk = \mathcal{O}([\mathsf{PubK}_k])$.
   (b) it generates a program ConK$_{k_S}$ as in Fig. 2 and sets $sk_S = \mathcal{O}([\mathsf{ConK}_{k_S}])$.

Finally, it returns $(sk_S, pk)$ to **A**.

3. The challenger receives an attack target $x_* \in S$ from **A**, it samples $b \leftarrow \{0,1\}$, if $b = 0$, it computes $t_* = F(k, x_*) = m_*\|r_*$ such that $m \in \{0,1\}^{p_1}$, $r \in \{0,1\}^{p_2}$, and returns $y_* = m_*$ to **A**; if $b = 1$ it returns $y_* \leftarrow \{0,1\}^{p_1}$.

4. Challenger receives **A**'s guess $b'$. The experiment outputs 1 if $b' = b$ else 0.

**Expt$_1$**: This experiment is identical to **Expt$_0$** except for step 2(a): the challenger generates $pk = \mathcal{O}([\mathsf{PubK}_{k,k_S}])$ where $\mathsf{PubK}_{k,k_S}$ is defined as in Fig. 3.

---

**Constants:** $k, k_S$;      **Inputs:** $x$
01 If $x \in S$, compute $t = F(k, x)$; If $x \notin S$, compute $t = \mathsf{CPRF.CEval}(k_S, x)$.
02 Parse $t = m\|r$ such that $m \in \{0,1\}^{p_1}$, $r \in \{0,1\}^{p_2}$. Output $c = \mathsf{Com}(m; r)$.

---

**Fig. 3.** The Program Description of $\mathsf{PubK}_{k,k_S}$

**Expt$_2$**: This experiment is identical to **Expt$_1$** except for step 2(a): the challenger samples a random function $R : \mathcal{X} \to \mathcal{Y}$, generates a program $\underline{\mathsf{PubK}_{k_S,R}}$ as in Fig. 4, sets $pk = \mathcal{O}([\mathsf{PubK}_{k_S,R}])$.

---

**Constants:** $k_S$;      **Inputs:** $x$
01 If $x \in S$, compute $\underline{t = R(x)}$; If $x \notin S$, compute $t = \mathsf{CPRF.CEval}(k_S, x)$.
02 Parse $t = m\|r$ such that $m \in \{0,1\}^{p_1}$, $r \in \{0,1\}^{p_2}$. Output $c = \mathsf{Com}(m; r)$.

---

**Fig. 4.** The Program Description of $\mathsf{PubK}_{k_S,R}$

For $i \in \{0, 1, 2\}$, we write $\Pr[\mathbf{Expt}_i(1^\lambda) = 1]$ as $\Pr[\mathbf{E}_i^1]$, $\mathsf{Adv}_i = |\Pr[\mathbf{E}_i^1] - \frac{1}{2}|$ as **A**'s advantage of the success in **Expt$_i$**.

**Lemma 1.** *If $\mathcal{O}$ is an indistinguishability obfuscator, $F$ satisfies the correctness of CPRF, there is $|\mathsf{Adv}_0 - \mathsf{Adv}_1| \leq negl(\lambda)$.*

*Proof.* If there exists a PPT adversary **A** whose advantage difference between the two experiments is non-negligible probability $\mu$ (i.e. $|\mathsf{Adv}_0 - \mathsf{Adv}_1| = \mu$), then we can construct adversaries (**Samp, D**) to attack IO's indistinguishability. (**Samp, D**) act as the challenger of **Expt$_0$** except for the generation of $pk$. In step 2(a), **Samp** generates two programs $(C_0, C_1, \sigma)$ where $C_0$ as in Fig. 1, $C_1$ as in Fig. 3, and $|C_0| = |C_1|$. Then **Samp** sends $(C_0, C_1)$ to the challenger who samples $e \leftarrow \{0,1\}$ and computes $\widehat{C_e} \leftarrow \mathcal{O}(\lambda, C_e)$ and returns $\widehat{C_e}$. **D** sets $pk = \widehat{C_e}$. At last, when **D** receives **A**'s guess $b'$ of $b$, it outputs $e' = b' \oplus b$.

Due to the correctness of CPRF, $C_0$ is functionality equivalent to $C_1$. Observed that, if $e = 0$, **D** outputs $e' = 1$ only if **A**'s guess $b' \neq b$ in **Expt$_0$**. if $e = 1$, **D** outputs $e' = 1$ only if **A**'s guess $b' \neq b$ in **Expt$_1$**. That's

$$\text{Adv}_D = \left| \Pr[\mathbf{D}(\sigma, \mathcal{O}(\lambda, C_0)) = 1] - \Pr[\mathbf{D}(\sigma, \mathcal{O}(\lambda, C_1)) = 1] \right|$$

$$= \left| (1 - \Pr[\mathbf{E}_0^1]) - (1 - \Pr[\mathbf{E}_1^1]) \right| \geq \left| \text{Adv}_0 - \text{Adv}_1 \right| = \mu$$

which contradicts the indistinguishability of IO, i.e. $|\text{Adv}_0 - \text{Adv}_1| \leq negl(\lambda)$.

**Lemma 2.** *If* Com *satisfies the hiding of commitment scheme, then* $|\text{Adv}_1 - \text{Adv}_2| \leq negl(\lambda)$.

*Proof.* From $\mathbf{A}$'s view, the difference between the two experiments lies in $pk$, when an input satisfies $x \in S$, $\mathbf{Expt}_1$ computes $t = F(k, x) = m_0 || r_0$ and outputs $c_0 \leftarrow \text{Com}(m_0; r_0)$, while $\mathbf{Expt}_2$ computes $t = R(x) = m_1 || r_1$ and outputs $c_1 \leftarrow \text{Com}(m_1; r_1)$. The hiding of Com says: for any $(\lambda, m_0, m_1 \in \{0, 1\}, r_0, r_1 \in \{0, 1\}^{p_2})$, there is $c_0$ is computationally indistinguishable from $c_1$. Hence the two experiments are also computationally indistinguishable, that's, for any PPT adversary $\mathbf{A}$, there is $|\text{Adv}_1 - \text{Adv}_2| \leq negl(\lambda)$.

**Lemma 3.** *If* $F$ *satisfies the single-key security of* CPRF, *then* $\text{Adv}_2 \leq negl(\lambda)$.

*Proof.* If there is a PPT adversary $\mathbf{A}$ whose advantage in $\mathbf{Expt}_2$ is non-negligible $\mu$, i.e. $\text{Adv}_2 = \mu$, then we can construct a PPT adversary $\mathbf{B}$ to attack CPRF single-key security. $\mathbf{B}$ acts as the challenge of $\mathbf{Expt}_2$ except the following: (1) after receiving the set $S$, $\mathbf{B}$ sends $S$ to the challenger to get a constrained key $k_S$. $\mathbf{B}$ uses $k_S$ to generate $sk_S$ and $pk$ (also using a random sample function $R$). (2) After receiving an attack target $x_* \in S$, $\mathbf{B}$ sends $x_*$ to the challenger who samples $e \leftarrow \{0, 1\}$ and computes $y_*^0 = F(k, x_*)$, $y_*^1 \leftarrow \{0, 1\}^{p_1}$ and returns $y_*^e$. Then, $\mathbf{B}$ sends $y_*^b = y_*^e$ to $\mathbf{A}$. (3) After receiving $\mathbf{A}$'s guess $b'$, $\mathbf{B}$ outputs $e' = b'$. $\mathbf{B}$'s probability of guessing $e' = e$ equals $\mathbf{A}$'s of guessing $b' = b$ in $\mathbf{Expt}_2$. That is, $\text{Adv}_2 = \text{Adv}_B \leq negl(\lambda)$.

Combining the above lemmas, Theorem 2 is proved.

## 5   Adaptively Secure CVRF

In order to achieve adaptively secure CVRF, we use the single-key technique and partition schemes. Let $\lambda \in \mathbb{N}$ be the security parameter, $F : \mathcal{K} \times \{0, 1\}^{n(\lambda)} \to \mathcal{Y}$ be CVRF, $\mathcal{S} = \{S_\lambda \subseteq 2^{\{0,1\}^{n(\lambda)}}\}_{\lambda \in \mathbb{N}}$ be a collection of circuit sets, the ingredients:

* CVRF$^{\text{Sig-key}}$: A value-function $F' : \mathcal{K} \times \{0, 1\}^{n'(\lambda)} \to \mathcal{Y}$, a proof-function $P' : \mathcal{K} \times \{0, 1\}^{n'(\lambda)} \to \mathcal{P}'$, a collection of circuit sets $\mathcal{S}' = \{S'_\lambda \subseteq 2^{\{0,1\}^{n'(\lambda)}}\}_{\lambda \in \mathbb{N}}$, algorithms $\Pi_{\text{Sig-key-CVRF}} = (\text{Setup}_{\text{key}}, \text{Eval}_{\text{key}}, \text{Cons}_{\text{key}}, \text{CEval}_{\text{key}}, \text{Verify}_{\text{key}})$.
* PAR: $\Pi_{\text{PAR}} = (\text{PAR.Enc}, \text{PAR.Gen})$, which are parameterized by $(n, n', \tau, \mathcal{S}')$: $n, n'$ are polynomially bounded functions, $\tau < 1$ is an inverse-polynomial function, and $\mathcal{S}'$ is a partition set.
* Indistinguishability obfuscator: $\mathcal{O}$.

There are the following algorithms: $\Pi_{\text{CVRF}} = (\text{Setup}, \text{Eval}, \text{Cons}, \text{CEval}, \text{Verify})$:

- **Setup**$(1^\lambda)$: On input the security parameter $1^\lambda$, the setup algorithm runs $(pk', sk') \leftarrow \mathsf{Setup_{key}}(1^\lambda)$ and outputs $(pk = pk', sk = sk')$.
- **Eval**$(sk, x)$: On input the secret key $sk = sk'$ and an input $x \in \{0,1\}^{n(\lambda)}$, the evaluation algorithm runs $(y, \pi) = \mathsf{Eval_{key}}(sk', \mathsf{PAR.Enc}(x))$ and outputs $(y, \pi) \in \mathcal{Y} \cup \mathcal{P}$. This algorithm is also written as a value-function $F\colon y = F(sk, x) = F'(sk', \mathsf{PAR.Enc}(x))$, and a proof-function $P\colon \pi = P(sk, x) = P'(sk', \mathsf{PAR.Enc}(x))$.
- **Cons**$(sk, S)$: On input of the secret key $sk = sk'$ and a set $S \in \mathcal{S}_\lambda$, the constrained algorithm first defines a circuit $C_{sk', S}(\cdot)$:

$$C_{sk', S}(x) = \begin{cases} \mathsf{Eval_{key}}(sk', \mathsf{PAR.Enc}(x)), & \text{if } x \in S \\ \bot, & \text{if } x \notin S \end{cases} \quad (1)$$

then runs $\widehat{C_{sk', S}} \leftarrow \mathcal{O}(\lambda, C_{sk', S})$, outputs an obfuscated circuit $sk_S = \widehat{C_{sk', S}}$.
- **CEval**$(sk_S, x)$: On input the constrained key $sk_S = \widehat{C_{sk', S}}$ and a value $x \in \{0,1\}^{n(\lambda)}$, the constrained evaluation algorithm outputs $(y, \pi) = \widehat{C_{sk', S}}(x)$ where $(y, \pi) \in \mathcal{Y} \times \mathcal{P} \cup \{(\bot, \bot)\}$.
- **Verify**$(pk, x, y, \pi)$: On input of the public key $pk = pk'$, $x \in \{0,1\}^{n(\lambda)}$, $y \in \mathcal{Y}$, and a proof $\pi$, the verify algorithm outputs what the algorithm $\mathsf{Verify_{key}}(pk', \mathsf{PAR.Enc}(x), y, \pi)$ outputs.

### 5.1  Proof of the Construction

**Theorem 3.** *If $F'$ is a single-key secure CVRF, PAR is a partition scheme, and $\mathcal{O}$ is an indistinguishability obfuscator, then the above construction on $F$ is an adaptively secure CVRF for any circuit set that contains $poly(\lambda)$-points.*

*Proof.* Provability: in the above construction, on any $S \in \mathcal{S}_\lambda$ and $x \in S$, no matter whether the value $y$ and proof $\pi$ are computed by the secret key or constrained key, $y$ is always the output of $\mathsf{Eval}(sk', \mathsf{PAR.Enc}(x))$. Uniqueness is from the uniqueness of the CVRF$^{\mathsf{Sig\text{-}key}}$. The constraint-hiding is satisfied because the proof generated by a secret key is the same as a constrained key. Pseudorandomness: we define some experiments in which the challenger interacts with a **valid** adversary **A** who always issues an attack target $x_*$ that has not been queried. Because in the PAR, the number of points that an adversary can query is a polynomial $Q$, thus here **A** can evaluate after polynomial queries of $\mathcal{O}_{\mathsf{Eval}}$ and $\mathcal{O}_{\mathsf{Cons}}$, which must also be $Q$, then the queried $S$ should contain $poly(\lambda)$-points. $\delta$ is the balance parameter of PAR.

**Expt$_0$**: This experiment emulates the real experiment in Definition 5.

1. The challenger generates $(pk', sk') \leftarrow \mathsf{Setup_{key}}(1^\lambda)$, sets $sk = sk'$ and $pk = pk'$, then sends $pk = pk'$ to **A**.
2. The challenger provides the following oracles to **A**:
    - $\mathcal{O}_{\mathsf{Eval}}$: On input $x \in \mathcal{X}$, it outputs $(y, \pi) \leftarrow \mathsf{Eval_{key}}(sk', \mathsf{PAR.Enc}(x))$.

- $\mathcal{O}_{\mathsf{Cons}}$: On input a set $S \in \mathcal{S}_\lambda$, it defines a circuit $C_{sk',S}(x)$ as Eq. 1 and outputs $sk_S := \widehat{C_{sk',S}} \leftarrow \mathcal{O}(\lambda, C_{sk',S})$.

3. The challenger receives an attack target $x_*$ from **A**, it samples $b \leftarrow \{0,1\}$, if $b = 0$, it returns $y_* = F'(sk', \mathsf{PAR.Enc}(x_*))$; if $b = 1$, it returns $y_* \leftarrow \mathcal{Y}$.
4. **A** outputs a guess $b'$ of $b$. $\mathbf{Expt}_0(1^\lambda)$ outputs 1 if $b' = b$; else outputs 0.

$\mathbf{Expt}_1$: This experiment is identical to $\mathbf{Expt}_0$ except for an additional abort operation. The challenger first runs $S' \leftarrow \mathsf{PAR.Gen}(\lambda, Q, \delta)$. Then for per query $x$ (or $S$, or $x_*$), the challenger checks whether $\mathsf{PAR.Enc}(x) \in S'$ (or there is $x$ such that $x \in S \land \mathsf{PAR.Enc}(x) \in S'$), or $\mathsf{PAR.Enc}(x_*) \notin S'$), if yes, the challenger aborts and outputs a random bit. Otherwise, it acts as in $\mathbf{Expt}_0$.

$\mathbf{Expt}_2$: This experiment is identical to $\mathbf{Expt}_1$ except for the replies made by the challenger in step 2. The challenger first generates $sk'_{S'} \leftarrow \mathsf{Cons}_{\mathsf{key}}(sk', S')$. When non-abort, for any query to $\mathcal{O}_{\mathsf{Eval}}$ and $\mathcal{O}_{\mathsf{Cons}}$, it uses the $sk_{S'}$ to evaluate the replies. Concretely, for $x \in \mathcal{X}$, it returns $(y, \pi) \leftarrow \mathsf{CEval}_{\mathsf{key}}(sk'_{S'}, \mathsf{PAR.Enc}(x))$; for $S$, it defines $C_{sk'_{S'},S}(x)$ as Eq. 2 and returns $sk_S := \widehat{C_{sk'_{S'},S}} \leftarrow \mathcal{O}(\lambda, C_{sk'_{S'},S})$.

$$C_{sk'_{S'},S}(x) = \begin{cases} \mathsf{CEval}_{\mathsf{key}}(sk'_{S'}, \mathsf{PAR.Enc}(x)), & \text{if } x \in S \\ \bot, & \text{if } x \notin S \end{cases} \quad (2)$$

For $i \in \{0,1,2\}$, we write $\Pr[\mathbf{Expt}_i(1^\lambda) = 1]$ as $\Pr[\mathbf{E}_i^1]$, $\mathsf{Adv}_i = |\Pr[\mathbf{E}_i^1] - \frac{1}{2}|$ as **A**'s advantage of success in $\mathbf{Expt}_i$. PAR is a partition scheme with a partition probability larger than $\tau = \tau(\lambda, Q, \delta^{-1})$.

**Lemma 1.** *If PAR is a partition scheme with a partition probability $\tau$, then* $\mathsf{Adv}_1 \geq \tau \cdot \mathsf{Adv}_0$.

*Proof.* The difference between $\mathbf{Expt}_0$ and $\mathbf{Expt}_1$ is that $\mathbf{Expt}_1$ adds an abort operation, defining it as "Abt". When Abt occurs, $\mathbf{Expt}_1(1^\lambda)$ will output a random bit in $\{0,1\}$; if not, it will act as $\mathbf{Expt}_0$. Hence there is

$$\Pr[\mathbf{E}_1^1] = \Pr[\mathsf{Abt}] \cdot \Pr[\mathbf{E}_1^1|\mathsf{Abt}] + \Pr[\overline{\mathsf{Abt}}] \cdot \Pr[\mathbf{E}_1^1|\overline{\mathsf{Abt}}]$$

$$= (1 - \Pr[\overline{\mathsf{Abt}}]) \cdot \frac{1}{2} + \Pr[\overline{\mathsf{Abt}}] \cdot \Pr[\mathbf{E}_0^1] = \frac{1}{2} + \Pr[\overline{\mathsf{Abt}}] \cdot (\Pr[\mathbf{E}_0^1] - \frac{1}{2})$$

That is, $|\Pr[\mathbf{E}_1^1] - \frac{1}{2}| = \Pr[\overline{\mathsf{Abt}}] \cdot |\Pr[\mathbf{E}_0^1] - \frac{1}{2}| \Rightarrow \mathsf{Adv}_1 \geq \tau \cdot \mathsf{Adv}_0$.

**Lemma 2.** *If $\mathcal{O}$ is an indistinguishability obfuscator, $F'$ satisfies the correctness of CVRF, then* $|\mathsf{Adv}_2 - \mathsf{Adv}_1| \leq \mathsf{negl}(\lambda)$.

*Proof.* The difference between the two experiments lies in the replies of the two oracles ($\mathcal{O}_{\mathsf{Eval}}, \mathcal{O}_{\mathsf{Cons}}$). Firstly, for simplicity, we define a transitional experiment $\mathbf{Expt}_{1.5}$ which is identical to $\mathbf{Expt}_1$ except that:

- $\mathcal{O}_{\mathsf{Eval}}$: On input $x \in \mathcal{X}$, the challenger checks if there is $\mathsf{PAR.Enc}(x) \in S'$, it aborts; Else, the challenger returns $(y, \pi) \leftarrow \mathsf{CEval}_{\mathsf{key}}(sk'_{S'}, \mathsf{PAR.Enc}(x))$.

From CVRF$^{\text{Sig-key}}$'s correctness, **Expt**$_{1.5}$ is identical to **Expt**$_1$, i.e. Adv$_1$ = Adv$_{1.5}$. We now prove **Expt**$_{1.5}$ is indistinguishable from **Expt**$_2$ by a standard hybrid technique. The hybrid experiments are (**Expt**$_{1.5}$ = Hyb$_0$, $\cdots$, Hyb$_j$, Hyb$_{j+1}$, $\cdots$, Hyb$_{Q_C}$ = **Expt**$_2$) where $Q_C$ is the number of **A**'s queries for $\mathcal{O}_{\text{Cons}}$. For $j \in [0, Q_C]$, Hyb$_j$ is identical to **Expt**$_2$ except that, for **A**'s first $i$'th query $S$, the challenger returns $sk_S$ using the circuit as Eq. 2, while the rest queries $S$, it returns $sk_S$ using the circuit as Eq. 1. If there exists a PPT adversary **A** such that $|$Adv$_{1.5}$ − Adv$_2|$ equals to a non-negligible probability $\mu$, then there must exist $j \in [0, Q_c]$ such $|$Adv$_j^{\text{Hyb}}$ − Adv$_{j+1}^{\text{Hyb}}|$ is also non-negligible, we can construct adversaries (**Samp, D**) to break the indistinguishability of IO.

Formally, (**Samp, D**) acts as the challenger of **Expt**$_2$ except $\mathcal{O}_{\text{Cons}}$'s response to $S_{j+1}$. When non-abort, **Samp** runs $(C_{sk',S}, C_{sk'_{S'},S}, S, \sigma) \leftarrow \mathbf{Samp}(1^\lambda)$ and sends $(C_0 = C_{sk',S}, C_1 = C_{sk'_{S'}})$ to the challenger who returns an obfuscated circuit $\mathcal{O}(\lambda, C_e)$ for a bit $e$, **D** then sends it to **A**. Until receiving **A**'s guess $b'$ of $b$, **D** outputs $e' = 1 − b' \oplus b$. If $e = 0$, the above experiment is Hyb$_j$; if $e = 1$, is Hyb$_{j+1}$. **D** outputs $e' = 1$ only if **A**'s guess $b' = b$ in the experiment.

$$\text{Adv}_D = \left| \Pr[\mathbf{D}(\sigma, \mathcal{O}(\lambda, C_{sk',S})) = 1] - \Pr[\mathbf{D}(\sigma, \mathcal{O}(\lambda, C_{sk'_{S'},S})) = 1] \right|$$

$$= \left| \Pr[\text{Hyb}_j] - \Pr[\text{Hyb}_{j+1}] \right| \geq \left| \text{Adv}_j^{\text{Hyb}} - \text{Adv}_{j+1}^{\text{Hyb}} \right|$$

which contradicts IO's indistinguishability, i.e. $|$Adv$_j^{\text{Hyb}}$ − Adv$_{j+1}^{\text{Hyb}}| \leq negl(\lambda)$. Similarly, $|$Adv$_2$ − Adv$_{1.5}| \leq negl(\lambda)$.

**Lemma 3.** *If $F'$ is a single-key secure CVRF, then $Adv_2 \leq negl(\lambda)$.*

*Proof.* If there is a PPT adversary **A** such that Adv$_2$ is a non-negligible probability $\mu$, we can construct adversary **B** to attack CVRF single-key security. **B** acts as the challenger of **Expt**$_2$ except (1) **B** sends $S'$ to its challenger for the replies $(sk'_{S'}, pk')$; (2) For the attack target $x_*$, **B** sends $x' = \text{PAR.Enc}(x_*)$ to the challenger who samples $e \leftarrow \{0, 1\}$, if $e = 0$, returns $y_*^0 = F'(sk', x')$; if $e = 1$, returns $y_*^1 \leftarrow \mathcal{Y}$. Then **B** returns $y_*^b = y_*^e$ to **A**. (3) After receiving **A**'s guess $b'$ of $b$, **B** outputs $e' = b'$. We find that $e' = e$ only if **A**'s guess $b' = b$, then $\mathbf{Adv_B} = \text{Adv}_2 = \mu$, which contradicts the single-key security of CVRF.

**Combining** the above lemmas and omitting some negligible functions for addition and subtraction, there is Adv$_2 \geq \tau \cdot$ Adv$_0$. Because $\tau$ is an inverse-polynomial function and thus incurs a polynomial level reduction loss. Theorem 3 is proved.

## 6     Possible Implications on Micro-Payment

In 2019, Liu et al. [27] briefly explained the possible application of CVRF in the micro-payment scenario in the abstract of their paper and showed that compared to just using VRF, the use of CVRF makes the public key infrastructure (PKI) for merchants' keys no longer needed. We will now retrospect this application, with appropriate expansions, and then analyze the possible implications of our adaptively secure construction on it.

In the process of probability-based micro-payments transaction, customers (**C**) pay some tiny cheques to merchants (**M**); **M** collects all cheques within a certain period and calculates the probability per cheque; according to the probability calculation result, **M** determines which cheque to complete once large payment and sends the cheque to the bank (**B**) for cash. Define CVRF's range $\mathcal{Y} \in [0,1]$, **B** saves the secret key $sk$ of CVRF in its own database, and exposes the public key $pk$. For each **M** with a unique identifying $id$, **B** generates a prefix-fixing set $S = \{s = id\|C : C \in \{0,1\}^{|C|}\}$ and sends the constrained key $sk_{id} \leftarrow \mathsf{Cons}(sk, S)$ to **M**. For each cheque $C$, **M** evaluates $(y, \pi) = \mathsf{CEval}(sk_{id}, id\|C)$ and checks whether $y \leq s$, here $s$ is a preset selection probability, if yes, $C$ is payable and **M** will send $(C, y, \pi)$ to **B** to finish the once large payment.

The use of CVRF has the following advantages: 1. There is one round of interaction between **M** and **C**. 2. Because of the uniqueness of CVRF, the result of the probabilistic calculation is fair, and no one can tamper with and falsify it. 3. The cheques $C$ are usually commitments to transaction information, due to CVRF constraints diversification, $C$ can be set to attach some fixed attribute concatenating after the commitment, such as timestamp, date, etc. Or even an explicit attribute related to the customers **C**. In short, **B** can customize different $S$ for different merchants and thus achieve more secure management. 4. With the semi-adaptive security of CVRF, **B** exposes the public key only after it triggers an attack target. While with adaptive security, **B** can disclose the public key at the beginning, and all merchants and customers can use the public key to repeatedly verify their forged cheques, probability calculation results and proof locally, then choose a beneficial target to attack. Systems that are resistant to the latter are more secure than the former. 5. Although the set $S$ contains $poly(\lambda)$-points, security, and feasibility can be satisfied if security parameters are appropriately selected in practice.

**Conclusion.** In addition to proving the existence of a single-key secure CVRF, we build an adaptively secure CVRF. There are some open problems, including constructing CVRF from the weaker cryptographic assumptions, or expanding its application scenarios to identity-based privacy-preserving authentication schemes [33] or password breach alerting services [31], and some others.

**Acknowledgments.** This work is supported by Beijing Natural Science Foundation (No. M22001).

# References

1. Abusalah, H., Fuchsbauer, G., Pietrzak, K.: Constrained PRFs for unbounded inputs. In: Sako, K. (ed.) CT-RSA 2016. LNCS, vol. 9610, pp. 413–428. Springer, Cham (2016). https://doi.org/10.1007/978-3-319-29485-8_24

2. Attrapadung, N., Matsuda, T., Nishimaki, R., Yamada, S., Yamakawa, T.: Adaptively single-key secure constrained PRFs for NC$^1$. In: Lin, D., Sako, K. (eds.) PKC 2019. LNCS, vol. 11443, pp. 223–253. Springer, Cham (2019). https://doi.org/10.1007/978-3-030-17259-6_8

3. Attrapadung, N., Matsuda, T., Nishimaki, R., Yamada, S., Yamakawa, T.: Constrained PRFs for NC1 in traditional groups. In: CRYPTO 2018, pp. 543–574 (2018)

4. Badertscher, C., Gazi, P., Kiayias, A., Russell, A., Zikas, V.: Ouroboros genesis: composable proof-of-stake blockchains with dynamic availability. In: CCS 2018, pp. 913–930 (2018)

5. Barak, B., et al.: On the (im)possibility of obfuscating programs. In: Kilian, J. (ed.) CRYPTO 2001. LNCS, vol. 2139, pp. 1–18. Springer, Heidelberg (2001). https://doi.org/10.1007/3-540-44647-8_1

6. Belenkiy, M., Chase, M., Kohlweiss, M., Lysyanskaya, A.: Compact E-cash and simulatable VRFs revisited. In: Shacham, H., Waters, B. (eds.) Pairing 2009. LNCS, vol. 5671, pp. 114–131. Springer, Heidelberg (2009). https://doi.org/10.1007/978-3-642-03298-1_9

7. Bitansky, N.: Verifiable random functions from non-interactive witness-indistinguishable proofs. J. Cryptol. **33**(2), 459–493 (2020)

8. Blum, M.: Coin flipping by telephone. In: Advances in Cryptology: A Report on CRYPTO 1981, IEEE Workshop on Communications Security, Santa Barbara, California, USA, 24–26 August 1981, pp. 11–15 (1981)

9. Bogdanov, A., Rosen, A.: Pseudorandom functions: three decades later. In: Tutorials on the Foundations of Cryptography. ISC, pp. 79–158. Springer, Cham (2017). https://doi.org/10.1007/978-3-319-57048-8_3

10. Boneh, D., Waters, B.: Constrained pseudorandom functions and their applications. In: Sako, K., Sarkar, P. (eds.) ASIACRYPT 2013. LNCS, vol. 8270, pp. 280–300. Springer, Heidelberg (2013). https://doi.org/10.1007/978-3-642-42045-0_15

11. Boyle, E., Goldwasser, S., Ivan, I.: Functional signatures and pseudorandom functions. In: Krawczyk, H. (ed.) PKC 2014. LNCS, vol. 8383, pp. 501–519. Springer, Heidelberg (2014). https://doi.org/10.1007/978-3-642-54631-0_29

12. Canetti, R., Chen, Y.: Constraint-hiding constrained PRFs for NC$^1$ from LWE. In: Coron, J.-S., Nielsen, J.B. (eds.) EUROCRYPT 2017. LNCS, vol. 10210, pp. 446–476. Springer, Cham (2017). https://doi.org/10.1007/978-3-319-56620-7_16

13. Chandran, N., Raghuraman, S., Vinayagamurthy, D.: Constrained pseudorandom functions: verifiable and delegatable. IACR Cryptology ePrint Archive 2014, 522 (2014)

14. Datta, P.: Constrained (verifiable) pseudorandom function from functional encryption. In: Su, C., Kikuchi, H. (eds.) ISPEC 2018. LNCS, vol. 11125, pp. 141–159. Springer, Cham (2018). https://doi.org/10.1007/978-3-319-99807-7_9

15. Datta, P.: Constrained pseudorandom functions from functional encryption. Theoret. Comput. Sci. **809**, 137–170 (2020)

16. Datta, P., Dutta, R., Mukhopadhyay, S.: Constrained pseudorandom functions for unconstrained inputs revisited: achieving verifiability and key delegation. In: Fehr, S. (ed.) PKC 2017. LNCS, vol. 10175, pp. 463–493. Springer, Heidelberg (2017). https://doi.org/10.1007/978-3-662-54388-7_16

17. Datta, P., Dutta, R., Mukhopadhyay, S.: Constrained pseudorandom functions for turing machines revisited: How to achieve verifiability and key delegation. Algorithmica **81**, 3245–3390 (2019)

18. Deshpande, A., Koppula, V., Waters, B.: Constrained pseudorandom functions for unconstrained inputs. In: Fischlin, M., Coron, J.-S. (eds.) EUROCRYPT 2016. LNCS, vol. 9666, pp. 124–153. Springer, Heidelberg (2016). https://doi.org/10.1007/978-3-662-49896-5_5

19. Fuchsbauer, G.: Constrained verifiable random functions. In: Abdalla, M., De Prisco, R. (eds.) SCN 2014. LNCS, vol. 8642, pp. 95–114. Springer, Cham (2014). https://doi.org/10.1007/978-3-319-10879-7_7

20. Garg, S., Gentry, C., Halevi, S., Raykova, M., Sahai, A., Waters, B.: Candidate indistinguishability obfuscation and functional encryption for all circuits. In: FOCS 2013, pp. 40–49 (2013)

21. Gilad, Y., Hemo, R., Micali, S., Vlachos, G., Zeldovich, N.: Algorand: scaling byzantine agreements for cryptocurrencies. In: Proceedings of the 26th Symposium on Operating Systems Principles, pp. 51–68. ACM (2017)

22. Goldreich, O., Goldwasser, S., Micali, S.: How to construct random functions (extended abstract). In: 25th Annual Symposium on Foundations of Computer Science, West Palm Beach, Florida, USA, 24–26 October 1984, pp. 464–479 (1984)

23. Jager, T.: Verifiable random functions from weaker assumptions. In: Dodis, Y., Nielsen, J.B. (eds.) TCC 2015. LNCS, vol. 9015, pp. 121–143. Springer, Heidelberg (2015). https://doi.org/10.1007/978-3-662-46497-7_5

24. Kiayias, A., Papadopoulos, S., Triandopoulos, N., Zacharias, T.: Delegatable pseudorandom functions and applications. In: In CCS 2013, pp. 669–684 (2013)

25. Liang, B., Li, H., Chang, J.: Constrained verifiable random functions from indistinguishability obfuscation. In: ProvSec 2015, pp. 43–60 (2015)

26. Liskov, M.: Updatable zero-knowledge databases. In: Roy, B. (ed.) ASIACRYPT 2005. LNCS, vol. 3788, pp. 174–198. Springer, Heidelberg (2005). https://doi.org/10.1007/11593447_10

27. Liu, M., Zhang, P., Wu, Q.: A novel construction of constrained verifiable random functions. Secur. Commun. Netw. **2019**, 4187892:1–4187892:15 (2019)

28. Lysyanskaya, A.: Unique signatures and verifiable random functions from the DH-DDH separation. In: Yung, M. (ed.) CRYPTO 2002. LNCS, vol. 2442, pp. 597–612. Springer, Heidelberg (2002). https://doi.org/10.1007/3-540-45708-9_38

29. Micali, S., Rabin, M., Vadhan, S.: Verifiable random functions. In: 40th Annual Symposium on Foundations of Computer Science, pp. 120–130 (1999)

30. Micali, S., Rivest, R.L.: Micropayments revisited. In: Preneel, B. (ed.) CT-RSA 2002. LNCS, vol. 2271, pp. 149–163. Springer, Heidelberg (2002). https://doi.org/10.1007/3-540-45760-7_11

31. Pal, B., et al.: Might I get pwned: a second generation compromised credential checking service. In: USENIX Security 2022, pp. 1831–1848 (2022)

32. Peikert, C., Shiehian, S.: Privately constraining and programming PRFs, the LWE way. In: Abdalla, M., Dahab, R. (eds.) PKC 2018. LNCS, vol. 10770, pp. 675–701. Springer, Cham (2018). https://doi.org/10.1007/978-3-319-76581-5_23

33. Wang, D., Cheng, H., He, D., Wang, P.: On the challenges in designing identity-based privacy-preserving authentication schemes for mobile devices. IEEE Syst. J. **12**(1), 916–925 (2018)

# A Robust Reversible Data Hiding Algorithm Based on Polar Harmonic Fourier Moments

Bin Ma[1], Zhongquan Tao[1], Jian Xu[2], Chunpeng Wang[1(✉)], Jian Li[1], and Liwei Zhang[3]

[1] Qilu University of Technology (Shandong Academy of Sciences), Jinan 250353, China
mpeng1122@163.com
[2] Shandong University of Finance and Economics, Jinan 250014, China
[3] Integrated Electronic Systems Lab Co., Ltd., Jinan 250000, China

**Abstract.** The Robust Reversible Data Hiding (RRDH) algorithm can recover both the secret data and the cover image entirely from an intact stego image, and can still restore the secret message comprehensively even if the stego image undergoes various attacks. While many current RRDH schemes possess a strong capability for resisting attacks, they often fail to withstand geometric deformation attacks such as rotation and scaling. In this paper, a new RRDH algorithm based on Polar Harmonic Fourier Moments (PHFMs) is presented to enhance its resistance to geometric transformation attacks. Firstly, leveraging the rotation invariance of PHFMs, a quantitation index modulation algorithm is designed to embed secret data into the coefficients of PHFMs. This approach achieves a high degree of anti-geometric transformation capability for the stego image while minimizing image distortion. Additionally, the differences between the original and restored image are employed as compensation information and embedded into the restored image. Moreover, a two-dimensional reversible data hiding scheme is adopted to embed the compensation information in order to minimize image distortion. The combination of PHFMs transformation and two-dimensional reversible data hiding enables the proposed RRDH algorithm to achieve high visual quality and strong resistance capability against geometric transformation attacks. Extensive experimental results demonstrate that the proposed RRDH algorithm outperforms other state-of-the-art techniques.

**Keywords:** Robust Reversible Data Hiding · Polar Harmonic Moments · Geometric deformation

Supported by National Natural Science Foundation of China (62272255); National key re-search and development program of China (2021YFC3340602); Shandong Provincial Natural Science Foundation Innovation and Development Joint Fund (ZR202208310038); Ability Im-provement Project of Science and Technology SMES in Shandong Province (2022TSGC2485); Project of Jinan Research Leader Studio (2020GXRC056); Project of Ji-nan Introducing Innovation Team (202228016); Project of Jinan City-School Integration De-velopment (JNSX2021030); Youth Innovation Team of Colleges and Universities in Shan-dong Province (2022KJ124); The "Chunhui Plan" Cooperative Scientific Research Project of Ministry of Education (HZKY20220482);Shandong Provincial Natural Science Foundation (ZR2020MF054).

© The Author(s), under exclusive license to Springer Nature Switzerland AG 2023
M. Yung et al. (Eds.): SciSec 2023, LNCS 14299, pp. 386–397, 2023.
https://doi.org/10.1007/978-3-031-45933-7_23

# 1 Introduction

Internet technology has advanced rapidly in the last decade, leading to its widespread use for communication and sharing of multimedia information. However, digital media is vulnerable to various types of attacks during transmission [1], making the need for ensuring multimedia information security a critical concern.

Various Reversible Data Hiding (RDH) techniques have been developed for the secure transmission of data and communications, including compression-based [2], Difference Expansion-based [3, 4], Histogram Shifting-based [5, 6], and Prediction-Error Expansion-based methods [7, 8]. However, these methods can cause irreversible distortions in the image, making them unsuitable for applications in sensitive fields such as medicine and military. Additionally, they do not account for robustness. To address these concerns, Robust Reversible Data Hiding (RRDH) techniques have been developed.

The RRDH methods can extract the secret data and the original image from an intact stego image and restore the secret data even if the stego image is attacked. RRDH methods can be broadly classified into two types.

The first type is histogram shifting-based RRDH methods. Ni [9] first proposed an RRDH method based on histogram shifting, which involves selecting robust features in the image to form the robust feature histogram and then forming the robust region by moving the histogram to embed the secret data. The larger the region interval, the more robust it is. Numerous researchers have since presented RRDH methods based on histogram shifting [10–13]. However, a major issue with histogram shifting-based RRDH methods is their sensitivity to geometric distortion. This is because the histogram embedded with robust data is related to the position of the blocks of the image, so even slight geometric distortion can cause significant distortion to the histogram. The second type of RRDH method is the two-stage embedding strategy approach proposed by Coltuc [14]. This approach divides the embedding process into two stages and embeds secret data in the first stage. In the second stage, the image distortion caused by the secret data embedding is reversibly embedded in the image as compensation information so that the image can be restored to its original state after the robust data is extracted at the receiving end. The design of this two-stage framework allows most robust and reversible data hiding methods to be incorporated to achieve robust reversible data hiding [15, 16].

Existing RRDH methods are vulnerable to geometric deformation, while image moment-based robust data hiding methods [17–19] can resist both geometric and non-geometric attacks with better visual quality. However, these methods suffer from irreversible image loss. To address this issue, this paper proposes a PHFMs-based RRDH algorithm.

The innovations of this paper are that: (1) This paper proposes a new quantization strategy that only quantizes the integer part of the Polar Harmonic Fourier moments, greatly reducing the impact of secret data embedding on image quality. (2) Another highlight is the computation method for compensation information, which divides the distortion of the image into two parts (quantization error and rounding error).

## 2  Proposed Method

This paper proposes a RRDH method, which comprises three stages: embedding of secret data, reversible embedding of compensation information, and decoding. In the embedding stage, secret data is embedded into the low-order coefficients of Polar Harmonic Fourier Moments (PHFMs) using the geometric invariance feature of PHFMs. The reversible embedding stage uses a two-dimensional reversible data hiding method to reversibly embed the difference between the stego image and the original image as compensation information for recovering the cover image. The extraction and recovery stage is the third and final stage of the RRDH method.

### 2.1  Robust Embedding Stage

Figure 1 shows a detailed flowchart, including the embedding process of secret data and the computation process of compensation information. In the first stage, PHFMs are first calculated from the cover image. The PHFMs are pseudo-randomly selected and normalized, and a new embedding strategy proposed in this paper is used to embed the secret data into the image. In the reversible embedding stage, the quantization error $d_q$ and rounding error $d_r$ are calculated and embedded into the secret image along with the generated image Hash value $H$ as compensation information.

A) *Calculation of PHFMs*

Unlike Zernike moments and pseudo-Zernik moments, Polar Harmonic Fourier moments do not do not suffer from deviation from orthogonality and can be accurately calculated [24]. Therefore, specifying the maximum order $N_{\max}$ and the repetition degree $m$ is sufficient for PHFMs. Since the PHFMs with repetition degree $m = 0$ are mutually conjugated, only moments with repetition degree $m \geq 0$ are selected for embedding. After information embedding, PHFMs at symmetrical positions are correspondingly modified to ensure the maintenance of the conjugate relationship. In summary, the order $n$ and the repetition degree $m$ should satisfy the following conditions:

$$0 \leq n \leq N_{\max}, m \geq 0 \tag{1}$$

To compute the PHFMs, we adopt the inner tangent circle as the reference unit circle. The PHFMs are then obtained by evaluating Eq. (2).

$$P_{nm} = \frac{2}{\pi} \sum_{p=0}^{N-1} \sum_{q=0}^{N-1} f(x_q, y_p) \overline{H_{nm}(r_{p,q}, \theta_{p,q})} \Delta x \Delta y \tag{2}$$

B) *Embed Secret Data into PHFMs*

Since images may be subject to scaling attacks during transmission, which can significantly change the size of PHFMs, normalization is used to avoid this from happening. And it can make the calculation easier.

To strengthen the security of the algorithm, we use the key $K_1$ to pseudo-randomly select $L$ PHFMs from the set $S$, denoted as set $P_{p,q} = \{P_{p_1,q_1}, P_{p_2,q_2}, ..., P_{p_i,q_i}\}$ except for $P_{00}$, based on the length $L$ of the secret information $W = w_i$, $i = 1, 2, ..., L$. In order to improve the image quality of secret images and reduce the compensation information's length, we preprocess set $P_{p,q}$ using the formula (3) to simplify subsequent calculations.

$$P^R_{p_i,q_i} = P_{p_i,q_i} \times K, \ i = 1, 2, ..., L \tag{3}$$

where $P^R_{p_i,q_i}$ denotes the normed moment after preprocessing and $K$ is a parameter set to $10^x$ ($x$ is a positive integer). Through a series of experiments, it has been observed that satisfactory result can be achieved when $x = 1$, $K = 10$.

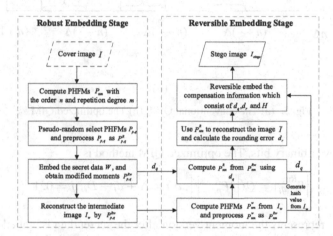

**Fig. 1.** New embedding strategy and computation of the compensation information

As illustrated in Fig. 2(a), embedding secret information using traditional quantization index modulation technology will cause an error between the secret image and the original image, denoted as quantization error $d_q$.

$$d_{qi} = \left| P^{Rw_j}_{p_i,q_i} \right| - \left| P^R_{p_i,q_i} \right| \tag{4}$$

The conventional quantization index modulation technique cannot control the quantization error to an integer, so more compensation information needs to be embedded to restore the original image. As a result, the conventional quantization index modulation technique is not appropriate for Robust Reversible Data Hiding. To solve this problem, a novel quantization index modulation technique is designed in this paper.

This paper proposes a novel technique for quantization index modulation (QIM) as illustrated in Fig. 2(b), which separates the amplitude of PHFMs into integer and fractional parts before embedding secret information. During the embedding process, only the integer part is quantized, while the fractional part $D_i = \left| P^R_{p_i,q_i} \right| - \left\lfloor \left| P^R_{p_i,q_i} \right| \right\rfloor$ is added to the quantized moment, forcing the quantization error to be an integer. In addition, in the optimized QIM, the embedding strength $\Delta$ is an even number greater than 0, that is, $\Delta = 2k, k \geq 1 \& k \in$ integer, which is beneficial to express quantization

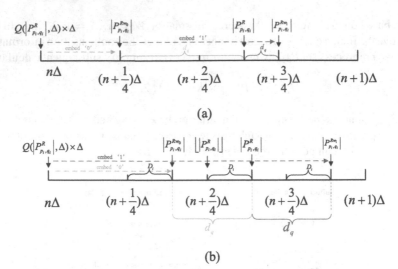

**Fig. 2.** (a) The conventional QIM; (b) The optimized QIM

errors with fewer binary bits. The optimized QIM method is

$$\left| P_{p_i,q_i}^{Rw_j} \right| = Q(\left\lfloor \left| P_{p_i,q_i}^{R} \right| \right\rfloor, \Delta) \times \Delta + d_j + D_i, \ j = 0, 1 \tag{5}$$

where $\left| P_{p_i,q_i}^{Rw_j} \right|$ represents the moment version after the secret data is embedded, $\Delta$ represents the embedding strength, and $Q(\cdot)$ is a quantizer, which can be expressed as:

$$Q(\left\lfloor \left| P_{p_i,q_i}^{R} \right| \right\rfloor, \Delta) = \left\lfloor \frac{\left\lfloor \left| P_{p_i,q_i}^{R} \right| \right\rfloor}{\Delta} \right\rfloor.$$

### C) *Robustly secret image Reconstruction*

During the process of embedding secret information, the pre-processed moments are quantized. If the image is to be reconstructed, an inverse processing operation is required, i.e.,

$$P_{p_i,q_i}^{w} = \frac{\left| P_{p_i,q_i}^{Rw} \right|}{\left| P_{p_i,q_i}^{R} \right|} \times P_{p_i,q_i} \tag{6}$$

Due to PHFMs are mutually conjugated with respect to repetition m = 0, the complex conjugate of $P_{p_i,q_i}$, i.e., $P_{-p_i,-q_i}$ needs to be quantized to the same magnitude. Then the reconstruction of the robust secret image $I_{Robust}$ can be performed.

$$I_{Robust} = I + \sum_{i=1}^{L} [((P_{p_i,q_i}^{R} - P_{p_i,q_i}) \times H_{p_i,q_i}) + ((P_{-p_i,-q_i}^{R} - P_{-p_i,-q_i}) \times H_{-p_i,-q_i})] \tag{7}$$

Reconstruction of $I_{robust}$ can be achieved by Eq. (9). However, direct addition and subtraction of pixel values during the reconstruction process can result in pixel values exceeding the range of [0,255]. And PHFMs compute discrete, the pixel values may be real. Therefore, a rounding operation will be performed so that the pixel value of the image is an integer in the range of [0,255]. The robust secret image $I_w$ is then generated.

### D) *Computation of the rounded error*

This subsection proposes a new strategy to represent the rounding error. Computing PHFMs from $I_w$, denote as $\widetilde{P}^w_{p_i,q_i}$. Then use the same preprocessing operation to get $\widetilde{P}^{Rw}_{p_i,q_i}$. And subtract the quantization error $d_q$ from $\widetilde{P}^{Rw}_{p_i,q_i}$, that is:

$$|\widetilde{P}^R_{p_i,q_i}| = |\widetilde{P}^{Rw}_{p_i,q_i}| + d_{q_i} \tag{8}$$

Because a rounding operation is performed after generating the robust secret image, $I_{robust}$ is not the same as $I_w$, so $\left|\widetilde{P}^R_{p_i,q_i}\right|$ is only approximate to $\left|P^R_{p_i,q_i}\right|$, not exactly the same as $\left|P^R_{p_i,q_i}\right|$. The inverse processing operation should continue.

$$\widetilde{P}_{p_i,q_i} = \frac{\left|\widetilde{P}^R_{p_i,q_i}\right|}{\left|\widetilde{P}^{Rw}_{p_i,q_i}\right|} \times \widetilde{P}^w_{p_i,q_i} \tag{9}$$

Similar to (7), use $\widetilde{P}_{p_i,q_i}$ to reconstruct the image to get the image $\widetilde{I}$ without the secret information. The difference between $\widetilde{I}$ and $I$ is caused by the rounding operation, so we regard this part as rounding error $d_r$ and use it as part of the compensation information to recover the original image.

$$d_r = I - \widetilde{I} \tag{10}$$

By now we obtained all the compensation information necessary for restoring the cover image, including quantization error $d_q$, rounding error $d_r$, and image Hash $H$. These pieces of information were embedded into the image using a two-dimensional histogram shifting algorithm based on Pairwise PEE [5], resulting in the final stego image $I_{stego}$.

## 2.2  Extraction of Secret Data and Recovery of the Image

When the recipient receives the stego image, their first step is to extract the compensation information. The compensation information includes the image hash value $H$, which the recipient uses to verify if the image has been tampered with during transmission. If the received image is lossless, the cover image can be restored losslessly after extracting the secret data. Otherwise, only the secret data can be extracted by Eq. (11)

$$w_i = \begin{cases} 0, & \text{if } \left\lfloor \left|\widetilde{P}^{Rw}_{p_i,q_i}\right| \right\rfloor - g(i) \leq \frac{1}{2}\Delta \\ 1, & \text{if } \left\lfloor \left|\widetilde{P}^{Rw}_{p_i,q_i}\right| \right\rfloor - g(i) \geq \frac{1}{2}\Delta \end{cases} \tag{11}$$

where $g(i) = \left\lfloor \left\lfloor \widetilde{P}^{Rw}_{p_i, q_i} \right\rfloor \div \Delta \right\rfloor \times \Delta$.

After the secret information extraction is completed, the original image can be recovered using the compensation information. The pre-recovered image $\widetilde{I}$ is obtained using the quantization error $d_q$ according to Eqs. (7)–(9), and then the original image is obtained by adding the rounding error $d_r$.

$$I_{original} = \widetilde{I} + d_r \tag{12}$$

## 3   Experimental Results

### 3.1   Design of Experiments

In this section, we evaluate the proposed RRDH method experimentally to showcase its feasibility and advantages. The *Misc* dataset, which includes four classic grayscale images (i.e., Lena, Goldhill, Peppers and Barbara), is selected for testing, as shown in Fig. 3. The purpose of the experiment is to assess the effectiveness of the proposed RRDH method by comparing it to other methods with regard to the robustness of the secret data and the length of the compensation information. The images utilized in the experiments are in grayscale and have dimensions of $512 \times 512$ pixels. All attack experiments on the images were performed using MATLAB 2018b.

(a)              (b)              (c)              (d)

**Fig. 3.** Four test images: (a) Lena, (b) Goldhill, (c) Peppers, (d) Barbara

The parameters of the algorithm proposed in this paper are embedding strength $\Delta = 6$, $K = 10$, and maximum order $N = 25$. To demonstrate the superiority of the proposed method, this paper is compared with four existing state-of-the-art methods such as Hu et al. [21], Hu et al. [22], Tang et al. [23] and Wang et al. [15]. The parameters of the methods for comparison are shown in Table 1.

### 3.2   Comparison of Data Embedding Capacity with Other Schemes

To ensure the reasonableness of the experiments, the experimental comparisons in this section are only compared with three Robust Reversible Data Hiding methods based on image moments in [21–23].

Based on Fig. 4(a), the proposed method requires the least amount of compensation information to restore the cover image, and the compensation information for all four test images does not exceed 20,000 bits. This indicates that the proposed method can

**Table 1.** The parameters of the compared methods and the proposed method

| Methods | Parameters settings |
|---|---|
| Hu's method [21] | $N = 15;\ \Delta = 40; T = 100$ |
| Hu's method [22] | $N = 31;\ \Delta = 18; T = 1000$ |
| Tang's method [23] | $M = 18;\ \Delta = 32; T_{start} = 2000; \Upsilon = 10$ |
| Wang's method [15] | block size $64 \times 16; \tau = 2.4$ |
| Proposed method | $N = 25;\ \Delta = 6; K = 10$ |

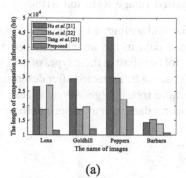 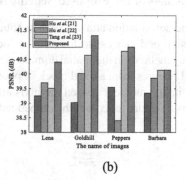

(a)                                    (b)

**Fig. 4.** (a) The length of compensation information in four test images; (b) The PSNR of the four test images

minimize the impact on the image after embedding secret data, and can achieve higher-quality images. This is also evident in Fig. 4(b). It can be concluded that the proposed method has better performance.

To further validate the superiority of the proposed method, we conducted an additional experiment using 45 images from the *Mis* dataset. We presented the average PSNR of the 45 images and the average amount of compensation information required to recover the images in Table 2. It can be obtained that the proposed method has the best image quality and only requires the least amount of compensation information to restore the cover image.

**Table 2.** Average PSNR and the average length of compensation information

| Method | Average of PSNR | Average length of compensation information |
|---|---|---|
| Hu's method [21] | 39.45 dB | 36861.18 bit |
| Hu's method [22] | 38.20 dB | 31456.13 bit |
| Tang's method [23] | 40.27 dB | 20878.11 bit |
| Proposed method | 40.39 dB | 18557.96 bit |

### 3.3 Comparison of Robustness with Other Schemes

The experiment in this paper employs Bit Error Rate (BER) as a metric to evaluate the robustness of the algorithm. The BER is defined as the frequency of error data occurrence, and is calculated by dividing the number of erroneous bits of the extracted secret data by the total number of bits of the secret data. Wang et al. [15] utilized the technique of HWT to embed secret data and differences into distinct regions. The method is capable of withstanding non-geometric attacks, but lacks robustness against geometric attacks. So, in the geometric attack experiments, we only compared the proposed method with the ones proposed in [21, 22], and [23].

### A) Robustness of the Proposed Scheme Against Image Rotation Attack

Geometric deformation attacks, such as rotation and scaling, are common during image transmission and can cause loss of the hidden data in the image. Some secret data embedding methods in the spatial domain may not be robust to these types of attacks since statistical features from a fixed block of pixels cannot be extracted after deformation. In contrast, image moments-based methods are more robust to geometric attacks and can be used as a feature to extract the embedded secret data even after such deformations.

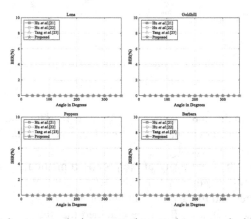

**Fig. 5.** Robustness to the image rotation attack compared with [21–23]

In the image rotation attack experiments, we varied the rotated angles from 0 to 360 with an interval of 20. In Fig. 5, it is apparent that all four methods are able to withstand the arbitrary angle rotation attack. Since image moment-based embedding methods have excellent rotational invariance.

### B) Robustness of the Proposed Scheme Against Image Scale Attack.

In the image scale attack experiments, the scale factors range from 0.5 to 2.0 with an interval of 0.1. As depicted in Fig. 6, the proposed method exhibits excellent robustness against scaling attacks. Notably, the BER of the proposed method remains below 2% for the four test images. And in the Barbara image, the secret data can be accurately extracted without error when facing any intensity of scaling attacks (Fig. 7).

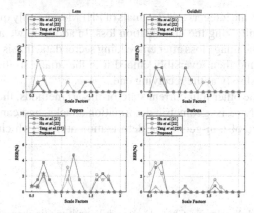

**Fig. 6.** Robustness to the image scale attack compared with [21–23]

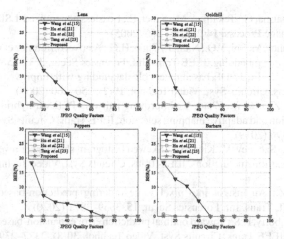

**Fig. 7.** Robustness to the JPEG compression attack compared with [15, 21–23]

## C) Robustness of the Proposed Scheme Against JPEG Compression Attack

Various types of image compression techniques such as JPEG compression and JPEG2000 compression are widely used. JPEG compression is a lossy compression technique that removes some insignificant data from the original image to save memory. During the experiment, we compressed four test images using JPEG compression with quality factors ranging from 10 to 100 at intervals of 10, it is evident that the proposed method provides complete resilience against JPEG compression attacks.

## 4   Conclusion

This paper proposes a robust reversible data hiding method based on Polar Harmonic Fourier Moments, which can effectively resist geometric and non-geometric deformation attacks, and can achieve good image quality. In this paper, we propose a new embedding strategy that separates the integers and decimals of PHFMs, and only quantizes the integer part of moments, minimizing the quantization loss. In addition, we also propose a new method to calculate the image loss after embedding secret data, that is, quantization error and rounding error, and then reversibly embed it in the image so that the cover image can be recovered losslessly at the receiving end.

Compared with the other four current state-of-the-art methods, the proposed method can almost completely resist geometric deformation attacks, and can still achieve good robustness in the face of non-geometric deformation attacks, which is better than the other four methods.

# References

1. Usman, M., Jan, M.A., He, X., Chen, J.: A mobile multimedia data collection scheme for secured wireless multimedia sensor networks. IEEE Trans. Netw. Sci. Eng. **7**(1), 274–284 (2020)
2. Celik, M.U., Sharma, G., Tekalp, A.M., Saber, E.: Lossless generalized-LSB data embedding. IEEE Trans. Image Process. **14**(2), 253–266 (2005)
3. Kim, H.J., Sachnev, V., Shi, Y.Q., Nam, J., Choo, H.G.: A novel difference expansion transform for reversible data embedding. IEEE Trans. Inf. Forensics Secur. **3**(3), 456–465 (2008)
4. Hu, Y., Lee, H.-K., Li, J.: DE-based reversible data hiding with improved overflow location map. IEEE Trans. Circuits Syst. Video Technol. **19**(2), 250–260 (2009)
5. Ou, B., Li, X., Zhang, W., Zhao, Y.: Improving pairwise PEE via hybrid-dimensional histogram generation and adaptive mapping selection. IEEE Trans. Circuits Syst. Video Technol. **29**(7), 2176–2190 (2019)
6. Kim, S., Qu, X., Sachnev, V., Kim, H.J.: Skewed histogram shifting for reversible data hiding using a pair of extreme predictions. IEEE Trans. Circuits Syst. Video Technol. **29**(11), 3236–3246 (2019)
7. He, W., Cai, Z.: An insight into pixel value ordering prediction-based prediction-error expansion. IEEE Trans. Inf. Forensics Secur. **15**, 3859–3871 (2020)
8. Roy, A., Chakraborty, R.S.: Toward optimal prediction error expansion-based reversible image watermarking. IEEE Trans. Circuits Syst. Video Technol. **30**(8), 2377–2390 (2020)
9. Ni, Z., Shi, Y.Q., Ansari, N., Su, W., Sun, Q., Lin, X.: Robust lossless image data hiding designed for semi-fragile image authentication. IEEE Trans. Circuits Syst. Video Technol. **18**(4), 497–509 (2008)
10. Zeng, X.-T., Ping, L.-D., Pan, X.-Z.: A lossless robust data hiding scheme. Pattern Recognit. **43**(4), 1656–1667 (2010)
11. Zong, T., Xiang, Y., Natgunanathan, I., Guo, S., Zhou, W., Beliakov, G.: Robust histogram shape-based method for image watermarking. IEEE Trans. Circuits Syst. Video Technol. **25**(5), 717–729 (2015)
12. Liang, X., Xiang, S.: Robust reversible audio watermarking based on high-order difference statistics. Signal Process. **173**, 107584 (2020)
13. Rajkumar, R., Vasuki, A.: Reversible and robust image watermarking based on histogram shifting. Cluster Comput **22**(Suppl. 5), 12313–12323 (2019)
14. Coltuc, D.: Towards distortion-free robust image authentication. In: Journal of Physics: Conference Series, vol. 77, p. 012005. IOP Publishing (2007)
15. Wang, X., Li, X., Pei, Q.: Independent embedding domain based two-stage robust reversible watermarking. IEEE Trans. Circuits Syst. Video Technol. **30**(8), 2406–2417 (2020)
16. Wang, W., Ye, J., Wang, T., et al.: Reversible data hiding scheme based on significant-bit-difference expansion. IET Image Proc. **11**(11), 1002–1014 (2017)

17. Yang, J., Lu, Z., Tang, Y.Y., Yuan, Z., Chen, Y.: Quasi fourier-mellin transform for affine invariant features. IEEE Trans. Image Process. **29**, 4114–4129 (2020)
18. Tarpau, C., Cebeiro, J., Nguyen, M.K., Rollet, G., Morvidone, M.A.: Analytic inversion of a radon transform on double circular arcs with applications in compton scattering tomography. IEEE Trans. Comput. Imaging **6**, 958–967 (2020)
19. Liu, X., Han, G., Wu, J., Shao, Z., Coatrieux, G., Shu, H.: Fractional krawtchouk transform with an application to image watermarking. IEEE Trans. Signal Process. **65**(7), 1894–1908 (2017)
20. Wang, C., Wang, X., Xia, Z., Ma, B., Shi, Y.-Q.: Image description with polar harmonic fourier moments. IEEE Trans. Circuits Syst. Video Technol. **30**(12), 4440–4452 (2020)
21. Hu, R., Xiang, S.: Cover-lossless robust image watermarking against geometric deformations. IEEE Trans. Image Process. **30**, 318–331 (2021)
22. Hu, R., Xiang, S.: Lossless robust image watermarking by using polar harmonic transform. Signal Process. **179**, 107833 (2021)
23. Tang, Y., Wang, S., Wang, C., Xiang, S., Cheung, Y.-M.: A highly robust reversible watermarking scheme using embedding optimization and rounded error compensation. IEEE Trans. Circuits Syst. Video Technol. **33**(4), 1593–1609 (2022)
24. Xin, Y., Liao, S., Pawlak, M.: Circularly orthogonal moments for geometrically robust image watermarking. Pattern Recognit. **40**(12), 3740–3752 (2007)

# Advanced Threat Detection Techniques and Blockchain

# VaultBox: Enhancing the Security and Effectiveness of Security Analytics

Devharsh Trivedi[✉] and Nikos Triandopoulos

Stevens Institute of Technology, Hoboken, USA
{dtrived5,ntriando}@stevens.edu

**Abstract.** Security tools like Firewalls, IDS, IPS, SIEM, EDR, and NDR effectively detect and block threats. However, these tools depend on the system, application, and event logs. Logs are the key ingredient for various purposes, including troubleshooting performance issues, satisfying compliance mandates, and monitoring and improving security. In addition, logs from multiple machines are collected and fed to the Security Information and Event Management (SIEM) system for further security analysis. Therefore, a SIEM system's efficiency and effectiveness depend heavily on the quality and quantity of logs provided. Unfortunately, logs are often targeted brutally and tampered with after a successful intrusion to cover the attack's traces. Thus it becomes critical to protect the confidentiality, integrity, availability, and authenticity of logs at rest or transit. This paper proposes a novel scheme to prevent logs from tampering, detect any tampering, and recuperate logs if lost or corrupt. Our scheme is forward-secure, replicated, randomized, and rate-less, aiming to help securely store and transmit logs to SIEM.

**Keywords:** Security analytics · Rateless encoding · Secure logging · SIEM security · LT codes · Secure coding

## 1 Introduction

Organizations worldwide are troubled by modern, advanced, and more organized threat actors. Ransomware and confidential data leaks are the major threats of the current times. In addition, businesses regularly face phishing, denial of service, URL poisoning, malware, and other types of cyberattacks. In a corporate environment with thousands of connected devices, it becomes crucial to monitor the network activity to stop possible breaches to secure personal user data and confidential business data, which can cause reputation and revenue damage to the organization. Devices can be small mobile tools like cell phones, tablets, laptops or big stationary machines like printers, desktops, or servers. With the vast scale and complex geographic distribution of users and devices in an organization, identifying and preventing security threats is a constant challenge. Therefore, monitoring and analyzing all devices' activities is imperative.

While it is proving impossible to keep up with all the vulnerabilities spanning the organization across continents, the next best thing to do is to identify

© The Author(s), under exclusive license to Springer Nature Switzerland AG 2023
M. Yung et al. (Eds.): SciSec 2023, LNCS 14299, pp. 401–422, 2023.
https://doi.org/10.1007/978-3-031-45933-7_24

Incidents of Compromise (IoC) early with the help of a 24x7 Security Operations Centre (SOC) and act on it. Information security technologies such as Intrusion Detection Systems (IDS), Intrusion Prevention Systems (IPS), and Security Information and Event Management (SIEM) systems are specifically designed to help fight cyberattacks and win this battle. Modern SOC uses SIEM, SOAR, XDR, and EDR tools that rely heavily on logs collected from endpoints. However, advanced threat actors use sophisticated methods to gain and maintain unauthorized access by manipulating their traces. Thus, protecting these logs' confidentiality, integrity, availability, and authenticity is critical.

SIEM uses system logs from various devices or SAS (Security Analytics Source) to monitor and identify security threats. They are vital for reconstructing an attack [1, 8, 17, 23, 28, 47, 50] once it is over which is somewhat different from the real-time IDS function. Log tampering sends investigators in the wrong direction, giving the attackers more time to cover their tracks. After intrusion, the logs are targeted, tampered with, or deleted to cover the attack's traces. Hence for SIEM to work better, the quality and quantity of the collected logs are paramount. If system logs are compromised, SIEM might fail to identify or prevent such attacks, which might go undetected. Besides security applications, systems logs are crucial for other business use cases like troubleshooting performance issues, providing user support, analyzing product usage patterns, and ensuring compliance and audit requirements. [40, 46] Therefore, it is paramount to protect the integrity and availability of these logs to be analyzed. Log generation tools like Rsyslog [45], Syslog-ng [43], Log4j [2], and OSSEC-HIDS [31] are prevalent, reliable, and best in the category of security and audit logs. However, there have been many vulnerabilities impacting log generation and management. Most engineers focus on the high-priority systems during threat modeling overlooking their low-priority systems like logging, which remains vulnerable.

There are vulnerabilities [10, 11] in logging applications disrupting service. Failure to transmit some logs from SAS endpoints to SIEM may result in missed intrusion detection. Thus a mechanism is required to detect such logging and transmission failures. There are other severe vulnerabilities [9] allowing an attacker to overwrite log messages. This is critical as the attacker can erase their tracks by modifying logs. Thus, a tamper-evident mechanism is needed. VaultBox aims to solve these critical issues.

This paper aims to protect logs' confidentiality, integrity, availability, and authenticity at rest or transit. We aim to prevent log tampering, detect log tampering, and recuperate logs if modified or deleted. Our proposed log protection scheme VaultBox is (i) forward-secure: each new log message in the data structure should be secured with Authenticated Encryption with a new key so that the attacker (probabilistic polynomial time (PPT) adversary) $\mathcal{A}$ can only learn the current key after compromise but does not learn about previous keys, (ii) replicated: multiple entries of the same log message in the data structure (secure buffer) $\mathcal{DS}$ should be added to provide redundancy in case of some entries being deleted, (iii) randomized: instead of merely appending the log messages in the

$\mathcal{DS}$ at the end, the location should be a different random index, and (iv) rate-less: it should be capable of generating an infinite number of encoding symbols for transmission.

**Contributions.** In this paper, we have demonstrated security vulnerabilities in security analytics sources, and to tackle these challenges, we have also proposed a solution which we refer to as VaultBox. We have specified security properties of correctness, stealth, and immutability for any SAS architecture in this design. We show a forward-secure, replicated, randomized, and rate-less scheme to prevent tampering, detect tampering, and recuperate if some logs are lost or corrupt while in transit or at rest. At rest, we rely on redundancy to recover data; if corrupted items are less than the redundancy (replication) factor, we can successfully recover the lost or corrupt message. In transit, we use fountain encoding for recovery.

We do not propose changes to the SAS design or offer new techniques for log generation. We do not address the efficiency and effectiveness of SAS. Components of SAS logs and how they are used in a SIEM to detect an anomaly or an intrusion are outside this paper's scope. Our scheme is agnostic to data design, how they are generated, and how they are used. Our scheme is highly parameterized - developers can enable or disable a component or fine-tune a parameter according to their requirements. Our scheme is data agnostic. E.g., it can secure HL7/DICOM data in a hospital network and system logs in a corporate network. VaultBox is a solution to securely store and relay the data and serve as a secure layer for any transmission channel $C$. VaultBox is general, and we do not require strong assumptions on the underlying operating system, disk, cache, and file system.

Our contributions to this paper can be summarized as follows:

- We propose a novel approach of using a periodic integrity checker at the sender to (immediately) spawn an alert if the secure buffer is compromised.
- We propose a novel approach of using rateless error-correcting codes (or fountain codes) for log security.
- We demonstrate the use of Uniform distribution and Robust soliton distribution for fountain codes degree distribution.
- We propose to use the Fisher-Yates shuffling to improve the performance of random shuffling used for fountain encoding.
- We propose a mechanism to (re-)verify the security of (previous) log messages at the receiver to detect modifications between transmissions.

**Organization.** This paper is organized as follows: threat model is described in Sect. 2, previous work is explained in Sect. 4, our scheme is described in detail in Sect. 3, security definitions and security analysis are described in Appendix A, Appendix B, and experimental results are shown in Sect. 6.

## 2    Threat Model

Our threat model considers three entities (i) the Security Analytics Source (SAS), which can be referred to as a client or endpoint or a host or a compromised machine, (ii) the SIEM server; and (iii) the attacker or Probabilistic Polynomial Time (PPT) adversary $\mathcal{A}$. SAS receives log messages $m_i \in \{0,1\}^*$ from logging agents (caused by events $\{e_i\}$) like Syslog or Windows Event Manager and writes to storage. SIEM downloads the log file from the SAS storage and checks the log's integrity. We consider SAS autonomous, and SIEM gets access to SAS logs only after some time.

At some point, the attacker compromises SAS. We assume that $\mathcal{A}$ at some time $T$ has gained complete control of the SAS system generating and transmitting the logs. The attacker can be the most potent possible adversary and modify, delete, forge, intercept, and delay messages at will. After successful intrusion, the attacker achieves a complete compromise of the client. It can learn about the machine's entire state, including all memory (RAM, disks) contents like cryptographic keys and buffered messages. It controls future behavior, including logging activity. After complete compromise at $T$, the attacker $\mathcal{A}$ can shut down the SAS or tamper with its outputs. Thus a client is assumed to produce trustworthy valid messages until the complete compromise. The intervening time interval is the critical window (Fig. 1) of the attack when intrusions are detectable, and we have to secure the logs before they are tampered with. We assume that the attacker $\mathcal{A}$ has complete knowledge of the workings of the SAS. It does not know about the host state, cryptographic keys, or the SAS anomaly signatures before the complete compromise.

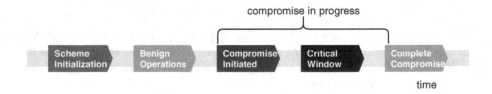

**Fig. 1.** Critical window

The attacker successfully compromises the host if any unmodified messages do not reach the SIEM server and if the server does not learn about the modification or suppression of messages. For a successful tampering attack without detection, $\mathcal{A}$ has to change the desired message stored at all locations in the buffer with a valid HMAC. At any time, these new unique keys are stored in the system: (i) Inner layer - for Encryption and HMAC of the encrypted message (Authenticated Encryption), and (ii) Outer layer - for Encryption and HMAC for symbols generated with fountain encoding. Therefore, to modify a previously stored message, $\mathcal{A}$ must regenerate these keys for a given message.

Our proposed scheme VaultBox provides security to logs at rest and in transit. At rest, we use hardware (optional) and software cryptography measures to protect the logs from $\mathcal{A}$. In transit, we use Falcon codes [21] (Falcon codes are authenticated Luby-Transform (LT) fountain codes. We use Falcon codes for enhanced security over LT codes.) for securely transmitting logs from SAS to SIEM. The goal of this paper is to (i) prevent adversarial modifications, (ii) detect malicious changes, and (iii) recuperate the lost/corrupt messages due to system crashes or malicious modifications at rest or in transit. In addition, we want to guarantee forward-secure integrity and confidentiality.

## 3   Architecture

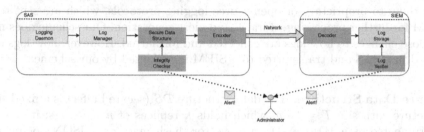

**Fig. 2.** VaultBox architecture and data flow. Green boxes show the VaultBox components. Encoder and Decoder represent Falcon encoding and decoding. (Color figure online)

Our scheme VaultBox proposes using software (and optionally hardware) based cryptography to protect logs at rest in a SAS endpoint and transmission to SIEM. We offer to leverage both software and hardware-based approaches to use cryptographically secure data structure inside a secure chip like Trusted Execution Environment, which operates at a higher privilege level, Ring 0, and thus protects against adversarial operations running at lower privilege levels, Ring 1, 2, and 3 by providing isolation. Alternatively, use cryptographically secure data structures inside a secure memory like ECC-RAM to detect and correct errors through hardware. We propose three levels of security for log protection. (i) To use VaultBox buffer secured by Authenticated Encryption to provide log security at rest. (ii) To use Falcon encoding to provide log protection in transit. (iii) To use (optionally) secure chips like Trusted Platform Module (TPM)/Trusted Execution Environment (TEE) and secure memory like Error Correcting Code - Random Access Memory (ECC-RAM) and Persistent Memory (P-Mem) for an extra layer of security. VaultBox is a general-purpose scheme agnostic to the type of data stored inside it and can be used for many applications besides system log protection (Fig. 2).

Asymmetric ciphers based on the Integer Factorization problem and Elliptic-curve Cryptography like RSA and DH are vulnerable to Quantum computers due to Shor's algorithm. In contrast, Symmetric ciphers (like AES-256 and Twofish-256) are considered quantum-safe. Thus, we propose using secure symmetric and secure asymmetric ciphers based on lattice-based cryptography.

## 3.1  VaultBox Components

**Logging Daemon.** One or more logging services (daemons) could be running on a SAS. These could be OS-specific tools like Event Manager (Windows) or Console (macOS) or third-party applications like a web server, a database, or custom tools. These are the processes that generate the log data.

**Log Manager.** Generally, there will only be a single Log Manager running inside a SAS, collecting logs from different applications it supports and formatting them to proprietary or open-source logging protocols, which are sent to a SIEM for further processing, storage, and alert generation. Our scheme does not propose changes to how logs are collected and formatted. However, how logs are stored in a SAS and transmitted to a SIEM is impacted by our scheme.

**Secure Data Structure.** The data structure $DS$ (secure buffer) is a fixed size structure with size $T = k \cdot n$, which holds $k$ replicas of $n$ messages; it can be instrumented as a 1D or N-D array, vector, hash map, or a JSON document. SAS endpoints send this data to SIEM every $x$ minutes; if SIEM does not receive it during this interval, it suggests an attack or a crash. Therefore, $DS$ should be large enough to hold all possible messages during that $x$ minutes interval.

Replicas of a message inside the $DS$ are stored at random locations generated by a PRNG. For example, consider a $DS$ that holds 1024 messages, each message being 32kB, then such a $DS$ would be of size 32MB. $DS$ is initialized with noise to avoid trivial detection from locations containing messages to locations without it. To protect the messages in transit, we use the Falcon encoding scheme. Our rate-less scheme assumes a highly adversarial channel $C$ where the attacker can eavesdrop or delete the messages. First, we break the sealed (encrypted) $DS$ into message blocks and apply LT encoding with a weak PRNG, and then we use the Authenticated Encryption scheme to generate output to transmit. The receiver at SIEM has to collect enough chunks (or symbols) for regenerating the scaled $DS$ and unsealing (decrypting) it to read and verify the logs inside it. The inner layer for individual messages and the outer layer for the buffer containing all messages provides two layers of encryption protection over plaintext.

**Integrity Checker.** After adding each message to $DS$ for the session, SAS updates and stores the hash value calculated and updated from each message for the session, which the Integrity Checker periodically compares with the hash value computed from the current contents of the buffer. An alert $\Psi$ is sent to an administrator if the hash values do not match.

**Encoder & Decoder.** These components convert the secure buffer messages to secure symbols (Falcon codes) and vice-versa. However, they do not modify the existing transmission channel $C$ (Fig. 3).

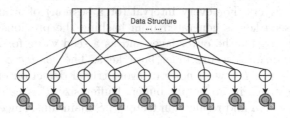

**Fig. 3.** Falcon encoding for $\mathcal{DS}$ in transit

**Log Verifier.** This component comprises a couple of security checks at a SIEM. First, the log Verifier alerts an administrator if Authenticated Decryption for any symbol or message fails for the current session. Second, *SequenceChecker* sends alert $\Psi$ for any gaps in the sequence numbers. Third, if replication factor $r > 1$, *IdentityChecker* is used to check if the two messages with the same sequence number are identical; if not, $\Psi$ is sent, indicating a potential compromise. Finally, since the cryptographic keys are stored for each index between sessions, Log Verifier also checks (re-verification for previous messages) the buffer contents if it includes any messages from prior sessions and sends $\Psi$ to an administrator if Authenticated Decryption fails for any of these messages.

**Log Storage.** How logs are internally stored (for security and compliance), how they are processed, and how the alerts are generated for IoC by a SIEM are out of the scope of our scheme.

---

**Algorithm 1.** Scheme setup phase

| | |
|---|---:|
| 1: seed key $K_S \leftarrow Gen(1^\lambda)$ | ▷ both sender and receiver |
| 2: Size $T = $ Messages $n \cdot$ Replicas $k$ | ▷ both sender and receiver |
| 3: Initialize $\mathcal{DS}[T]$ with random bytes | ▷ only sender |
| 4: $Count_{Global} \leftarrow 0$ | ▷ only sender |
| 5: $M_{prev} \leftarrow \emptyset$ | ▷ only receiver |
| 6: $M_{seq} \leftarrow 0$ | ▷ only receiver |
| 7: $IndexKeys[T] \leftarrow \emptyset$ | ▷ only receiver |

### 3.2  Initialization

The scheme is initialized between a *sender* $S$ and a *verifier* $V$ with a shared security parameter to generate the seed key $K_S$. Both parties share a seed or master key for generating forward-secure Authenticated Encryption keys for the inner and outer layers. For every interval (or a session) of $x$ minutes, a new forward-secure seed key is generated from hashing the previous seed key and used as the new key for the outer layer and as a seed value for $KeyGen(K_S)$, which generates forward-secure inner layer keys. The *sender* $S$ (SAS) initializes the secure buffer $DS$ with noise to avoid trivial detection of indexes with and without messages in an empty buffer. Buffer size $T$ is calculated by the agreed message size $m$ and replication factor $k$. $S$ maintains, since initialization, a global state of (1) incremental counter for total messages sent and (2) $DS$, which may contain noise or messages from previous sessions. The *verifier* $V$ (SIEM) maintains a global state of (1) a list or an empty buffer of size $T$ to store the latest forward-secure keys (using an FS-PRNG) for each index of the secure buffer $DS$ to reverify the authenticity and integrity of a particular message at a corresponding position and (2) the latest values for the last message sequence number (initialized with 0) and its value (initialized with null).

### 3.3  Adding Logs

SAS service providers are logging agents such as Syslog-NG [43], Rsyslog [45], Log4j [2], FileBeat, WinBeat, and AuditBeat [44] or a custom service that collects log and send them to the Log Manager for storage. For example, PRNG with seed $s$ denoted as PRNG($s$) shuffles the indexes of $DS$ and stores it in a list. This is the predefined random order of messages stored in the data structure for the current session. Seed $s$ is randomly generated, and indexes are shuffled for each transmission (session). The receiver on SIEM needs to replay these random coins for shuffling, as the messages can be recovered using a sequence number.

Session keys for the inner layer are generated from $KeyGen(K_S)$ using seed. These keys are then used as a seed to generate authenticated encryption keys for each message. Each message is encrypted with the newly generated forward-secure key, and the previous key is deleted. The encrypted message is stored at the shuffled random index. Any previous message (or noise) at that location is overwritten. When a message $A_j$ arrives, it is stored in the $DS$ at the index value specified in the list. For example, the following $A_{j+1}$ is stored in the $DS$ at the index value specified after the previous value from the list. A monotonically increasing counter is maintained, tracking the number of messages added to $DS$ for each transmission and sent to the receiver.

### 3.4  Retrieving Logs

The entire encrypted buffer is sent for each session (which may contain logs from previous sessions). The number of messages added in that session is sent along

**Algorithm 2.** Adding messages to VaultBox buffer

```
 1: for every session do
 2:     Count_Session ← 0
 3:     orders[T] ← {0, 1, ..., T − 1}                    ▷ for random storage location
 4:     K_S ← Hash(K_S)
 5:     Nonce ← PRNG(K_S)
 6:     orders ← ShuffleIndices(orders, Nonce)
 7:     K_M ← KeyGen(K_S)
 8:     for every message M do
 9:         M ← ApplyPadding(M)                ▷ ensure all messages are of equal size
10:         M ← Count_Global||M
11:         for r = ReplicationFactor do
12:             K_M ← Hash(K_M)
13:             k_ENC, k_HMAC ← K_M
14:             M_AE ← AuthEnc(M, k_ENC, k_HMAC)
15:             DS[orders[Count_Session]] ← M_AE
16:             Count_Session ← Count_Session + 1
17:         end for
18:         Count_Global ← Count_Global + 1
19:     end for
20:     k_ENC, k_HMAC ← K_S
21:     orders[T] ← {0, 1, ..., T − 1}
22:     for every symbol S do
23:         if generating first symbol then
24:             degree ← 1                         ▷ for at least one message with degree = 1
25:         else
26:             degree ← Distribution(T)
27:         end if
28:         orders ← ShuffleIndices(orders)               ▷ for random XOR sequence
29:         S ← DS[orders[0]]
30:         for l ∈ {1, 2, ..., degree − 1} do
31:             S ← XOR(S, DS[orders[l]])
32:         end for
33:         S_AE ← AuthEnc(S, k_ENC, k_HMAC)
34:     end for
35: end for
```

with the buffer. The receiver updates the seed key for the session, overwriting the previous one and decrypting and authenticating the symbols received. The outer layer key for all symbols to decode in that session remains the same and is only updated for the next session, unlike the inner layer key, which is different for all the messages in the buffer. Symbols contain the encrypted XOR-ed data and the neighboring indices for that symbol. Authenticated Decryption for the inner layer starts after successfully decoding enough symbols to get all the messages.

**Algorithm 3.** Retrieving Messages from Symbols

```
1: for every session do
2:      receive a value Count_Session and a set {S_AE} from Sender
3:      retrievedIndices[T] ← ∅
4:      K_S ← Hash(K_S)
5:      k_ENC, k_HMAC ← K_S
6:      for every secure symbol S_AE do
7:          if AuthDec(S_AE, k_ENC, k_HMAC) = ⊥ then
8:              § send alert Ψ to administrator §
9:          else
10:             S ← AuthDec(S_AE, k_ENC, k_HMAC)
11:             data, neighbors ← S
12:             degree ← sizeof(neighbors)
13:             if degree == 1 then
14:                 retrievedIndices.push(neighbors[0])
15:                 DS[neighbors[0]] ← data
16:             else
17:                 DecodeSymbols(DS, data, neighbors, degree, retrievedIndices)
18:             end if
19:         end if
20:     end for
21:     Nonce ← PRNG(K_S)
22:     K_M ← KeyGen(K_S)
23:     orders ← {0, 1, ..., T − 1}
24:     orders ← ShuffleIndices(orders, Nonce)
25:     for r ∈ Count_Session do
26:         K_M ← Hash(K_M)
27:         k_ENC, k_HMAC ← K_M
28:         if AuthDec(M_AE, k_ENC, k_HMAC) = ⊥ then
29:             § send Ψ to administrator §
30:         else
31:             M ← AuthDec(M_AE, k_ENC, k_HMAC)
32:             DS[orders[r]] ← M
33:             IndexKeys[orders[r]] ← K_M
34:             if SequenceChecker(DS[orders[r]], M_prev, M_seq) = ⊥ then
35:                 § send Ψ to administrator §
36:             end if
37:         end if
38:     end for
39:     if Count_Session < T then
40:         for w ∈ {Count_Session, ..., T − 1} do
41:             K_M ← IndexKeys[orders[w]]
42:             k_ENC, k_HMAC ← K_M
43:             if AuthDec(DS[orders[w]], k_ENC, k_HMAC) = ⊥ then
44:                 § send Ψ to administrator §
45:             end if
46:         end for
47:     end if
48: end for
```

The receiver maintains a list of indices for the buffer and the decryption key for that indices, and it is updated every session. Using the seed key for the current session, the receiver forward-securely generates as many keys as per the counter value of the number of messages added. It updates the list with the latest key for that index. This list of decryption keys across sessions helps to reverify the confidentiality, integrity, and authenticity of messages and detect corruption. This feature works when adding fewer messages to the buffer than capacity. An alert $\Psi$ is sent if there is an error during Authenticated Decryption for any message.

To retrieve logs, the receiver has to collect enough symbols and knowledge of the number of messages added to the buffer for that session from the sender. First, the receiver initializes the buffer of retrieved indices for the session to null. Then, the receiver updates the 'master' key for the session's outer layer $K_S$, which decodes symbols. Every symbol is recovered using Authenticated Decryption by the same Encryption and HMAC keys derived from the outer key. When the symbol is decoded successfully, and the message is retrieved, it is copied to the retrieved indices buffer. This buffer checks which messages are already decoded to help reduce the degrees of other symbols. After all the symbols are decoded, the receiver generates Nonce from a PRNG using the $K_S$ and the 'master' key for the inner layer (Authenticated Decryption of the buffer) $K_M$. This Nonce is used to replay the random coins for the shuffled index. The receiver then forward-securely updates the key $K_M$ for the number of messages for that session. If the Authenticated Decryption fails for any message, then an alert $\Psi$ is sent to an administrator, suggesting a possible compromise of the scheme. If the Authenticated Decryption is successful, it is placed at the predefined shuffled index for that message. Thus the order in which the messages were added to the buffer is maintained. Log verifier also runs a subroutine $SequenceChecker$ to extract the sequence ID from the message and send $\Psi$ to the administrator if a "gap" is identified. It also sends $\Psi$ if the two replicas of a message with the same sequence ID do not match.

## 3.5  Key Evolution

The outer layer key for symbols, $K_S$, is forward-securely updated with a hash function for every session. $K_S$ generates a Nonce value using a PRNG to shuffle the indices of messages to be XORed. The inner layer key for the session, $K_M$, is forward-securely generated using a KeyGen function $KeyGen(K_S)$. Every message in the buffer $K_M$ is hashed to generate message-specific keys $k_{ENC}$ and $k_{HMAC}$ for Authenticated Encryption. After all the messages are encrypted and stored in the buffer, $k_{ENC}$ and $k_{HMAC}$ are generated using $K_S$ for Authenticated Encryption of symbols (Fig. 4).

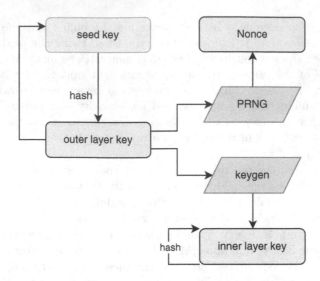

**Fig. 4.** Forward-secure key generation in VaultBox

### 3.6 Fisher-Yates Shuffle

The Fisher-Yates approach [13,14] is a shuffling algorithm for generating a random permutation of a finite sequence. In plain terms, the algorithm shuffles the sequence. It produces an unbiased permutation: every permutation is equally likely. It takes time proportional to the number of shuffled items and shuffles them in place.

## 4    Related Work

Securing logs is a widely studied topic in both the industry and academics. Extensive work has been carried out to secure SAS logs. There are many schemes for secure logging using cryptography [5,16–18,24–26,28,29,33,42] or using hardware like TPM, Hypervisor, and TEE [19,22,32,39,41]. Modern cryptography-based solutions emerged when Bellare and Yee [3,4] introduced the concept of forward integrity for secure logs. Schneiner and Kelsey [37,38] introduced the concept of Forward-secure MACs. Schemes by Goodrich [15], and Mitzenmacher and Pontarelli et al. [35] introduced data structures with redundancy. Blass and Noubir [6] extended the redundancy with crash recovery while providing forward-integrity and forward-confidentiality. Most state-of-the-art solutions work on a client-server-based model where multiple SAS endpoints generate logs collected by a central SIEM server.

---

**Algorithm 4.** Forward-secure Key Generation

---

1: **for** every session **do**
2:     $K_S \leftarrow Hash(K_S)$
3:     $Nonce \leftarrow PRNG(K_S)$
4:     $K_M \leftarrow KeyGen(K_S)$
5:     **for** every message $M$ **do**
6:         $K_M \leftarrow Hash(K_M)$
7:         $k_{ENC}, k_{HMAC} \leftarrow K_M$                          ▷ inner layer key
8:         $M_{AE} \leftarrow AuthEnc(M, k_{ENC}, k_{HMAC})$
9:     **end for**
10:    $k_{ENC}, k_{HMAC} \leftarrow K_S$                              ▷ outer layer key
11:    **for** every symbol $S$ **do**
12:        $S_{AE} \leftarrow AuthEnc(S, k_{ENC}, k_{HMAC})$
13:    **end for**
14: **end for**

---

**Algorithm 5.** Fisher-Yates Shuffle Algorithm

---

1: **for** $i$ **from** $n - 1$ **downto** 1 **do**
2:     $j \leftarrow$ random integer such that $0 \leq j \leq i$
3:     exchange $a[j]$ and $a[i]$
4: **end for**

---

Our paper is an extension of work carried out in PillarBox [7]. PillarBox uses forward-secure cryptographic keys and is a black box for securely storing and relaying data. PillarBox is a valuable tool for tamper detection but does not recover lost or corrupt messages. Another solution provided by Forward Integrity and Crash Recovery [6] is an encoding scheme that offers forward-secure integrity and confidentiality. It can detect tampering and provide a recovery mechanism if up to $\delta$ of data gets deleted or corrupted. While this scheme is adequate to detect and recover from tapering, it lacks a protection mechanism to prevent such tampering by an attacker in the first place. SGX-Log [22] demonstrates such a protection mechanism using a Trusted Execution Environment (TEE) viz. Intel SGX prevents malicious behavior but does not provide a recovery mechanism. Our scheme VaultBox aims to use a PillarBox-like black box or a buffer to secure data with forward-secure cryptography, which can optionally run inside secure hardware (chip and memory) and provides log recovery through replication at rest. We use Falcon [21] encoding to securely transmit the data from a SAS client to a SIEM server in a corrupt adversarial transmission channel $C$ to provide log recovery in transit (Table 1).

**Table 1.** Comparing our scheme VaultBox with other state-of-the-art log protection schemes. Encoding refers to Falcon encoding in our scheme.

| Features | [7] | [16] | [5] | [22] | [6] | [19] | [32] | [33] | Ours |
|---|---|---|---|---|---|---|---|---|---|
| Confidentiality | ✓ | | ✓ | ✓ | ✓ | ✓ | | | ✓ |
| Integrity | ✓ | ✓ | ✓ | ✓ | ✓ | ✓ | ✓ | ✓ | ✓ |
| Availability | | | | | | | ✓ | | ✓ |
| Non-repudiation | ✓ | | ✓ | ✓ | ✓ | | ✓ | | ✓ |
| Crash Detection | | ✓ | | ✓ | | | | | ✓ |
| Fault Tolerance | | | | ✓ | | | ✓ | | ✓ |
| Aegis at Rest | ✓ | | ✓ | ✓ | ✓ | | ✓ | ✓ | ✓ |
| Aegis at Transit | ✓ | ✓ | ✓ | ✓ | ✓ | ✓ | ✓ | ✓ | ✓ |
| Aegis at Execution | | | | ✓ | | ✓ | ✓ | ✓ | ✓ |
| Encoding | | | | | | ✓ | | | ✓ |
| Software-crypto | ✓ | ✓ | ✓ | | | ✓ | ✓ | | ✓ |
| Hardware-crypto | | | | ✓ | | ✓ | ✓ | ✓ | ✓ |
| Centralized | ✓ | | | ✓ | ✓ | | ✓ | | ✓ |
| Decentralized | | | | | | | ✓ | | ✓ |
| Quantum-safe | | | | | | | | | ✓ |

## 5  Security Evaluation

Our scheme employs a buffer $T$ and transmission time interval $\mu$ that are both fixed; each message is also of fixed size or padded to that size. While transmitting, the entire fixed-size buffer, not just new messages, is sent. This fixed communication pattern defeats **traffic analysis**.

Adversarial buffer modification (or destruction) is a detectable attack. It causes the server to receive a symbol $\perp$ while recovering messages, indicating a cryptographic integrity-check failure. The log verifier detects both **buffer overwriting attacks** and **buffer dropping attacks**. It looks for lost messages, as indicated by a gap in message sequence numbers. That is, a gap alert is issued when the sequence numbers $s_j$ and $s_{j'}$ of two successively received buffers $j$ and $j'$ are such that $s_{j'} - s_j \geq T$.

An attacker cannot undetectably delete arbitrary events from the secure buffer. Removing a subset of events from the buffer will fail log verification with Authenticated Encryption. In addition, an attacker can only insert or modify messages into the buffer, invalidating its integrity proof. An attacker also **cannot re-order** messages because it will fail Authenticated Encryption when the Verifier replays the random coins to regenerate the order sequence.

VaultBox satisfies **third-party verifiability** by having each client publish its key. Any verifier can use this key to verify any logs from a particular client. Our scheme can be modified to support **fine-grained audits** by encrypting the entire buffer and storing them with a block ID.

We achieve *log recovery* at rest by applying a RAID-1 (mirroring) like mechanism with multiple copies of a message, where each message is analogous to a file and a buffer index as a disk in a RAID-1 style recovery. In transit, we use *Falcon* encoding, which is authenticated rateless LT codes. Suppose the original data consists of $k$ input symbols (messages). In that case, they can be recovered from any $k + O(\sqrt{k}\ln^2(k/\delta))$ of the encoding symbols with probability $1 - \delta$ by on average $O(k \cdot \ln(k/\delta))$ symbol operations.

We advocate leveraging hardware-based security mechanisms (secure chips like TPM or TEE, and secure memory like ECC-RAM [27,49] or Persistent Memory (P-Mem) [20,34] if supported by the systems) in conjunction with a software-based security scheme to provide an extra layer of security. In our scheme, we propose using *hardware-based* solutions like a TEE that provide isolation. The secure buffer can be placed inside a TEE, providing an extra layer of security and making log tampering difficult. In addition, we propose using ECC-RAM to fix corruptions and Persistent-Memory (P-Mem), a fusion between DRAM and NAND storage, to protect against shutdown attacks. ECC-RAM is server commodity hardware, while P-Mem is a relatively new security mechanism.

We strongly encourage the use of *quantum-safe cryptography* [36] schemes like AES-256, ChaCha20-256, Twofish-256, Camellia-256, SHA512, SHA3-512 and lattice-based schemes like NTRU.

Security definitions and analysis are described in Appendix A and B.

# 6   Experimental Analysis

Our scheme is available at [12] as a header-only C++ library. It is free to use under GPLv3 licensing. It has about 500 lines of code and is about 20kB in size. It depends on Crypto++ [48] for cryptographic properties (other providers like OpenSSL [30] or custom cryptography library can also be used). Test.cpp is provided as a dummy application. We experimented on a client-server C++ application over a TCP socket running locally for proof-of-concept. Experiments were performed on a MacBook Pro with a 2.4 GHz Quad-Core Intel Core i5 processor and 8 GB 2133 MHz LPDDR3 memory.

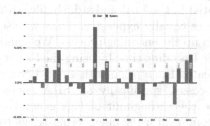

**Fig. 5.** User and System CPU time differences in macOS

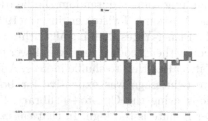

**Fig. 6.** User CPU time differences in Kali Linux VM

First, we measured the overhead of symbol generation (XOR operation) and shuffling operations. This experiment measured the difference between average user-CPU time and average system-CPU time for the logs of different sizes without XOR and shuffle operations and with XOR and shuffle operations. For each measurement, the log replication factor was set to 3, and the symbols generated were three times as of the total messages. E.g., for example, there were 15 replicas in the buffer for five messages added to logs, with each message replicated three times. Furthermore, for these 15 messages, 45 symbols were generated. The symbol size were 45, 90, 135, 180, 225, 270, 315, 360, 405, 450, 900, 2250, 4500, and 9000. The experiments show that there is no significant impact on macOS user-CPU time and system-CPU overhead are under 5% on average. This test was also done in a Kali Linux VM inside the MacBook Pro. We observed 0% overhead for system-CPU and user-CPU time overhead under 5% on average (Figs. 5 and 6).

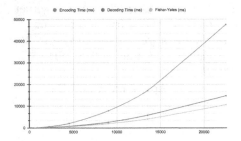
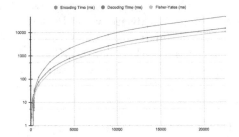

**Fig. 7.** Encoding comparison (time-ms)     **Fig. 8.** Encoding comparison (log-scale)

Next, we measured the performance for encoding and decoding the authenticated encryption symbols for size 45, 90, 135, 180, 225, 270, 315, 360, 405, 450, 900, 2250, 4500, 9000, 13500 and 22500 (Figs. 7 and 8). It is similar to the previous experiment, where five messages generate $5 \times 3 = 15$ replicas and $15 \times 3 = 45$ symbols. We discovered that the Encoding time (ms) was much higher than the Decoding. This was due to the expensive shuffle operations of indices. Shuffling using std::random_shuffle (deprecated in C++14) and std::shuffle with the Mersenne Twister engine (mt19937) were equally expensive. Thus we use the Fisher-Yates shuffling algorithm to speed up the encoding process. We could reduce the Encoding time even lower than the Decoding time by using the Fisher-Yates algorithm to shuffle indices. Authenticated Encryption with a new key for every message was expensive for both processes, specifically for SHA operations using the Crypto++ library. Suppose the throughput of logs generated per minute is too high. In that case, we recommend using engineering tricks to improve performance, e.g., shuffle and update keys for every x'th amount of logs instead of every message.

We also show the use of Uniform distribution for Falcon codes instead of Robust soliton distribution for degree distribution of neighbors to generate a symbol. Generally, it is good to use Uniform distribution when generating symbols for a few messages and Robust soliton distribution when generating symbols for many messages. For this experiment, we compared the distribution frequency for the random numbers generated by std::rand(), std::random_device(), and Robust soliton distribution. Random numbers were generated for $N = \{25, 50, 100, 250, 500, 1000\}$. We capped the degree range from 0 to $N/2$ to make it practical to use uniform distribution.

# 7   Conclusion

This paper proposes a novel approach to secure logs at rest and in transit. Quantum-safe probabilistic encryption schemes and Hash-based Message Authentication Codes provide Confidentiality and Integrity. Error correction and recovery are achieved by replicating logs in memory at rest and Falcon codes (authenticated LT codes) in transit. Furthermore, we propose a mechanism for the sender to detect log tampering in memory and for the receiver to detect log tampering between the transmissions. Additionally, we propose improvements through secure hardware.

Secure chips like TPM and TEE can provide a layer of security for logs. ECC-RAM, generally used for applications where high reliability is necessary, like in a server-grade system or scientific computation, can be used for log error correction. Availability of logs can be improved with additional usage of permanent write-once-read-many (WORM) disks or a Persistent Memory (P-Mem) in case of a malicious or benign shutdown. Finally, our scheme is data agnostic, making it suitable for other use cases besides log security, like health data security.

However, our scheme may need to be more efficient for a web server generating millions of log messages per minute due to forward-secure key requirements and the resource usage for Authenticated Encryption for all the messages. Also, it imposes storage overhead to provide error recovery through replication at rest and Falcon encoding at transit.

**Acknowledgements.** We want to thank Matthew Butler (Laurel Lye LLC) for his valuable feedback on the paper.

# A   Security Definitions

A logging protocol is an algorithm running on a logging device $D_i$ that receives messages from *sender* $S$ and writes entries to its logging device $L_i$. The log on logging device $L_i$ is a finite but increasingly long sequence of entries. An essential part of each secure logging protocol is a verifier protocol. A verifier protocol is an algorithm run by the *verifier* $V$ that accesses some log $L_i$, extracts the log entries stored on $L_i$, and each log entry computes whether to accept or reject the entry. The decision of the Verifier is the basis for defining security properties below. For all properties, we assume that there is a "ground truth", i.e., a sequence $< e_1, e_2, ... >$ of events that happened. Our scheme VaultBox aims to achieve the following security properties.

**Definition 1 (Correctness).** *A logging protocol satisfies correctness for log $L_k$ if and only if a corresponding event actually happened for each entry in the log of $L_k$ that the Verifier accepts. Correctness dictates that under normal operation, any sequence of messages of size at most $T$ added to channel$C$ by sender $S$ can be correctly read by verifier $V$ in an order-preserving way; in particular, the $T$ most recent messages of $C$ and their exact order can be determined by $V$. More formally, let $< e_1, e_2, ... >$ denote the event sequence that actually happened and $< l_1, l_2, ... >$ be the sequence of log entries in $L_k$ accepted by the verifier. Then the following condition should hold:*

$$\forall i : l_i \Rightarrow \exists e_j : e_j \sim l_i$$

**Definition 2 (Immutability).** *A logging protocol satisfies immutability for log $L_k$ if $\mathcal{A}$ cannot undetectably forge the messages that are contained in channel $C$ at the time of compromise. Although $\mathcal{A}$ takes full control of sender $S, \mathcal{A}$ may only make two uses of the channel:*

- *$\mathcal{A}$ may either write in $C$ legitimate messages in $L$, or*
- *$\mathcal{A}$ may destroy the delivery of chosen messages in $C$, but in a detectable way, since a failure message, $\perp$ is substituted.*

**Definition 3 (Stealth).** *A logging protocol satisfies stealth for log $L_k$ if $\mathcal{A}$ can learn nothing about the contents of $C$ that is not explicitly added to it by $\mathcal{A}$.*

- *At the pre-compromise state, $\mathcal{A}$ cannot learn anything about the contents of the channel: at all times, $C$ itself perfectly hides its contents. Not only can messages in the channel be written such that the $\mathcal{A}$ never learns them, but $\mathcal{A}$ does not even learn of their existence.*
- *At the point of compromise, $\mathcal{A}$ cannot learn the messages present in the channel or even learn if such messages exist in the channel. $\mathcal{A}$ can only learn the current slot in $C$, but since this index is initially randomized, it conveys no information about the usage of $C$ thus far.*
- *At the post-compromise state, $\mathcal{A}$ completely controls all the messages added to the channel and the positions of the messages that $V$ will fail to produce.*

# B     Security Analysis

We briefly provide the intuition behind the above valuable properties.

**Correctness.** Messages are included in the buffer at a random position in order of their arrival. Under normal operation of the system, it is always possible for the receiver to replay the random coins to generate the corresponding secret keys for Authenticated Decryption and reconstruct the exact sequence of the most recent messages written in the buffer.

**Immutability.** Since any message added in the buffer is encrypted and verifiable through Authenticated Encryption scheme, individual messages cannot be undetectably tampered with. Additionally, since the seed to replay the random coins is generated forward-securely, messages in an individual buffer cannot be undetectably received out of order, across different buffers, or within the buffer.

**Stealth.** Since both individual messages and buffers are encrypted through the inner layer and outer layer Authenticated Encryption and since the buffer is of fixed size $T$, observing buffers before the compromise or actual buffer contents after the compromise reveals no information about the actual messages previously added in the channel or about whether messages have been ever added.

# References

1. Allen, J., et al.: Mnemosyne: an effective and efficient postmortem watering hole attack investigation system. In: Proceedings of the 2020 ACM SIGSAC Conference on Computer and Communications Security, CCS 2020, pp. 787–802. Association for Computing Machinery, New York (2020). https://doi.org/10.1145/3372297.3423355
2. Apache Log4j 2. https://logging.apache.org/log4j/2.x/
3. Bellare, M., Yee, B.: Forward integrity for secure audit logs. Technical report, Citeseer (1997)
4. Bellare, M., Yee, B.: Forward-security in private-key cryptography. In: Joye, M. (ed.) CT-RSA 2003. LNCS, vol. 2612, pp. 1–18. Springer, Heidelberg (2003). https://doi.org/10.1007/3-540-36563-X_1
5. Blass, E.O., Noubir, G.: Secure logging with crash tolerance. In: 2017 IEEE Conference on Communications and Network Security (CNS), pp. 1–10 (2017). https://doi.org/10.1109/CNS.2017.8228649
6. Blass, E.O., Noubir, G.: Forward integrity and crash recovery for secure logs. Cryptology ePrint Archive, Report 2019/506 (2019). https://ia.cr/2019/506
7. Bowers, K.D., Hart, C., Juels, A., Triandopoulos, N.: PillarBox: combating next-generation malware with fast forward-secure logging. In: Stavrou, A., Bos, H., Portokalidis, G. (eds.) RAID 2014. LNCS, vol. 8688, pp. 46–67. Springer, Cham (2014). https://doi.org/10.1007/978-3-319-11379-1_3
8. Buyukkayhan, A.S., Oprea, A., Li, Z., Robertson, W.: Lens on the endpoint: hunting for malicious software through endpoint data analysis. In: Dacier, M., Bailey, M., Polychronakis, M., Antonakakis, M. (eds.) RAID 2017. LNCS, vol. 10453, pp. 73–97. Springer, Cham (2017). https://doi.org/10.1007/978-3-319-66332-6_4
9. CVE-2011-0343. https://cve.mitre.org/cgi-bin/cvename.cgi?name=CVE-2011-0343
10. CVE-2011-1951. https://cve.mitre.org/cgi-bin/cvename.cgi?name=CVE-2011-1951
11. CVE-2014-3683. https://cve.mitre.org/cgi-bin/cvename.cgi?name=CVE-2014-3683
12. Devharsh: GitHub - devharsh/VaultBox: VaultBox is a static C++ library for secure storage and transmission. https://github.com/devharsh/VaultBox
13. Fisher yates shuffle algorithm. https://www.ahtcloud.com/fisher-yates-shuffle-algorithm

14. Fisher, R.A., Yates, F.: Statistical tables for biological, agricultural and medical research, 6th edn. https://hdl.handle.net/2440/10701
15. Goodrich, M.T., Mitzenmacher, M.: Invertible bloom lookup tables. In: 2011 49th Annual Allerton Conference on Communication, Control, and Computing (Allerton), pp. 792–799 (2011). https://doi.org/10.1109/Allerton.2011.6120248
16. Hartung, G.: Secure audit logs with verifiable excerpts. In: Sako, K. (ed.) CT-RSA 2016. LNCS, vol. 9610, pp. 183–199. Springer, Cham (2016). https://doi.org/10.1007/978-3-319-29485-8_11
17. Hassan, W.U., Bates, A., Marino, D.: Tactical provenance analysis for endpoint detection and response systems. In: 2020 IEEE Symposium on Security and Privacy (SP), pp. 1172–1189 (2020). https://doi.org/10.1109/SP40000.2020.00096
18. Holt, J.E.: Logcrypt: forward security and public verification for secure audit logs. In: Proceedings of the 2006 Australasian Workshops on Grid Computing and E-Research, ACSW Frontiers 2006, vol. 54, pp. 203–211. Australian Computer Society Inc, AUS (2006)
19. Homoliak, I., Szalachowski, P.: Aquareum: a centralized ledger enhanced with blockchain and trusted computing (2020)
20. Intel®optane™ dc persistent memory: A major advance in memory and storage architecture. https://www.intel.com/content/www/us/en/developer/articles/technical/optane-dc-persistent-memory-a-major-advance-in-memory-and-storage-architecture.html
21. Juels, A., Kelley, J., Tamassia, R., Triandopoulos, N.: Falcon codes: fast, authenticated LT codes (or: making rapid tornadoes unstoppable). In: Proceedings of the 22nd ACM SIGSAC Conference on Computer and Communications Security, CCS 2015, pp. 1032–1047. Association for Computing Machinery, New York (2015). https://doi.org/10.1145/2810103.2813728
22. Karande, V., Bauman, E., Lin, Z., Khan, L.: SGX-log: securing system logs with SGX. In: Proceedings of the 2017 ACM on Asia Conference on Computer and Communications Security, ASIA CCS 2017, pp. 19–30. Association for Computing Machinery, New York (2017). https://doi.org/10.1145/3052973.3053034
23. Kwon, Y., Wang, W., Jung, J., Lee, K.H., Perdisci, R.: C2SR: cybercrime scene reconstruction for post-mortem forensic analysis. In: Proceedings 2021 Network and Distributed System Security Symposium (2021)
24. Ma, D.: Practical forward secure sequential aggregate signatures. In: Proceedings of the 2008 ACM Symposium on Information, Computer and Communications Security, ASIACCS 2008, pp. 341–352. Association for Computing Machinery, New York (2008). https://doi.org/10.1145/1368310.1368361
25. Ma, D., Tsudik, G.: A new approach to secure logging. ACM Trans. Storage 5(1) (2009). https://doi.org/10.1145/1502777.1502779
26. Marson, G.A., Poettering, B.: Even more practical secure logging: tree-based seekable sequential key generators. In: Kutyłowski, M., Vaidya, J. (eds.) ESORICS 2014. LNCS, vol. 8713, pp. 37–54. Springer, Cham (2014). https://doi.org/10.1007/978-3-319-11212-1_3
27. Memory RAS technologies for HPE proliant/synergy/blade gen10 servers with intel xeon scalable processors. https://psnow.ext.hpe.com/doc?id=4aa4-3490enw.pdf
28. Michael, N., Mink, J., Liu, J., Gaur, S., Hassan, W.U., Bates, A.: On the forensic validity of approximated audit logs. In: Annual Computer Security Applications Conference, ACSAC 2020, pp. 189–202. Association for Computing Machinery, New York (2020). https://doi.org/10.1145/3427228.3427272

29. Noura, H.N., Salman, O., Chehab, A., Couturier, R.: Distlog: a distributed logging scheme for IoT forensics. Ad Hoc Netw. **98**, 102061 (2020). https://doi.org/10.1016/j.adhoc.2019.102061. https://www.sciencedirect.com/science/article/pii/S1570870519306997
30. OpenSSL: GitHub - OpenSSL/OpenSSL: TLS/SSL and crypto library. https://github.com/openssl/openssl
31. OSSEC. https://www.ossec.net/
32. Paccagnella, R., et al.: Custos: practical tamper-evident auditing of operating systems using trusted execution. In: NDSS (2020)
33. Paccagnella, R., Liao, K., Tian, D., Bates, A.: Logging to the danger zone: race condition attacks and defenses on system audit frameworks. In: Proceedings of the 2020 ACM SIGSAC Conference on Computer and Communications Security, CCS 2020, pp. 1551–1574. Association for Computing Machinery, New York (2020). https://doi.org/10.1145/3372297.3417862
34. Persistent memory. https://www.micron.com/campaigns/persistent-memory
35. Pontarelli, S., Reviriego, P., Mitzenmacher, M.: Improving the performance of invertible bloom lookup tables. Inf. Process. Lett. **114**(4), 185–191 (2014). https://doi.org/10.1016/j.ipl.2013.11.015. https://www.sciencedirect.com/science/article/pii/S0020019013002950
36. Quantum-safe cryptography. https://cryptobook.nakov.com/quantum-safe-cryptography
37. Schneier, B., Kelsey, J.: Cryptographic support for secure logs on untrusted machines. In: 7th USENIX Security Symposium (USENIX Security 1998), San Antonio, TX. USENIX Association (1998). https://www.usenix.org/conference/7th-usenix-security-symposium/cryptographic-support-secure-logs-untrusted-machines
38. Schneier, B., Kelsey, J.: Secure audit logs to support computer forensics. ACM Trans. Inf. Syst. Secur. **2**(2), 159–176 (1999). https://doi.org/10.1145/317087.317089
39. Shepherd, C., Akram, R.N., Markantonakis, K.: EmLog: tamper-resistant system logging for constrained devices with TEEs. In: Hancke, G.P., Damiani, E. (eds.) WISTP 2017. LNCS, vol. 10741, pp. 75–92. Springer, Cham (2018). https://doi.org/10.1007/978-3-319-93524-9_5
40. Siem compliance requirements and standards. https://www.peerspot.com/articles/siem-compliance-requirements-and-standards
41. Sinha, A., Jia, L., England, P., Lorch, J.R.: Continuous tamper-proof logging using TPM 2.0. In: Holz, T., Ioannidis, S. (eds.) Trust 2014. LNCS, vol. 8564, pp. 19–36. Springer, Cham (2014). https://doi.org/10.1007/978-3-319-08593-7_2
42. Soriano-Salvador, E., Guardiola-Múzquiz, G.: Sealfs: storage-based tamper-evident logging. Comput. Secur. **108**, 102325 (2021). https://doi.org/10.1016/j.cose.2021.102325. https://www.sciencedirect.com/science/article/pii/S0167404821001498
43. syslog-ng. https://www.syslog-ng.com/
44. The beats family. https://www.elastic.co/beats/
45. The rocket-fast syslog server. https://www.rsyslog.com/
46. Using siem for regulatory compliance: Importance, best practices, use cases. https://logsentinel.com/blog/using-siem-for-regulatory-compliance-importance-best-practices-use-cases/
47. Wang, Q., et al.: You are what you do: hunting stealthy malware via data provenance analysis. In: NDSS (2020)
48. Weidai: GitHub - weidai11/cryptopp: free C++ class library of cryptographic schemes. https://github.com/weidai11/cryptopp

49. What is ECC memory? https://www.crucial.com/products/memory/server/ecc
50. Yuan, X., Setayeshfar, O., Yan, H., Panage, P., Wei, X., Lee, K.H.: Droidforensics: accurate reconstruction of android attacks via multi-layer forensic logging. In: Proceedings of the 2017 ACM on Asia Conference on Computer and Communications Security, ASIA CCS 2017, pp. 666–677. Association for Computing Machinery, New York (2017). https://doi.org/10.1145/3052973.3052984

# Two-Stage Anomaly Detection in LEO Satellite Network

Yipeng Wang[1]($\boxtimes$), Peixian Chen[1], Shan Ai[1], Weipeng Liang[1], Binjie Liao[1], Weichuan Mo[1], and Heng Wang[2]

[1] Institute of Artificial Intelligence and Blockchain, Guangzhou University, Guangzhou, Guangdong, China
2112106219@e.gzhu.edu.cn, 1369740347@qq.com, aishan@gzhu.edu.cn, 2909847341@qq.com, liaobinjie123@163.com, moweichuan@foxmail.com
[2] CASIC Space Engineering Development Co., Ltd., Wuhan 430416, Hubei, China
whbeny@sina.com

**Abstract.** We introduce a novel two-stage method for detecting anomaly signal in Low Earth Orbit (LEO) satellite network in response to increasing clutter signals and interference. First, a convolutional neural network (CNN) classifier is trained on real sensor signals and synthesized wireless modulated signals. Then, an anomaly detector is used to detect the classified signal. To address the limited computing resources on the satellite, we utilize transfer learning to reduce the scale of the classifier and anomaly detector. Our proposed method consists of a multi-class CNN model that reliably detects the modulation methods used in a specific satellite environment by I/Q signals and a recurrent neural network model that identifies anomalies when events significantly deviate from expected or predicted values. Experimental results show the effectiveness of our proposed method.

**Keywords:** Anomaly Detection · Satellite Network · Deep Learning

## 1 Introduction

In the real world, satellite technology is extensively used, and communication technology plays a crucial role in the growth of globalization. Therefore, research on satellite communication and privacy [1] has become a critical factor in the rapid development of this era. Wireless signals are also essential for satellite communications and sensing. While several wireless and radar communication systems have been extensively designed and optimized, the increasing complexity of new hardware, communication, or access technologies poses challenges to spectrum management in satellite networks. Anomaly detection aims to identify patterns that are inconsistent with expected patterns, which are defined as anomalies or outliers [2].

This work was funded by the Key Research and Development Program of Guangzhou (No. 202103050003).

© The Author(s), under exclusive license to Springer Nature Switzerland AG 2023
M. Yung et al. (Eds.): SciSec 2023, LNCS 14299, pp. 423–438, 2023.
https://doi.org/10.1007/978-3-031-45933-7_25

Wireless signal anomalies can be characterized in several ways, such as the presence of unlicensed signals in the target band or the absence of target signals. Compared with other anomaly detection tasks, radio spectrum anomaly detection in wireless communication has many challenges [3]. For example, due to the complex physical environment (e.g., electrical signals), it is challenging to extract handcrafted features for anomalies, and it is also difficult to annotate models due to the diversity of anomalies. Some new approaches can easily make communication devices misuse unlicensed spectrum by observing the relevant hardware and spectrum usage. These attacks have prompted some countries to take new steps to defend against them [4]. Satellite networks need viable and robust tools to detect faults and misbehavior when using spectrums, which we refer to as spectrum anomalies in this paper. However, since the normal data in the original data is far more than the anomalous data, and these data may be complex, it may prevent the anomaly detection model from learning well.

Detecting anomalies in electromagnetic communication environment data, which can be considered big data, presents significant challenges. When dealing with mixed electromagnetic signals, common types of electromagnetic interference signals include pulse electromagnetic interference signals, white noise, and periodic narrowband interference signals [5]. Pulse electromagnetic interference signals can be divided into random and periodic types. Periodic pulse interference arises primarily from Silicon Controlled Rectifiers, while random pulse interference is typically associated with lightning and relays. White noise interference comprises various random noises, including those caused by transmission lines and thermal noise from windings. Periodic narrowband interference primarily results from electromagnetic interference signals used in radio and carrier communication.

Existing literature primarily assumes the presence of a single signal in the receiving channel when recognizing modulation in communication signals. However, in practice, the widespread use of radio communication in military and civilian fields has led to an increase in the number of signals within the radio communication frequency band. Consequently, even ordinary narrowband communication receivers sharing the same channel face challenges in avoiding the presence of multiple signals. This phenomenon is particularly common in specialized broadband receivers, such as Low Earth Orbit (LEO) satellites.

We propose a novel supervised, adaptive deep learning-based anomaly detection model to address the challenges mentioned above. Deep learning has been extensively studied for specific detection tasks such as image detection, speech detection, and so on. Since the signals propagating in the environment are unpredictable, the wide area spectrum received by the device is complex, and traditional methods may not learn these complex features well. On the other hand, deep learning models are known for automatically learning complex features in target data. Recent studies have demonstrated the benefits of using deep learning to model the spectrum [6]. However, the large number of handcrafted features required in the data preprocessing stage makes it difficult to handle large-scale data, and performance is limited for more complex patterns.

Recurrent neural networks (RNN), such as gated recurrent unit (GRU) [7] and long short-term memory (LSTM) [8], are extensively studied and used due to their powerful ability to automatically extract high-quality features and handle long-term dependencies. Moreover, through the attention mechanism, these deep learning methods can capture the common features between sequences to handle the classification or prediction of time series. In this paper, we design an anomaly detection system based on deep learning to address the anomaly detection issues of multi-signals in satellite networks.

Overall, our paper makes the following contributions:

- We propose a two-stage anomaly detection framework (as show in Fig. 1) consisting of a multi-class Convolutional Neural Network (CNN) model and a Recurrent Neural Network (RNN) model. The multi-class CNN model is designed to accurately detect the modulation method used by In-phase/Quadrature-phase (I/Q) signals in LEO satellite network. By employing this model for pre-classification, we effectively reduce data complexity and enhance predictor performance. Additionally, we train an RNN model that identifies anomalies by detecting significant deviations from expected or predicted events. To detect anomalous signals, we employ a similarity metric algorithm. Experimental validation of our system is conducted on LEO satellite network signals.
- We reduce the training time of the model used in satellite networks by employing transfer learning. This approach helps calibrate the model with a smaller number of samples and reduces the computational resource consumption at each LEO satellite and ground station.

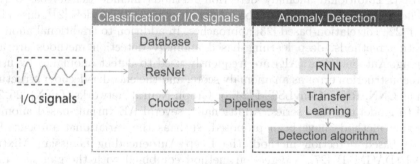

**Fig. 1.** Our proposed framework with two components of classifier (green part) and detector (blue part). (Color figure online)

The rest of this paper is structured as follows: In Sect. 2, we present related works on signal modulation recognition, time series anomaly detection, and transfer learning. In Sect. 3, we propose a LEO signal detection algorithm and provide some definitions. We introduce our framework, which comprises a multi-class classifier and an anomaly detector in Sect. 4. We report the results of our experiments in Sect. 5, and we make a conclusion in Sect. 6.

## 2   Related Work

**Signal Modulation Recognition.** Traditional Automatic Modulation Recognition (AMR) algorithms can be divided into likelihood function-based and signal feature-based methods. [9] proposed a maximum likelihood algorithm based on phase parameters to classify the modulation of multiple linearly modulated signals. However, the computational complexity of the likelihood function is high, and offline computing of fast likelihood functions [10], nonparametric likelihood functions (NPLF) [11], and other methods have been proposed to address this issue. [12] used statistical features of signals to distinguish signal modulation types, such as ASK, PSK, and QAM. The constellation diagram corresponds to the Euclidean distance of the geometric shape, thus [13] used them as a classification feature of the signal. Wavelet transforms have also been employed as a classification feature in the works of [14] and [15].

More recently, deep learning methods have been extensively studied in signal modulation recognition. [16] used several different CNN backbones, such as VGGNet and ResNet, to identify multiple modulated signals. [17] combined the Inception ResNet V2 model with transmission adaptation to extract constellation diagram features by unsupervised learning and then inputs these features into support vector machines for classification. These deep learning methods have shown promise in improving the accuracy of signal modulation recognition, but there are still challenges to be addressed, such as limited training data and model interpretability.

**Time Series Anomaly Detection.** Time series data is often unlabeled, which means that anomaly detection methods for time series are typically unsupervised. Traditional anomaly detection methods include density-based [18], distance-based [19], linear-model based [20], classification models [21], ensemble-based [22], correlation-based [23] approaches. In addition to traditional anomaly detection methods, deep learning-based anomaly detection methods are also thriving. Autoencoders (AE) are frequently used to detect anomalies by using the reconstruction error as an anomaly score, and the encoder-decoder structure, such as CNN, RNN, ConvLSTM [24] or Graph Neural Networks (GNN) [25,26], used to model the time series. Additionally, several AE variant-based anomaly detection methods have been proposed, such as the variational autoencoders (VAE) based detection method, the Deep Autoencoding Gaussian Mixture Model (DAGMM) [27], a detection method combined with the gaussian mixture model, and so on.

**Transfer Learning.** In the past, transfer learning was mostly studied in traditional machine learning methods. For example, [28] reweighted some label data in the source domain before using it in the target domain. Other work used some unsupervised approaches for unlabeled data in the source domain and target domain, such as dimensionality reduction [29], clustering [30], or density estimation. However, with the advent of deep learning, more studies are focusing on effectively transferring knowledge using deep neural networks, which is called deep transfer learning. [31] proposed a metric transfer learning framework

(MTLF) to learn instance weights in the source domain. [32] mapped instances of the source and target domains to a new data space, which is suitable for deep networks. [33] transferred the network structure or connection parameters pretrained in the source domain to another network in the target domain. Moreover, inspired by generative adversarial networks (GAN), [34] and [35] employed several adversarial technologies to find transferable representations that are applicable to both the source domain and the target domain. Overall, these works demonstrate the versatility of transfer learning and its potential in various applications.

## 3    Problem Formulations

For the purpose of clarity, we introduce the notations and definitions that are used throughout the algorithm in Table 1.

Table 1. Notations

| Notations | Description |
|---|---|
| $T$ | the time series |
| $m$ | the length of $T$ |
| $C$ | the subsequence of time series |
| $Q$ | another subsequence of time series |
| $\bar{C}$ | the mean of $C$ |
| $\bar{Q}$ | the mean of $Q$ |
| $\tilde{C}$ | the variance of $C$ |
| $\tilde{Q}$ | the variance of $Q$ |
| $\mathrm{Dist}(C, Q)$ | the distance between $C$ and $Q$ |
| $\mathrm{Sim}(C, Q)$ | the similarity between $C$ and $Q$ |
| $\mathrm{Dist}(\bar{C}, \bar{Q})$ | the distance between $\bar{C}$ and $\bar{Q}$ |
| $\mathrm{Sim}(\tilde{C}, \tilde{Q})$ | the similarity between $\tilde{C}$ and $\tilde{Q}$ |

### 3.1    Time Series Anomaly Detection

In machine learning, it is common practice to divide a time series into multiple subseries and represent each subsequence by its average value. However, this technique can lead to a loss of important information, making it difficult to distinguish between different subseries. For instance, some subsequences may have distinct shapes despite having the same mean and variance (as shown in Fig. 2(a) and Fig. 2(b)). To overcome this limitation, we use a similarity metric to evaluate the similarity between two subsequences.

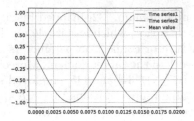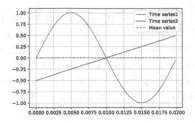

**Fig. 2.** Two examples of different time series with the same mean value.

The similarity metric is a crucial concept in machine learning that determines the degree of similarity between two objects, such as images, words, or subsequences. It is commonly used in various machine learning and deep learning applications, including clustering and recommendation algorithms. There are many similarity measurement methods, such as Euclidean distance, Manhattan distance, and Chebyshev distance. For the purposes of this paper, we employ the Euclidean distance (defined in Definition 3) to measure the similarity between two sequences. We also incorporate the mean and variance of these two sequences to calculate their similarity.

Typically, the closest match to any subsequence is located in one or two subsequences before or after it. Therefore, when calculating the similarity between a subsequence and other subsequences, it is important to exclude these highly similar subsequences to avoid nonsensical results. In this paper, we refer to this matching pattern as a non-self match (defined in Definition 5).

### 3.2 Definitions

**Definition 1.** *Time series are ordered sets of real-valued observations, $T = t_1, t_2, ..., t_m$, where $m$ is the length of the time series. Depending on the dimension of observations, time series can be classified into univariate time series (where $t_m \in \mathbb{R}$), or multivariate time series (where $t_m \in \mathbb{R}^d$).*

**Definition 2.** *We can define a subsequence of a time series, $C = t_p, t_{p+1}, ..., t_{p+n-1}$, as a sampling of length $n \leq m$ of contiguous positions from $T$ starting at position $p$ for $1 \leq p \leq m - n + 1$.*

**Definition 3.** *The distance between two subsequences $C$ and $Q$, both of length $n$, can be calculated using the Euclidean distance metric, as follows:*

$$Dist(C, Q) = \sqrt{\sum_{i=1}^{n}(c_i - q_i)^2} \tag{1}$$

*where $c_i, q_i$ are the observations of two subsequences $C$ and $Q$, respectively.*

---

**Algorithm 1.** Anomaly signal detection algorithm

---

**Input:** Time series $\{T_{t-1}^q\}_{t\in m}$ from $k$ clients
**Parameter:** learning rate $\eta$, model parameters $\theta$
**Output:** Anomaly series $C_t$

1: **for** $j = 1$ to $k$ **do**
2:    Randomly sample $T_{t-1}^q$
3:    ResNet model computers and sends classification results $R$ to detector $X$
4: **end for**
5: **for** $r = 1$ to $R$ **do**
6:    RNN model computers and predict time series $T$
7:    Calculate similarity
8:    send abnormal series $C_t$
9: **end for**
10: **return** abnormal series $C_t$

---

**Definition 4.** *To calculate the similarity between two subsequences $C$ and $Q$, we combine the mean and variance of the subsequences. Specifically, the similarity between $C$ and $Q$ is defined as:*

$$Sim(C, Q) = \sqrt{Dist(\bar{C}, \bar{Q})^2 + Dist(\tilde{C}, \tilde{Q})^2} \tag{2}$$

*where $\bar{C}$ and $\bar{Q}$ are the means of $C$ and $Q$, respectively; $\tilde{C}$ and $\tilde{Q}$ are the variance of $C$ and $Q$, respectively.*

**Definition 5.** *Given a time series $T$ of length $m$, containing a subsequence $C$ of length $n$ starting at position $p$, and a matching subsequence $M$ starting at position $q$. If $|p - q| \geq n$, then subsequence $M$ is a non-self match to subsequence $C$ at a distance of $Dist(M, C)$, where $|p - q|$ is the absolute distance between the starting positions of subsequences $C$ and $M$.*

**Definition 6.** *In signal anomaly detection, a discord of a time series $T$ is a subsequence $D$ that deviates significantly from other subsequences in the time series. It is defined as a subsequence with the largest distance from its non-self match subsequence $M_D$. In other words, for any subsequence $C$ of $T$ and its non-self match subsequence $M_C$, we have:*

$$\min(Dist(D, M_D)) > \min(Dist(C, M_C)) \tag{3}$$

## 4  Anomaly Signal Detection Method

### 4.1  Background

[36] presented a scenario for a communications satellite company where the operation team aims to detect anomalies or outliers in thousands of signals. The team used multiple spacecraft, each containing 37,000 signals from 9 different subsystems. Each signal is a univariate time series collected at the microsecond level and has been tracked for over ten years.

The team employed a traditional approach to anomaly detection, based on setting and adjusting thresholds to flag anomalous intervals. On average, 20 alarms were reported daily, most of which were false alarms and could be resolved within a few hours. However, the team faced some challenges during this process: 1) Setting and adjusting thresholds can be time-consuming and may be requested by the customer, necessitating selection of signal parts using domain-specific knowledge. 2) The anomaly detection model often mislabels some unusual signals, even if they are not anomalous. 3) The machine learning model often flags unusual signals, even if they do not necessarily have problems.

Motivated by these challenges, we designed a two-stage method for effectively classifying and detecting anomalous signals in LEO satellite networks (more details in Sect. 4.2).

## 4.2   Two-Stage Framework

Our proposed two-stage framework (illustrated in Fig. 1) includes two modules: signal classification and anomaly detection. Input data is classified by the sequence classification module before being labeled and passed through to the next module. The data is then analyzed for anomalies and stored accordingly. Lightweight transfer learning can be used to calibrate the anomaly detection model when sufficient data is available. Algorithm 1 presents the complete anomaly signal detection algorithm.

Wireless communication signals exhibit specific patterns in terms of center frequency, bandwidth, spectral density, and changing trends. To classify the input I/Q signals, we utilize the CNN model to extract key features, and the ResNet architecture backbone serves as the backbone for the classification module, as shown in Fig. 3.

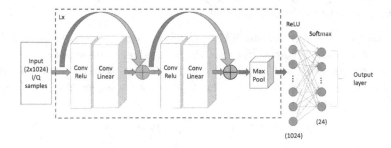

**Fig. 3.** The ResNet architecture. We use it to classify the I/Q signals.

Due to limited computing resources, satellites may face difficulties training complex deep anomaly detection models. Edge computing [37] and transfer learning are two approaches that have been employed to overcome this issue. In this paper, we adopt a lightweight deep transfer learning approach that involves pre-training a base network on the source dataset, freezing the weights of the backbone, and fine-tuning the network with the target dataset.

# 5 Experiments

In this section, we demonstrate that our framework is able to handle anomaly detection on two datasets. In addition, we show that our framework has excellent anomaly detection capabilities under classification and prediction task settings.

## 5.1 Datasets

In this paper, we evaluate the performance of our model using two datasets: 1) a modulated signal dataset generated by GNU Radio software, and 2) a weather dataset provided by the UCR time series classification archive [38]. The modulated signal dataset contains several types of signals necessary to evaluate the model performance in a controlled environment. Some signals are shown in Fig. 4 and Fig. 5, respectively.

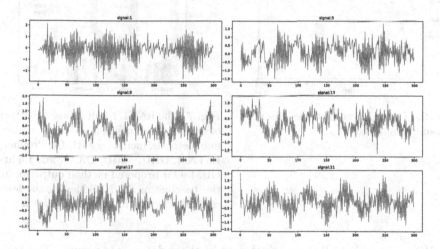

**Fig. 4.** Some input I/Q signals with SNR of 5 dB.

All the used datasets are divided into two subsets: a training subset and a testing subset. We use a seed to generate a random mutually exclusive array index, which is then used to split the data into two parts to determine that the training and testing sets are completely different. The UCR dataset contains hourly air quality information collected from 12 meteorological stations in Beijing between March 2013 and February 2017.

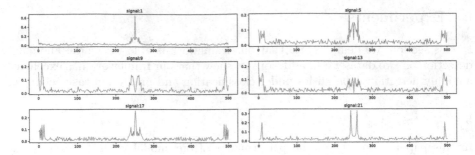

**Fig. 5.** Apply FFT (Fast Fourier Transform) on input I/Q signals with SNR of 5 dB.

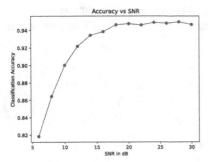

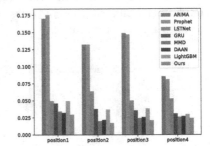

**Fig. 6.** The accuracy of the ResNet classification model.

**Fig. 7.** The RMSE result (lower is better) of our proposed method and several baseline methods, including ARIMA, Prophet, LSTNet, GRU, MMD, DAAN and Light-GBM. Our proposed method outperforms the baseline methods, achieving the lowest RMSE value in all four positions.

We assume that all the original data in these datasets is normal. To obtain anomalous signals, we intercepted a part of the data from a certain class of normal signals and replaced it with signals from other classes in the dataset.

## 5.2   Implementation Details

Our two-stage method for detecting anomalous signals in LEO satellite networks is implemented using the PyTorch framework on a server equipped with a single NVIDIA Tesla V100S GPU. In the first stage, We trained a ResNet classification model on a dataset of 100,000 samples of 24 modulated signals generated by GNU Radio. We used the Adam [46] optimizer with a dynamic learning rate (from 0.01 to 0.0001) to optimize our classifier model. The ResNet model achieved an accuracy of 92.3% on the entire modulated signals dataset (illustrated in Fig. 6) after training for approximately 16 h. In the second stage, we optimized the anomaly detector using the Adam optimizer with a fixed learning rate of 0.005 and the batch size is 36.

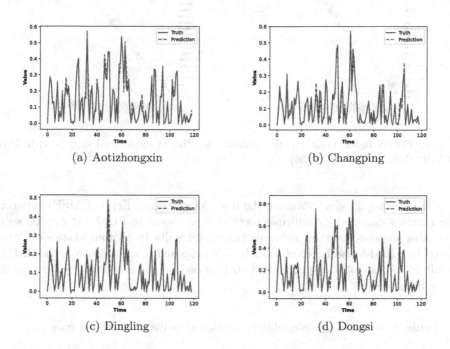

(a) Aotizhongxin

(b) Changping

(c) Dingling

(d) Dongsi

**Fig. 8.** Some prediction results in different locations.

### 5.3 Classification Result

Figure 9 presents the confusion matrix for the ResNet classifier, which was trained on 100,000 samples of 24 modulated signals generated by GNU Radio. These signals are relatively similar, and the classification accuracy of the existing methods is significantly reduced under high noise conditions. The main classification errors occur with 16 or 32-order phase shift keying (PSK), 64, 128, or 256-order quadrature amplitude modulation (QAM), and AM modulation. However, as shown in Fig. 6, we observe that our ResNet model maintains high classification accuracy when the signal-to-noise ratio (SNR) is high. This result confirms the observations in Sect. 5.4.

### 5.4 Anomaly Detection Result

To evaluate the effectiveness of our proposed anomaly signal detection model, we conducted a comparative analysis against several baseline methods. The baseline methods considered in our study include ARIMA [40], Prophet [41], Light-GBM [42], GRU [7], LSTNet [43], MMD [44], and DAAN [45]. On the normal

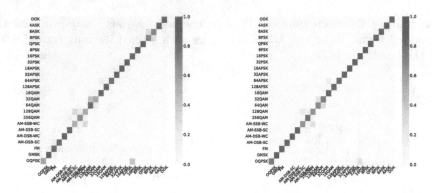

**Fig. 9.** The confusion matrix of the classifier for the 24 modulated signals with SNR of 5 dB (left) and 10 dB (right).

weather sensing data, we obtained the Root Mean Square Error (RMSE) between the stations detected by multiple receivers (as shown in Fig. 7). It can be seen that the proposed method achieves the best results in all four stations. Compared to ARIMA and LSTNet models, our proposed model improves the RMSE result by 14% and 4%, respectively. Our proposed model is able to predict signals more accurately.

**Table 2.** Received signals quality evaluation (The data is obtained from [39]).

| EIRP/dBW | Gr/dBi | Ls/dB | B/MHz | Margin/dB | SNR/dB |
|----------|--------|-------|-------|-----------|--------|
| 1 | 5 | 157.3 | 0.05 | 5 | −1.3 |
| 1 | 5 | 157.3 | 2 | 5 | −17.3 |
| 1 | 5 | 163.6 | 0.05 | 5 | −7.6 |
| 1 | 5 | 163.6 | 2 | 5 | −23.6 |
| 1 | 35 | 157.3 | 0.05 | 5 | 28.7 |
| 1 | 35 | 157.3 | 2 | 5 | 12.7 |
| 1 | 35 | 163.6 | 0.05 | 5 | 22.4 |
| 1 | 35 | 163.6 | 2 | 5 | 6.4 |

To evaluate the effectiveness of our anomaly detection model on partial time series, we generated anomalous signals from normal data and analyzed the results. Figure 8 shows some of the prediction results of our model, which accurately predicts signals received at different locations, such as Aotizhongxin, Changping, and Dingling. We then calculated the similarity between predicted and ground truth signals and set a threshold of 0.1 to detect anomalies. As illustrated in Fig. 10, when the similarity exceeds this threshold, we identify an anomaly.

Table 2 summarizes the received signal quality in different scenarios. We found that satellites with an antenna gain exceeding 35 dB can receive signals with a higher SNR than the recognition sensitivity and handle most narrow-band and microstrip signals effectively. Since the noise figure (NF) of the signal receiver on the satellite is about 2 dB, a propagation attenuation coefficient (Ls) of approximately 5 dB should be reserved in our system. Our experimental results demonstrate that our signal detection model performs well even when the SNR exceeds 10 dB(as shown in Fig. 9).

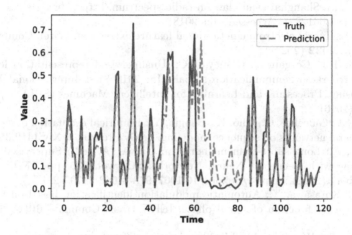

**Fig. 10.** The detection of anomaly sensor sequences using similarity measurement.

## 6  Conclusion

In this paper, we propose a deep multi-class prediction architecture, which is used for multi-signal anomaly detection in satellite communication networks and telemetry data. And it uses ResNet model as a multi-class classifier that relies on frequency spectrum features of the signal. Experiments show that our model is able to automatically extract useful features from the data without any additional manual selection.

Furthermore, we verify the viability of frequency spectrum change detection and offer some workable tricks for enhancing traditional time series change detection techniques. And through transfer learning, we can effectively reduce the amount of data required by lightweight devices in satellite networks. As a future work, we would like to expand this work in terms of algorithms. We plan to explore distributed machine learning in satellite networks, which may improve the performance of our system. Besides, an end-to-end anomaly detector can simplify the training process and make training easier.

# References

1. Li, J., Hu, X., Xiong, P., Zhou, W., et al.: The dynamic privacy-preserving mechanisms for online dynamic social networks. IEEE Trans. Knowl. Data Eng. **34**, 2962–2974 (2020)
2. Chandola, V., Banerjee, A., Kumar, V.: Anomaly detection: a survey. ACM Comput. Surv. (CSUR) **41**(3), 1–58 (2009)
3. Feng, Q., Zhang, Y., Li, C., Dou, Z., Wang, J.: Anomaly detection of spectrum in wireless communication via deep auto-encoders. J. Supercomput. **73**(7), 3161–3178 (2017). https://doi.org/10.1007/s11227-017-2017-7
4. J, K. (n.d.). Shanghai wants law on radio spectrum. https://www.shine.cn/news/metro/1803061282/. Accessed Mar 2018
5. Dong, J.: Hybrid electromagnetic signal feature extraction. Wirel. Commun. Mob. Comput. **2022** (2022)
6. O'Shea, T.J., Corgan, J., Clancy, T.C.: Unsupervised representation learning of structured radio communication signals. In: 2016 First International Workshop on Sensing, Processing and Learning for Intelligent Machines (SPLINE), pp. 1–5. IEEE (2016)
7. Chung, J., Gulcehre, C., Cho, K., Bengio, Y.: Empirical evaluation of gated recurrent neural networks on sequence modeling. arXiv Preprint arXiv:1412.3555 (2014)
8. Graves, A.: Long short-term memory. In: Graves, A. (ed.) Supervised Sequence Labelling with Recurrent Neural Networks. SCI, vol. 385, pp. 37–45. Springer, Heidelberg (2012). https://doi.org/10.1007/978-3-642-24797-2_4
9. Shi, Q., Karasawa, Y.: Automatic modulation identification based on the probability density function of signal phase. IEEE Trans. Commun. **60**(4), 1033–1044 (2012)
10. Xu, J.L., Su, W., Zhou, M.: Likelihood-ratio approaches to automatic modulation classification. IEEE Trans. Syst. Man Cybern. Part C (Appl. Rev.) **41**(4), 455–469 (2010)
11. Zhu, Z., Nandi, A.K.: Blind digital modulation classification using minimum distance centroid estimator and non-parametric likelihood function. IEEE Trans. Wirel. Commun. **13**(8), 4483–4494 (2014)
12. Liedtke, F.: Computer simulation of an automatic classification procedure for digitally modulated communication signals with unknown parameters. Signal Process. **6**(4), 311–323 (1984)
13. Mobasseri, B.G.: Digital modulation classification using constellation shape. Signal Process. **80**(2), 251–277 (2000)
14. Ho, K., Prokopiw, W., Chan, Y.: Modulation identification of digital signals by the wavelet transform. IEE Proc.-Radar Sonar Navig. **147**(4), 169–176 (2000)
15. Wang, L., Guo, S., Jia, C.: Recognition of digital modulation signals based on wavelet amplitude difference. In: 2016 7th IEEE International Conference on Software Engineering and Service Science (ICSESS), pp. 627–630. IEEE (2016)
16. Lin, Y., Tu, Y., Dou, Z., Wu, Z.: The application of deep learning in communication signal modulation recognition. In: 2017 IEEE/CIC International Conference on Communications in China (ICCC), pp. 1–5. IEEE (2017)
17. Jiang, K., Zhang, J., Wu, H., Wang, A., Iwahori, Y.: A novel digital modulation recognition algorithm based on deep convolutional neural network. Appl. Sci. **10**(3), 1166 (2020)
18. Breunig, M.M., Kriegel, H.-P., Ng, R.T., Sander, J.: LOF: identifying density-based local outliers. In: Proceedings of the 2000 ACM SIGMOD International Conference on Management of Data, pp. 93–104 (2000)

19. Angiulli, F., Pizzuti, C.: Fast outlier detection in high dimensional spaces. In: Elomaa, T., Mannila, H., Toivonen, H. (eds.) PKDD 2002. LNCS, vol. 2431, pp. 15–27. Springer, Heidelberg (2002). https://doi.org/10.1007/3-540-45681-3_2

20. Shyu, M.-L., Chen, S.-C., Sarinnapakorn, K., Chang, L.: A novel anomaly detection scheme based on principal component classifier, Technical report, Miami University, Coral Gables Fl Department of Electrical and Computer Engineering (2003)

21. Schölkopf, B., Platt, J.C., Shawe-Taylor, J., Smola, A.J., Williamson, R.C.: Estimating the support of a high-dimensional distribution. Neural Comput. **13**(7), 1443–1471 (2001)

22. Lazarevic, A., Kumar, V.: Feature bagging for outlier detection. In: Proceedings of the Eleventh ACM SIGKDD International Conference on Knowledge Discovery in Data Mining, pp. 157–166 (2005)

23. Kriegel, H.-P., Kröger, P., Schubert, E., Zimek, A.: Outlier detection in arbitrarily oriented subspaces. In: 2012 IEEE 12th International Conference on Data Mining, pp. 379–388. IEEE (2012)

24. Zhang, C., et al.: A deep neural network for unsupervised anomaly detection and diagnosis in multivariate time series data. In: Proceedings of the AAAI Conference on Artificial Intelligence, vol. 33, pp. 1409–1416 (2019)

25. Deng, A., Hooi, B.: Graph neural network-based anomaly detection in multivariate time series. In: Proceedings of the AAAI Conference on Artificial Intelligence, vol. 35, pp. 4027–4035 (2021)

26. Jiang, N., Jie, W., Li, J., Liu, X., Jin, D.: GATrust: a multi-aspect graph attention network model for trust assessment in OSNs. IEEE Trans. Knowl. Data Eng. (2022)

27. Zong, B., et al.: Deep autoencoding Gaussian mixture model for unsupervised anomaly detection. In: International Conference on Learning Representations (2018)

28. Dai, W., Xue, G.-R., Yang, Q., Yu, Y.: Transferring Naive Bayes classifiers for text classification. In: AAAI (2007)

29. Wang, Z., Song, Y., Zhang, C.: Transferred dimensionality reduction. In: Daelemans, W., Goethals, B., Morik, K. (eds.) ECML PKDD 2008. LNCS (LNAI), vol. 5212, pp. 550–565. Springer, Heidelberg (2008). https://doi.org/10.1007/978-3-540-87481-2_36

30. Dai, W., Yang, Q., Xue, G.-R., Yu, Y.: Self-taught clustering. In: ICML 2008 (2008)

31. Xu, Y., et al.: A unified framework for metric transfer learning. IEEE Trans. Knowl. Data Eng. **29**, 1158–1171 (2017)

32. Tzeng, E., Hoffman, J., Zhang, N., Saenko, K., Darrell, T.: Deep domain confusion: maximizing for domain invariance. arXiv abs/1412.3474 (2014)

33. Oquab, M., Bottou, L., Laptev, I., Sivic, J.: Learning and transferring mid-level image representations using convolutional neural networks. In: 2014 IEEE Conference on Computer Vision and Pattern Recognition, pp. 1717–1724 (2014)

34. Ajakan, H., Germain, P., Larochelle, H., Laviolette, F., Marchand, M.: Domain-adversarial neural networks. arXiv abs/1412.4446 (2014)

35. Luo, Z., Zou, Y., Hoffman, J., Fei-Fei, L.: Label efficient learning of transferable representations across domains and tasks. arXiv abs/1712.00123 (2017)

36. Liu, D., Alnegheimish, S., Zytek, A., Veeramachaneni, K.: MTV: visual analytics for detecting, investigating, and annotating anomalies in multivariate time series. arXiv preprint arXiv:2112.05734 (2021)

37. Tianqing, Z., Zhou, W., Ye, D., Cheng, Z., Li, J.: Resource allocation in IoT edge computing via concurrent federated reinforcement learning. IEEE Internet Things J. **9**(2), 1414–1426 (2021)

38. Dau, H.A., et al.: The UCR time series archive. IEEE/CAA J. Automatica Sinica **6**(6), 1293–1305 (2019)
39. Zhou, X., Xiao, Y., Hu, M., Liu, L.: Wireless signal recognition based on deep learning for LEO constellation satellite. In: Yu, Q. (ed.) SINC 2019. CCIS, vol. 1169, pp. 275–285. Springer, Singapore (2020). https://doi.org/10.1007/978-981-15-3442-3_23
40. Zhang, G.: Time series forecasting using a hybrid ARIMA and neural network model. Neurocomputing **50**, 159–175 (2003)
41. Taylor, S., Letham, B.: Forecasting at scale. Am. Stat. **72**, 37–45 (2018)
42. Ke, G., et al.: LightGBM: a highly efficient gradient boosting decision tree. In: Advances In Neural Information Processing Systems, vol. 30 (2017)
43. Lai, G., Chang, W.-C., Yang, Y., Liu, H.: Modeling long-and short-term temporal patterns with deep neural networks. In: The 41st International ACM SIGIR Conference on Research & Development in Information Retrieval, pp. 95–104 (2018)
44. Li, H., Pan, S., Wang, S., Kot, A.: Domain generalization with adversarial feature learning. In: Proceedings of the IEEE Conference on Computer Vision and Pattern Recognition, pp. 5400–5409 (2018)
45. Yu, C., Wang, J., Chen, Y., Huang, M.: Transfer learning with dynamic adversarial adaptation network. In: 2019 IEEE International Conference On Data Mining (ICDM), pp. 778–786 (2019)
46. Kingma, D.P., Ba, J.: Adam: a method for stochastic optimization. arXiv preprint arXiv:1412.6980 (2014)

# Hydra: An Efficient Asynchronous DAG-Based BFT Protocol

Zhuo An[1,2(✉)] [ID], Mingsheng Wang[1,2(✉)], Dongdong Liu[1,2], Taotao Li[3(✉)], and Qiang Lai[1,2]

[1] State Key Laboratory of Information Security, Institute of Information Engineering, Beijing 100093, China
{anzhuo,wangmingsheng,liudongdong,laiqiang}@iie.ac.cn
[2] School of Cyber Security, University of Chinese Academy of Sciences, Beijing 100049, China
[3] School of Software Engineering, Sun Yat-sen University, Zhuhai 529478, China
ltt93@mail.sysu.edu.cn

**Abstract.** The directed acyclic graph (DAG) technique applied in asynchronous Byzantine fault tolerance (BFT) consensus has tremendously improved the system throughput in practice. However, there are still two critical limitations on performance. First, most work relies on the transaction batching technique to achieve high throughput while maintaining low amortized communication complexity, which places a huge bandwidth burden on the relevant single party.

Second, due to the concurrency of the DAG-based multi-leader, it is inevitable for them to mix duplicate transactions in their proposal blocks. This paper presents Hydra, an efficient asynchronous DAG-based BFT protocol that is the first work to address these problems. The core design of Hydra is to leverage a request transaction pre-processing scheme, which fairly distributes the transactions to different node for proposing to eliminate the transaction duplication. And an AVID-M protocol with a *dispersal-then-retrieval* workflow to split the bulky proposal blocks into tiny segments for dispersing and recovering, which achieves optimal amortized $O(N)$ communicate complexity and alleviates the bandwidth pressure on a single party.

**Keywords:** DAG-based blockchain consensus · Asynchronous BFT protocol · State machine replication

## 1 Introduction

Recently, the surge of the blockchain [18] development and its application [3] in the distributed system has fueled research in the traditional state machine replication (SMR) field. A key challenge in the blockchain system is building a robust system that can maintain a consistent state among a large number of distributed nodes and tolerate attacks and unknown bugs. This problem, known as the consensus problem in blockchain systems and the Byzantine general problem

© The Author(s), under exclusive license to Springer Nature Switzerland AG 2023
M. Yung et al. (Eds.): SciSec 2023, LNCS 14299, pp. 439–459, 2023.
https://doi.org/10.1007/978-3-031-45933-7_26

in SMR [14], has garnered significant interest from researchers. The first practical Byzantine tolerant protocol (PBFT) [4] was introduced 20 years ago, and since then, a series of BFT consensus protocols [13,24] have been proposed. However, many of these protocols are based on predefined assumptions of synchronous or partially synchronous network conditions, which can result in significant performance degradation and even security threats when network fluctuations occur.

Given the unpredictable network conditions in the real world, it is important to eliminate the impact of network conditions. This has prompted researchers to investigate the design of consensus protocols in an asynchronous environment, which allows nodes to operate without any restrictions imposed by the network.

**Practical Solutions to SMR in the Asynchronous Network:** In recent times, an important work of HoneyBadger (HB) [17] protocol has been proposed by Miller. The HB protocol is the first practical asynchronous BFT-consensus work. It turns to treats consensus as a Byzantine atomic broadcast (BAB) problem, and introduces a new pattern that combines the reliable broadcast (RBC) protocol with Byzantine agreement (BA) protocol to accomplish the BAB in asynchronous conditions. However, the Byzantine agreement (BA) is costly and requires $O(N)$ process operating rounds to terminate in the worst case.

As a comparison work, Keidar proposed DAG-rider [12], a new framework of consensus that introduces a DAG (directed acyclic graph) structure as the communication layer. With this approach, replicas can commit blocks by depending on their local DAG duplicates without contacting to others. This reduces the communication cost in the agreement phase and tremendously improves the throughput and latency. However, due to the concurrency of multi-leader, duplicate transactions are inevitable between proposals, leading to unpredictable delay attacks on correct transactions and affecting the performance of the protocols. Therefore, there is still room for further improvement.

**Removing the Restrictions and Improving the Efficiency:** In this paper, we present Hydra, which leverages the strengths of DAG-rider while addressing its limitations. To our knowledge, Hydra is the first DAG-based asynchronous BFT consensus protocol that eliminates the duplication of transactions and achieves optimal amortized $O(N)$ communication complexity with only slightly increased bandwidth pressure on relevant nodes. More specifically, we make the following contributions:

- We present a scheme to preprocess request transactions that effectively eliminates the issue of transaction overlapping between different proposal blocks, resulting in improved throughput quality and indirectly enhancing the practical throughput.
- We introduce the AVID-M protocol with a carefully designed two-stage broadcast structure, which improves system concurrency and achieves significantly lower communication cost for a single node. Each node only needs $O(|v|+\lambda N)$ bits to broadcast a batch of transactions of size $|v|$ along with the $\lambda$ bits of publicly verifiable proof.
- We optimize a *Bufferpool* to enable reliable dissemination and storage of transactions among $N$ parties, along with a pre-defined global partitioning

mechanism to grant replicas to individually select distinct transactions to be placed in their proposal block. It feasts a solid and effective foundation for upper-layer protocols.
- We elevate the latency of the DAG-based asynchronous consensus from 4 rounds to 3 rounds for each commitment and allow for pipelining.

## 2    Related Work

In this section, we introduce the additional related work and their detail features. This paper focuses mainly on the asynchronous BFT-based protocols, which achieve state machine replication among widely distributed parties without making harsh assumptions about the network conditions.

The seminal HoneyBadger [17] protocol was the first practical asynchronous BFT protocol developed from previous theoretical studies [2,5]. It is built upon an Byzantine atomic broadcast (BAB) abstraction which enables to reach a common agreement on the order of the input transactions among replicas. Moreover, it simplifies the atomic broadcast primitive into $N$ parallel RBC instances, combined with Byzantine agreement (BA) instances to output an asynchronous common set (ACS) that contains almost every input transaction. HoneyBadger incurs $O(N^2|v| + \lambda N^3 \log N)$ communication complexity and $O(1)$ time in the best case. However it requires ample bandwidth for the replicas and causes excessive time delays in the presence of failures, limiting its practical performance.

To address this issue, some scholars have focused on the use of DAG technology in blockchain systems. The application of DAG technology in blockchain has been ongoing for some time, as it inherently matches the demands of concurrency. However, previous works like [1,15,19,20] have mostly focused on the POW-based or POS-based consensus in the synchronous network. Aleph [9] is the first protocol that implements the DAG on the asynchronous protocol. Aleph innovatively proposes a DAG structure as a communication layer that allows replicas to accumulate transactions and utilizes RBC to ensure a high probability of reaching ACS among honest nodes. However, most of their efforts were spent designing a decentralized randomness beacon to provide randomness in the absence of a trust broker, thus achieving liveness in asynchronous conditions. And it has $O(|v|N^2 + RN^2 \log N)$ communication complexity, where $|v|$ is the transaction batch size and $R$ is the protocol rounds for consensus, which can increase significantly in the worst-case scenario.

The recent work of Idit proposed DAG-rider [12], which provides a comprehensive explanation of DAG technology and presents an asynchronous consensus protocol based on the DAG structure. Specifically, it builds on the framework of Aleph [9], which deploys an independent communication layer with DAG structure, reduces the BA process, and relies only on a common coin for liveness. The main focus is on explaining and providing rigorous security and feasibility proofs of how atomic broadcast can be achieved in a practical way by attaching RBC instances to causal DAG structured histories. However, scalability is limited due to the requirement for each node to have unbounded memory and ample

**Table 1.** A comparison between Hydra with different related protocols.

|  | Communication Complexity | Disperse cost for single node | No duplication |
|---|---|---|---|
| HoneyBadger | amortized $O(N)$ | $O(N\|v\| + \lambda N^2 \log N)$ | ✗ |
| Dumbo | amortized $O(N)$ | $O(N\|v\| + \lambda N^2 \log N)$ | ✗ |
| DAG-rier | amortized $O(N)$ | $O(N\|v\| + \lambda N^2 \log N)$ | ✗ |
| Tusk | amortized $O(N))$ | $O(N\|v\| + \lambda N^2 \log N)$ | ✗ |
| Dumbo-NG | amortized $O(N)$ | $O(\|v\| + \lambda N)$ | ✗ |
| Hydra | amortized $O(N)$ | $O(\|v\| + \lambda N)$ | ✓ |

\* Here |v| denotes the size of proposal block for each node to disperse.

bandwidth. As an extension of DAG-rider, Tusk [6] takes the modularity of protocol one step further by proposing an independent modular *Mempool* to envelop reliable data transmission and storage processes. This *Mempool* allows honest nodes to share the same transaction pool, generates quorum-based certificates of availability and validity that are bound with unique hashes of transactions, and serves as a foundation to sustain the upper consensus layer in a more efficient manner. Efforts are made to promote the concurrency of sub-processes and reduce latency, thereby improving system scalability. Bullshark [21] is another work that focuses on studying the fast path of DAG-rider in the synchronous network.

In contrast to Hydra, the series of works focused on optimizing the protocol structure and achieving an optimal amortized $O(N)$ communication complexity. However, they overlook the key issue of transaction overlapping and the bandwidth burden on the sole node that truly impacts practical performance.

There also exist some works in the field of asynchronous BFT consensus protocols, such as BEAT [7] and Dumbo [10]. BEAT retains the main framework of HoneyBadger but replaces some major components for slight performance optimization. Dumbo reduces the BA instances from $N$ to a constant number of $k$ by combining the $k$ size committee with a reformed MVBA protocol. This reduction results in a significant improvement in latency, but the protocol still has high communication complexity. Dumbo-NG [8] is another recent work that separates the broadcast and agreement processes into individual steps. It uses a sequence broadcast protocol to disseminate transactions with predefined sequence numbers and provides certificates as proof of availability, which can also satisfy the external-validity requirement of the next MVBA process. Furthermore, Dumbo-NG emphasizes the potential for pipeline work due to the concurrency of the two independent processes, leading to high throughput and low latency. Compared to previous works, Dumbo-NG reduces communication complexity and achieves better performance. However, none of these works address the issue of transaction duplication. We give a comparison to related works in the Table 1.

# 3  Preliminaries

## 3.1  System Model

We assume a system with a set of $N$ parties denoted as $\{P_i\}_{i \in [N]}$, where $[N]$ represents $\{1, ..., N\}$ in this paper. Each party $\{P_i\}_{i \in [N]}$ has the same status and acquires its identity from the initial trust setup. We further assume the existence of an *adversary* who can corrupt at most $f$ parties in the system, with a constrain of $(3f + 1) \leq N$. Once a party is corrupted, we call it a *Byzantine* node in this paper. The other parties, who constantly follow the protocols, are deemed *honest*. The *adversary* takes over the identities and entire states of the corrupted parties, and obligates or even colludes with them to perform arbitrary behaviors that can undermine the system. Note the *adversary* is considering computation-bounded and unable to broke the cryptography primitives in this paper.

Our system runs on a fully asynchronous network which means that we will not make any assumptions about the message delays other than that the messages will eventually arrive. Specifically, we consider pairs of honest nodes to be connected with authenticate communication channels. These channels are established under the control of *adversary*, which means it can dominate the message delays between nodes, but it is incapable to forge or drop messages between correct processes. This ensures that the communication between honest nodes is secured and not compromised by the *adversary*.

## 3.2  Building Blocks

**Definition 1 (Reliable Broadcast (RBC) protocol):** RBC is an abstraction of communication protocols in which messages is guaranteed to broadcast to all nodes in a distributed system even with some failed or unavailable nodes. And provided these properties:

- **Validity:** If there has an honest process $P_i$ invoked $r\_broadcast_i\{m, r\}$ to broadcast a message $m$ with a sequence number $r \in N$, then all honest processes $P_{j \in [N/i]}$ eventually output $r\_deliver_j\{m, r, p_i\}$;
- **Integrity:** For each $r \in N$ and party $P_{s \in [N]}$, honest process $P_i$ output $r\_deliver_i\{m, r, p_s\}$ at most once regardless of $m$;
- **Agreement:** If an honest process $P_i$ has outputted $r\_deliver_i\{m, r, p_s\}$, then all honest processes $P_j$ eventually output $r\_deliver_j\{m, r, p_s\}$.

**Definition 2 (Verifiable Transaction Partition):** Verifiable transaction partition is a scheme used to randomly distribute the request transactions from clients to the proposal blocks of different *validators*, thus solving the transaction duplication problem, and guaranteed these properties:

- **No-duplication:** If a request transaction $q$ has distributed to a specific area $\alpha$ in the buffer of nodes, then only the *validator* $P_i$ has assigned with area $\alpha$ can contain $q$ in its proposal block $b$;

– **Agreement:** Any correct *Validaotr* has the same view about the blocks distribution.

**Definition 3 (Common Coin):** Common coin is a scheme to provide randomness to select a *leader* which is unpredictable to the *adversary*, and have the following properties:

– **Agreement:** Each honest node $P_i$ calls *choose_leader*() to active the *common coin*, and for $(\forall i, j \in N, i \neq j)$, $P_i$ always get a return value the same to $P_j$;
– **Termination:** If at least $f+1$ honest nodes active the *choose_leader*(), then they eventually get a result;
– **Unpredictability:** The probability of the *adversary* to predict the return value of *choose_leader*() before honest nodes is at most $1/n$;
– **Fairness:** The probability of each node to be the returned value of the *choose_leader*() is equal to $1/n$.

We implement the RBC protocols and *common coin* based on a PKI and a $(2f + 1, N)$-threshold signature scheme [16]. Utilizing them to construct a *common coin* and achieve *liveness* in the asynchronous network is a well-studied pattern, as demonstrated in previous work [9,12]. Regarding the RBC protocols, we introduce AVID-M, which incorporates a $(f + 1, N)$-erasure code [11] and a Merkle hash tree to improve efficiency. Additionally, we utilize a hash function as the VRF to ensure efficiency and resistance to duplication. Further details will be presented in later sections.

### 3.3  Problem Definition

This paper focuses on solving the problem of Byzantine Atomic Broadcast (BAB), which is a transformed consensus problem. The goal is to achieve a consistent state among nodes in State Machine Replication (SMR), even when there are some malicious nodes.

**Definition 4 (Byzantine Atomic Broadcast (BAB)):** Byzantine Atomic Broadcast (BAB) is a protocol that ensures all honest parties deliver messages in the same order. For any honest party $P_i$, BAB provides two functions: $a\_broadcast(m, r)$ and $a\_deliver(m, r, j)$, representing the broadcast and delivery programs of BAB, respectively. Here, $m$ is a message, $r \in N$ is a sequence number, and $P_j$ is another process. Furthermore, each party can broadcast an infinite number of messages with consecutive sequence numbers. A BAB protocol must satisfy the properties of reliable broadcast, and an additional property:

– **Total Order:** If there is a correct process $P_i$ output $a\_deliver(m, r, j)$ before $a\_deliver(m', r', j')$, then no correct process in the system should output $a\_deliver(m', r', j')$ before $a\_deliver(m, r, j)$.

Furthermore, Hydra also strengthen the validity to ensure the all transactions proposed by correct parties are eventually ordered with probability 1, and provides the following properties:

- **No duplication:** In any time, if correct nodes output $a\_deliver(m, r, i)$ and $a\_deliver(m', r', j)$, then the transactions in $m$ from $P_i$ is completely different to $m'$ from $P_j$.
- **Liveness:** If a client send request transaction to $f + 1$ correct nodes, then it is eventually contained in the output $a\_deliver(m, r, i)$ by every correct node.
- $\frac{1}{2}$-**Chain quality:** For each honest party $P_i$ in any view of the system, at least $\frac{1}{2}$ of the consecutive blocks it hold are from honest parties.

## 4  Overview

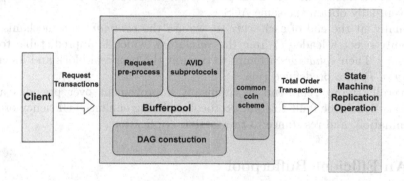

**Fig. 1.** The framework of Hydra with modular designs

In this section, we present an overview of Hydra. Hydra is designed with a modular architecture, inheriting the design of DAG-rider [12], it progresses in *wave* $w \in N$, but with one difference: each *wave* only has three *round* $r \in N$. Additionally, it introduces new components to improve efficiency and provide new features. Figure 1 depicts the refined framework of Hydra.

Firstly, Hydra employs the request pre-processing technique in *Definition* 2 to partition requests into separate *baskets* to ensure that each validator proposes a block with disjoint transactions. This enhances the efficiency of multi-leader working mode and provides duplication-resistance.

Secondly, instead of using RBC protocols in *Definition* 1 as a black-box, such as the previous work [6,12], we present a new AVID-M protocol with an innovative workflow of *dispersal-then-retrieval* while providing the same properties as RBC protocols. This facilitates data dissemination, reduces overhead, and improves the efficiency of individual parties.

Then we develop a new module called the *Bufferpool* upon the above protocols that integrates the request transactions handling, fair distribution of transactions to *validators* for proposal, and dissemination of proposals among nodes. This module also generates quorum-based certificates (QCs) that serve as available proofs of transactions in proposals. However, it should be noted that the QC of one *Bufferpool* block only represents the current state of a single *validator* in the local round.

Furthermore, to achieve the same ACS in the asynchronous network, we establish a causality between blocks in succeeding rounds. Once a *validator* $P_i$ receives at least $(2f+1)$ QCs in the current round $r$, it has guaranteed access to the matching $(2f+1)$ valid blocks, which contains at least $f+1$ proposals from honest *validators*. Subsequently, it moves to the next round $r+1$ and proposes a new block that references these QCs to blocks from the previous round $r$. By performing three consecutive rounds of operations in a *wave w*, the quorum intersection of each round blocks will yield an ACS that contains almost all blocks in this *wave*, thus preventing censorship attacks from *adversaries*. Additionally, due to the agreement property of the AVID-M protocol, each party will eventually obtain the same ACS.

Finally, at the end of each *wave w*, we use the *common coin* mechanism to randomly select a leader $P_l$ from the *validators*, which is unpredictable to the *adversary*. Then *validators* attempt to commit the proposal block and its entire history of $P_l$ in a predefined rule.

Overall, Hydra represents a significant improvement over previous asynchronous Byzantine fault-tolerant protocols, offering enhanced efficiency of data dissemination, and resistance to transaction duplication.

## 5   An Efficient Bufferpool

In this section, we will provide a detailed explanation of the core components of the *Bufferpool*, which is a fundamental element that facilitates fairly partitioning transaction requests, distributing data reliably among $N$ parties, and providing certificates of availability.

### 5.1   Request Preprocessing

First, we present the specific request preprocessing scheme for verifiable transaction partition (*Definition* 2) before they are proposed by the *validators* in Hydra. The request pre-processing scheme, illustrated in Fig. 2, aims to eliminate overlapping transactions between the blocks of proposals in the multi-leader or leaderless consensus protocols. To execute the pre-processing, we divide the request hash space into equally sized *baskets*, then evenly allocate them to the *validators* for proposal. This approach was first introduced in [22]. We argue that this scheme is suitable for porting to Hydra because it does not depend on any specific network condition. We provide the details in Algorithm 1.

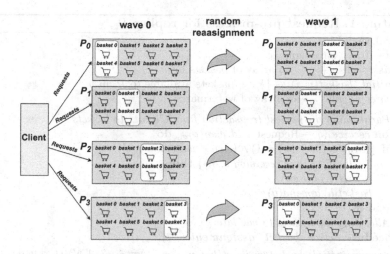

**Fig. 2.** The request pre-process.

**Request Partition** (line 1–5): In each *wave w*, a client *c* puts its request *o* along with an assigned sequence number *c.sn* and its identifier *c.id* in a specific message $\langle \mathbf{Request}, c.id, c.sn, o \rangle_{\sigma_c}$ and broadcasts it to all replicas. The sequence number *c.sn* is an incremental integer starting from 0, and a dynamic *watermark* $w_L < c.sn \le w_H$ is used to control the request rate of the client, in order to resist denial-of-service (DOS) attacks. The *watermark* is tied to the system state and only moves forward whenever the system completes a deterministic state transfer, which occurs when the honest nodes reach an agreement. Then, the request partition function $basket_{h \in N} = Hash(c.id \| c.sn)$ is applied, which relies only on the identity of the client *c* and the unique sequence number *sn* to place the request in the matching $basket_h$. This thwarts *adversaries* from biasing fairness by altering the request payload *o*. Finally, a *validator* will only accept requests in *baskets* if the following conditions are met: (i) the signature $\sigma_c$ of request is correctly signed by the accordant client *c*; (ii) it is the first time receiving the sequence number *c.sn* from client *c*, and *c.sn* is in the rang of current *watermark*;

**Assign Baskets to Validators** (line 6–7): For the *validators*, in each *wave w*, they receive requests from clients and store them in their local buffers. Once the requests from clients have been partitioned into different *baskets* and the buffer of *validators* has been converted into these disjoint *baskets*, we need to reassign them to different *validators* for proposing blocks. The *baskets* are denoted by $basket_h, h \in (0, ..., |baskets[w]| - 1)$, and $|baskets[w]|$ is the number of *baskets* available for allocation.

To determine the distribution, we directly use a modulo operation on the size of the *validators*:

$$Baskets(w, i) = \{(h + w + s_{w-1}) \equiv i \ mod \ N| \ h \in |baskets[w]|\} \qquad (1)$$

---

**Algorithm 1. Request pre-process for replica $\mathcal{P}_i$**

---

**Local variables:**

$w \leftarrow 0 - -$ wave number

$basket[w] \leftarrow \{\} - -$ an array of baskets in $w$

$initial\_basket(w, i) \leftarrow \{\} - -$ baskets available to $\mathcal{P}_i$ in $w$

$watermark[w] - -$an array of watermarks in $w$

// *Partitioning the request transaction from $Client_c$ to baskets*

1: **Upon** $receive \langle q \leftarrow Request, c.id, c.sn, o \rangle_{\sigma_c}$ **do**

2:     **if** $\{(SigVer(\sigma_c, c.p_k) = 1) \wedge (c.sn \in watermark[c])$

        $\wedge$ *(is the first time receiving c.sn)*$\}$ **then**

3:         $h = H(c.id\|c.sn)$

4:         $basket[h].append(q)$

5:     **end if**

// *Allocating the baskets to validators*

6: **Procedure** $active\_basket\_assignment(w, s_{w-1}, i)$

7:     **return** $initial\_basket(w, i) = ((h + w + s_{w-1}) \equiv i \mod N)|(h \in |baskets[w]|)$

---

The result $Baskets(w, i)$ represents the *baskets* assignment for *validator* $P_i$ in *wave* $w$, and $s_{w-1}$ is a randomly generated seed from the *common coin* process of previous *wave* $w - 1$, which is a pre-defined number in the genesis *wave* 0.

It is worth noting that while we require clients to broadcast their requests to $N$ nodes in the system, in practice it is feasible to send requests to a smaller subset of nodes (such as $f + 1$ nodes). Furthermore, the *baskets* in $Baskets(w, i)$ of *validator* $P_i$ will be randomly reassigned during the agreement phase of *wave* $w$ before entering the next *wave* $w + 1$. This approach also helps to prevent censorship attacks from the Byzantine replicas.

## 5.2    The AVID-M Protocol

Hydra choose to use AVID-M protocol [23] as a RBC protocol to construct *Bufferpool*. AVID-M achieves linear amortized communication complexity and provides a more affable broadcast strategy for individual party. It does this by modifying the broadcast process to a *dispersal-then-retrieval* workflow as shown in Fig. 3.

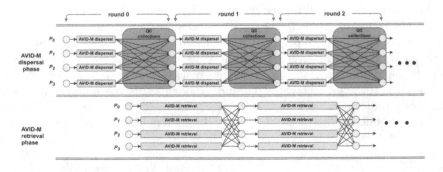

**Fig. 3.** The AVID-M process.

- **Dispersal phase:** The *dispersal* phase aims to proficiently transmit data to others, we show the details in Algorithm 2.

  Firstly, the sender party $P_{i \in [N]}$ selects a batch of transactions with $|v|$ bits size from their own baskets in $Baskets(w, i)$, and a list of certificates $CL_{r-1}$ from the previous round, as their proposal in round $r$. They then invoke $AVID - M\_Dispersal(ID)$ with a unique ID associated with the current round, and use $(f + 1, N)$-erasure code to encode the proposal block into an array of $N$ segments $S[N]$. They also construct a Merkle tree upon these segments, and compute a proof $\rho_j$ with the Merkle tree root that can prove the segments belong to the $j$-th position of the block.

  Then $P_i$ sends a specific message $Dispersal$ to distribute the segments $seg_j$ corresponding to node $P_j$. Waiting for responses from other parties $P_{j \in [N/i]}$ to ensure that at least $2f + 1$ nodes have received the corresponding segments. Once a node $P_j$ receives the $Dispersal$ message from $P_i$, it verifies its validity. If correct, $P_j$ delivers it and signs its partial secret key $sk_j$ of the $(N - f, N)$-threshold signature, then sends a particular message $Saved$ to $P_i$. Upon receiving at least $2f + 1$ valid $Saved$ responses from different nodes, the sender $P_i$ combines them into a full signature $\sigma_i$ as $QC_i$, and broadcasts $Done$ to the other nodes. If a node $P_j$ receives the $Done$ message, it checks the validity of the $QC_i$ and saves it in its local certificates list $CL_r$, then invokes $\langle Retrieval, ID, QC_i \rangle$ to retrieve the segments from the proposal block.

  Additionally, a party can invoke $abort(r)$ once it has collected enough certificates to step to the next round, or if it suspects the sender $P_i$.
- **Retrieval phase:** For the retrieval phase, it responsible to retrieve segments from different parties if it receive a request contains a valid certificate.

The $P_i$ invokes $\langle Retrieval, ID, QC_k \rangle$ to retrieve associate segments from other parties. If $P_i$ has collected $f + 1$ segments, it tries to recover the proposal block and verify the correctness use the associated proof of Merkle tree root $tree.root$. We show the details in Algorithm 3.

Note the *retrieval* process runs in parallel with the accumulation of certificate list $CL_r$ for each *validator* $P_i$ as shown in Fig. 3. This approach is more efficient and convenient for a single party, as it only needs to request small segments.

**Security Intuitions:** The QC certificates indicate that at least $2f + 1$ nodes have submitted the relevant segments of the dispersed proposal. This ensures that the original value can be retrieved since there are no fewer than $f + 1$ honest nodes involved. Thus, the *Bufferpool* guarantees the validity and availability of the proposals. Additionally, each proposal includes a certificate list $CL_{r-1}$ that refers to at least $2f + 1$ blocks from the previous round $r - 1$. Therefore, the *Bufferpool* provides causality among blocks from consecutive rounds and ensures that at least $f + 1$ of these blocks are from honest nodes, which is also known as $1/2$ chain quality.

**Algorithm 2. AVID-M_Dispersal subprotocol with identifier ID and sender $\mathcal{P}_i$**

Local variables:
$r \leftarrow 0$ – –the current round
$res[\ ] \leftarrow \{\}$ – –an array for saving responses in $r$
$CL_r[\ ] \leftarrow \{\}$ – – an array for saving QC in $r$
$Stop \leftarrow 0$ – – the flag of termination

// Dispersing the proposal of $\mathcal{P}_i$ to others
1: **Upon** receive an input $\mathcal{V}$ **do**
2:    $S[\ ] \leftarrow Encode(\mathcal{V})$ // $S$ is a vector with $N$ segments
3:    $tree \leftarrow Merkle(S)$
4:    **for** $j \leftarrow (1$ to $N)$ **do**
5:        $\rho_j \leftarrow tree.get\_proof(seg_j)$
6:        **let** $Save = \langle tree.root, seg_j, \rho_j, r \rangle$
7:        **send** $\langle Dispersal, ID, CL_{r-1}, Save \rangle$ to $\mathcal{P}_j$
8:    **end for**
9:    **wait** until $(|res[r]| = 2f + 1)$ **do**
10:        $\sigma_i = Combine_{(2f+1)}(\sigma_{(i,j)} \in res[r])$
11:        **let** $QC_i = \langle \sigma_i, tree.root, r \rangle$
12:        **multicast** $\langle Done, ID, QC_i \rangle$

// Collecting the response from others
13: **Upon** receive $\langle Saved, ID, \sigma_{i,j}, r \rangle$ from $\mathcal{P}_j$ for the first time **do**
14:    **if** VerShare$(\langle Saved, ID, tree.root \rangle, (j, \sigma_{(i,j)})) = 1) \wedge (Stop = 0)$ **then**
15:        $res[j] = \sigma_{(i,j)}$
16:    **end if**

// Validating the segment of proposal from $\mathcal{P}_i$
17: **Upon** receive $\langle Dispersal, ID, Save \rangle$ from sender $\mathcal{P}_i$ for the first time **do**
18:    **if** VerSig$(\langle i, Save \rangle) = 1 \& Stop = 0$ **then**
19:        **deliver** $Save$ **and** parse $Save$ as $\langle tree.root, seg_j, \rho_j, r \rangle$
20:        $\sigma_{(i,j)} \leftarrow PartialSign(sk_j, \langle Saved, ID, tree.root \rangle)$
21:        **Send** $\langle Saved, ID, \sigma_{(i,j)}, r \rangle$ to $\mathcal{P}_i$
22:    **end if**

// Collecting QC in round $r$
23: **Upon** receive $\langle Done, ID, QC_s \rangle$ from sender $\mathcal{P}_i$ for the first time **do**
24:    **if** Validate$(\langle ID, tree.root, \sigma_i \rangle) = 1 \& Stop = 0$ **then**
25:        **Save** $CL_r[\ ] \leftarrow \langle ID, QC_i \rangle$
26:        **wait** until $(|CL_r[\ ]| = 2f + 1)$ **do**
27:            $abort(r)$
28:            $r = r + 1$
29:    **end if**

// Terminating the process in round $r$
30: **Procedure** $abort(r)$:
31:    **return** $Stop \leftarrow 1$

**Algorithm 3. AVID-M_Retrieval subprotocol with identifier ID, for each party $\mathcal{P}_i$**

---

Local variables:
  $\mathcal{S}[\ ] \leftarrow \{\}$ – – the *segments* vector

1: **Upon** receive $\langle Retrieve, ID, QC_k \rangle$ **do**
2:    **if** $ValidCert\langle ID, QC_k \rangle = 1$ **then**
3:      **multicast** $\langle Retrieve, ID, Save \rangle$ to all
4:      parse $QC_k$ as $\langle tree.root, \sigma_k \rangle$
5:      **wait** until $|\mathcal{S}[seg]| = f + 1$ **do**
6:        $\mathcal{V} \leftarrow Decode(\mathcal{S}[seg])$
7:        **if** $Merkle(Encode(\mathcal{V})).getroot() = tree.root$ **then**
8:          **return** $\mathcal{V}$
9:        **else**
10:          **return** $\perp$
11:        **end if**
12:    **end if**

---

# 6    The Hydra Protocol

In this section, we introduce the complete version of Hydra. Hydra is an asynchronous consensus protocol, it shares a similar DAG structure with DAG-rider as the communication layer. In this paper, we implement Hydra based on the efficient *Bufferpool* of Sect. 5, that extends several novel features to improve the time latency and reduce the storage requirements.

**DAG Construction:** Hydra designs to operate in *waves*, and each *wave* $w \in N$ consists of three consecutive *round* $r \in N$. Algorithm 4 specifies the DAG basic structure and utilities for replica $P_i$ (line 1–10). For each replica $P_i$, it keeps a DAG-structure duplicate $DAG_i[]$ in local, which is comprised of sets of *vertices* and *edges* based on the *Bufferpool*. Actually the $DAG_i[]$ also represents the local view of $P_i$ to the global state, so it can be interpreted into sequentially different fragments $DAG_i[0...r]$, $i, r \in N$ according to the round number $r$. Each $DAG_i[r]$ is an array contains at least $2f + 1$ *vertices* that represent to the proposal blocks $P_i$ previously received from *source* $P_j$ in round $r$. And each *vertex* in $DAG_i[r \geq 1]$ has at least $2f + 1$ *edges* link to the previous *vertices* of $DAG_i[r-1]$, which denotes the certificate list $CL_{r-1}$ contained in the corresponding proposal block. Thus these *edges* establish a *"happened-before"* relationship among the relevant vertices from $DAG_i[r]$ to $DAG_i[0]$, which is also called causality. The function $path(u, v)$ checks if there exist a path comprised of connecting *edges* from *vertex* $u$ to *vertex* $v$. Then Algorithm 4 describes the DAG construction function for replica $P_i$ (line 11–32). Each party $P_i$ consistently checks its local buffer to examine if there exist a *vertex* $v$ that satisfies the conditions to be added to $DAG_i[]$. Once $P_i$ collects at least $2f + 1$ vertices with the same *round* $r$ from different sources, it advances to next *round* $r + 1$ and invokes $create\_new\_vertex(r+1)$ through the upper layer BAB protocols. And $P_i$ puts the proposals it has delivered through the BAB processes as *vertices*

---

**Algorithm 4. DAG basic structure and construction for replica $\mathcal{P}_i$**

---

   **Local variables:**
     $r \leftarrow 0$ – – round number, start at 0
     $buffer \leftarrow \{\}$ – – vertices received from other *validators*
     struct vertex $v$:
       $v.round$– –the round of $v$ in the DAG
       $v.source$– –the replica that broadcast $v$
       $v.block$– –a batch of transactions
       $v.edges$– –a set of vertices in $v.round - 1$

     $DAG_i[\ ]$: – – an array of sets of vertices
       $DAG_i[0] \leftarrow$ predefined hardcoded set of $2f + 1$ "genesis" vertices
       $DAG_i[j] \leftarrow [\ ]$, for $\forall j \geq 1$
     $initial\_baskets(w, i) \leftarrow$ a queue, initially empty, $\mathcal{P}_i$ enqueues valid
       transactions from clients

1: **Procedure** $path(v, u)$
2:     $v_1 = v, v_k = u$
3:     **return**$((\exists k \in N, v_1, v_2, ..., v_k) : v_i \in \bigcup_{r \geq 1} DAG_i[r]) \wedge (v_i \in v_{i-1}.Edges())$

4: **Procedure** $create\_new\_vertex(round)$
5:     **wait until** $\neg initial\_baskets(w, i).empty()$
6:     $v.round \leftarrow round$
7:     $v.source \leftarrow p_i$
8:     $v.block \leftarrow initial\_baskets(w, i).dequeue(|v|)$
9:     $v.Edges \leftarrow DAG[round - 1]$
10:    **return** $v$

    // Collecting valid vertices from algorithm 3
11: **Upon** $r\_deliver_i(v, round, p_k)$ **do**
12:    $v.source \leftarrow p_k$
13:    $v.round \leftarrow round$
14:    **if** $|v.Edges| \geq 2f + 1$ **then**
15:       $buffer \leftarrow buffer \cup \{v\}$
16:    **end if**
    // Constructing $DAG_i[r]$ of round $r$
17: **while** $\neg$ $buffer.IsEmpty()$ **do**
18:    **for** $v \in buffer: v.round \leq r$ **do**
19:       **if** $\forall v' \in v.Edges: v' \in DAG[v.round - 1]$ **then**
20:          $DAG[v.round] \leftarrow DAG[v.round] \cup \{v\}$
21:          $buffer \leftarrow buffer \setminus \{v\}$
22:       **end if**
23:    **end for**
24:    **if** $|DAG[r]| \geq 2f + 1$ **then**
    // Ending of a wave $w$, call for ordering algorithm 5
25:       **if** $r \bmod 3 = 0$ **then**
26:          $wave\_end(r/3)$
27:       **end if**
28:       $r \leftarrow r + 1$
29:       $v \leftarrow create\_new\_vertex(r)$
    // Calling algorithm 2 for proposing new vertex
30:       $r\_bcast_i(v, r)$
31:    **end if**
32: **end while**

---

to its buffer. Note each party $P_i$ only allows to create one *vertex* $v$ per round $r$, and each *vertex* $v$ is delivered through the *Bufferpool* with validity and non-ambiguity.

**DAG Ordering:** Although the DAG structure allows replicas to order the blocks in local, achieving consensus on the total order still requires a *common coin* scheme in *Definition* 3 to provide randomness for choosing a replica $P_l$ and committing its causal history. This bypasses the FLP theory and achieves *liveness* in an asynchronous environment, while also resisting censorship attacks from *adversaries*. The Algorithm 5 demonstrate the consensus details for replica $P_i$.

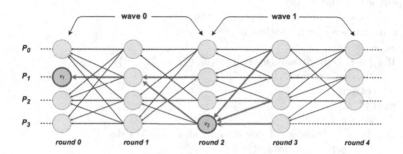

**Fig. 4.** The commit process.

As previous mentioned, each *validator* $P_i$ can interpret its $DAG_i[]$ by dividing it into *waves*, each of which includes three rounds. During each round of a *wave* $w$, *validators* propose their blocks along with QCs for proposal blocks from the previous round. In the third round of a *wave* $w$, *validators* $P_i$ invoke the *choose_leader*($w$) process to get a randomly selected leader *validator* $P_l$, and try to commit its proposal block in $DAG_i[1st\ round\ of\ w]$ along with the entire history. Note the *choose_leader*($w$) process is based on the *common coin* scheme, it outputs an identical value to each replica only if there has at least $2f + 1$ replicas invoked the process. Therefore the compute-bounded *adversary* is incapable to predict or manipulate the result with at least $f + 1$ honest parties involved.

To commit a leader *vertex* $v$ in *wave* $w$, as shown in Fig. 4, a *validator* $P_i$ checks if there are at least $2f + 1$ locally stored $path(u, v)$, which is from *vertices* $u \in DAG_i[3rd\ round\ of\ w]$ to leader *vertex* $v \in DAG_i[1st\ round\ of\ w]$. This condition aims to assure that the leader *vertex* $v$ will eventually committed by all *validators*. However, *validators* may not commit $v$ at the same time since they perhaps have different local views in their $DAGs$. To address this issue, when a *validator* $P_i$ decides to commit a *vertex* $v$, it pushes $v$ into the $order\_stack(v)$ to be ordered next. When $P_i$ try to order a *vertex* $v$ from the $order\_stack()$, it first checks if there is a path leading to an uncommitted leader *vertex* $v'$ in the previous *wave* $w'$. If such a path exists, it orders $v'$ before $v$

**Algorithm 5. Byzantine Atomic Broadcast based on DAG for replica $\mathcal{P}_i$**

---

**Local Variables:**
  $decidedWave \leftarrow 0$ $--$ the wave of $\mathcal{P}_i$ committed last leader
  $deliveredVertices \leftarrow \{\}$ $--$ the vertices $\mathcal{P}_i$ has committed
  $leadersStack \leftarrow$ initialize empty stack with isEmpty(), push(),
    and pop() functions

// Collecting transactions from algorithm 1
1: **Upon** $a\_bcast_i(b, r)$ **do**
2:     $baskets.get\_basket().enqueue(b)$

// Trying to commit a leader when wave $w$ completed
3: **Upon** $wave\_end(w)$ **do**
4:     $v \leftarrow choose\_leader(w)$
  // Random seed for algorithm 1
5:     $s_w \leftarrow w$
6:     **if** $(|\{v' \in DAG_i[round(w, 3)]||path(v', v)\}| < 2f + 1))$
7:       $\vee(v = \bot)$ **then**
8:         **return**
9:     **end if**
10:     $leadersStack.push(v)$
11:     **for** wave $w'$ from $w - 1$ down to $decidedWave + 1$ **do**
12:         $v' \leftarrow get\_wave\_leader(w')$
13:         **if** $v' \neq \bot \wedge strong\_path(v, v')$ **then**
14:             $leadersStack.push(v')$
15:             $v \leftarrow v'$
16:         **end if**
17:     **end for**
18:     $decidedWave \leftarrow w$
19:     $try\_to\_order(leadersStack)$

// Choosing a leader randomly through the *common coin*
20: **Procedure** $choose\_leader(w)$
21:     $p_j \leftarrow get\_wave\_leader_i(w)$
22:     **if** $\exists v \in DAG[round(w, 1)]$ s.t. $v.source = p_j$ **then**
23:         **return** $v$
24:     **end if**
25:     **return** $\bot$
  // Trying to order the vertices locally
26: **Procedure** $try\_to\_order(leadersStack)$
27:     **while** $\neg leadersStack.isEmpty()$ **do**
28:         $v \leftarrow leadersStack.pop()$
29:         **if** $\{(v' \in \bigcup_{r>0} DAG_i[r]) \wedge (path(v, v')) \wedge (v' \notin deliveredVertices)\}$ **then**
30:             $verticesToDeliver \leftarrow v'$
31:         **end if**
32:         **for** $\forall v' \in verticesToDeliver$ in a deterministic order **do**
33:             **output** $a\_deliver_i(v'.block, v'.round, v'.source)$
34:             $deliveredVertices \leftarrow deliveredVertices \cup \{v'\}$
35:         **end for**
36:     **end while**

---

and sets $v'$ to the $order\_stack(v')$. This process is repeated recursively until $P_i$ meets the previously committed leader $vertex$ $v''$. This approach enables that all $validators$ eventually commit the same leader $vertex$ $v$ with identical orders.

Note Hydra is building on the refined $Bufferpool$, so it naturally eradicates the redundant overlapping transactions in the DAG duplicates of each node, and further reduce the bandwidth cost of a single node. Which results in higher concurrency performance and also the time delay without comprise the security guarantees. We will give a more detailed analysis of Hydra in the Appendix.

## 7  Conclusion

We present Hydra, an efficient asynchronous DAG-based BFT consensus protocol. It achieves high throughput and optimal linear amortized communication complexity with low bandwidth overhead on a single party. Additionally, Hydra guarantees no-duplication and censorship-resistance.

There still remains some open problems in the asynchronous DAG-based BFT consensus domain, such as exploring the garbage collection without relying on a third party and the strictly order fairness of the $vertices$ in the DAG structure.

**Acknowledgments.** This work was supported by projects of Huawei Technologies Co., Ltd (E2V1061112) and Academy of Mathematics and Systems Science (E150131112).

## A  Appendix

We will prove the correctness, and then discuss the time and communication complexity of the Hydra.

### A.1  Correctness

**Theorem 1.** *The Hydra protocol satisfies Validity.*

*Proof.* Assume an honest replica $P_i$ propose a new $vertex$ $v$ by invoking $r\_broadcast(b, r)$ function. The AVID-M protocol in Algorithm 2 will encode it into $N$ segments and construct a proof for each segments, which are then dispersed to all parties. Since there are at most $f$ $Byzantine$ nodes, a QC can be formed upon responses from $2f + 1$ parties with at least $f + 1$ of them are honest. This QC can be broadcasted to all honest replicas, allowing them to access the $vertex$ $v$ and eventually add to their $DAG[r]$. Therefore, the Hydra protocol satisfies validity.

**Theorem 2.** *The Hydra protocol satisfies Agreement.*

*Proof.* Consider an honest party $P_i$ *commit* a *vertex* $v$ in round $r$ of *wave* $w$. Either $v$ is the leader returned by the *common coin*, or there exists a *path*$(v', v)$ from the leader *vertex* $v'$ in *wave* $w' \geq w$ to the *vertex* $v$. Due to the fact that we imply a *common coin* requires at least $2f + 1$ participants to randomly choose a leader *vertex* and return to all parties, all the honest replicas will *commit* the same leader and its total history. Therefore, *vertex* $v$ will be committed and ordered by all the honest parties at some point.

**Theorem 3.** *The Hydra protocol satisfies No-duplication.*

*Proof.* This follows directly from the *request pre-processing* scheme that reckons on the collusion-resistant property of the hash function to distribute transactions into different *baskets*, forcing replicas to propose distinct blocks from the locally allocated *baskets*, and ultimately eradicating the transaction duplication among replicas, whether intentional or accidental.

**Theorem 4.** *The Hydra protocol satisfies Causality.*

*Proof.* When a replica $P_i$ proposes a new *vertex* $v$ in round $r$, it must have collected at least $2f + 1$ QCs in the local $CL_{r-1}$ for the *vertices* in previous round $r - 1$ except in the genesis round. These QCs can act as links between the *vertices* of $DAG_i[r]$ and $DAG_i[r - 1]$, creating a *"happened-before"* relationship between *vertices* in $DAG_i[r \in N]$. Since a *wave* $w$ comprises of three continuous rounds, and each round $r$ can have at most $3f + 1$ *vertices*. The accumulation feature ensures that at the end of $w$, each party $P_i$ can form a general ACS which includes almost all blocks in the previous round with the *"happened-before"* relationship, thus build the causal order of the *vertices*.

**Theorem 5.** *The Hydra protocol satisfies* $\frac{1}{2}$ *chain quality.*

*Proof.* Let $P_i$ be a *validator* in round $r$. From **Theorem** 4, we know that any *validator* $P_i$ in round $r$ must have at least $2f + 1$ links to vertices in round $r - 1$. Since at most $f$ of the links may be to a dishonest vertex, $P_i$ must have at least $f + 1$ links to honest vertices in round $r - 1$.

Thus, in each round $r$, $P_i$ must include at least $f + 1$ honest vertices in its own $DAG_i$, this means that on average, at least $\frac{f+1}{2f+1} > \frac{1}{2}$ of the vertices in any given round will be honest.

**Theorem 6.** *The Hydra protocol satisfies Total order.*

*Proof.* Let $P_i$ be a replica trying to order the *vertices* in its local $DAG_i$. Upon receiving a leader *vertex* $v$ of *wave* $w$, $P_i$ examines if there exists a *path*$(v, v')$ leading to a *vertex* $v'$ in the previous *wave* $w' < w$. If so, $P_i$ pushes $v'$ before $v$ in the local *leader_stack*, repeating this until the last committed leader *vertex* $v''$ in the *wave* $w''$. To order the vertices, $P_i$ pops a *vertex* $v$ from the *leader_stack* and verifies if it has enough votes from the *vertices* in $DAG_i[r]$ of the current round. If it has, $P_i$ commits $v$ successfully and moves on to the next *vertex* until the conditions are not met.

By **Theorem** 3, each *wave* $w$ allows replicas to commit only one specific leader *vertex* $v$ and its causal history. By **Theorem** 2 and 4, all the honest nodes will eventually have the same causal history of the leader *vertex* $v$. Thus, all correct replicas will acquire identical DAGs and the total order of *vertices*.

**Theorem 6 (Liveness).** *The Hydra protocol will make progress even in the asynchronous network.*

*Proof.* We begin by noting that the system has $3f + 1$ nodes, with at most $f$ *Byzantine* nodes. Therefore, in each round, there will be at least $2f + 1$ *vertices* from honest parties. By **Theorem** 1, all honest parties will receive every *vertex* from other honest parties. Thus, each honest replica $P_i$ can proceed to the next round immediately after collecting $2f + 1$ *vertices* from the current round.

Furthermore, the Hydra protocol uses a *common coin* to choose a leader in the third round of each *wave* $w$. This ensures that all honest replicas will commit the same leader *vertex* $v$ and its entire history. By the commit rule, a replica $P_i$ can commit a *vertex* $v$ in the third round of a wave $w$ only if there are at least $2f + 1$ paths in $DAG_i[3rd\ round\ of\ w]$ leading to $v$. By **Theorem** 4 and 5, soon or later each replica $P_i$ will accumulate an identical ACS that contains almost all proposals, including at least $2f + 1$ associated *vertices* in $DAG_i[1st\ round\ of\ w]$, with at least $f + 1$ of them from honest parties.

Therefore, $P_i$ has at least $\frac{1}{3}$ probabilities of committing a correct leader *vertex* $v$. Hence, the Hydra protocol expects to commit a leader *vertex* $v$ in three waves in the worst case. It is important to note that when a leader block is committed, its entire history will also be committed. This allows the protocol to achieve high performance even in a fluctuating environment.

## A.2    Communication and Time Complexity

The communication overhead of Hydra mainly focuses on the *Bufferpool* and the *common coin* scheme.

The *Bufferpool*'s main costs are concentrated in the AVID-M protocol, which uses a *disperse-then-retrieve* pattern and erasure coding to reduce communication costs. The communication complexity of AVID-M's dispersal phase is $O(|v| + \lambda N)$ bits, and the retrieval phase has a communication complexity of $O(N|v| + \lambda N^2)$ bits.

The *common coin* scheme uses a threshold signature mechanism to output a verifiable value, with a communication complexity of $O(\lambda N^2)$ bits.

So the whole system communication complexity is $O(N|v| + \lambda N^2)$, and time complexity is $O(1)$ followed by **Theorem** 6, and we adapted batch processing technique hence achieve the optimal amortized communication cost $O(N)$ as known so far.

## References

1. Baird, L.: The swirlds hashgraph consensus algorithm: fair, fast, Byzantine fault tolerance (2016)

2. Bracha, G.: Asynchronous Byzantine agreement protocols. Inf. Comput. **75**(2), 130–143 (1987). https://doi.org/10.1016/0890-5401(87)90054-X. https://www.sciencedirect.com/science/article/pii/089054018790054X
3. Buterin, V., et al.: A next-generation smart contract and decentralized application platform. White Paper **3**(37), 2–1 (2014)
4. Castro, M.: Practical Byzantine fault tolerance. In: USENIX Symposium on Operating Systems Design and Implementation (1999)
5. Cristian, F., Aghili, H., Strong, H.R., Dolev, D.: Atomic broadcast: from simple message diffusion to Byzantine agreement. In: Twenty-Fifth International Symposium on Fault-Tolerant Computing, 'Highlights from Twenty-Five Years', p. 431 (1995)
6. Danezis, G., Kogias, E.K., Sonnino, A., Spiegelman, A.: Narwhal and tusk: a DAG-based mempool and efficient BFT consensus. https://arxiv.org/abs/2105.11827
7. Duan, S., Reiter, M.K., Zhang, H.: Beat: asynchronous BFT made practical. In: Proceedings of the 2018 ACM SIGSAC Conference on Computer and Communications Security, pp. 2028–2041. ACM (2018). https://doi.org/10.1145/3243734.3243812
8. Gao, Y., Lu, Y., Lu, Z., Tang, Q., Xu, J., Zhang, Z.: Dumbo-NG: fast asynchronous BFT consensus with throughput-oblivious latency (2022). https://arxiv.org/abs/2209.00750
9. Gągol, A., Leśniak, D., Straszak, D., Świętek, M.: Aleph: efficient atomic broadcast in asynchronous networks with Byzantine nodes (2019). https://arxiv.org/abs/1908.05156
10. Guo, B., Lu, Z., Tang, Q., Xu, J., Zhang, Z.: Dumbo: faster asynchronous BFT protocols. In: Proceedings of the 2020 ACM SIGSAC Conference on Computer and Communications Security, pp. 803–818 (2020)
11. Guruswami, V., Wootters, M.: Repairing Reed-Solomon codes. IEEE Trans. Inf. Theory **63**, 5684–5698 (2015)
12. Keidar, I., Kokoris-Kogias, E., Naor, O., Spiegelman, A.: All you need is DAG. https://arxiv.org/abs/2102.08325
13. Kotla, R., Alvisi, L., Dahlin, M., Clement, A., Wong, E.L.: Zyzzyva: speculative Byzantine fault tolerance. ACM Trans. Comput. Syst. **27**, 7:1–7:39 (2007)
14. Lamport, L.: Time, clocks, and the ordering of events in a distributed system. Commun. ACM **21**, 558–565 (1978)
15. Li, C., Li, P., Zhou, D., Xu, W., Long, F., Yao, A.: Scaling Nakamoto consensus to thousands of transactions per second (2018). https://arxiv.org/abs/1805.03870
16. Libert, B., Joye, M., Yung, M.: Born and raised distributively: fully distributed non-interactive adaptively-secure threshold signatures with short shares. In: Proceedings of the 2014 ACM Symposium on Principles of Distributed Computing, pp. 303–312. ACM (2014). https://doi.org/10.1145/2611462.2611498. https://dl.acm.org/doi/10.1145/2611462.2611498
17. Miller, A., Xia, Y., Croman, K., Shi, E., Song, D.: The honey badger of BFT protocols. In: Proceedings of the 2016 ACM SIGSAC Conference on Computer and Communications Security, pp. 31–42. ACM (2016). https://doi.org/10.1145/2976749.2978399. https://dl.acm.org/doi/10.1145/2976749.2978399
18. Nakamoto, S.: Bitcoin: a peer-to-peer electronic cash system. White Paper (2008)
19. Sompolinsky, Y., Lewenberg, Y., Zohar, A.: Spectre: a fast and scalable cryptocurrency protocol. https://eprint.iacr.org/undefined/undefined
20. Sompolinsky, Y., Zohar, A.: Secure high-rate transaction processing in bitcoin. In: Financial Cryptography (2015)

21. Spiegelman, A., Giridharan, N., Sonnino, A., Kokoris-Kogias, L.: Bullshark: DAG BFT protocols made practical. https://arxiv.org/abs/2201.05677
22. Stathakopoulou, C., David, T., Pavlovic, M., Vukolić, M.: Mir-BFT: high-throughput robust BFT for decentralized networks. https://arxiv.org/abs/1906.05552
23. Yang, L., Park, S.J., Alizadeh, M., Kannan, S., Tse, D.: DispersedLedger: high-throughput Byzantine consensus on variable bandwidth networks. https://arxiv.org/abs/2110.04371
24. Yin, M., Malkhi, D., Reiter, M.K., Gueta, G.G., Abraham, I.: Hotstuff: BFT consensus in the lens of blockchain. arXiv, Distributed, Parallel, and Cluster Computing (2018)

# Redactable Blockchain
# in the Permissioned Setting

Chunying Peng[1,2], Haixia Xu[1,2(✉)], Huimei Liao[1,2], Jinling Tang[1,2],
and Tao Tang[1,2]

[1] State Key Laboratory of Information Security, Institute of Information
Engineering, CAS, Beijing 100093, China
{pengchunying,xuhaixia}@iie.ac.cn
[2] School of Cyber Security, University of Chinese Academy of Sciences,
Beijing 100049, China

**Abstract.** As a momentous attribute of blockchains, the immutability
ensures the integrity and credibility of historical data, but it is inevitably
abused to spread illegal content and does not meet certain requirements of relevant privacy protection laws and regulations such as the
General Data Protection Regulation (GDPR). In this paper, we focus
on the redactable blockchain, which can break the immutability in a
safe and controllable way without affecting the normal operation of the
blockchain. We propose a redactable blockchain based on the aggregation signature in the permissioned setting. This scheme supports democratic instant modification and accountability, in which every user has
the right to propose editing requests, and the credible balloting committee is responsible for reviewing and voting on the redaction. In addition,
in order to achieve accountability, we introduce the concept of a witness
chain to ensure that every revision can be traced.

**Keywords:** Redactable blockchain · Aggregation signature · Schnorr
signature · Democratic modification · Accountability

## 1 Introduction

Blockchain as the underlying core technology for achieving transparent,
autonomous, and tamper-proof decentralized ledgers, is widely used to build
credible data distribution platforms. It was pioneered by Satoshi Nakamoto [25]
in 2008. It reduces the reliance on intermediaries and introduces a new trust
structure into the trading systems, so that entities from all over the world can
trade directly. The immutability as a key attribute of blockchains, can ensure
that multiple stakeholders build trust without relying on trusted third parties.
It ensures the integrity, immutability, and traceability of historical transaction
data, making the audit process more effective and transparent. But there are two
sides to everything, in certain scenarios and even legal requirements, we require
it to break the immutability in a controlled manner.

© The Author(s), under exclusive license to Springer Nature Switzerland AG 2023
M. Yung et al. (Eds.): SciSec 2023, LNCS 14299, pp. 460–477, 2023.
https://doi.org/10.1007/978-3-031-45933-7_27

For example, current blockchain applications pay more attention to the storage security of data, while ignoring the security and legitimacy of block content, resulting in a lot of illegal and harmful information on the chain [28]. The immutability of the blockchain makes it an effective way to disseminate illicit content. What's more, the "General Data Protection Regulations (GDPR)" [30] issued by the European Union stipulated that users have the "right to be forgotten". However, the transparent and immutable nature of blockchain data contradicts this regulation and violates users' privacy rights. In addition, the essence of the blockchain is a decentralized database. It is a new technology that is constantly innovating and developing, so there may be erroneous data and vulnerabilities caused by subjective or objective reasons. There is an imperative need for safe and controllable editing technology to support the modification of wrong data and the update of content. Therefore, a redactable blockchain that is compatible with practical applications and legal obligations is beneficial to the sustainable and healthy development of blockchain technology.

**Our Contributions.** In this paper, we present a redactable blockchain based on the aggregation signature in the permissioned setting that simultaneously supports democratic instant modification and accountability. It does not rely on complex cryptographic primitives such as chameleon hashing and attribute-based encryption (ABE), which is conducive to the compatibility of the solution with the current blockchain technology. Specifically, we have two technical contributions.

- *Redactable blockchain protocol.* We put forward a redactable blockchain protocol for permissioned systems, which enables democratic instant modification and provides accountability for editing operations. Any user can make an editing request. The credible balloting committee is responsible for reviewing and voting. The editing request can only be executed after the blockchain policy has been approved. Here, the blockchain policy is to get the consent of three-quarters of the balloting committee members. Miners take charge of uploading and rewriting the transaction as well as maintaining the witness chain. In addition, any redaction in the chain can be publicly verified through the witness chain, which provides accountability for editing operations.
- *A new consensus-based voting mechanism.* We leverage a Schnorr-based multiple signature aggregation scheme as the basis of the consensus mechanism. This aggregation signature scheme supports signature compression and fast verification. The balloting committee's signature on the candidate block is regarded as a vote. If the total number of signatures exceeds three-quarters of the committee, it will be considered as adopted. Then the aggregation of all signatures (votes) will be written on the redactable block as the consensus witness. At the same time, all single signatures will be maintained on a separate witness chain, which provides accountability for editing operations.

**Related Work.** In recent years, scholars have made some achievements in the emerging field of redactable blockchain. There are mainly two categories of redaction. One is the cryptography-based redaction and the other is non-cryptography-based redaction approaches [36].

For *cryptography-based redaction*, there are three main types: chameleon-hash-based redaction, polynomial-based redaction and RSA-based redaction [36].

In 2017, Ateniese et al. [2] first proposed the concept of chameleon-hash-based redactable blockchain. The holder of the trapdoor key can effectively calculate the collision to revise the block content at the block level. In order to make the revision more fine-grained, Derler et al. [9] proposed a policy-based chameleon hash (PCH) in 2019, which combined access control and chameleon hashing, people with enough privileges satisfying the policy can redact the single transaction. But this scheme lacks accountability and verifiability, and there may be the risk of authorizing illegal users because of the existence of central authority in CP-ABE. To solve the accountability problem, Xu et al. [34] proposed a new design of PCH with blackbox accountability (PCHA). And Ma et al. [22] proposed the decentralized policy-based chameleon hash (DPCH), which solved the authorization problem. In addition, Huang et al. [17] proposed the redactable consortium blockchain (RCB), which realized accountability and effectively solved the corruption problem in the redaction process through multi-party key management. Similarly, Xu et al. [35] proposed the k-time modifiable and epoch-based redactable blockchain (KERB), achieving the accountability through deposit locking and CA authorization. Besides, some researchers have made some achievements in other aspects. For example, Ashritha et al. [1] proposed an enhanced chameleon hash function and applied it to the modifiable blockchain. Hou et al. [16] proposed a fine-grained controllable editable blockchain, in which harmful information can be forcibly deleted, Peng et al. [26] put forward a redactable blockchain with fine-grained autonomy and transaction rollback, Jia et al. [18] proposed an editable blockchain that supports supervision and self-management, Li et al. [19] introduced the novel concept of non-interactive chameleon hash (NITCH) and presented Wolverine, a scalable and transaction-consistent redactable permissionless blockchain, and so on.

Cheng et al. [7] first put forward the polynomial-based blockchain in 2019, in which the block was represented by a polynomial, and the block header as well as the transactions was defined by coordinates respectively. However, PRB is not compatible with the mainstream blockchain, and the structure of the original blockchain needs to be greatly changed when it is implemented. The RSA-based redaction was proposed by Grigoriev and Shplrain [15], which focused on the private blockchain. This scheme provides strong security based on cryptography, but it may also suffer from moderator circumvention attacks and reversion attacks.

For *non-cryptography-based redaction*, it can be mainly divided into: consensus-based redaction, data-appending-based redaction, framework and protocol layer redaction.

In 2019, Deuber et al. [10] proposed the first consensus-based redactable blockchain in the permissionless setting. It was based on the voting consensus and did not need to resort to heavy cryptographic primitives or additional trust assumptions. The block could be revised only if the modification request got adequate votes within the specified time. It is inefficient because the voting period is consecutive 1024 blocks. In the same year, Marsalek and Zefferer [23] introduced a new editable blockchain with double-chain structure, which contained two chains (a *Standard Chain* and a *Redaction Chain*). This solution solves the limitations of data editing. However, this scheme is also voted on the blockchain, which is inefficient. In 2021, Li et al. [20] proposed an instantly editable blockchain protocol in permissionless setting. Compared with the scheme of Deuber et al. [10], Li et al.'s scheme has better security, faster revision speed and provides public verifiability. Our work is also based on the voting consensus, compared with the above three schemes, our voting doesn't depend on the block generation time and the credible balloting committee is small in scale, so the redaction is secure and can be as fast as the underlying blockchain. In the meantime, we introduce the concept of a witness chain, which makes the accountability and audit of the revision process possible. Through the *queryBlock* interface, users can audit the revision process without storing the witness chain, which reduces the storage requirements. We now give a comparison of our work with the most relevant prior schemes, as shown in Table 1.

**Table 1.** A comparison of redactable methods between relevant schemes and ours

| Scheme | Type | Main idea | Authorization | Consensus approach | Accountability |
|--------|------|-----------|---------------|--------------------|----------------|
| Deuber et al. [10] | permissionless | two hash links | miners | on-chain vote | ✓ |
| Marsalek and Zefferer [25] | \ | standard chain and redaction chain | miners | on-chain vote | ✗ |
| Li et al. [22] | permissionless | one chain | committee | off-chain vote | ✗ |
| Ours | permissioned | redaction chain and witness chain | committee | off-chain vote | ✓ |

The data-appending-based redaction was pioneered by Peddu et al. [27] in 2017, called $\mu$chain. In this work, they introduced the concept of mutations to replace the data records, which are controlled by policy, enforced by consensus, and can be validated as routine transactions. However, the illicit data is still on the blockchain, and transaction sets waste a lot of storage space, which affects the scalability of the blockchain.

For the framework and protocol layer redaction, Thyagarajan et al. [32] proposed Reparo in 2021, which was a publicly verifiable repair layer on the blockchain. It stored the content without changing the block structure by introducing an external data structure, and all versions of records were stored in the database. In addition, Dousti et al. [11] proposed a definitional framework with an efficient construct for moderated redactable blockchains, which was based on the signature scheme and had been proved to be correct and secure.

Besides, there are also some non-mainstream redacting methods, such as hard fork [33] and soft fork [21], which will not be described in detail here.

## 2    Preliminaries

In this section, we first describe the notations and notions that will be used in the later work, then introduce the cryptographic building blocks and basic knowledge of blockchain.

### 2.1    Notation

In this paper, we use $\lambda$ to denote the main cryptographic security parameter and $\mathbb{Z}_q$ to denote the set of integers modulo an integer $q \geq 1$. For a nonnegative integer $k$, $[k]$ denotes $\{1, \cdots, k\}$.

### 2.2    Algebraic Group Model

The algebraic group model (AGM) was first proposed by Fuchsbauer et al. [12] in 2018. It is a model that lies in between the standard model and the generic group model (GGM) [24,31]. In AGM, the adversary is required to provide a representation of any group elements it outputs as the product of the elements it received. For instance, let $Y_1, ..., Y_k$ be group elements of a multiplicative group provided to the adversary as inputs or from the oracle. For any group element $X$ it outputs or queries to the oracle, it gives a representation of $X$: a vector $(\gamma_1, ..., \gamma_k)$ satisfying $X = \prod_{i=1}^{k} Y_i^{\gamma_i}$.

### 2.3    Aggregate Signature

The concept of aggregate signature (AS) was first put forward by Boneh et al. [4]. It is a digital signature that supports aggregation: An unspecified aggregator is allowed to compress multiple signatures into a short aggregation. The AS scheme, whose advantages are saving storage space and accelerating verification, is applicable to applications where many signatures need to be stored, transmitted or verified together, such as certificate chains and electronic voting. Next, we will briefly review the definition of the aggregate signature scheme.

**Definition 1** (Aggregate Signature [6]). *An aggregate signature (AS) scheme AS is a tuple of algorithms as follows:*

$KG(1^\lambda)$ : *the key generation algorithm that given a security parameter $\lambda$ and outputs a pair of public/private key $(pk, sk)$;*

$Sign(sk, m)$ : *an algorithm which, on input the secret key sk and a message m, outputs a signature $\sigma$;*

$Verify(pk, m, \sigma)$ : *the verification algorithm that, on input the public key pk, a message m and a signature $\sigma$, outputs accept or reject;*

$AggSig((pk_1, m_1, \sigma_1), ..., (pk_n, m_n, \sigma_n)) \rightarrow \sigma_{agg}$ : *the aggregation algorithm that, on input set of n triplets–public key, message and signature, outputs an aggregate signature $\sigma_{agg}$.*

$AggVerify((pk_1, m_1), ..., (pk_n, m_n), \sigma_{agg}) \rightarrow \{accept/reject\}$ : *the AggVerify algorithm takes as inputs public keys $pk_1, ..., pk_n$, messages $m_1, ..., m_n$ and an aggregate signature $\sigma_{agg}$, outputs accept or reject.*

**Security.** In literature [4], Boneh et al. formalized the security of aggregate signature schemes. They brought the existential unforgeability under chosen-message attacks (EUF-CMA) of normal signature schemes [14] into the aggregate chosen-key security model, and proposed CK-AEUF-CMA [6]. In this security model, the adversary $\mathcal{A}$ is given a target public key, and his goal is to give a forgery of an aggregate signature. The adversary has the ability to select all public keys except the challenge public key. He also has access to the signing oracle on the challenge key. The CK-AEUF-CMA game consists of the following three stages.

**Setup.** The forger $\mathcal{A}$ is given a randomly generated public key $pk^*$.

**Queries.** $\mathcal{A}$ adaptively requests signatures with $pk^*$ on messages of its choice.

**Response.** The forger $\mathcal{A}$ outputs a set of public keys $(pk_1, ..., pk^*, ...pk_n)$, messages $(m_1, ..., m^*, ...m_n)$, and an aggregate signature $\sigma$, where $n$ is a game parameter not exceeding $N$.

We say the forger $\mathcal{A}$ wins the game if the aggregate signature $\sigma$ is a valid aggregate signature on messages $(m_1, ..., m^*, ...m_n)$ under keys $(pk_1, ..., pk^*, ...pk_n)$. And $\mathcal{A}$ has not queried $m^*$ to the signing oracle. His advantage is defined to be the probability of success in this game.

**Theorem 1.** *A forger $\mathcal{A}$ $(t, q_{H_1}, ...q_{H_k}, q_S, \epsilon)$-breaks the CK-AEUF-CMA security of an aggregate signature scheme AS in the ROM if: $\mathcal{A}$ runs in time at most $t$; $\mathcal{A}$ makes at most $q_{H_1}, ..., q_{H_k}$ queries respectively to the random oracles $H_1, ...H_k$ modeling the hash functions used in AS and at most $q_S$ queries to the signing oracle; his advantage is at least $\epsilon$. An aggregate signature scheme is $(t, q_{H_1}, ...q_{H_k}, q_S, \epsilon)$-secure against existential forgery in the ROM if no forger $(t, q_{H_1}, ...q_{H_k}, q_S, \epsilon)$-breaks it.*

### 2.4 Schnorr Signature

The schnorr signature was first introduced by C. P. Schnorr in [29], here we briefly review this signature as follows.

Let $G$ be a cyclic group of prime order $q$, g is the generator of G. Pick a random value $x \in \mathbb{Z}_q$ and compute $y = g^x$. Then the public key is $y$, the private key is $x$. $\mathcal{H} : \{0, 1\}^* \to \mathbb{Z}_q$ is a cryptographic hash function.

- **Signature generation:** To sign message $m$ with the private key $x$ perform the following steps:
  1. Pick a random number $r \in \mathbb{Z}_q$, compute $R = g^r$.
  2. Compute $c = \mathcal{H}(R, m)$.
  3. Compute $S = r + cx$ and output the signature $(R, S)$.
- **Signature verification:** To verify the signature $(R, S)$ for message $m$ with public key $y$, compute $v = g^S$ and check that $v = Ry^c$. A signature $(R, S)$ is accepted if it withstands verification.

## 2.5   Blockchain Basics

We recall the notation in the work of Gray et al. [13] to describe a blockchain.

Let $H : \{0,1\}^* \rightarrow \{0,1\}^k$ and $G : \{0,1\}^* \rightarrow \{0,1\}^k$ be the collision-resistant hash functions. A block is a triple of the form $B_i := <s_i, x_i, ctr_i>$, $s_i \in \{0,1\}^k$, denotes the state of the previous block; $x_i \in \{0,1\}^*$, represents the block data and $ctr_i \in \mathbb{N}$, is the difficulty level of the block. The block $B_i$ is valid iff

$$\text{validBlock}_q^T(B_i) := (H(ctr_i, G(s_i, x_i)) < T) \wedge (ctr_i \leq q)$$

Here, the parameter $T \in \mathbb{N}$ is the difficulty level of the block, and $q \in \mathbb{N}$ represents the maximum allowed number of hash queries.

The blockchain, or simply a chain is a sequence of blocks, that we call $C$. We call the rightmost block the head of the chain, denoted by $\text{Head}(C)$. A chain $C$ with $\text{Head}(C) := <s, x, ctr>$ can be extended to a new longer chain $C' := C \| B'$ by attaching a (valid) block $B' := <s', x', ctr'>$ that satisfies $s' = H(ctr, G(s, x))$; the head of the new chain $C'$ is $\text{Head}(C') = B'$.

# 3   Schnorr-Based Multiple Signature Aggregation Scheme

In this section, we put forward a Schnorr-based multiple signature aggregation scheme based on the redactable blockchain application scenario, which is a variant of a half-aggregate scheme for Schnorr signatures proposed by Chalkias et al. [5] (named *ASchnorr*).

## 3.1   Our Schnorr-Based Multiple Signature Aggregation Scheme

Our modified Schnorr-based multiple signature aggregation scheme, denoted by $MSAgg$ in all the following. The scheme is parameterized by group parameters $(\mathbb{G}, g, q)$. $\mathbb{G}$ is a group generated by $g$, whose order is prime $q$, $q$ is a $n$-bit prime. It makes use of cryptographic hash functions $\mathcal{H}_1, \mathcal{H}_2 : \{0,1\}^* \rightarrow \mathbb{Z}_q$. We use $\text{H}_2$ to denote the range of $\mathcal{H}_2$. The specific algorithm is shown as follows.

- **Parameters generation:** The parameters generation algorithm $\text{PG}(n)$ sets up a group $\mathbb{G}$ of order $q$ with generator $g$, and output $par \leftarrow (\mathbb{G}, g, q)$.
- **Key generation:** The key generation algorithm $KG(par)$ chooses a random value $x \in \mathbb{Z}_q$ as the secret key $sk$ and compute the public key $pk : Y = g^x$. Output $(pk, sk)$.
- **Signing:** To sign a message $m$, the signer $i$ with private key $x_i$ chooses a random number $r_i \in \mathbb{Z}_q$ and computes $R_i = g^{r_i}$. Let $c_i = \mathcal{H}_1(R_i, Y_i, m)$ and $S_i = r_i + c_i x_i$. Then the signature is $(R_i, S_i)$.
- **Aggregation:** At first, the aggregator verifies every signature $(R_i, S_i)$. It means that check $g^{S_i} = R_i Y_i^{c_i}$. After all the signatures have been authenticated, let $L = \{pk_1, ..., pk_n\}$ and $a_i = \mathcal{H}_2(L, pk_i)$, computes $S = \sum_{i=1}^{n} a_i S_i$. Finally, outputs the aggregate signature $\sigma \leftarrow (\{R_1, ..., R_n\}, S)$.
- **Verification:** The verification algorithm calculates the correlation coefficients and outputs 1 iff $g^S = \prod_{i=1}^{n} (R_i Y_i^{c_i})^{a_i}$.

## 3.2  Security

We refer to the CK-AEUF-CMA security proof of *ASchnorr* [5] in AGM+ROM proposed by Chen et al. [6], and give a similar CK-AEUF-CMA security proof to *MSAgg* in AGM+ROM, which is based on the EUF-CMA security of *Schnorr*. We model the hash function $\mathcal{H}_2$ as a random oracle, the forger has access to the signing oracle *Sign* and the random oracle $\mathcal{H}_2$.

**Theorem 2.** *Assume that there exists a forger that $(t, Q_{\mathcal{H}_2}, Q_S, \varepsilon)$-breaks the CK-AEUF-CMA security of MSAgg in the AGM+ROM with $\mathcal{H}_2$ modeled as a random oracle, then there exists an algorithm that $(t', Q_S, \varepsilon')$-breaks the EUF-CMA security of Schnorr with*

$$t' = \mathcal{O}(t)$$

$$and \quad \varepsilon' \geq \varepsilon - \frac{Q_{\mathcal{H}_2} + 1}{|\mathrm{H}_2|}$$

*Proof.* Here, we briefly generalize the main idea of the proof. We suppose that $\mathcal{F}$ is the forger that $(t, Q_{\mathcal{H}_2}, Q_S, \varepsilon)$-breaks the CK-AEUF-CMA security of $MSAgg$, then our goal is to construct an algorithm $\mathcal{A}$ to break the EUF-CMA security of *Schnorr*. Compared with [6], our scheme has two differences. One is the coefficient $c_i$, and the other is the coefficient $a_i$. For the coefficient $c_i$, our $\mathcal{H}_1$ is computed as $\mathcal{H}_1(R_i, Y_i, m)$ rather then not binding to the public key, which can avoid malleability attack (related content is shown in the "Malleability paragraph" on page 7 of the EdDSA paper [3]). Therefore, in the signing queries, the calculation of $\hat{c}_j$ should be changed accordingly, i.e., $\hat{c}_j = \mathcal{H}_1\left(\hat{R}_j, Y^*, m\right)$, where the target public key is $Y^*$. In addition, our $L = \{pk_1, ..., pk_n\}$ and $a_i = \mathcal{H}_2(L, pk_i)$, which are different from the scheme of Chen et al. [6], but the simulation of random oracle $\mathcal{H}_2$ is the same, all of which are simulated by $\mathcal{A}$. The rest of the proofs are similar to [6], for more details, please refer to [6].

# 4  Blockchain Redacting Protocol

In this section, we will present the details of our protocol, clarifying how to leverage cryptography technology to obtain a redactable blockchain system that supports democratic instant modification and accountability.

## 4.1  Blockchain Protocol

With reference to [10] and [13], we use $\Gamma$ to represent an immutable blockchain protocol, which is characterized by a set of global parameters and a set of common verification rules. The protocol $\Gamma$ has the following basic interfaces, which are described in the Table 2.

We build our redactable blockchain protocol $\Gamma'$ by modifying and extending the above standard immutable blockchain protocol $\Gamma$. The protocol $\Gamma'$ has all the basic blockchain functionalities exposed by $\Gamma$ through the above interfaces,

**Table 2.** The basic interfaces description of protocol $\Gamma$

| Interface | Description |
|---|---|
| $\{C', \bot\} \leftarrow \Gamma.\text{updateChain}$ | returns a longer and valid chain $C'$ in the network, otherwise returns $\bot$ |
| $\{0,1\} \leftarrow \Gamma.\text{validateChain}(C)$ | checks the validity of the chain, inputs a chain $C$ and returns 1 iff the chain is valid according to the public rules |
| $\{0,1\} \leftarrow \Gamma.\text{validateBlock}(B)$ | inputs a block B that needs to be verified and returns 1 iff the block is valid according to the public rules, otherwise returns 0 |
| $\Gamma.\text{broadcast}(x)$ | broadcasts the massage $x$ to the whole network |

in which algorithms *validateChain* and *validateBlock* are adjusted to adapt to the revision operation. Furthermore, the $\Gamma'$ protocol also provides the following interfaces.

$B_i^* \leftarrow \Gamma'.\text{proposeRedact}(C, i, x^*)$: on input chain $C$, the index $i$ of the block to edit and data $x^*$, returns a candidate block $B_i^*$.

$\{0,1\} \leftarrow \Gamma'.\text{validateCand}(B_i^*, C)$: on input a candidate block $B_i^*$ and the chain $C$, returns 1 iff the candidate block $B_i^*$ is valid.

$W_{j+1} \leftarrow \Gamma'.\text{proposeWitblock}(W_C, i, S)$: on input the witness chain $W_C$ of length $j$, the index $i$ of the block to edit in chain $C$ and votes set $S$, returns a witness block $W_{j+1}$.

$\{W_i, \bot\} \leftarrow \Gamma'.\text{queryBlock}(W_C, i)$: on input the witness chain $W_C$, the index $i$ of the witness block to query, returns the queried block $W_i$. If it does not exist, returns $\bot$.

## 4.2   Protocol Description

Our protocol extends the immutable blockchain of Garay et al. [13] to adapt to the editing operations in the following manner: we extend the block structure to have a capacity of the modification flag. In spirit of [8] and [13], we make the block structure described above to be of the form $B_i := <m_i, header_i, x_i>$, where $header_i = <s_i, G(x_i), ctr_i>$, as shown in Fig. 1.

In this structure, $s_i$ is the hash of the previous block header, denoted by $H(header_{i-1})$; $x_i$ denotes the block data, $ctr_i$ is the consensus witness of the block, i.e., the proof of work of block $B_i$, and $G(x_i)$ is the Merkle root of the block data. To maintain the link relationship between the edited block and its adjacent blocks, we have added a new modification flag $m_i$. When the block is revised, the modification flag is rewritten from 0 to 1, and the index of the corresponding witness block is also written in the flag section. In addition, the redaction policy of the chain is also considered, which determines the constraints and requirements for approving editing operations. To edit a block in the chain, our protocol performs the steps as follows.

1) The user first submits an editing request to the system. The request includes the index of the block he wants to edit and the candidate block to replace it, where the candidate block contains the "PreHash" field of the next block and the modification flag of the candidate block is 1.

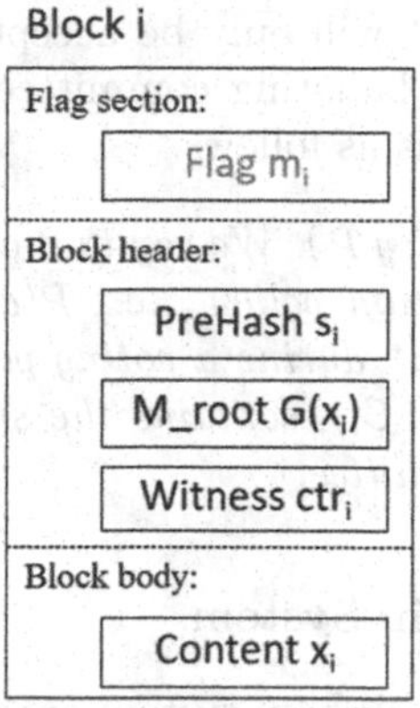

Block i

Flag section:

Flag $m_i$

Block header:

PreHash $s_i$

M_root $G(x_i)$

Witness ctr$_i$

Block body:

Content $x_i$

**Fig. 1.** Block structure of the redactable blockchain.

2) When the members of the balloting committee in the network receive an editing request, they first verify the validity of the candidate block and check whether the following conditions hold: a) it contains the correct information of the previous block; b) it contains the correct consensus; c) it will not invalidate the next block in the chain. If the candidate block is valid, the balloting committee can vote for it during the voting period by signing the hash of the request. The unforgeability of the signature ensures the authenticity of the votes.

3) After the user has collected sufficient votes that meet the redaction policy within the specified time, he aggregates all the signatures into an aggregate signature and writes the aggregate signature into the candidate block as the consensus witness. In addition, he creates a witness block including these signatures and the index of the block he wants to edit. In the end, he sends the revised candidate block and the corresponding witness block to the miners.

4) The miner verifies the correctness of the candidate block and the corresponding witness block. If the verification passes, he adds the index of the witness block to the candidate block. Finally, the edit operation is performed by replacing the original block with the candidate block and the corresponding witness block is attached to the witness chain.

To verify an edited chain, miners verify each block, just as in an immutable protocol. If the modification flag of a block is 1, then miners need to ensure that the edited block has collected ample votes and been approved according to the redaction policy of the chain.

**Redaction Policy.** We introduce the concept of redaction policy $\mathcal{P}$, which determines whether an edit to the blockchain $\mathcal{C}$ should be approved or not. The redaction policy $\mathcal{P}$ is a function that takes as input the blockchain $\mathcal{C}$, a witness chain $\mathcal{C}_W$ which contains the set of signatures on the candidate block $\mathcal{S} := (\sigma_1, ..., \sigma_s)$ and a candidate block $B^*$, outputs accept or reject. In other words, the candidate block $B^*$ will only be accepted by $\mathcal{P}$ if it was effectively voted by the majority of the balloting committee members within the voting period. The formal definition is as follows.

**Definition 2** *(Redaction Policy $\mathcal{P}$). We say that a candidate block $B^*$ generated in round $r$ satisfies the redaction policy, i.e., $\mathcal{P}(chain, B^*, \mathcal{C}_W, r) = accept$, if the number of valid votes on $B^*$ during a voting period is more than a threshold value, where the witness chain $\mathcal{C}_W$ contains the signatures set $\mathcal{S} := (\sigma_1, ..., \sigma_s)$ on $B^*$ from the balloting committee.*

### 4.3   Redactable Blockchain System

The roles in our redactable blockchain system are composed of a balloting committee, users and miners. The balloting committee is an authoritative organization trusted by users and miners, which is mainly responsible for reviewing and voting for redaction. Users can send transactions and propose redactions. Miners take charge of reviewing, uploading and rewriting the transaction. Also, miners need to maintain the witness chain. A simple system model is shown in Fig. 2.

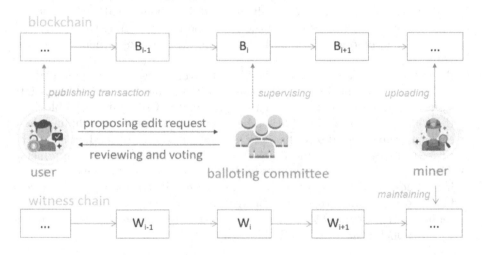

**Fig. 2.** System model of our redactable blockchain.

**Technique Overview.** We have recourse to the techniques in [9] and Schnorr-based multiple signature aggregation scheme. We also introduce the concept of a witness chain. The hash values of all signatures in a witness block constitute the leaf nodes of Merkle tree, and its root node is stored in this witness block header. The nature of the signature scheme guarantees the authenticity of the votes. The witness chain that provides accountability is shown in Fig. 3.

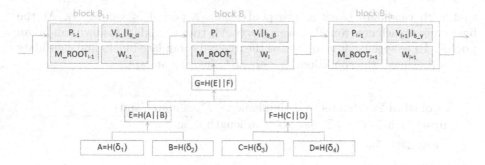

**Fig. 3.** Witness chain.

In Fig. 3, a block $B_i$ contains a Merkle root, which accumulates four signatures: $\sigma_1$, $\sigma_2$, $\sigma_3$ and $\sigma_4$. They are all immutable transactions. Note that the $P_i$ denotes the hash value of the previous block, $W_i$ is the consensus witness of the block, $V_i$ is the version of the witness block, and $I_{B_\beta}$ is the index of the corresponding redacted block. This index implements the correspondence between the redacted block and the witness block, providing support for accountability.

**Block Information Redaction.** In our system, every user has the right to propose revision requests. Specifically, the revision process of a block is shown below. Let the chain be $\mathcal{C} = (B_1, ..., B_n)$.

**1) Proposing a redaction.** Any user in the system can propose an edit to block $B_i$ for a particular data to be removed or replaced, he first calls the algorithm *proposeRedact* described in **Algorithm** 1 to generate the candidate block $B_i^*$. The *proposeRedact* algorithm takes as input chain $C$, the index $i$ of the block to edit and new data $x_i^*$, returns $B_i^* := <m_i', s_{i+1}, header_i, x_i^*>$, where $header_i = <s_i, G(x_i^*), ctr_i>$. Then he submits the candidate block to the balloting committee for review, which requires payment of a transaction fee.

---

**Algorithm** 1: *proposeRedact* (implements$\Gamma'$.proposeRedact)

**input** : Chain $\mathcal{C} = (B_1, \cdots, B_n)$ of length n, an index $i \in [n]$, and the new data $x_i^*$.

**output**: A candidate block $B_i^*$.

1: Parse $B_i := \langle m_i, header_i, x_i \rangle$, $header_i = <s_i, G(x_i), ctr_i>$;

2: Build the candidate block $B_i^* := \langle m_i', s_{i+1}, header_i, x_i^* \rangle$, $header_i = <s_i, G(x_i^*), ctr_i>$;

3: **return** $B_i^*$;

---

**2) Updating the candidate pool.** Upon receiving the candidate block $B_i^*$ from the network, every member of the balloting committee $\mathcal{M}_i$ first calls the algorithm *validateCand* described in **Algorithm** 2 to validate whether $B_i^*$ is a valid candidate block, and stores it in his own candidate pool $\mathcal{P}$ if it is. Notice that each candidate block in the pool $\mathcal{P}$ has a period of validity $t_l$. At the beginning of each new round $r$, every party $\mathcal{M}_i$ tries to update his own candidate pool $\mathcal{P}$. To be specific, it is to discard the expired candidate blocks and candidate blocks that meet the condition $\mathcal{P}(chain, B^*, \mathcal{C}_W, r) = accept$.

---

**Algorithm 2:** validateCand (implements $\Gamma'$.validateCand)

**input** : Chain $\mathcal{C} = (B_1, \cdots, B_n)$ of length $n$, and a

   candidate block $B_j^*$ for an edit.

**output**: $\{0, 1\}$.

1: Parse $B_j^* := \langle m_j', s', header_j, x_j^* \rangle$ where $header_j = <s_j, G(x_j^*), ctr_j>$.

2: Validate data $x_j^*$, if invalid return 0;

3: Parse $B_{j-1} := \langle m_{j-1}, s_{j-1}, G(x_{j-1}), ctr_{j-1}, x_{j-1} \rangle$

4: Parse $B_{j+1} := \langle m_{j+1}, s_{j+1}, G(x_{j+1}), ctr_{j+1}, x_{j+1} \rangle$

5: if $m_j' = 1 \wedge s' = s_{j+1} \wedge s_j = H\left(s_{j-1}, G(x_{j-1}), ctr_{j-1}\right)$

   then **return** 1

---

**3) Voting for candidate block.** After updating the candidate pool, for each candidate block $B_i^*$ in $\mathcal{P}$, $\mathcal{M}_i$ computes the hash value of $B_i^*$, and broadcasts the signature $\sigma_i$ on $H(B_i^*)$ as his vote.

**4) Collecting votes.** The user who has initiated the revision request collects the signatures (votes) from the balloting committee. If he has collected adequate valid votes that meet the policy within the voting period, then he computes the aggregate signature of these signatures and writes the final aggregate signature into his candidate block $B_i^*$ as the consensus witness.

**5) Creating a new witness block.** Upon completion of the votes collection work, the user calls the algorithm *proposeWitblock* described in **Algorithm 3** to generate a witness block $W_t^* := <s_t, I_{B_i}, x_t, w_t>$, where $s_t$ is the hash of the previous block, $I_{B_i}$ denotes the index of the corresponding redacted block, $x_t$ is the content of the witness block (i.e., the signatures on $H(B_i^*)$), and $w_t$ is the consensus witness of the block. Finally, he sends the revised candidate block and the corresponding witness block to the miners.

---

**Algorithm 3:** proposeWitblock (implements $\Gamma'$.proposeWitblock)

**input** : Chain $\mathcal{C}_W = (W_1, \cdots, W_{t-1})$ of length t-1, an index $i \in [n]$ of

the redactable block in chain $\mathcal{C}$ (length is n), and the votes set

$\mathcal{S} := (\sigma_1, ..., \sigma_s)$.

**output**: A witness block $W_t^*$.

1: Build the candidate block $W_t^* := <s_t, I_{B_i}, x_t, w_t>$, where $x_t$ is $\mathcal{S}$.

3: **return** $W_t^*$;

---

**6) Redacting the chain.** The miner verifies the correctness of the candidate block and the corresponding witness block. If the verification passes, he adds the index of the witness block to the candidate block. Finally, the edit operation is performed by replacing the original block with the candidate block and the corresponding witness block is attached to the witness chain.

**Block Redaction Accountability.** In this scheme, we introduce the witness chain to achieve accountability, in which every revision process can be reviewed. The concrete structure of the witness chain is shown in Fig. 3. Our protocol provides a *queryBlock* interface described in **Algorithm** 4, through which the specific witness block can be obtained by inputting the index $i$, without storing the whole witness chain, which balances the contradiction between storage cost and accountability. We can see all the votes (signatures) of the corresponding block in the consensus stage through the witness block.

---

**Algorithm** 4: queryBlock (implements $\Gamma'$.queryBlock)

**input** : the witness chain $W_C$, index i of the witness block to query.

**output**: $\{W_i, \bot\}$.

1: Validate index $i$, if invalid return $\bot$ ;

2: else **return** $W_i$;

---

### 4.4 Analysis

In this section, we will analyse our redactable blockchain protocol $\Gamma'$ from the perspectives of security and efficiency. We use $\Gamma$ to represent an immutable blockchain protocol, as described in Sect. 4.1, that satisfies the properties of *chain growth*, *chain quality* and *common prefix* [13]. The proof line behind our security analysis is that if $\Gamma$ satisfies the above properties, then $\Gamma'$ will meet the same properties (where the common prefix property is a variant). And we will discuss these three aspects respectively.

**Chain Growth.** Our redactable blockchain $\Gamma'$ will automatically inherit the property of chain growth form $\Gamma$, because the redaction does not allow deleting blocks or affecting the growth of the chain. Next, we will give a formal definition.

**Definition 3** (Chain Growth Property [13]). *The chain growth property $Q_{cg}$ with parameters $\delta \in \mathbb{R}$ and $t \in \mathbb{N}$ states that for any honest party $P$ that has a chain $C$ at two slots $sl_1$ (corresponding to chain $C_1$), $sl_2$ (corresponding to chain $C_2$) with $sl_1$ at least $t$ slots behind of $sl_2$, then it holds that $C_2$ is at least $\delta \cdot t$ blocks longer than $C_1$, where $\delta$ is the speed coefficient.*

**Theorem 3.** *If $\Gamma$ satisfies $(\delta, t)$-chain growth, then $\Gamma'$ satisfies $(\delta, t)$-chain growth for any redaction policy $\mathcal{P}$.*

*Proof.* Our redactable blockchain $\Gamma'$ is an extension of $\Gamma$, for any redaction policy $\mathcal{P}$, we can't delete the existing blocks from the chain, but replace them with candidate blocks, which means that the editing operation will not shorten the chain. Therefore, we conclude that when $\Gamma$ satisfies the property of $(\delta, t)$-chain growth, $\Gamma'$ also satisfies it.

**Definition 4** (Chain Quality Property [13]). *The chain quality property $Q_{cq}$ with parameters $\sigma \in \mathbb{R}$ and $\ell \in \mathbb{N}$ states that for any honest party $P$ with a chain $\mathcal{C}$, the ratio of adversarial blocks in any $\ell$ consecutive blocks of the chain $\mathcal{C}$ is at most $\sigma$, where $0 < \sigma \leq 1$ is the chain quality coefficient.*

**Theorem 4.** *Let $\mathcal{H}$ be a collision-resistant hash function. If $\Gamma$ satisfies $(\sigma, \ell)$-chain quality, then $\Gamma'$ satisfies $(\sigma, \ell)$-chain quality.*

*Proof.* Note that if there is no redaction in the chain, $\Gamma'$ behaves exactly like the underlying immutable protocol $\Gamma$. In other words, the only difference between $\Gamma$ and $\Gamma'$ is that the blocks in $\Gamma'$ can be modified. If an adversary $\mathcal{A}$ can replace honest blocks with malicious blocks (e.g., containing illicit content), he can increase the proportion of malicious blocks and finally achieve the purpose of destroying the chain quality property of the chain. We prove in the following that $\mathcal{A}$ has a negligible probability to violate the chain quality property.

We assume that $\mathcal{A}$ propose a malicious candidate block $B_i^*$ to redact an honest block $B_i \in \mathcal{C}$. As the blockchain policy is to get the consent of three-quarters of the credible balloting committee members, the adversary can only attempt to create a candidate block $B_i^{*\prime} \neq B_i^*$ that appears "honest" (e.g., without illicit content) satisfying $H\left(B_i^{*\prime}\right) = H\left(B_i^*\right)$, trying to cheat the committee during the voting phase; the committee could endorse the candidate block $B_i^*$ during the voting phase, and the adversary would instead replace the candidate block with the malicious block $B_i^{*\prime}$. However, the probability that the adversary finds a candidate block $B_i^*$ where $H\left(B_i^{*\prime}\right) = H\left(B_i^*\right)$ is negligible, because the hash function $\mathcal{H}$ is collision-resistance. At the same time, our Schnorr-based multiple signature aggregation scheme $MSAgg$ has CK-AEUF-CMA security, so the votes cannot be forged. Therefore, we conclude $\Gamma'$ satisfies $(\sigma, \ell)$-chain quality.

**Definition 5** (Common Prefix Property [13]). *The common prefix property $Q_{cp}$ with common prefix parameter $\mu \in \mathbb{N}$ states that for two chains $C_1, C_2$, possessed by two honest parties at rounds $\mathrm{r}_1 \leq \mathrm{r}_2$, it holds that $C_1^{\lceil \mu} \preceq C_2$.*

However, due to the redaction, our protocol $\Gamma'$ will not directly meet the above definition, so we introduce a variant of common prefix property, which called redactable common prefix.

**Definition 6 (Redactable Common prefix).** *The redactable common prefix property $Q_{Rcp}$ with common prefix parameter $\mu \in \mathbb{N}$ states that for two chains $C_1$ (with length $L_1$), $C_2$ (with length $L_2$) possessed by two honest parties at rounds $\mathrm{r}_1 \leq \mathrm{r}_2$, it meets one of the following conditions:*

*a) $C_1^{\lceil \mu} \preceq C_2$, or*

b) for $i \in [L_1 - \mu]$, if there exists $B_i^* \in C_2^{\lceil (L_2 - L_1) + \mu}$ but $B_i^* \notin C_1^{\lceil \mu}$, it must be the case that $\mathcal{P}(chain, B_i^*, \mathcal{C}_W, r) = accept$,

where $C_1^{\lceil \mu}$ represents the chain in which the last $\mu$ blocks are removed from $\mathcal{C}_1, \mathcal{P}$ denotes the chain policy and $r$ denotes the voting round.

**Theorem 5.** *Let $\mathcal{H}$ be a collision-resistant hash function. If $\Gamma$ satisfies $\mu$-common prefix, then $\Gamma'$ satisfies $\mu$-redactable common prefix for chain policy $\mathcal{P}$.*

*Proof.* We know that if no redaction occurred in the chain, $\Gamma'$ behaves exactly like the underlying immutable protocol $\Gamma$. And in the case of a redaction, we assume that $\mathcal{A}$ proposes a candidate block $B_i^*$ to redact a block $B_i \in C_2$, which is later edited by an honest miner at round $r_2$. Due to the hash function $\mathcal{H}$ is collision-resistance, the probability that adversary $\mathcal{A}$ finds a candidate block $B_i^{*\prime} \neq B_i^*$ where $H\left(B_i^{*\prime}\right) = H\left(B_i^*\right)$ is negligible. In addition, the unforgeability of the signature scheme $MSAgg$ ensures the authenticity of the votes. Therefore, for the honest miner and adopted redaction $B_i^*$, it must satisfy the chain policy $\mathcal{P}$. This concludes the proof. ∎

**Efficiency.** From the perspective of efficiency, our solution collects votes from the small-scale credible balloting committee off the blockchain within a specified time. Compared with voting on the blockchain, such as Deuber et al.'s scheme [10] (their voting period requires collecting votes in consecutive 1024 blocks) and Marsalek and Zefferer's scheme [23], our voting is independent of the block generation time, and we only need to collect three-quarters of the votes from a small balloting committee, which is efficient. As a result, our protocol provides faster speed and is more in line with practical application requirements.

## 5    Conclusion

In this paper, we present a blockchain redactable protocol that simultaneously supports democratic instant modification and accountability. We introduce a block structure having a capacity of the modification flag to facilitate the verification and accountability of the revision by system members. Our scheme enables the blockchain to meet the dual needs of chain audit and data modification. The introduction of a witness chain not only reduces the storage overhead of redaction in the main chain, but also ensures the complete revision traceability and provides accountability. The scheme can achieve instant redaction by collecting votes from the small-scale credible balloting committee off the blockchain within a specified time. In comparison with the existing related work, our protocol has faster speed and provides public accountability while ensuring security.

## References

1. Ashritha, K., Sindhu, M., Lakshmy, K.: Redactable blockchain using enhanced chameleon hash function. In: 2019 5th International Conference on Advanced Computing & Communication Systems (ICACCS), pp. 323–328. IEEE (2019)

2. Ateniese, G., Magri, B., Venturi, D., Andrade, E.: Redactable blockchain-or-rewriting history in bitcoin and friends. In: 2017 IEEE European Symposium on Security and Privacy (EuroS&P), pp. 111–126. IEEE (2017)
3. Bernstein, D.J., Duif, N., Lange, T., Schwabe, P., Yang, B.Y.: High-speed high-security signatures. J. Cryptogr. Eng. **2**(2), 77–89 (2012)
4. Boneh, D., Gentry, C., Lynn, B., Shacham, H.: Aggregate and verifiably encrypted signatures from bilinear maps. In: Biham, E. (ed.) EUROCRYPT 2003. LNCS, vol. 2656, pp. 416–432. Springer, Heidelberg (2003). https://doi.org/10.1007/3-540-39200-9_26
5. Chalkias, K., Garillot, F., Kondi, Y., Nikolaenko, V.: Non-interactive half-aggregation of EdDSA and variants of Schnorr signatures. In: Paterson, K.G. (ed.) CT-RSA 2021. LNCS, vol. 12704, pp. 577–608. Springer, Cham (2021). https://doi.org/10.1007/978-3-030-75539-3_24
6. Chen, Y., Zhao, Y.: Half-aggregation of Schnorr signatures with tight reductions. In: Atluri, V., Di Pietro, R., Jensen, C.D., Meng, W. (eds.) ESORICS 2022. LNCS, vol. 13555, pp. 385–404. Springer, Cham (2022). https://doi.org/10.1007/978-3-031-17146-8_19
7. Cheng, L., Liu, J., Su, C., Liang, K., Xu, G., Wang, W.: Polynomial-based modifiable blockchain structure for removing fraud transactions. Futur. Gener. Comput. Syst. **99**, 154–163 (2019)
8. David, B., Gaži, P., Kiayias, A., Russell, A.: Ouroboros praos: an adaptively-secure, semi-synchronous proof-of-stake blockchain. In: Nielsen, J.B., Rijmen, V. (eds.) EUROCRYPT 2018. LNCS, vol. 10821, pp. 66–98. Springer, Cham (2018). https://doi.org/10.1007/978-3-319-78375-8_3
9. Derler, D., Samelin, K., Slamanig, D., Striecks, C.: Fine-grained and controlled rewriting in blockchains: chameleon-hashing gone attribute-based. IACR Cryptology ePrint Archive, p. 406 (2019)
10. Deuber, D., Magri, B., Thyagarajan, S.A.K.: Redactable blockchain in the permissionless setting. In: 2019 IEEE Symposium on Security and Privacy, pp. 124–138. IEEE (2019)
11. Dousti, M.S., Küpçü, A.: Moderated redactable blockchains: a definitional framework with an efficient construct. In: Garcia-Alfaro, J., Navarro-Arribas, G., Herrera-Joancomarti, J. (eds.) DPM/CBT -2020. LNCS, vol. 12484, pp. 355–373. Springer, Cham (2020). https://doi.org/10.1007/978-3-030-66172-4_23
12. Fuchsbauer, G., Kiltz, E., Loss, J.: The algebraic group model and its applications. In: Shacham, H., Boldyreva, A. (eds.) CRYPTO 2018. LNCS, vol. 10992, pp. 33–62. Springer, Cham (2018). https://doi.org/10.1007/978-3-319-96881-0_2
13. Garay, J., Kiayias, A., Leonardos, N.: The bitcoin backbone protocol: analysis and applications. In: Oswald, E., Fischlin, M. (eds.) EUROCRYPT 2015. LNCS, vol. 9057, pp. 281–310. Springer, Heidelberg (2015). https://doi.org/10.1007/978-3-662-46803-6_10
14. Goldwasser, S., Micali, S., Rivest, R.L.: A digital signature scheme secure against adaptive chosen-message attacks. SIAM J. Comput. **17**(2), 281–308 (1988)
15. Grigoriev, D., Shpilrain, V.: RSA and redactable blockchains. Int. J. Comput. Math Comput. Syst. Theory **6**(1), 1–6 (2021)
16. Hou, H., Hao, S., Yuan, J., Xu, S., Zhao, Y.: Fine-grained and controllably redactable blockchain with harmful data forced removal. Secur. Commun. Netw. **2021** (2021)
17. Huang, K., et al.: Building redactable consortium blockchain for industrial internet-of-things. IEEE Trans. Industr. Inf. **15**(6), 3670–3679 (2019)

18. Jia, Y., Sun, S.F., Zhang, Y., Liu, Z., Gu, D.: Redactable blockchain supporting supervision and self-management. In: ACM Asia Conference on Computer and Communications Security, ASIA CCS 2021, pp. 844–858 (2021)
19. Li, J., Ma, H., Wang, J., Song, Z., Xu, W., Zhang, R.: Wolverine: a scalable and transaction-consistent redactable permissionless blockchain. IEEE Trans. Inf. Forensics Secur. **18**, 1653–1666 (2023)
20. Li, X., Xu, J., Yin, L., Lu, Y., Tang, Q., Zhang, Z.: Escaping from consensus: instantly redactable blockchain protocols in permissionless setting. IACR Cryptology ePrint Archive, p. 223 (2021)
21. Lin, I.C., Liao, T.C.: A survey of blockchain security issues and challenges. Int. J. Netw. Secur. **19**(5), 653–659 (2017)
22. Ma, J., Xu, S., Ning, J., Huang, X., Deng, R.H.: Redactable blockchain in decentralized setting. IEEE Trans. Inf. Forensics Secur. **17**, 1227–1242 (2022)
23. Marsalek, A., Zefferer, T.: A correctable public blockchain. In: 2019 18th IEEE International Conference on Trust, Security and Privacy in Computing and Communications, pp. 554–561. IEEE (2019)
24. Maurer, U.: Abstract models of computation in cryptography. In: Smart, N.P. (ed.) Cryptography and Coding 2005. LNCS, vol. 3796, pp. 1–12. Springer, Heidelberg (2005). https://doi.org/10.1007/11586821_1
25. Nakamoto, S.: Bitcoin: a peer-to-peer electronic cash system. Decentralized Bus. Rev. 21260 (2008)
26. Peng, C., Xu, H.: Redactable blockchain with fine-grained autonomy and transaction rollback. In: Su, C., Sakurai, K., Liu, F. (eds.) SciSec 2022. LNCS, vol. 13580, pp. 68–84. Springer, Cham (2022). https://doi.org/10.1007/978-3-031-17551-0_5
27. Puddu, I., Dmitrienko, A., Capkun, S.: $\mu$chain: how to forget without hard forks. IACR Cryptology ePrint Archive, p. 106 (2017)
28. Schellekens, M.: Does regulation of illegal content need reconsideration in light of blockchains? Int. J. Law Inf. Technol. **27**(3), 292–305 (2019)
29. Schnorr, C.P.: Efficient signature generation by smart cards. J. Cryptol. **4**, 161–174 (1991)
30. Shabani, M., Borry, P.: Rules for processing genetic data for research purposes in view of the new EU general data protection regulation. Eur. J. Hum. Genet. **26**(2), 149–156 (2018)
31. Shoup, V.: Lower bounds for discrete logarithms and related problems. In: Fumy, W. (ed.) EUROCRYPT 1997. LNCS, vol. 1233, pp. 256–266. Springer, Heidelberg (1997). https://doi.org/10.1007/3-540-69053-0_18
32. Thyagarajan, S.A.K., Bhat, A., Magri, B., Tschudi, D., Kate, A.: Reparo: publicly verifiable layer to repair blockchains. In: Borisov, N., Diaz, C. (eds.) FC 2021. LNCS, vol. 12675, pp. 37–56. Springer, Heidelberg (2021). https://doi.org/10.1007/978-3-662-64331-0_2
33. Tsankov, P., Dan, A., Drachsler-Cohen, D., Gervais, A., Buenzli, F., Vechev, M.: Securify: practical security analysis of smart contracts. In: Proceedings of the 2018 ACM SIGSAC Conference on Computer and Communications Security, pp. 67–82 (2018)
34. Xu, S., Huang, X., Yuan, J., Li, Y., Deng, R.H.: Accountable and fine-grained controllable rewriting in blockchains. IEEE Trans. Inf. Forensics Secur. **18**, 101–116 (2022)
35. Xu, S., Ning, J., Ma, J., Huang, X., Deng, R.H.: K-time modifiable and epoch-based redactable blockchain. IEEE Trans. Inf. Forensics Secur. **16**, 4507–4520 (2021)
36. Ye, T., Luo, M., Yang, Y., Choo, K.K.R., He, D.: A survey on redactable blockchain: challenges and opportunities. IEEE Trans. Netw. Sci. Eng. (2023)

# Workshop Session

# A Multi-level Sorting Prediction Enhancement-Based Two-Dimensional Reversible Data Hiding Algorithm for JPEG Images

Bin Ma[1,2], Songkun Wang[1,2], Jian Xu[3(✉)], Chunpeng Wang[1,2], Jian Li[1,2], and Xiaolong Li[4]

[1] Shandong Provincial Key Laboratory of Computer Networks, Jinan 250000, China
[2] Qilu University of Technology (Shandong Academy of Sciences), Jinan 250000, China
[3] Shandong University of Finance and Economics, Jinan 250000, China
sddxmb@126.com
[4] Beijing Jiaotong University, Beijing 100044, China

**Abstract.** To improve the performance of reversible data hiding on JPEG images, a multi-level sorting prediction enhancement-based two-dimensional reversible data hiding algorithm for JPEG images is proposed in this paper. Firstly, the smoothness of each DCT block is evaluated by counting the number of non-zero AC coefficients in each DCT block. Then, the gradient of the object block is calculated with the DC coefficients of its surrounding blocks, and the lowest gradient direction blocks are employed to achieve the prediction of object block AC coefficients. The AC coefficient in the object block is accurately predicted by AC coefficients in the same position of its adjacent blocks. These chosen adjacent blocks are located in the lowest gradient direction. Then, the obtained DCT coefficient prediction errors are paired by a similar prediction-error pairing strategy to construct a two-dimensional prediction-error histogram. Finally, the secret data are imperceptibly embedded into the AC coefficients by using the two-dimensional histogram shifting algorithm. The performance multi-level sorting prediction enhancement-based two-dimensional reversible data hiding algorithm was validated on different classical images, and six representative

This work was partially supported by National Natural Science Foundation of China (62272255); National key research and development program of China (2021YFC3340602); Shandong Provincial Natural Science Foundation Innovation and Development Joint Fund (ZR202208310038); Ability Improvement Project of Science and Technology SMES in Shandong Province (2022TSGC2485); Project of Jinan Research Leader Studio (2020GXRC056); Project of Jinan Introducing Innovation Team (202228016); Project of Jinan City-School Integration Development (JNSX2021030); Youth Innovation Team of Colleges and Universities in Shandong Province (2022KJ124) The "Chunhui Plan" Cooperative Scientific Research Project of Ministry of Education (HZKY20220482) Shandong Provincial Natural Science Foundation (ZR2020MF054).

© The Author(s), under exclusive license to Springer Nature Switzerland AG 2023
M. Yung et al. (Eds.): SciSec 2023, LNCS 14299, pp. 481–495, 2023.
https://doi.org/10.1007/978-3-031-45933-7_28

images were elaborately chosen to demonstrate the superiority of the proposed scheme. Extensive experimental results show that, compared to other state-of-the-art reversible data hiding algorithms, the proposed algorithm can achieve higher marked image visual quality while maintaining low JPEG image file size increments.

**Keywords:** Reversible data hiding · Multi-level sorting · Prediction error · Two-dimensional embedding

# 1    Introduction

With the fast development of internet technology, more and more multi-media files need to be transmitted through the network. However, these transmitted multimedia files always face a series of security risks, such as tampering, theft, infringement, illegal use, and so on. In order to solve these problems, people often hide the secret data in some carrier images, ensuring the security of the information. However, in some special fields that require high carrier image integrity, such as military, medical, and digital forensics, the receiving party needs to extract secret messages and restore the original carrier without loss. Reversible date hiding, as an important branch of messages hiding, not only allows the receiving party to extract secret messages, but also restores the carrier image accurately and correctly. Given this feature, reversible data hiding technology has received widespread attention and become a research hotspot in the field of information security nowadays.

At present, reversible data hiding algorithms are mainly divided into two categories according to the embedding domain: the spatial domain and the transform domain. The spatial domain reversible data hiding algorithms are mainly divided into three kinds: lossless compression, difference expansion (DE), histogram shifting (HS). Fridrich et al. [1] proposed a reversible data hiding algorithm based on the Joint Bi-Level Image Experts Group (JBIG) lossless compression. Tian [2] first presented a reversible data hiding algorithm based on difference expansion by extending the difference between a pair of pixels for data embedding. Ni et al. [3] employed histograms for data embedding and proposed a reversible data hiding algorithm based on histogram shifting.

Moreover, Thodi and Rodriguez et al. [4] proposed a reversible data hiding algorithm based on prediction error expansion, which could achieve the purpose of embedding secret messages by expanding the prediction errors. In recent years, people have further improved the prediction error expansion algorithm by modifying the high-dimensional and multi-dimensional histograms. Ou et al. [5] proposed a two-dimensional reversible mapping algorithm is adapted to specially distributed prediction error pairs, which extends the pixel-valued sorting embedding to a two-dimensional form by integrating paired prediction error expansions and improves the data hiding performance significantly compared to other pixel-valued sorting algorithms. Since then, Ou et al. [6] chose to remove the rough pixels from the pairing and pair the smooth pixels so that the entropy of the two-dimensional prediction error histogram obtained is smaller. Compared with the previous paired prediction error algorithm, this method is a significant improvement. Recently, Hu et al. [7] proposed a new mapping method that can adaptively change the size of the mapping area based on the frequency of error pairs, improving the performance of data hiding.

In addition, with the fast growth of digital imaging and image compression technology, JPEG has become the most popular way to record images in recent years. Considering the file format characteristics of compressed images, some traditional reversible data hiding algorithms are difficult to apply directly to JPEG images. According to the coding principle of JPEG images, at present, reversible data hiding algorithms for JPEG images are mainly divided into three kinds: reversible data hiding algorithms based on modified quantization table [8–10], reversible data hiding algorithms based on modified Huffman table [11, 12], and reversible data hiding algorithms based on modified quantization DCT coefficients [13–22]. In recent years, significant progress has been made in other fields of data hiding [23–31].

Among them, algorithms based on modifying quantized DCT coefficients have received widespread attention. Huang et al. selected the AC coefficients of DCT with the value of "$\pm 1$" as embedding points, while keeping the "0" coefficient unchanged. The algorithm adaptively selects the data embedding image blocks based on the AC coefficients. Compared with existing JPEG reversible data hiding algorithms, this algorithm has better performance. Wedaj et al. proposed a new data embedding method that followed Huang's block ordering strategy and proposed a frequency selection method according to the data embedding capabilities of each quantized AC coefficient at different frequency positions. Subsequently, Hou et al. estimated the data embedding frequency by quantifying the AC coefficients at different frequency positions in DCT blocks, chosen the set of embedding frequencies using the optimized search method, and selected the to be embedded DCT blocks according to the analog embedding distortion values. Di et al. put forward a new measurement strategy to select the most suitable coefficient position for embedding by measuring the distortion of each coefficient. This strategy employed the "0" coefficient for data embedding, which improved the embedding performance and reduced the image distortion caused by secret messages embedding. He et al. proposed a JPEG image coding scheme based on improved block prediction and frequency selection, they accurately estimated the possible distortion caused by data embedding in each block according to the DCT coefficients distribution and prioritized the use of block embedding data with less distortion, which obtained a better performance in terms of both image quality and file size retention. Li et al. paired adjacent non-zero AC coefficients and designed a two-dimensional reversible mapping rule to shift the generated coefficient pairs to embed messages. This algorithm advances the visual quality and better solves the file increment problem. Chen et al. paired the AC coefficients located in different blocks according to the similarity of DCT blocks, and optimized the two-dimensional map to achieve a better experimental performance.

In recent years, Xiao et al. [32] have proposed a reversible data hiding algorithm for JPEG images based on multiple histogram modifications and rate distortion optimization. This algorithm not only achieves adaptive data embedding but also greatly improves embedding performance. Weng et al. [33] designed a general reversible data hiding framework for JPEG images with multiple two-dimensional histograms. This framework could adaptively select different two-dimensional histogram mappings for embedding. An improved discrete particle swarm optimization algorithm is used to speed up the optimization process, which improves the performance of data embedding. Moreover, Weng et al. [34] have further improved the algorithm, reducing embedding distortion by block smoothness estimator and band smoothness estimator.

The algorithms proposed in recent years have made full use of various sorting strategies to improve the performance of reversible data hiding. However, the correlation of the DCT coefficients is still not fully utilized and the generated marked image eventually have problems such as low visual quality and excessive file increment. If the visual quality of the marked image is relatively low or the file increment is high, it is easy for criminals to detect the image during network transmission, and it is difficult to guarantee the security of secret messages. The multi-level sorting algorithm proposed in this paper tries to make full use of the correlation among DCT coefficients to embed secret messages into the prediction errors and uses a two-dimensional embedding algorithm to greatly improve the performance of data hiding. To evaluate the performance of the proposed scheme, two aspects, i.e., visual quality and file increment are considered. The innovation points of this paper are following:

a) This paper proposes an error prediction algorithm based on multi-level sorting of DCT blocks and coefficients. The algorithm first comprehensively sorts the DCT blocks' complexity according to their last non-zero coefficient position in each DCT block. And the DC difference can further evaluate the close correlation of the current block with its neighbors in case that the non-zero coefficients of blocks are equal to each other.

b) The gradient of adjacent blocks of the object block are compared to found the lowest gradient direction. A DCT coefficient prediction model is constructed using coefficients in the same position of its adjacent DCT blocks, and the optimal coefficient gradient direction of the adjacent blocks is chosen to predict the value of the object coefficients, and thus improve the error prediction accuracy greatly.

c) An approximately prediction-error pairing strategy is proposed only to pair prediction errors whose values are "0" or "−1", and the two dimensional prediction-error plane with a steep prediction error distribution is achieved. An improved two-dimensional histogram shifting algorithm is applied to embed secret data into the chosen prediction errors. The combination of approximately prediction-error pairing strategy and two-dimensional histogram shifting algorithm minimize the image distortion caused by data embedding, enhancing both data embedding capacity and visual quality of the data-embedded image.

The structure of the rest of this paper is as follows. The second part introduces in detail the two-dimensional reversible data hiding algorithm of JPEG image based on multi-level sorting and the two-dimensional data embedding algorithm based on prediction error pairing method is also illustrated in this part. In the third part, the experimental results of the algorithm and the comparison results with other RDH methods based on JPEG images are shown comprehensively. And in the last part of the paper, the algorithm proposed in this paper is summarized.

## 2   Propose Scheme

### 2.1   Inter-block DCT Coefficients Error Prediction

The DCT inter-block coefficient prediction error algorithm based on multi-level sorting proposed in this paper will analyze the distribution characteristics of DCT coefficients in JPEG images. Due to the large correlation between DCT blocks, an inter-block prediction model is constructed. In order to improve the performance of the algorithm, the algorithm also designs an optimization strategy, which uses DC coefficients to build a two-level sorting model, that is, complexity sorting and gradient sorting. The complexity sorting first calculates the smoothing coefficient of each DCT block (the number of non-zero AC coefficients in each DCT block), and then embeds the secret messages into the DCT blocks with higher smoothness. The gradient sorting first calculates the gradient values in the horizontal and vertical directions of the target block according to the DC coefficients of the adjacent blocks of the target, then selects the direction of the smaller gradient value as the gradient direction of the target block, and finally uses the adjacent blocks in the gradient direction to predict the coefficients of the target block.

The distribution of DCT coefficients in each block of JPEG image correlates whether in a single block (intra-block correlation) or between two blocks (inter-block correlation). In each 8*8 DCT coefficient block, the AC coefficient decreases with the increase of frequency, which indicates that adjacent DCT coefficient within the same DCT block have some correlation. As we all know that the correlation within the DCT block decreases with the increase of frequency. Compared with intra-block correlation, the coefficients in the current DCT block are still related to the coefficients in the same location of its adjacent blocks, which are generally called local (or intersubband) correlation or inter-block correlation. There is a significant correlation between the coefficient values on a DCT block and those at the exact location in the adjacent block. In general, the inter-block correlation is stronger than the intra-block correlation. However, with the increase of the frequency in the current DCT block and the distance from the target block, the inter-block correlation is gradually weakening. Therefore, using the inter-block correlation of DCT coefficients is crucial in the DCT coefficient prediction process.

In the process of data embedding, in order to improve the prediction performance of the predictor, the algorithm adopts a double-layer sorting scheme.

After the double-layer sorting, the predictor can predict more embeddable prediction error values, thus reducing the distortion caused by invalid displacement, and the embedding capacity is also improved.

**Sorting of DCT Block Complexity.** The primary purpose of complexity sorting is to find a smoother block for the image and embed it first. As we all know that in the smooth area of the image, the absolute values of most non-zero AC coefficients are minimal, and more zero values and smaller prediction error values can be obtained after coefficient prediction. In complex texture areas of the image, the absolute value of non-zero coefficients will be more significant, and predictions of AC coefficients will be more dispersed. For this feature, this algorithm uses the number of non-zero coefficients within each block as the smoothness measure for each block and then ranks each block's smoothness, giving priority to the smoothness block for embedding in the data embedding process. By selecting the last non-zero coefficient in each block as the base point, the algorithm keeps the point unchanged during the embedding process to ensure the algorithm's reversibility. Zigzag scanning is performed on each DCT block. Through scanning, the quantized DCT coefficients are changed from two-dimensional to one-dimensional, the last non-zero coefficient in one-dimensional data is set as the reference point, and its position is set as the threshold T. The DCT blocks are sorted from small to large according to the threshold T, and the blocks with smaller T values are embedded preferentially. At the same time, in order to keep the algorithm reversible during the embedding process, the DCT coefficient value corresponding to the position of T will not be changed.

Among them, the distortion caused by data embedding is not only related to the DCT coefficients within the block, but also related to the correlation of adjacent blocks. Therefore, if there are two or more DCT blocks with the same threshold, the DCT blocks are sorted twice by calculating the correlation between the current block and adjacent blocks. Firstly, eight DCT blocks with four directions: horizontal, vertical, 45° and 135°, are selected, and the correlation is calculated using the difference between the current DC coefficients and the DC coefficients of adjacent blocks. Formulas for calculating correlation are (1):

$$C_{i,j} = \left| \sum_{m=-1}^{1} \sum_{n=-1}^{1} \left( D_{i,j} - D_{i-m,j-m} \times \frac{1}{8} \right) \right| \tag{1}$$

where $C_{i,j}$ represents the mean of the correlation with the neighboring blocks and $D_{i-m,j-n}$ represents the DC coefficients of the eight adjacent blocks. If the value of $C_{i,j}$ is larger, it indicates that the correlation between the block and the adjacent block is smaller. Otherwise, the correlation is larger. By complexity sorting, the message is first embedded in DCT blocks with smaller T values and greater correlation with adjacent blocks.

**Sorting of DCT Blocks Gradient.** There is correlation between each adjacent blocks in JPEG images, but the correlation of adjacent blocks in each direction is also different. In order to make the predicted coefficient values more accurate, use the DC coefficient of each block to calculate the gradient change trend of the target block and adjacent blocks, and select the optimal gradient direction of the target block. By calculating and comparing DC coefficients, the gradient values in the horizontal and vertical directions of the block are calculated, and the current block is predicted according to the optimal gradient direction so as to improve the accuracy of prediction.

The specific methods are as follows: assuming that the current block is $x_{i,j}$, first calculate the gradients $D_1$ and $D_2$ of the current block and adjacent blocks $x_{i,j-1}$ and $x_{i,j+1}$ in the horizontal direction and adjacent blocks $x_{i+1,j}$ and $x_{i-1,j}$ in the vertical direction, respectively. The formulas are as follows:

$$D_1 = \frac{(|N - NW| + |N - NE| + |X - W| + |X - E| + |S - SW| + |S - SE|)}{6} \quad (2)$$

$$D_2 = \frac{(|W - NW| + |W - SW| + |X - N| + |X - S| + |E - NE| + |E - SE|)}{6} \quad (3)$$

$$D = min\,(D_1, D_2) \quad (4)$$

The symbols in formula (2) and formula (3) correspond to adjacent blocks in the corresponding direction of the current block, and the minor gradient direction of the two gradient directions of the current block $D_1$ and $D_2$ is taken as the optimal gradient direction $D$ of the current block. After the gradient direction of the current block is selected, the AC coefficients at the exact position of adjacent blocks in the same gradient direction are averaged. The obtained value is the predicted value of the AC coefficient at the same position as the current block. The formula is as follows:

$$\bar{x}_{i,j} = \begin{cases} \left\lfloor \frac{(W+E)}{2} \right\rfloor, & if\ D = D_1 \\ \left\lceil \frac{(N+S)}{2} \right\rceil, & if\ D = D_2 \end{cases} \quad (5)$$

where $\bar{x}_{i,j}$ is the predicted value of the DCT coefficient. It should be noted that in this algorithm, the DC coefficients of all DCT blocks remain unchanged during the embedding process. Therefore, the receiver can obtain the same DC coefficients as the embedding process for data extraction and image restoration.

## 2.2  Two-Dimensional Embedding Algorithm Based on Pairing Principle

**Prediction Error Pairing Method.** According to the adaptive pixel matching strategy proposed by Ou et al. [6], in order to increase the similarity between a pair of pixels, this strategy chooses that rough do not participate in the matching process, only smooth pixels will be paired, and the entropy of the two-dimensional prediction error histogram will be reduced. After the message is embedded, the result will be further improved compared with the previous algorithm.

Inspired by the adaptive pixel matching strategy proposed by Ou et al., this paper proposes a new prediction error matching strategy based on the distribution of the one-dimensional prediction error histogram: the prediction error is paired with two pairs according to the size of the value, a large number of experiments have proved that the prediction error with the value equal to "0, −1" is selected for pairing, and the data hiding performance obtained is optimal.

For ease of understanding, we will give an example explaining the pairing principle of prediction errors in more detail. As shown in Fig. 1, assume that the one-dimensional prediction error sequence is "0, 0, 2, 1, −1, −2, 0, −4, −1, 0". According to the proposed pairing principle of prediction errors, the prediction errors of the green markers in the figure are paired prediction errors, and the red markers are unpaired prediction errors. The index values of the one-dimensional prediction errors are 1, 2, 5, 7, N − 1 and N corresponding to the prediction errors of "0, 0, −1, 0". The selected six-bit prediction error values are paired sequentially, and the resulting one-dimensional prediction error sequence is paired. The available pairs of prediction error coefficients are "(0, 0), (−1, 0), (−1, 0)", and the remaining prediction error coefficients will not be changed during the message embedding process. The coefficient pair "(0, 0)" will be shifted in the messages embedding process according to the embedded secret messages. In contrast, the other coefficient pairs will make room for secret messages embedding by shifting.

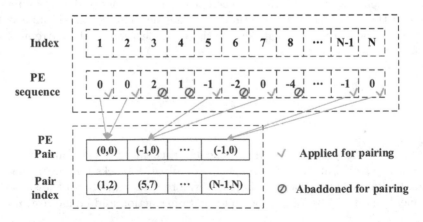

**Fig. 1.** Prediction error pairing diagram.

**Two-Dimensional Reversible Data Hiding Algorithm.** Based on the analysis in the previous section, this paper chooses to embed a secret message on the peak of the two-dimensional prediction error histogram. Inspired by the two-dimensional mapping of non-zero AC coefficient pairs proposed by Li et al. [22], this paper improves the two-dimensional mapping proposed by Li et al. [22] because the algorithm in this paper pairs the prediction errors with values of "0, −1". Because the secret message is embedded in the peak points of the two-dimensional prediction error histogram composed of internal regions, the two-dimensional map proposed by Li et al. is modified accordingly, and the moving rules of the remaining quadrants are deleted, leaving only the third quadrant. Figure 2 is an improved two-dimensional mapping in which coordinate points are divided into two types, type A points for embedding secret message and type B points for translating to make room for secret message embedding.

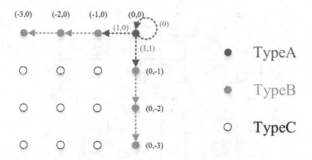

**Fig. 2.** Two-dimensional embedding mapping.

# 3   Experimental Results

In the following sections, the experiments are conducted on standard $512 \times 512$ sized grayscale JPEG images from the *Misc* dataset, including most popular images *Lena, Baboon, Barbara, Bridge*, and *Splash*. All experiments are implemented on the Matlab2021b platform.

In order to improve the data embedding performance, the prediction errors are further paired according to their value size to form a two-dimensional prediction error coordinate. Only the prediction errors with a value of "−1, 0, 1" are paired to test how well different matching principles are used. After pairing the obtained prediction errors according to their values, a number of coefficient prediction error pairs is formed.

Figure 3 and Fig. 4 are the two-dimensional prediction error histograms formed by pairing the predicted errors with the values of "0, 1" and "0, −1" when the image *Lena, Baboon, Barbara, Peppers, Bridge*, and *Splash* quality factors are 70 respectively. Among them, the prediction error in red color is the peak of the two-dimensional histogram. The prediction error pair "(0, 0)" gains the most frequent distribution. That is, the proposed predictor can estimate the value of the object coefficient very accurately, and the prediction error paring

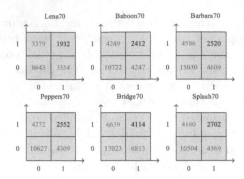

**Fig. 3.** Two-dimensional distribution of prediction error combinations of 0 and 1.

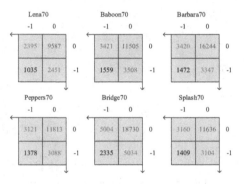

**Fig. 4.** Two-dimensional distribution of prediction error combinations of 0 and −1.

strategy further enables to obtain more prediction error pairs around the origin of coordinate. According to the principle of two dimensional RDH, the more prediction error pair "(0, 0)" is, the more secret data can be embedded and the higher the data embedded image quality is maintained.

The results also demonstrate that the distribution of the histogram formed by pairing the prediction errors of "0, −1" is steeper than that formed by pairing the prediction error values with "0, 1". That is, the number of prediction error pair "(0, 0)" obtained by paring the prediction errors of "0, −1" is much larger than that obtained with the scheme of paring the prediction errors of "0, 1". The small number of other prediction errors also would reduce the image distortion in the process of reversible data hiding.

Therefore, this paper chooses to pair the prediction errors with values of "0, −1" to form a two-dimensional prediction error histogram. The prediction error values other than "0, −1" are shifted to make room for secret data embedding.

Tested the experimental performance of six representative experimental images with quality factors of 70, 80, and 90 under different conditions, we will show the visual quality and file increment performance of the original JPEG

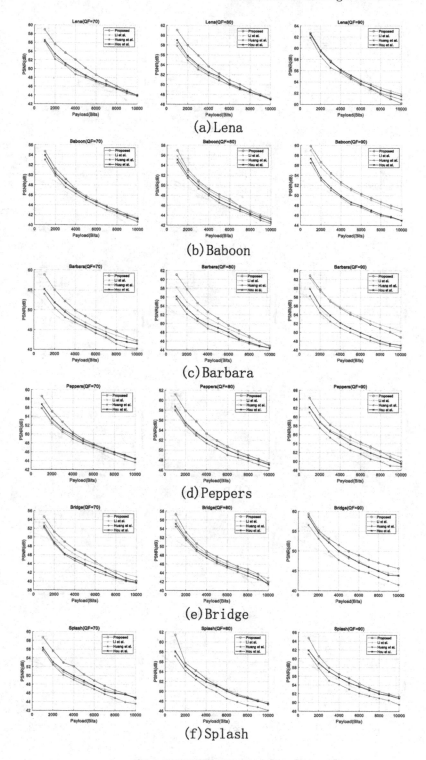

**Fig. 5.** PSNR Test Results.

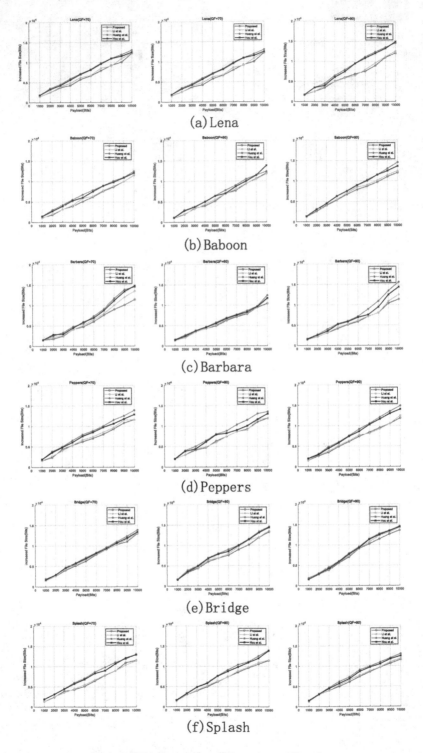

**Fig. 6.** File Incremental Experiment Results.

image and the encrypted JPEG image, respectively. They were compared with the schemes of Hou et al. [18], Huang et al. [16] and Li et al. [22], respectively. The following experimental results were marked as "Huang et al.", "Hou et al." and "Li et al." in Fig. 5 PSNR Test Results and in Fig. 6 File Incremental Experiment Results. The experimental results show that the PSNR is improved compared with other algorithms in the process of embedding capacity from 1000 bits to 10000 bits. The smaller the embedding capacity, the more pronounced the PSNR is, and about 3 dB can increase the maximum PSNR. The algorithm proposed in this paper can effectively reduce the file size expansion of the image when embedding secret messages of the same size, compared with the method of Huang et al. and Hou et al. when embedding secret messages of the same size with the same embedding capacity. The proposed method can better maintain the file size, the file increment of the scheme proposed by Li et al. is almost the same as that of the scheme proposed by Li et al., and even some images can maintain smaller file increments.

## 4   Conclusion

In this paper, a new RDH scheme for JPEG images based on a multi-level block sorting and two-dimensional data embedding algorithms is presented. The DCT blocks are firstly sorted to found flatten areas for data embedding, and then, the object coefficient of current block is estimated with the coefficients at the same position of its adjacent blocks in the lowest gradient direction. Moreover, a prediction error paring strategy is designed to pair only the prediction error values of "0, −1", making the two-dimensional histogram distributes steep. An improved two-dimensional histogram shifting algorithm is further applied to improve the performance of RDH for JPEG images, which can obtain high visual quality of the data embedded image and low file size increment while at large data is embedded. Extensive experiments including error prediction precision, data embedding performance and file size increment are conducted and compared with other state-of-the-art JPEG image reversible data hiding schemes and the result shows the superiority of the proposed scheme is apparent.

## References

1. Fridrich, J., Goljan, M., Du, R.: Lossless data embedding for all image formats, vol. 4675, pp. 572–583 (2002)
2. Tian, J.: Reversible data embedding using a difference expansion. IEEE Trans. Circuits Syst. Video Technol. **13**(8), 890–896 (2003)
3. Ni, Z., Shi, Y.-Q., Ansari, N., Su, W.: Reversible data hiding. IEEE Trans. Circuits Syst. Video Technol. **16**(3), 354–362 (2006)
4. Thodi, D.M., Rodríguez, J.J.: Expansion embedding techniques for reversible watermarking. IEEE Trans. Image Process. **16**(3), 721–730 (2007)
5. Ou, B., Li, X., Wang, J.: High-fidelity reversible data hiding based on pixel-value-ordering and pairwise prediction-error expansion. J. Vis. Commun. Image Represent. **39**, 12–23 (2016)

6. Ou, B., Li, X., Zhang, W., Zhao, Y.: Improving pairwise pee via hybrid-dimensional histogram generation and adaptive mapping selection. IEEE Trans. Circuits Syst. Video Technol. **29**(7), 2176–2190 (2018)

7. Hu, R., Xiang, S.: Efficient 2D mapping for reversible data hiding. IEEE Signal Process. Lett. **30**, 329–333 (2023)

8. Fridrich, J., Goljan, M., Du, R.: Invertible authentication watermark for JPEG images. In: Proceedings International Conference on Information Technology: Coding and Computing, pp. 223–227. IEEE (2001)

9. Chang, C.-C., Lin, C.-C., Tseng, C.-S., Tai, W.-L.: Reversible hiding in DCT-based compressed images. Inf. Sci. **177**(13), 2768–2786 (2007)

10. Wang, K., Lu, Z.-M., Hu, Y.-J.: A high capacity lossless data hiding scheme for JPEG images. J. Syst. Softw. **86**(7), 1965–1975 (2013)

11. Mobasseri, B.G., Berger, R.J., Marcinak, M.P., NaikRaikar, Y.J.: Data embedding in JPEG bitstream by code mapping. IEEE Trans. Image Process. **19**(4), 958–966 (2009)

12. Wu, Y., Deng, R.H.: Zero-error watermarking on JPEG images by shuffling Huffman tree nodes. In: Visual Communications and Image Processing (VCIP), pp. 1–4. IEEE (2011)

13. Xuan, G., Shi, Y.Q., Ni, Z., Chai, P., Cui, X., Tong, X.: Reversible data hiding for JPEG images based on histogram pairs. In: Kamel, M., Campilho, A. (eds.) ICIAR 2007. LNCS, vol. 4633, pp. 715–727. Springer, Heidelberg (2007). https://doi.org/10.1007/978-3-540-74260-9_64

14. Sakai, H., Kuribayashi, M., Morii, M.: Adaptive reversible data hiding for JPEG images. In: International Symposium on Information Theory and Its Applications, pp. 1–6. IEEE (2008)

15. Nikolaidis, A.: Reversible data hiding in JPEG images utilising zero quantised coefficients. IET Image Proc. **9**(7), 560–568 (2015)

16. Huang, F., Qu, X., Kim, H.J., Huang, J.: Reversible data hiding in JPEG images. IEEE Trans. Circuits Syst. Video Technol. **26**(9), 1610–1621 (2015)

17. Wedaj, F.T., Kim, S., Kim, H.J., Huang, F.: Improved reversible data hiding in JPEG images based on new coefficient selection strategy. EURASIP J. Image Video Process. **2017**, 1–11 (2017)

18. Hou, D., Wang, H., Zhang, W., Yu, N.: Reversible data hiding in JPEG image based on DCT frequency and block selection. Signal Process. **148**, 41–47 (2018)

19. Di, F., Zhang, M., Huang, F., Liu, J., Kong, Y.: Reversible data hiding in JPEG images based on zero coefficients and distortion cost function. Multimed. Tools Appl. **78**, 34541–34561 (2019)

20. Chen, Y., He, J., Xian, Y.: Reversible data hiding for JPEG images based on improved mapping and frequency ordering. Signal Process. **198**, 108604 (2022)

21. He, J., Pan, X., Wu, H.-T., Tang, S.: Improved block ordering and frequency selection for reversible data hiding in JPEG images. Signal Process. **175**, 107647 (2020)

22. Li, N., Huang, F.: Reversible data hiding for JPEG images based on pairwise nonzero AC coefficient expansion. Signal Process. **171**, 107476 (2020)

23. Ma, B., Shi, Y.Q.: A reversible data hiding scheme based on code division multiplexing. IEEE Trans. Inf. Forensics Secur. **11**(9), 1914–1927 (2016)

24. Ma, B., Chang, L., Wang, C., Li, J., Wang, X., Shi, Y.-Q.: Robust image watermarking using invariant accurate polar harmonic Fourier moments and chaotic mapping. Signal Process. **172**, 107544 (2020)

25. Ma, B., Hou, J.-C., Wang, C.-P., Wu, X.-M., Shi, Y.-Q.: A reversible data hiding algorithm for audio files based on code division multiplexing. Multimed. Tools Appl. **80**, 17569–17581 (2021)

26. Ma, B., Fu, Y., Wang, C., Li, J., Wang, Y.: A high-performance insulators location scheme based on YOLOv4 deep learning network with GDIoU loss function. IET Image Proc. **16**(4), 1124–1134 (2022)

27. Xiong, L., Han, X., Yang, C.-N., Zhang, X.: Reversible data hiding in shared images based on syndrome decoding and homomorphism. IEEE Trans. Cloud Comput. (2023)

28. Fan, M., Zhong, S., Xiong, X.: Reversible data hiding method for interpolated images based on modulo operation and prediction-error expansion. IEEE Access **11**, 27290–27302 (2023)

29. Yang, X., Huang, F.: New CNN-based predictor for reversible data hiding. IEEE Signal Process. Lett. **29**, 2627–2631 (2022)

30. Yang, Y., He, H., Chen, F., Yuan, Y., Mao, N.: Reversible data hiding in encrypted images based on time-varying Huffman coding table. IEEE Trans. Multimed. (2023)

31. Hua, Z., Wang, Z., Zheng, Y., Chen, Y., Li, Y.: Enabling large-capacity reversible data hiding over encrypted JPEG bitstreams. IEEE Trans. Circuits Syst. Video Technol. **33**, 1003–1018 (2022)

32. Xiao, M., Li, X., Ma, B., Zhang, X., Zhao, Y.: Efficient reversible data hiding for JPEG images with multiple histograms modification. IEEE Trans. Circuits Syst. Video Technol. **31**(7), 2535–2546 (2020)

33. Weng, S., Zhou, Y., Zhang, T., Xiao, M., Zhao, Y.: General framework to reversible data hiding for JPEG images with multiple two-dimensional histograms. IEEE Trans. Multimed. (2022)

34. Weng, S., Zhou, Y., Zhang, T., Xiao, M., Zhao, Y.: Reversible data hiding for JPEG images with adaptive multiple two-dimensional histogram and mapping generation. IEEE Trans. Multimed. (2023)

# Research on Encrypted Malicious 5G Access Network Traffic Identification Based on Deep Learning

Zongning Gao[1,2](✉) and Shunliang Zhang[1,2](✉)

[1] Institute of Information Engineering, Chinese Academy of Sciences,
Beijing 100093, China
{gaozongning,zhangshunliang}@iie.ac.cn
[2] School of Cyber Security, University of Chinese Academy of Sciences,
Beijing 100049, China

**Abstract.** The 5G small stations are widely deployed to improve the capacity of the 5G communication system. However, increasing malicious attacks are hidden in the traffic of small stations. Therefore, it is of great significance to identify the encrypted malicious traffic in the 5G access network. However, the traffic transferred through the backhaul of 5G small base station is usually encrypted. To this end, a deep learning-based method to identify the signaling hijacking traffic on the access network is proposed. Firstly, a 5G signaling hijacking system is developed to address the vulnerabilities of small stations and generate practical malicious traffic. To identify encrypted malicious traffic from 5G backhaul links, a 1D-CNN recognition model based on data packets is constructed. Finally, the 1D-CNN model is tested and validated multiple dimensions. The extensive experiment results reveal that the proposed method can achieve a recognition accuracy of over 99.95%.

**Keywords:** 5G Smallcell · Access Network · Deep learning · IPSec · Encrypted Malicious Traffic Identification

## 1 Introduction

5G has been widely utilized in transportation, medicine, and other areas of life, which brings tremendous convenience to people's lives. More than 2.31 million 5G base stations were built in China by the end of 2022. The total number of base stations accounts for over 60% of the world, with 561 million 5G mobile phone users [1]. There has been a explosive growth in 5G mobile communication traffic.

As an effective carrier of information transmission, traffics of mobile communication networks contain a large amount of valuable information. Meanwhile, network attacks, such as hijackings and interpolation, can be carried out by web

This work was supported by the National Key R&D Program of China No. 2021YFB 2910105.

© The Author(s), under exclusive license to Springer Nature Switzerland AG 2023
M. Yung et al. (Eds.): SciSec 2023, LNCS 14299, pp. 496–512, 2023.
https://doi.org/10.1007/978-3-031-45933-7_29

attackers. Therefore, encryption technologies such as IPsec and SSL are widely used to ensure the secure transmission of mobile communication. More than 80% of network traffic was encrypted in 2019. However, attackers can hidden malicious traffic by encryption technologies, which makes it difficult to detect malicious traffic encapsulated by encrypted tunnels. As the link between mobile communication users and the core network, the access network plays a crucial role in mobile communication. To our knowledge, encrypted traffic classification of 5G access network is rarely discussed in the literature, we are the first to classify encrypted 5G access network using 1D-CNN. Therefore, conducting encrypted malicious traffic identification on access network is of great significance to ensure the security communication of 5G.

The identification of signaling hijacking attack from encrypted 5G access network traffics is carried out in this paper. Firstly, 5G Smallcell access network environment is built. Secondly, the signaling hijacking system is designed. Then we capture the 5G access network traffic and construct the database including normal encrypted traffic and encrypted signaling hijacking traffic. By exploiting the special IPsec frame structures, we build a traffic identification framework based on deep learning to identify encrypted signaling hijacking traffic. The experiment result shows that signaling hijacking attacks can be recognized effectively on the access network without decrypting the packets, and the recognition accuracy is above 99.96%. Moreover, we conducted comparative experiments with different numbers of classifications including binary class, three class, nine class, and twelve class. The results verify the feasibility and superiority of the proposed model to identify the encrypted traffic on the access network.

The remainder of this paper is organized as follows. Section 2 reviews related work. Section 3 describes the signaling hijacking system. Section 4 presents the details of 1D-CNN and 2D-CNN models. Section 5 conducts comparison experiments and evaluates model performances. Section 6 concludes the work.

## 2    Related Work

### 2.1    Traffic Identification Methods

Traditional traffic identification methods mainly include port-based detection and deep packet inspection. With the widespread use of encryption technology, malicious traffic hidden in encrypted traffic poses a new threat to network security. Since the end of the last century, research institutions and scholars at home and abroad have conducted a lot of work on the identification and detection of malicious encrypted traffic. Traffic detection methods can be classified into four categories: port-based detection, deep packet inspection, machine learning detection using statistical features, and deep learning detection methods [2].

Most research on encrypted traffic classification focus on SSL traffic classification. SSL encrypts traffic above the transport layer, which is convenient for the purpose of classification because researchers can divide packets into session flows based on IP quintuples (source address, source port, destination address, destination port, transport layer protocol). Divided packets are then used to extract model features including the famous rule-based network intrusion detection system Snort [3], encrypted malicious traffic detection and confirmation method based on secure two-party computation [4], the Stacking method [5] for encrypted malicious traffic detection based on multi-feature recognition, and the RMETD-MF method [6].

This paper uses the convolutional neural network (CNN) method for feature extraction and model training of original encrypted traffic on the access network, aiming to achieve the goal of identifying Smallcell encrypted malicious traffic.

## 2.2    Smallcell Introduction

In the early stages of 5G network construction, network coverage focused on breadth. However, with the increase in the number of users, there was a surge in traffic in high-traffic areas during peak periods, leading to a deterioration in user experience. According to the latest data from the Ministry of Industry and Information Technology, as of the end of 2022, China has built 2.312 million 5G base stations, accounting for over 60% of the global total. With the advent of the 5G era, people demand higher transmission speed of wireless communication. However, the deployment of 5G in high-frequency bands has led to larger path losses in 5G macro station signal transmission and inadequate indoor 5G signal coverage due to building blockages.

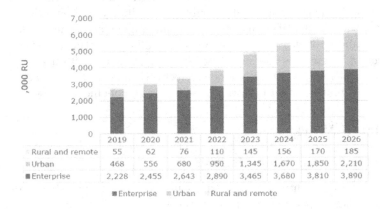

**Fig. 1.** New deployments and upgrades of small cells by environment 2019-2026.

In 2022, ABI Research, a professional research institution, predicted that when massive MIMO macro stations cannot meet 5G capacity demands, the deployment of outdoor smallcells will start to increase around 2025. Fei Liu, an industry analyst for 5G mobile network infrastructure at ABI Research, pointed out that 5G smallcells and macro stations complement each other, improving network capacity and expanding coverage in densely populated areas with weak or unavailable signals. As an effective supplement to 5G macro stations, the use of 5G smallcells is becoming more and more common. According to the

Smallcell Forum (SCF) in their SCF Market Status Report July 2020 [7], the number of smallcells is predicted to maintain an annual growth rate of 13% from 2019 to 2026, as shown in Fig. 1, increasing from 2.7 million to 6.3 million, with a cumulative deployment of 38.3 million. Among them, telecom operators' smallcell deployments in urban environments will grow at an average annual rate of 24%, with 80% for enterprises and schools. Enterprise deployments of smallcells will grow at an average annual rate of 9%, accounting for 68% of the total; smallcell deployments in rural and remote areas will maintain an annual growth rate of 19%, mainly serving industrial and IoT enterprises, and will install 957,000 by 2026.

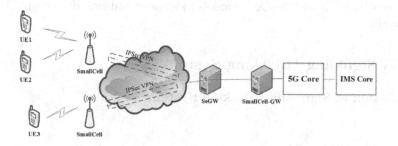

**Fig. 2.** IPSec tunnel in Smallcell architecture.

Considering the characteristics of the Smallcell access network and IPsec tunnel, as shown in Fig. 2, once the IPsec tunnel is successfully established, the addresses and port numbers at both ends of the tunnel between the Smallcell base station and the security gateway generally remain unchanged during the rekey period. This means that the source and destination addresses of the data packets of all mobile devices accessing the network through the Smallcell base station remain unchanged, and both the source and destination port numbers are 4500. Traditional methods [8–10] for distinguishing data packets into sessions or streams based on the five-tuple are no longer applicable in the Smallcell network architecture. Based on this, this paper proposes extracting features from 5G traffic packets using IP data packets as units.

## 2.3 Convolutional Neural Network

Convolutional Neural Networks (CNNs) are a class of feedforward neural networks that include convolutional computations and have a deep structure, representing one of the key algorithms in deep learning. CNNs have the ability to learn representations and perform translation invariant classification according to their hierarchical structure, hence they are also referred to as "translation invariant artificial neural networks". A typical CNN consists of three parts: the convolutional layer, the pooling layer and the fully connected layer. The convolutional layer is responsible for extracting local features of objects to be classified;

The pooling layer significantly reduces the number of parameters (dimensionality reduction); and the fully connected layer, similar to a traditional neural network, is used to output the desired result. Convolutional neural networks are among the most successful applications of deep learning algorithms. They include one-dimensional, two-dimensional, and three-dimensional convolutional neural networks. One-dimensional CNNs are primarily used for sequence data processing, two-dimensional CNNs are commonly used for image and text recognition [11,12], and three-dimensional CNNs are mainly used for medical image and video data recognition. As a deep learning model, CNNs have been successfully applied in multiple fields [13] in recent years, such as image classification, object localization and detection, face recognition, and autonomous driving. This paper uses 1D-CNN and 2D-CNN models to identify and classify encrypted malicious traffic.

# 3    5G Signaling Hijacking System

## 3.1    Design of Signaling Hijacking System

**Fig. 3.** The architecture of smallcelll signaling hijacking system.

Smallcell have security risks that can be exploited by attackers, we selected a commercial Smallcell base station, cracked it, and obtained root access to the station board. Mobile communication traffics were obtained at the signaling level, from the TCP three-way handshake to DNS resolution and then to the HTTP process, and a real-life signaling hijacking scenario was developed. As shown in Fig. 3, By modifying the response signaling in the mobile internet process and sending it through the base station to the mobile terminal, the mobile terminal interface is redirected to the attacker's specified website, achieving the goal of signaling hijacking. Under normal circumstances, the mobile phone connects to the 5G core network directly through the Smallcell base station via an IPsec tunnel. In this article, a signal hijacking system is added behind the Smallcell base station to act as a man-in-the-middle attack, simulating the access network side signaling hijacking attack scenario.

## 3.2   Implementation of Signaling Hijacking System

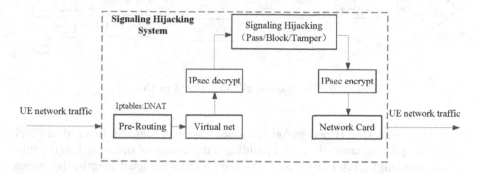

**Fig. 4.** The data processing flow diagram of signaling hijacking system.

As shown in Fig. 4, after the signaling hijacking system obtains the UE internet data uploaded by the Smallcell base station, it sets IPtables to perform destination address translation for the data packets at the pre-routing hook point, and processes the internet data packets in the virtual network card, i.e., decrypting the IPsec data packets, locating specific signals, and modifying the corresponding signaling according to the set hijacking mode (pass, block, tamper). Afterwards, the modified data packets are encrypted and encapsulated before being sent out through the physical network card. The same process is applied to the downlink data.

Through testing and verification, the signaling hijacking system can achieve passing, blocking, and tampering with the UE data packets uploaded by the access network, providing a foundation for subsequent malicious encrypted traffic targeting signaling hijacking.

# 4   Encrypted Malicious Traffic Identification Framework

The detection process of encrypted malicious traffic from signaling hijacking mainly consists of three stages: data collection and processing, model construction, classification and recognition, as shown in Fig. 5.

## 4.1   Data Collection and Processing

1. We set up a Smallcell network environment to fetch IPSec encrypted traffic data, detailed in Fig. 9. We selected nine major mobile applications, including social media, mobile payment, shopping, audio and video, as well as three types of malicious traffic from signaling hijacking system, totaling 12 encrypted traffic types that make up the raw dataset.
2. Extract IPSec packets from raw data.

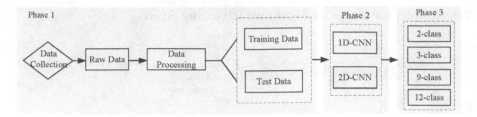

**Fig. 5.** The framework of proposed method.

3. We then converted IPSec packet data into a format that can be used as CNN model input, by truncating and padding data frames of different lengths into the same length(126 bytes). Since we need a unified packet length, the excess part will be trimmed, the part less than 126 bytes will be filled with zeros. The dataset is randomly divided into training and testing sets at an 8:2 ratio.

To visualize, 12 types of traffic are transformed into a two-dimensional gray image, as shown in Fig. 6.

**Fig. 6.** Grayscale map of 12 types of encrypted traffic.

From Fig. 6, we can see that different types of data packets have different grayscale images, and these differences also provide the possibility for feature extraction of CNN model.

## 4.2    Model Construction

In recent years, popular representation learning methods [14,15] have learned data features directly from raw data, eliminating the tedious process of manually selecting features. Deep learning algorithms, in particular, have shown significant advantages in image classification and speech recognition domains. For comparison and verification, we construct a 1D-CNN model and 2D-CNN model. For the sake of observation, the 2D-CNN model structure is shown in Fig. 7.

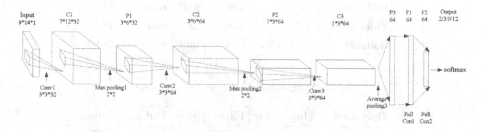

**Fig. 7.** The model structure of 2D-CNN.

The CNN reads pixel values of traffic images with dimensions of 9x14x1 from the dataset. In the first convolutional layer (C1), a 3x3 convolution kernel is used for convolution with 32 channels, generating 32 feature maps of size 7x12. Then, a 2x2 max-pooling operation is performed in layer P1, producing 32 feature maps of size 3x6. In the second convolutional layer (C2), the convolution kernel size is also 3x3, but with 64 channels, generating 64 feature maps of size 3x6. Then, a 2x2 max-pooling operation is performed in layer P2, producing 64 feature maps of size 1x3. In the third convolutional layer (C3), the same 3x3 convolution with 64 channels is used, generating 64 feature maps of size 1x3. This is followed by average pooling and two fully connected layers, converting the data size to 64. Finally, a softmax function is used to output the probabilities of each class. To reduce overfitting, dropout is applied before the output layer. Detailed parameters of 1D-CNN model and 2D-CNN model are shown in Table 1 and Table 2.

## 4.3    Classification Experiments Design

To verify the effectiveness of the proposed model, two application scenarios are set up, with a total of four classification experiments: 2-class, 3-class, 9-class, and 12-class classification. As shown in Fig. 8.

Scenario A (2-class, 3-class, 9-class): In practical applications of encrypted traffic classification, the most common requirement is to identify malicious encrypted traffic from mixed traffic, which is also the main goal of general intrusion detection systems. If there is a further need, classification can be performed

**Table 1.** 1D-CNN Model Parameter List

| Num. | Operation | Input Size | Filter | Step | Padding | Output |
|------|-----------|------------|--------|------|---------|--------|
| 1 | Conv1 | 126*1 | 32*3*1 | 2 | same | 32*124*1 |
| 2 | Max pool1 | 32*124*1 | 3*1 | 2 | same | 32*62*1 |
| 3 | Conv2 | 32*62*1 | 64*3*1 | 2 | same | 64*60*1 |
| 4 | Max pool2 | 64*60*1 | 2*1 | 2 | same | 64*30*1 |
| 5 | Conv3 | 64*30*1 | 64*3*1 | 2 | same | 64*28*1 |
| 6 | Average pool | 64*28*1 | null | 2 | same | 64 |
| 7 | Full connect1 | 64 | null | null | none | 64 |
| 8 | Full connect2 | 64 | null | null | none | 2/3/9/12 |
| 9 | Softmax | 2/3/9/12 | null | null | none | 2/3/9/12 |

**Table 2.** 2D-CNN Model Parameter List

| Num. | Operation | Input Size | Filter | Step | Padding | Output |
|------|-----------|------------|--------|------|---------|--------|
| 1 | Conv1 | 1*9*14 | 32*3*3 | 2 | same | 32*7*12 |
| 2 | Max pool1 | 32*7*12 | 2*2 | 2 | same | 32*3*6 |
| 3 | Conv2 | 32*3*6 | 64*3*3 | 2 | same | 64*3*6 |
| 4 | Max pool2 | 64*3*6 | 2*2 | 2 | same | 64*1*3 |
| 5 | Conv3 | 64*1*3 | 64*3*3 | 2 | same | 64*1*3 |
| 6 | Average pool | 64*1*3 | null | 2 | same | 64 |
| 7 | Full connect1 | 64 | null | null | none | 64 |
| 8 | Full connect2 | 64 | null | null | none | 2/3/9/12 |
| 9 | Softmax | 2/3/9/12 | null | null | none | 2/3/9/12 |

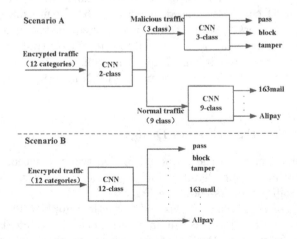

**Fig. 8.** Classification experiments under different scenarios.

separately within the already distinguished malicious encrypted traffic and normal encrypted traffic. Specifically, in this scenario, 12 types of traffic are mixed together. First, a 2-class experiment is conducted to classify malicious encrypted traffic and normal encrypted traffic. Then, a 3-class experiment is conducted within the identified malicious encrypted traffic, and a 9-class experiment is conducted within the normal encrypted traffic.

Scenario B (12-class): In practical applications, sometimes there is a need to classify all traffic types at once, which places higher demands on the classifier. In this scenario, the 12 types of traffic are mixed together and a single 12-class classification experiment is conducted.

## 5    Experiment and Result Analysis

### 5.1    Experiment Environment Setup

As shown in Fig. 9, two links are set up: Link 1 and Link 2. Link 1 consists of UE1 and Smallcell1, representing the encrypted normal traffic link. Link 2 consists of UE2, Smallcell2 and the signaling hijacking system, representing the encrypted malicious traffic link. To reflect traffic diversity, Link 1 captures 9 types of commonly used traffic, while Link 2 captures 3 types of traffic from the signaling hijacking system. A total of 1.41G of mixed traffic from Links 1 and 2 is captured through the mirror switch, providing data for subsequent experiments.

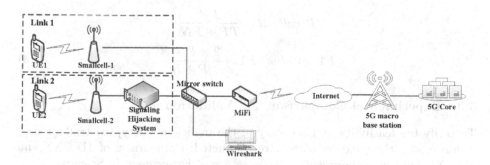

**Fig. 9.** Experiment environment diagram.

### 5.2    Experimental Evaluation Metrics

Network encrypted traffic recognition usually employs accuracy-related metrics to evaluate the quality of the proposed model. Accuracy are based on the confusion matrix, which is defined in Table 3. The confusion matrix consists of True Positive, False Positive, False Negative and True Negative. True Positive (TP) represents a positive sample that has been determined by the model to be positive. False Positive (FP) refers to negative samples that are judged positive by the model. False Negative (FN) represents a positive sample that is judged by the model to be negative. True Negative (TN) represents a negative sample that has been determined by the model to be negative.

**Table 3.** Confusion Matrix

| Label | Predict | |
|---|---|---|
| | Positive | Negative |
| True | TP | FN |
| False | FP | TN |

Accuracy-related metrics mainly include accuracy (A), precision (P), recall (R), and F1-Value (F1). Accuracy is used to evaluate the overall classification performance of the model, while precision, recall, and F1-Value are used to evaluate the recognition performance of each traffic category. The calculation formulas are as follows:

$$Accuracy : A = \frac{TP + TN}{TP + TN + FP + FN} \tag{1}$$

$$Precision : P = \frac{TP}{TP + FP} \tag{2}$$

$$Recall : R = \frac{TP}{TP + FN} \tag{3}$$

$$F1 - Value : F1 = \frac{2 \times P \times R}{P + R} \tag{4}$$

### 5.3 Experimental Comparison and Validation

To verify the feasibility and accuracy of the malicious encrypted traffic recognition model, three experiments were conducted: Comparison of 1D-CNN and 2D-CNN Models, Experiment in Scenario A and Experiment in Scenario B.

**Comparison of 1D-CNN and 2D-CNN Models.** Most research in the field of encrypted traffic classification converts traffic data into 2D images in a certain format and then uses a 2D convolutional neural network to identify and classify the images. However, mobile communication network traffic is essentially a type of time series data, organized in a hierarchical structure of 1D byte streams from low to high, including bytes, frames, sessions, and overall traffic. This structure is similar to the structure of letters, words, sentences, and paragraphs in natural language processing. Recently, 1D convolutional neural networks have been widely applied in various tasks such as text classification and sentiment analysis in natural language processing. Therefore, this experiment compares the 1D-CNN model with the 2D-CNN model, we show the results in Fig. 10. Detailed performance tuning was conducted to ensure the best results for each model. Detailed performance tuning was conducted to ensure the best results for each model. The precision and recall of the 12 traffic types under Scenario B were selected for comparison, as shown in Fig. 11 and Fig. 12.

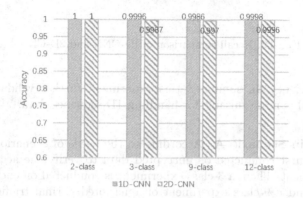

**Fig. 10.** Accuracy comparison of lD-CNN and 2D-CNN.

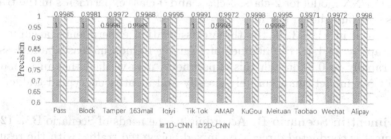

**Fig. 11.** Precision comparison of lD-CNN and 2D-CNN.

As shown in Fig. 10, the accuracy of both neural networks in the four classification experiments is above 99.86%. Among them, the accuracy of 1D-CNN is higher than that of 2D-CNN, with the maximum difference being 0.09%. As seen in Fig. 11, for 11 out of 12 types of encrypted traffic in Scenario B, the precision of 1D-CNN is higher than 2D-CNN. From Fig. 12, in the 12 types of encrypted traffic under Scenario B, the recall of 8 types of encrypted traffic in 1D-CNN is higher than that of 2D-CNN, with an average increase of 0.16%, while the remaining 4 types are roughly the same.

This experiment further demonstrates the superiority of the 1D-CNN model in processing mobile communication data streams, allowing it to directly process raw time-series data, eliminating the need for conversion to 2D images and the additional process of using 2D-CNN for recognition, thus improving model

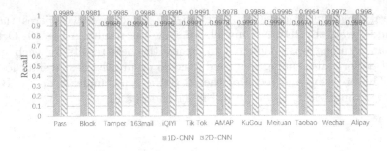

**Fig. 12.** Recall comparison of lD-CNN and 2D-CNN.

processing efficiency. In summary, this experiment further validates that the 1D-CNN model is more suitable for handling 1D sequence data.

**Experiment in Scenario A.** According to the needs of Scenario A, a 2-class experiment is first conducted on encrypted data to verify the accuracy of the 2-class experiment. Then, a 3-class experiment is conducted on encrypted malicious traffic, and a 9-class experiment on encrypted normal traffic. Given the superiority of the 1D-CNN model over the 2D-CNN model, the following experiments will be presented using the 1D-CNN model. The accuracy and loss curves of the 1D-CNN model for 2-class, 3-class, and 9-class experiments in the training and test sets are shown below.

As shown in Fig. 13, in the three classification experiments, accuracy converges rapidly to 1 as the number of training rounds increases, and loss converges rapidly to 0. Through this experiment, it is verified that in scenario A, the 1D-CNN model can accurately classify and identify encrypted traffic.

**Experiment in Scenario B.** According to the needs of Scenario B, a 12-class experiment is conducted directly on mixed encrypted traffic, with the results as follows.

As shown in Fig. 14, the accuracy of the 12-class experiment for encrypted traffic under Scenario B converges rapidly to 1, and the loss converges rapidly to 0, achieving the expected experimental results. Figure 15 shows that over 99% of the data in the confusion matrix can be classified correctly on the diagonal line. Through this experiment, it is verified that in scenario B, the 1D-CNN model can accurately classify and identify encrypted traffic.

## 5.4    Experimental Results and Evaluation

The effects of four classifiers were tested in both Scenario A and Scenario B. The overall accuracy data is shown in Table 4, and the precision, recall, and F1-Value for each traffic type are shown in Table 5, Table 6, and Table 7. Since the accuracy of the 2-class classifier reached 100%, there is no need to list the classification data for each class in the 2-class classifier. As seen in Table 6, the

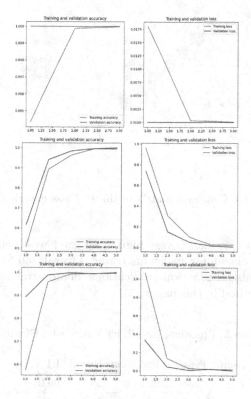

**Fig. 13.** Accuracy and Loss curves for 2/3/9-Class experiments.

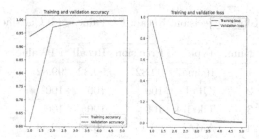

**Fig. 14.** Accuracy and Loss curves of the 12-Class experiment.

**Table 4.** The Accuracy of 4 types classifier(%)

| Average | 4 Classifier | | | |
|---|---|---|---|---|
| | 2 Class | 3 Class | 9 Class | 12 Class |
| | | malicious | normal | |
| 99.95 | 100 | 99.96 | 99.86 | 99.98 |

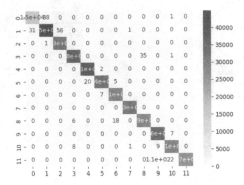

**Fig. 15.** Confusion matrix of the 12-Class experiment.

accuracy of the four classifiers is very high, with the lowest being 99.86% for the 9-class classifier in classifying malicious traffic. Overall, all four classifiers have met the requirements for practical applications, demonstrating the effectiveness of the proposed method in this paper.

**Table 5.** Precision, Recall, F1 Values of 3 Classifier

| Num. | Type | Precision | Recall | F1 Value |
|------|------|-----------|--------|----------|
| 1 | Pass | 99.91 | 99.91 | 99.91 |
| 2 | Block | 100 | 100 | 100 |
| 3 | Tamper | 99.96 | 99.96 | 99.96 |

**Table 6.** Precision, Recall, F1 Values of 9 Classifier

| Num. | Type | Precision | Recall | F1 Value |
|------|------|-----------|--------|----------|
| 1 | 163mail | 99.82 | 99.82 | 99.82 |
| 2 | iQIYI | 100 | 100 | 100 |
| 3 | Tik Tok | 100 | 100 | 100 |
| 4 | AMAP | 99.45 | 99.44 | 99.44 |
| 5 | KuGou | 100 | 100 | 100 |
| 6 | Meituan | 100 | 100 | 100 |
| 7 | Taobao | 100 | 100 | 100 |
| 8 | Wechat | 99.56 | 99.56 | 99.56 |
| 9 | Alipay | 99.94 | 99.94 | 99.94 |

**Table 7.** Precision, Recall, F1 Values of 12 Classifier

| Num. | Type | Precision | Recall | F1 Value |
|------|------|-----------|--------|----------|
| 1 | Pass | 100 | 100 | 100 |
| 2 | Block | 100 | 100 | 100 |
| 3 | Tamper | 99.98 | 99.98 | 99.98 |
| 4 | 163mail | 99.89 | 99.89 | 99.89 |
| 5 | iQIYI | 100 | 100 | 100 |
| 6 | Tik Tok | 100 | 100 | 100 |
| 7 | AMAP | 99.98 | 100 | 99.99 |
| 8 | KuGou | 100 | 99.99 | 99.99 |
| 9 | Meituan | 99.98 | 99.98 | 99.98 |
| 10 | Taobao | 100 | 100 | 100 |
| 11 | Wechat | 100 | 100 | 100 |
| 12 | Alipay | 100 | 100 | 100 |

# 6 Conclusion

We propose a deep learning encrypted malicious traffic recognition method for access networks, and use CNN to validate this idea: we create an access network dataset, design and develop a Smallcell access network signaling hijacking system, and construct a CNN model. We verified in two different application scenarios that the models have accuracy ranging from 99.86% to 100%, which is close to the level of practical application. This method can identify and classify traffic on the access network, thus it can detect hidden encrypted malicious traffic early and eliminate security risks early.

We propose the following areas for future research exploration:

Data set: currently, there is no publicly available IPSec data set for mobile communication networks. Researchers often collect their own data, making it difficult to compare works among different researchers.

Model generalizability: the model proposed in this paper is aimed at the classification of 5G access network traffic in Smallcell architecture. However, whether the model can be generalized in classifying encrypted traffic in other mobile network system remains to be explored.

# References

1. https://www.miit.gov.cn/zwgk/zcjd/art/2023/art_9f5022af3cdf48789484117d9d a03c58.html
2. Deng, W.: Network traffic classification based on deep learning. Xidian University (2020)
3. Roesch, M.: Network Intrusion Detection & Prevention System (Snort) (2017). https://www.snort.org

4. He, G., Wei, Q., Xiao, X., et al.: Confirmation method for the detection of malicious encrypted traffic with data privacy protection. J. Commun. **2**, 156–170 (2022)
5. Wang, T., Ding, Y.: Network malicious encryption traffic identification method based on stacking. Commun. Technol. **7**, 935–942 (2022)
6. Li, H., Zhang, S., Song, H., et al.: Robust malicious encrypted traffic detection based with multiple features. J. Cyber Secur. **2**, 129–142 (2021)
7. https://www.smallcellforum.org/scf-market-forecast/
8. Lin, W., Huamin, F., Biao, L., et al.: SSL VPN encrypted traffic identification based on hybrid method. Comput. Appl. Softw. **36**(2), 315–322 (2019)
9. Guo, S., Su, Y.: Encrypted traffic classification method based on data stream. J. Comput. Appl. **41**(5), 1386-1391 (2021)
10. Kang, P., Yang, W., Ma, H.: TLS malicious encrypted traffic identification research. Comput. Eng. Appl. **12**, 1–11 (2022)
11. Pan, Y., Zhang, X., Jiang, H., et al.: A network traffic classification method based on graph convolution and LSTM. IEEE Access **9**, 158261–158272 (2021)
12. Zhao, J., Li, Q., Liu, S., et al.: Towards traffic supervision in 6G: a graph neural network-based encrypted malicious traffic detection method. Sci. China (Inf. Sci.) **52**, 270–286 (2022)
13. Zhou, Y., Liu, F., Wang, Y.: IPSec VPN encrypted traffic identification based on hybrid method. Comput. Sci. **48**(4), 295–302 (2021)
14. Li, C., Dong, C., Niu, K., et al.: Mobile service traffic classification based on joint deep learning with attention mechanism. IEEE Access. **9**, 74729–74738 (2021)
15. Chakraborty, I., Kelley, B.M., Gallagher, B.: Industrial control system device classification using network traffic features and neural network embeddings. Array **12**, 100081 (2021)

# A Design of Network Attack Detection Using Causal and Non-causal Temporal Convolutional Network

Pengju He, Haibo Zhang$^{(\boxtimes)}$, Yaokai Feng, and Kouichi Sakurai

Graduate School of Information Science and Electrical Engineering,
Kyushu University, Fukuoka, Japan
haiboz0105@gmail.com, fengyk@ait.kyushu-u.ac.jp,
sakurai@inf.kyushu-u.ac.jp

**Abstract.** Temporal Convolution Network(TCN) has recently been introduced in the cybersecurity field, where two types of TCNs that consider causal relationships are used: causal TCN and non-causal TCN. Previous researchers have utilized causal and non-causal TCNs separately. Causal TCN can predict real-time outcomes, but it ignores traffic data from the time when the detection is activated. Non-causal TCNs can forecast results more globally, but they are less real-time. Employing either causal TCN or non-causal TCN individually has its drawbacks, and overcoming these shortcomings has become an important topic.

In this research, we propose a method that combines causal and non-causal TCN in a contingent form to improve detection accuracy, maintain real-time performance, and prevent long detection time. Additionally, we use two datasets to evaluate the performance of the proposed method: NSL-KDD, a well-known dataset for evaluating network intrusion detection systems, and MQTT-IoT-2020, which simulates the MQTT protocol, a standard protocol for IoT machine-to-machine communication. The proposed method in this research increased the detection time by about 0.1ms compared to non-causal TCN when using NSL-KDD, but the accuracy improved by about 1.5%, and the recall improved by about 4%. For MQTT-IoT-2020, the accuracy improved by about 3%, and the recall improved by about 7% compared to causal TCN, but the accuracy decreased by about 1% compared to non-causal TCN. The required time was shortened by 30ms (around 30%), and the recall was improved by about 7%.

**Keywords:** Network security · Causal Temporal Convolutional Network · Non-causal Temporal Convolutional Network · Intrusion Detection System

## 1 Introduction

In this section, the background of this study and related work are briefly introduced, followed by our main contributions.

© The Author(s), under exclusive license to Springer Nature Switzerland AG 2023
M. Yung et al. (Eds.): SciSec 2023, LNCS 14299, pp. 513–523, 2023.
https://doi.org/10.1007/978-3-031-45933-7_30

## 1.1 Background

With the increasing use of network technology, networks are becoming more widespread and complex. Unfortunately, cyber crimes are also on the rise, as there are more potential targets and the network market continues to grow. To protect against these threats, it is important to implement strong security measures like intrusion detection system(IDS) [9,10,16]. IDSs are security tools that monitor networks or computer systems for signs of unauthorized access or malicious activity. They are designed to identify security threats and alert administrators or security personnel when such activity is detected. AI technology is being widely implemented in various fields, making AI-IDS a hot topic for researchers. AI-based IDS can be particularly effective in detecting new or previously unknown threats since they are capable of learning and adapting to changing patterns of network activity over time [1,3,4,17]. Deep learning IDS, a subset of AI-IDS, has gained popularity. Neural networks are used to learn and model complex patterns in data. Previously, Multi-Layer Perceptron (MLP) was used for image recognition tasks, but its accuracy was low, and the model size was large. Convolutional Neural Network (CNN) is highly used in image recognition fields and is also being used in the attack detection field [18,19]. However, CNN cannot accept time-series input. In the field of AI-IDS, the original datasets consist of packets that are collected by machines. During the implementation phase, the machines continue to collect data over time, which is then utilized as input for the IDS. Therefore, when using time-series input data, other models such as Recurrent Neural Network (RNN) or Long short-term memory (LSTM) are used. RNN and LSTM are often used in tasks that accept time-series input because they can remember the previous input. When implementing them in attack detection fields, researchers use their ability to perform time-series prediction, such as stock prediction [27,28], and similarly in attack detection fields [20–22,25,26]. However, the model size of RNN or LSTM is larger than CNN generally. Temporal convolutional network(TCN), which is a type of deep learning model that is specifically designed to process sequential data. It is often used for tasks such as natural language processing, time series forecasting, speech recognition and also for AI-IDS [6,11–13]. In the field of AI-IDS, as the input data is usually time-series input and TCN can handle variable-length sequences, TCN is useful in context of IDS where data may vary in length and complexity.

## 1.2 Related Work

Lea [15] firstly proposed TCN to do action segmentation and detection in 2016. Bai [2] firstly introduced TCN and implement it on CNN using various fields such as image recognition and natural language processing to prove the ability of sequential data processing of TCN. As TCN is implemented in different fields, researchers implemented it into other fields such as speech recognition. Different from deep learning model which can also accept time-series input such as RNN, the model size of TCN is much smaller, and TCN has no information leakage from the past.

Fu [8] implemented causal-TCN into AI-IDS, the proposed model is an end-to-end IDS consists of character-level preprocessing and causal-TCN model. Comparing with previous model such as CNN or RNN, the model size is smaller and accuracy is higher.

Yao [30] based on attention mechanism combined improved TCN and bidirectional long short-term memory (BiLSTM) to extract the input temporal features. In their research, they implemented an attention mechanism into TCN. The attention mechanism is used to extract features.

However, there are still problems remained. Causal TCN uses data up to the present time to do prediction, while non-causal TCN uses not only data up to now, but also data for some time from now. Either of them has advantages and disadvantages. Besides, previously used dataset is not collected by IoT devices or embedded devices, and the dataset has already been made up by researchers, many features of the data packet were lost. Finally, the researches lack real-time benchmarks and performance. Research[8] did not test the real-time benchmark of TCN. As NSL-KDD dataset [14] don't have feature of timestamp, we could not know how exactly time cost for collecting data.

### 1.3  Contribution

We propose the combined model of causal and non-causal TCN, to gain the advantage of both and implement it into both NSL-KDD dataset [14] and MQTT-IoT-2020 dataset [29]. We compare the general benchmark metric based on causal, non-causal, and proposed combined models using the NSL-KDD dataset. And we first implement TCN in the MQTT-IoT-2020 dataset and we compare general benchmark metrics based on the causal, non-causal, and proposed combined model, besides, we also compare real-time benchmarks and show the pros and cons of these models. And in MQTT-IoT-2020, there is meta packet data that is not preprocessed, so we can easily compute the time cost which consists of both the time for every prediction and the time to deliver the packet in the real environment. As shown in [14] the combined model improves the accuracy and recall, and the MQTT-IoT-2020 dataset improves the recall and accuracy compared with causal TCN, recall, and time cost in the non-causal TCN.

This paper is organized as follows. Firstly, we introduce the background and challenging issues in this topic, then we state the terminologies of this research, then show the proposed model and the experiment results, and finally, we conclude this paper and discuss the future work.

## 2  Terminology

### 2.1  Temporal Convolutional Network

Temporal convolution network(TCN) is based on the convolutional neural network(CNN), which is an efficient model in many fields, it implements causal convolution, dilated convolution, and residual block to improve performance compared with the original CNN. There are also some differences between these two models. General CNNs are primarily designed for spatial data, such as images, while TCNs are designed for handling sequential data, like time series or natural language. To achieve this, TCNs implement 1-dimensional(1D) convolution while general CNNs implement 2-dimensional(2D) convolution. Figure 1 shows the relation between CNN and TCN.

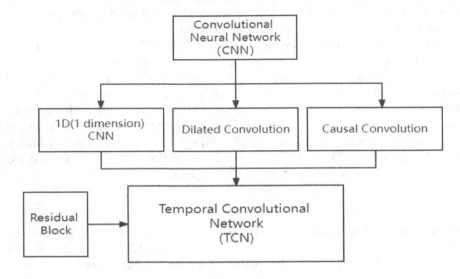

**Fig. 1.** TCN in CNN

Dilated convolution is a variant of the standard convolution operation that involves inserting gaps or dilations between the elements of the convolutional kernel as it is applied to the input data. The purpose of dilated convolution is to increase the receptive field of the convolutional filters, which is the area of the input data that each filter processes.

Considering the structure of TCN, there is causal TCN and non-causal TCN. Causal TCN uses data up to the present time to do prediction, while non-causal TCN uses not only data up to now, but also data for some time from now. Figure 2 and Fig. 3 here are examples of both of causal and non-causal TCN.

The receptive field of TCN can be seen as the memory of the model, if the receptive field is large, the memory is long. It can be calculated using Eq. (1).

$$Receptive\_size = 1 + 2 * (K_{size} - 1) * N_{stack} * \sum_i d_i, \tag{1}$$

where $k_{size}$ kernel size, which is also the size of convolution. Here, $N_{stack}$ is 1, means that there is only one block.

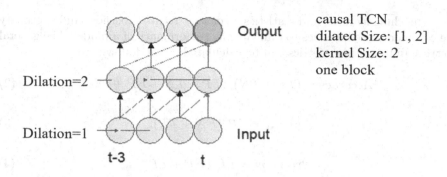

**Fig. 2.** Causal TCN

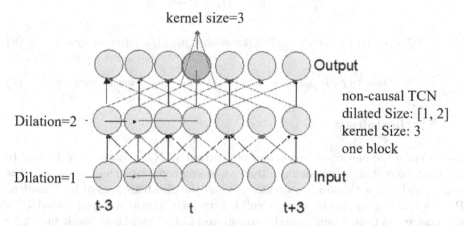

**Fig. 3.** Non-Causal TCN

## 2.2 Benchmarks Metrics

To compare with the performance of models, we use general benchmark metrics and real-time performance metrics.

1. TP: The row of the data is attack and regarded as attack.
2. TN: The row of the data is benign and regarded as benign.
3. FP: The row of the data is benign but regarded as attack.
4. FN: The row of the data is attack but regarded as benign.

In machine learning and statistics, accuracy, recall, precision, and specificity are commonly used metrics to evaluate the performance of a model. The general metrics and real-time metrics can be calculated as the followings.

$$Accuracy = (TP + TN)/(TP + TN + FP + FN) \tag{2}$$

$$Recall = TP/(FN + TP) \tag{3}$$

$$Precision = TP/(FP + TP) \tag{4}$$

$$Specificity = TN/(FP + TN) \tag{5}$$

$$Time\ cost\ per\ prediction = (total\ cost\ time)/prediction\ cases \tag{6}$$

$$Time\ to\ wait\ pakcets\ in = (total\ wait\ time)/wait\ cases \tag{7}$$

## 3    Methodology

Considering the benefit of causal and non-causal TCN, we hope to find a way to get both advantages. It is definite that when you combined the two models, the new model will both gain the advantages and the disadvantages of both models. However, in the real world, the tradeoff is extremely important. Non-causal TCN will cost much time (training and waiting) and lack of real-time, but it may have higher performance because of its longer step of dilation. But when using it in a real-time system, we have a tolerance of 20–200 ms delay, so if we can control the delay, we may improve the performance and can keep the real-time of the whole system.

As Fig. 4 shows, $T_0$ packet or data is firstly captured and sent into the causal TCN model to judge if it is benign. In the real world, we cannot capture the Tn packet or data at the same time, the system has to wait for the Tn packet or data in. So, there is time to wait from $T_0$ to $T_n$, which depends on the gap between the two packets or the piece of processed data.

Intuitively, if most of the packets are attack packets, the model can improve the accuracy or other performance compared with the original packets compared with the causal TCN, if most of the packets are benign packets, the model can improve the real-time performance compared with the non-causal TCN.

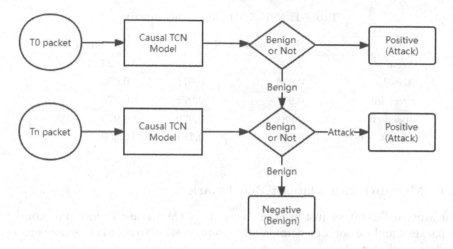

**Fig. 4.** General description of proposed combined TCN

## 4    Experiment

### 4.1    Datasets and Experiment Environment

The experiment consists of two parts: first, we use the NSL-KDD dataset [14] to compare benchmarks using accuracy, recall, precision, and specificity, as the NSL-KDD dataset doesn't contain the gap between every two pieces of data. Second, we use the Mqtt-IoT-2020 dataset to compare benchmarks using accuracy, recall, precision, specificity, and time to cost for every prediction, which contains the time to wait and the time to execute.

The experiment environment is the following. AMD Ryzen 5900x, RTX 3080ti, 32g memory. Python 3.9, Tensorflow2.6.

### 4.2    NSL-KDD Dataset Benchmark

The NSL-KDD dataset consists of preprocessed data, which may result in longer processing times compared to unprocessed, real-world data. However, it remains a suitable choice for performance comparison as long as the evaluation does not take time into account. The NSL-KDD dataset is a refined version of the KDD-cup99 dataset, NSL-KDD dataset consists of different kinds of features, generally, they can be divided into 4 categories.

First is basic features, which are almost the same as the packet data or data flow extracted from packets. Second is the content-related features of each network connecting vector. Third is time-related traffic features. The last is host-based traffic features in network connection features.

In this experiment, the non-causal kernel size of TCN is 3 and the causal kernel size is 2. The dropout rate is 0.2, and the dilation of TCN is (1, 2, 4, 8). Table 1 is

Table 1. NSL-KDD dataset benchmark

|                  | Non-causal TCN | Causal TCN | Combined TCN |
|------------------|----------------|------------|--------------|
| Accuracy         | 0.809          | 0.809      | 0.823        |
| Recall           | 0.658          | 0.671      | 0.696        |
| Precision        | 0.977          | 0.975      | 0.975        |
| Specificity      | 0.981          | 0.979      | 0.978        |
| Time Cost/Per(ms)| 0.193          | 0.181      | 0.300        |

### 4.3   Mqtt-IoT-2020 Dataset Benchmark

For Mqtt-IoT-2020, we just use the metadata of this dataset, which only consists of packets, and we don't do complicated preprocessing to avoid the whole process being out of control.

The Mqtt-IoT-2020 dataset is generated by some tools like MQTT-PWN which is used for the MQTT-brute-force attack. From the description of the Mqtt-IoT-2020 dataset, the attacker is on the same local network as the camera feed server, so the attacker can easily hack in and attack.

In this experiment, as we just used a packets dataset, so we don't have columns like *num_packets* because we hope the data collected by the host or broker is instant, and it may be hard for devices to do more processing when doing data transmission. The kernel size of causal TCN is 2, the kernel size of non-causal TCN is 3, the dropout rate is 0.2, and the dilation of TCNs is both (1, 2, 4, 8) (Table 2).

Table 2. Mqtt-IoT-2020 dataset benchmark

|                 | Non-causal TCN        | Causal TCN            | Combined TCN          |
|-----------------|-----------------------|-----------------------|-----------------------|
| Accuracy        | 0.700                 | 0.660                 | 0.691                 |
| Recall          | 0.476                 | 0.478                 | 0.540                 |
| Precision       | 0.863                 | 0.751                 | 0.773                 |
| Specificity     | 0.924                 | 0.842                 | 0.842                 |
| Time Cost/Per(s)| $5.922 * 10^{-5}$     | $4.392 * 10^{-5}$     | $8.670 * 10^{-5}$     |
| Time to wait(s) | 0.1037                | 0                     | 0.0707                |

### 4.4   Discussion

In these experiments, we compared the causal, non-causal and combined cases. From the first experiment, we found that when using extracted features, single causal TCN or non-causal TCN may perform almost the same in some benchmarks because even in the same number of receptive fields, the extracted feature may help the causal TCN model learn something which they cannot catch using packets data or not fully extracted data as we mentioned before. We can make the size of receptive fields larger, but it may cost much more time and computational resources compared with before. But we can use the proposed combined TCN to make use of their advantages to improve the performance by using a bit more computational resources. In the experiment using the Mqtt-IoT-2020 dataset, we showed that we could also save time and improve the performance compared with non-causal TCN using combined TCN, and improve recall, accuracy, and precision compared with causal TCN. So, using different kinds of datasets and compared with different models, we can get different performance improved.

## 5    Conclusion and Future Work

In this study, after a brief explanation of causal and noncausal TCNs, some weaknesses when applying them to detect network attacks respectively are discussed. Then, for the first time, a new method for network attack detection using a combination of causal and non-causal TCN models is proposed. Two public datasets of Mqtt-IoT-2020 and NSL-KDD were used to examine behaviors of our proposal. The discussion and experimental results indicate that our proposal can, to some extent, improve the detection performance and suppress the weaknesses of a single model. In the future, the performance of our proposal will be further verified using other parameters and other datasets.

**Acknowledgement.** This work is partially supported by JSPS international scientific exchanges between Japan and India (Bilateral Program DTS-JSPS).

## References

1. Cheng, P., Xu, K., Li, S., et al.: TCAN-IDS: intrusion detection system for internet of vehicle using temporal convolutional attention network. Symmetry **14**(2), 310 (2022). https://doi.org/10.3390/sym14020310
2. Bai, S., Kolter, J.Z., Koltun, V.: An empirical evaluation of generic convolutional and recurrent networks for sequence modeling. arXiv preprint arXiv:1803.01271 (2018)
3. Ahmad, Z., Shahid Khan, A., Wai Shiang, C., et al.: Network intrusion detection system: a systematic study of machine learning and deep learning approaches. Trans. Emerg. Telecommun. Technol. **32**(1), e4150 (2021). https://doi.org/10.1002/ett.4150
4. Kim, A., Park, M., Lee, D.H.: AI-IDS: application of deep learning to real-time Web intrusion detection. IEEE Access **8**, 70245–70261 (2020). https://doi.org/10.1109/ACCESS.2020.2986882
5. Gopali, S., Abri, F., Siami-Namini, S., et al.: A comparative study of detecting anomalies in time series data using LSTM and TCN models. arXiv preprint arXiv:2112.09293 (2021)
6. Thill, M., Konen, W., Bäck, T.: Time series encodings with temporal convolutional networks. In: Filipič, B., Minisci, E., Vasile, M. (eds.) BIOMA 2020. LNCS, vol. 12438, pp. 161–173. Springer, Cham (2020). https://doi.org/10.1007/978-3-030-63710-1_13

7. Derhab, A., Aldweesh, A., Emam, A.Z., et al.: Intrusion detection system for internet of things based on temporal convolution neural network and efficient feature engineering. Wirel. Commun. Mob. Comput. **2020**, 1–16 (2020)

8. Fu, N., Kamili, N., Huang, Y., et al.: A novel deep intrusion detection model based on a convolutional neural network. Aust. J. Intell. Inf. Process. Syst. **15**(2), 52–59 (2019)

9. Ashoor, A.S., Gore, S.: Importance of intrusion detection system (IDS). Int. J. Sci. Eng. Res. **2**(1), 1–4 (2011)

10. Liao, H.J., Lin, C.H.R., Lin, Y.C., et al.: Intrusion detection system: a comprehensive review. J. Netw. Comput. Appl. **36**(1), 16–24 (2013)

11. Liu, Y., Dong, H., Wang, X., et al.: Time series prediction based on temporal convolutional network. In: 2019 IEEE/ACIS 18th International Conference on Computer and Information Science (ICIS), pp 300–305. IEEE (2019)

12. Xu, B., Lu, C., Guo, Y., et al.: Discriminative multi-modality speech recognition. In: Proceedings of the IEEE/CVF Conference on Computer Vision and Pattern Recognition, pp. 14433–14442 (2020)

13. He, Y., Zhao, J.: Temporal convolutional networks for anomaly detection in time series. In: Journal of Physics: Conference Series, vol. 1213, no. 4, p. 042050. IOP Publishing (2019)

14. Ghulam Mohi-ud-Din, December 29, 2018, "NSL-KDD", IEEE Dataport. https://doi.org/10.21227/425a-3e55

15. Lea, C., Vidal, R., Reiter, A., Hager, G.D.: Temporal convolutional networks: a unified approach to action segmentation. In: Hua, G., Jégou, H. (eds.) ECCV 2016. LNCS, vol. 9915, pp. 47–54. Springer, Cham (2016). https://doi.org/10.1007/978-3-319-49409-8_7

16. Raiyn, J.: A survey of cyber attack detection strategies. Int. J. Secur. Appl. **8**(1), 247–256 (2014)

17. Kaloudi, N., Li, J.: The AI-based cyber threat landscape: a survey. ACM Comput. Surv. (CSUR) **53**(1), 1–34 (2020)

18. Ding, Y., Zhai, Y.: Intrusion detection system for NSL-KDD dataset using convolutional neural networks. In: Proceedings of the 2018 2nd International Conference on Computer Science and Artificial Intelligence, pp. 81–85 (2018)

19. Belgrana, F.Z., Benamrane, N., Hamaida, M.A., et al.: Network intrusion detection system using neural network and condensed nearest neighbors with selection of NSL-KDD influencing features. In: 2020 IEEE International Conference on Internet of Things and Intelligence System (IoTaIS), pp. 23–29. IEEE (2021)

20. Chaibi, N., Atmani, B., Mokaddem, M.: Deep learning approaches to intrusion detection: a new performance of ANN and RNN on NSL-KDD. In: Proceedings of the 1st International Conference on Intelligent Systems and Pattern Recognition, pp. 45–49 (2020)

21. Muhuri, P.S., Chatterjee, P., Yuan, X., et al.: Using a long short-term memory recurrent neural network (LSTM-RNN) to classify network attacks. Information **11**(5), 243 (2020)

22. Tang, T.A., Mhamdi, L., McLernon, D., et al.: Deep recurrent neural network for intrusion detection in SDN-based networks. In: 2018 4th IEEE Conference on Network Softwarization and Workshops (NetSoft), pp. 202–206. IEEE (2018)

23. Xu, W., Jang-Jaccard, J., Singh, A., et al.: Improving performance of autoencoder-based network anomaly detection on NSL-KDD dataset. IEEE Access **9**, 140136–140146 (2021)

24. Chen, Z., Yeo, C.K., Lee, B.S., et al.: Autoencoder-based network anomaly detection. In: 2018 Wireless Telecommunications Symposium (WTS), pp. 1–5. IEEE (2018)
25. Yin, C., Zhu, Y., Fei, J., et al.: A deep learning approach for intrusion detection using recurrent neural networks. IEEE Access **5**, 21954–21961 (2017)
26. Althubiti, S.A., Jones, E.M., Roy, K.: LSTM for anomaly-based network intrusion detection. In: 2018 28th International Telecommunication Networks and Applications Conference (ITNAC), pp. 1–3. IEEE (2018)
27. Selvin, S., Vinayakumar, R., Gopalakrishnan, E.A., et al.: Stock price prediction using LSTM, RNN and CNN-sliding window model. In: 2017 International Conference on Advances in Computing, Communications and Informatics (ICACCI), pp. 1643–1647. IEEE (2017)
28. Pawar, K., Jalem, R.S., Tiwari, V.: Stock market price prediction using LSTM RNN. In: Rathore, V.S., Worring, M., Mishra, D.K., Joshi, A., Maheshwari, S. (eds.) Emerging Trends in Expert Applications and Security. AISC, vol. 841, pp. 493–503. Springer, Singapore (2019). https://doi.org/10.1007/978-981-13-2285-3_58
29. Hindy, H., Bayne, E., Bures, M., Atkinson, R., Tachtatzis, C., Bellekens, X.: Machine learning based IoT intrusion detection system: an MQTT case study (MQTT-IoT-IDS2020 dataset). In: Ghita, B., Shiaeles, S. (eds.) INC 2020. LNNS, vol. 180, pp. 73–84. Springer, Cham (2021). https://doi.org/10.1007/978-3-030-64758-2_6
30. Yao, C., Yang, Y., Yang, J., et al.: A network security situation prediction method through the use of improved TCN and BiDLSTM. Math. Probl. Eng. **2022** (2022)

# Author Index

© The Editor(s) (if applicable) and The Author(s), under exclusive license
to Springer Nature Switzerland AG 2023
M. Yung et al. (Eds.): SciSec 2023, LNCS 14299, pp. 525–526, 2023.
https://doi.org/10.1007/978-3-031-45933-7